THE REGULATION OF GENETICALLY MODIFIED ORGANISMS: COMPARATIVE APPROACHES

The Regulation of Genetically Modified Organisms: Comparative Approaches

Edited by
LUC BODIGUEL
and
MICHAEL CARDWELL

UNIVERSITY PRESS

OXFORD
UNIVERSITY PRESS

Great Clarendon Street, Oxford OX2 6DP

Oxford University Press is a department of the University of Oxford.
It furthers the University's objective of excellence in research, scholarship,
and education by publishing worldwide in

Oxford New York

Auckland Cape Town Dar es Salaam Hong Kong Karachi
Kuala Lumpur Madrid Melbourne Mexico City Nairobi
New Delhi Shanghai Taipei Toronto

With offices in

Argentina Austria Brazil Chile Czech Republic France Greece
Guatemala Hungary Italy Japan Poland Portugal Singapore
South Korea Switzerland Thailand Turkey Ukraine Vietnam

Oxford is a registered trade mark of Oxford University Press
in the UK and in certain other countries

Published in the United States
by Oxford University Press Inc., New York

© The several contributors, 2010

The moral rights of the author have been asserted
Database right Oxford University Press (maker)

Crown copyright material is reproduced under Class Licence
Number C01P0000148 with the permission of OPSI
and the Queen's Printer for Scotland

First published 2010

All rights reserved. No part of this publication may be reproduced,
stored in a retrieval system, or transmitted, in any form or by any means,
without the prior permission in writing of Oxford University Press,
or as expressly permitted by law, or under terms agreed with the appropriate
reprographics rights organization. Enquiries concerning reproduction
outside the scope of the above should be sent to the Rights Department,
Oxford University Press, at the address above

You must not circulate this book in any other binding or cover
and you must impose this same condition on any acquirer

British Library Cataloguing in Publication Data
Data available

Library of Congress Cataloging in Publication Data
Data available

Typeset by MPS Limited, A Macmillan Company
Printed in Great Britain
on acid-free paper by
CPI Antony Rowe

ISBN 978-0-19-954248-2

1 3 5 7 9 10 8 6 4 2

Acknowledgements

The idea for this edited collection first arose in the region of the Guérande in Brittany, whose natural beauty provided inspiration. Since then, the editors have been indebted to a great number of people in seeing the project through to a conclusion, and not least the contributors. Their scholarship has been matched by their enthusiasm and attention to detail, and at all stages this played a key role in maintaining momentum. In addition, material assistance on economic aspects was provided by Lee Ann Jackson and Chantal Pohl Nielsen, while Melaku Desta, Louise Ellison, Charlotte O'Brien, and Ole Windahl Pedersen provided constructive advice on matters of law. Gill Ewing kindly translated the final chapter, Ludivine Petetin helped to compile the Tables of Cases and Legislation, and again a debt of gratitude is owed to those at Leeds who carried the heavier burden during study leave.

The staff at Oxford University Press have given tremendous support since the book was first mooted with John Louth. In particular, many thanks are extended to Chris Champion, Glynis Dyson, Alex Flach, Emma Hawes, and Natasha Knight, all of whom showed not only great expertise, but also great patience. Further, the anonymous referees of the book proposal suggested many wise amendments, which we hope have been properly accommodated.

Ce travail a été réalisé avec passion sous l'œil attentif de nos proches, grands et plus petits— les parents de Michael, Adèle, Mariette, Yves, et Odile, à qui nous dédions cet ouvrage.

Luc Bodiguel Michael Cardwell
Nantes York
January 2010

Contents

List of Contributors	ix
Table of Cases	xiii
Table of Legislation	xix
Table of Treaties and Conventions	xxxi
List of Abbreviations	xxxv

Introduction   1
Michael Cardwell

I. THE PUBLIC, ECONOMICS, AND RISK

1. Genetically Modified Organisms and the Public: Participation, Preferences, and Protest   11
Luc Bodiguel and Michael Cardwell

2. The Importance and Limits of Cost-Benefit Analysis in the Regulation of Genetically Modified Organisms   37
Christophe Charlier and Egizio Valceschini

3. Genetically Modified Organisms and Risk   54
Karen Morrow

II. THE EUROPEAN UNION

4. The European Union Regulatory Regime for Genetically Modified Organisms and its Integration into Community Food Law and Policy   79
Marine Friant-Perrot

5. Multi-level Governance of Genetically Modified Organisms in the European Union: Ambiguity and Hierarchy   101
Maria Lee

6. Coexistence of Genetically Modified, Conventional, and Organic Crops in the European Union: The Community Framework   123
Margaret Rosso Grossman

7. Coexistence of Genetically Modified, Conventional, and Organic Crops in the European Union: National Implementation   163
Luc Bodiguel, Michael Cardwell, Ana Carretero García, and Domenico Viti

8. Implementing the Community Environmental Liability Directive: Genetically Modified Organisms and the Problem of Unknown Risk   198
Christopher Rodgers

III. REGULATION BEYOND THE EUROPEAN UNION

9. Genetically Modified Organisms in Africa: Regulating a Threat or an Opportunity?   227
Fikremarkos Merso Birhanu

10. The Coexistence of Genetically Modified and Non-genetically Modified Agriculture in Canada: A Courtroom Drama   254
Jane Matthews Glenn

11. Genetically Modified Organisms in MERCOSUR   274
Rosario Silva Gilli

12. Genetically Modified Crops and Food in the United States: The Federal Regulatory Framework, State Measures, and Liability in Tort   299
Margaret Rosso Grossman

IV. INTERNATIONAL IMPLICATIONS

13. The *EC—Biotech* Decision: Another Missed Opportunity?   337
Joseph McMahon

14. The Regulation of Genetically Modified Organisms and International Law: A Call for Generality   355
Duncan French

Conclusion   375
Luc Bodiguel

Index   387

List of Contributors

Fikremarkos Merso Birhanu
Fikremarkos Merso Birhanu gained his LLB from Addis Ababa University, LLM from the University of Pretoria, and PhD from the Osaka School of International Public Policy, Osaka University (the subject of his PhD being the issue of intellectual property in plant-related innovations). Prior to taking his PhD, he was a lecturer and Assistant Dean at the Faculty of Law, Addis Ababa University, where he now serves as Dean and Assistant Professor. He has published several articles and produced a number of research papers with focus on international economic law, intellectual property (including TRIPS), and human rights. In addition, Fikremarkos has served as an adviser to various national, regional, and intergovernmental organizations and he has taught widely as visiting lecturer.

Luc Bodiguel
Dr Luc Bodiguel is Chargé de Recherche with the Centre National de la Recherche Scientifique (CNRS) and teaches agricultural law at the University of Nantes. His published work includes *L'Entreprise Rurale: Entre Activités Économiques et Territoire Rural* (Harmattan, 2002) and a wide range of articles on the Common Agricultural Policy, agricultural law, and environmental law. He has participated in several national research projects (addressing, *inter alia*, the governance of water resources, multifunctionality, and Natura 2000). At present, he is Scientific Director of the University of Nantes Diploma on the Environment and Sustainable Development of Land and he serves on the executive committee of the Comité Européen de Droit Rural.

Michael Cardwell
Michael Cardwell is Professor of Agricultural Law at the School of Law, University of Leeds, which he joined in 1990 after working in legal practice with Burges Salmon, Bristol. His early research was directed to agricultural tenancies and European Community quota regimes. More recently, he has addressed also broader legal issues generated by Common Agricultural Policy reform, together with the regulation of agriculture within the WTO. His publications include *Milk Quotas: European Community and United Kingdom Law* (OUP, 1996), *Agriculture and International Trade: Law, Policy and the WTO* (CAB International, 2003) (co-edited with Peggy Grossman and Chris Rodgers), and *The European Model of Agriculture* (OUP, 2004).

Ana Carretero García
Ana Carretero García is Professor of Civil Law at the University of Castilla-La Mancha and Vice-Dean of the Facultad de Ciencias Jurídicas y Sociales. Her doctoral thesis was awarded both national and international prizes and her research encompasses agricultural law, consumer law, and civil responsibility. She has published extensively, including the monographs *Empresa Agraria y Profesionales de la Agricultura en Derecho Español y Comunitario* (Comares, 2003) and *Origen, Evolución, Situación Actual y Retos de la Política Agraria Común* (Esperia, 2003).

List of Contributors

Christophe Charlier
Christophe Charlier is an economist and Maître de Conférences at the University of Nice—Sophia Antipolis. His main areas of interest are the harmonization of national regulatory frameworks in a globalizing world and the study of traceability and labelling at a time of vertical integration in the food chain. He has addressed both these topics in the specific context of GMOs as coordinator of a recent and multidisciplinary research project (GICOGM—French National Research Agency).

Duncan French
Duncan French is Professor of International Law at the University of Sheffield. He is the author of *International Law and Policy of Sustainable Development* (Manchester University Press, 2005) and has written numerous papers on various aspects of international environmental law. He is also co-rapporteur of the International Law Association (ILA) Committee on International Law on Sustainable Development.

Marine Friant-Perrot
Marine Friant-Perrot is Maître de Conférences in the Faculté de Droit at the University of Nantes, where she is Director of the Masters Programme. Her research interests and publications cover various aspects of food law, with recent emphasis on sustainable consumption and consumer law (these publications including 'La Consommation Durable et la Protection des Consommateurs' in *Production et Consommation Durables: de la Gouvernance au Consommateur Citoyen* (Éditions Yvon Blais, 2008) (edited by Geneviève Parent)). In December 2005 she co-organized an international congress to celebrate the Twenty-fifth Anniversary of the Conseil National de l'Alimentation (CNA) and is a co-editor of its proceedings. During 2007–2009 she was a member of the CNA Working Group on Food Information, Advertising, and Education.

Jane Matthews Glenn
Jane Matthews Glenn is Emeritus Professor at the Faculty of Law and School of Urban Planning, McGill University, where she is also a Member of the Institute of Comparative Law and an Associate Member of the McGill School of Environment. She was elected an Associate Member of the International Academy of Comparative Law in 1995. Professor Glenn has a long-standing research interest in agricultural law and has been a Member of the Union Mondiale des Agraristes Universitaires since 1992 and an Associate Member of the Comité Européen de Droit Rural since 1996. In 2005 she was named Honorary Member of the Comité Americano de Derecho Agrario. Her published work has recently focussed on the regulation of GMOs in North America.

Margaret Rosso Grossman
Margaret (Peggy) Rosso Grossman is Professor and Bock Chair in Agricultural Law in the Department of Agricultural and Consumer Economics, University of Illinois. She is author of numerous US and European law review articles and book chapters on agricultural and environmental law topics. She is editor of *Agriculture and the Polluter Pays Principle* (BIICL, 2009) and co-editor of *Agriculture and International Trade: Law, Policy and the WTO* (CAB International, 2003). Professor Grossman has received three Fulbright Senior Scholar Awards and a German Marshall Fund Research Fellowship to support her research in Europe. In 1991 she was President of the American Agricultural Law Association (AALA). She received the AALA Distinguished Service Award in 1993

and the AALA Professional Scholarship Award in 2006 and 2008. The Comité Européen de Droit Rural awarded her the Silver Medal in 1999. Since 1986, Professor Grossman has been a frequent visiting research professor in the Law and Governance Group at Wageningen University, The Netherlands. In Spring 2008, she was Visiting Professor at the University of Copenhagen (Denmark) and at Newcastle Law School (England).

Maria Lee
Maria Lee is Professor of Law at University College London and a member of the Royal Commission on Environmental Pollution. She has written widely on the environmental law of the EU and on risk regulation, including *EU Environmental Law: Challenges, Change and Decision-making* (Hart Publishing, 2005), EU Regulation of GMOs: Law and Decision making for a New Technology (Edward Elgar, 2008), and 'Beyond Safety? The Broadening Scope of Risk Regulation' in *Current Legal Problems 2009* (OUP, 2010) (edited by Colm O'Cinneide).

Joseph McMahon
Joseph McMahon is Professor of Commercial Law at University College Dublin. After his undergraduate study at the Queen's University of Belfast, he undertook doctoral research at the University of Edinburgh on European trade policy in agricultural products. He writes extensively on international agricultural trade issues and the Common Agricultural Policy; and he is currently completing a book on the negotiations for a new WTO Agreement on Agriculture (to be published by Kluwer). He has also previously been a member of staff at the Victoria University of Wellington (New Zealand), the University of Leicester, and the Queen's University Belfast.

Karen Morrow
Karen Morrow is Professor of Environmental Law at Swansea University. She has published widely on environmental law, including (with Sean Coyle) *A Philosophical Foundation for Environmental Law* (Hart Publishing, 2004), which won a Society of Legal Scholars Peter Birks Prize for Outstanding Legal Scholarship in that year. She is Deputy Convenor of the Environment Panel of the Society of Legal Scholars; and she is also an associate member of the Monash European and EU Centre, a visiting member of the Faculty of Law at Leuven, and a member of the International Working Group on Property, Community and Social Entrepreneurism, University of Syracuse, USA. In addition, she is co-editor of the Journal of Human Rights and Environment and of the UCN Academy Environmental Law Journal.

Christopher Rodgers
Christopher Rodgers is Professor of Law at Newcastle University. He has lectured widely on environmental and agricultural law in the United Kingdom, Europe, and the USA, and is Editor-in-Chief of the Environmental Law Review. He is currently Principal Investigator of a major research project on 'Contested Common Land—Environmental Governance, Law and Sustainable Land Management 1600–2006', funded by the Arts and Humanities Research Council as part of its thematic Landscape and Environment Programme. He is also the author of numerous journal articles and books on agricultural law and the environment, including *Agricultural Law* (Tottel Publishing, 3rd edition, 2008).

Rosario Silva Gilli
Rosario Silva Gilli, Dr en Derecho y Ciencias Sociales, has for some time worked in a socio-legal research institute at Universidad de la Republica, Uruguay; and she is currently member both of a committee to monitor regional transformation in South America (based at Universidad Brasilia, CNRS-CIRAD) and an academic group which addresses MERCOSUR issues (based at Universidad de Rosario, Argentina). In addition, she has written many publications (including international publications) on agricultural law and the environment, with particular reference to MERCOSUR integration.

Egizio Valceschini
Dr Egizio Valceschini, PhD (Agricultural Eonomics), is Research Director at the Institut National de la Recherche Agronomique (INRA). He has published numerous research articles on the economics of quality and strategies for differentiation within the European agrifood sector. He also works on the traceability and labelling of GM products and has managed several national and European research projects. He is adviser to the French Ministry of Agriculture on matters of regulation and public policy. Since 2006, he has been the French representative for the Ministry of Research on the Standing Committee on Agricultural Research (SCAR). He is currently coordinator of the ERANET RURAGRI project 'Facing Sustainability: New Relationships between Rural Areas and Agriculture in Europe (2009–2013)'.

Domenico Viti
Domenico Viti is Professor of Agricultural and Food Law at the University of Foggia, where he is Delegate of the Rector for Environmental and Energy Policy. He is also a Member of the Executive Board of the Italian Agricultural Law Association and has written extensively on agri-environmental law, biosafety, and agricultural co-operatives.

Table of Cases

BRAZIL

Case 1998-00.36170-4 280
Case 1998-34-00.027682-8 280

CANADA

Bernèche v Canada (Procureur Général) 2007 QCCS 2945 (CanLII) 269, 270
Bernèche v Canada (Procureur Général) 2008 QCCA 1581 (CanLII);
    2008 QCCS 2815 (CanLII) 269
Bernèche v Canada (Procureur Général) 2008 QCCS 2248 (Can LII) 271
Canadian National Rly v Norsk Pacific Steamships Co [1992] 1 SCR 1021 267
Childs v Desormeaux [2006] 1 SCR 643 264
Cooper v Hobart [2001] 3 SCR 537 .... 261, 263, 264, 265
Edwards v Law Society of Upper Canada [2001] 3 SCR 562 ..... 261, 263, 264, 265
Eliopoulos v Ontario (2006) 82 Ontario Rep (3d) 321 (Ont CA) 264
Harvard College v Canada (Commissioner of Patents) [2004] 4 SCR 45 258–9
Hercules Management v Ernst & Young [1997] 2 SCR 165 263
Hoffman and Beaudoin v Monsanto Canada and Bayer Cropscience Inc (2007)
    283 DLR (4th) 190 (Sask CA); [2005] 7 Western Weekly Rep
    665 (Sask QB) 6, 21, 256, 257, 259–63, 267, 268, 269, 270, 271, 272
Holland et al v Saskatchewan (2004) 258 Saskatchewan Rep 243 (Sask QB) 270–71
Holland v Saskatchewan 2008 SCC 42; (2007) 281 DLR (4th) 349
    (Sask CA); (2006) 277 Saskatchewan Rep 131 (Sask QB) 270–1, 272
Hollick v Toronto (City) [2001] 3 SCR 158 261
Just v British Columbia [1989] 2 SCR 1228 265
McMillan v Canada Mortgage and Housing Corporation [2008] 3
    Western Weekly Rep 505 (BC Sup Ct) 264
Monsanto Canada Ltd v Schmeiser [2004] 1 SCR 902; (2002) 218 DLR
    (4th) 31 (Federal CA); (2001) 12 Canadian Patent Rep (4th) 204
    (Federal Ct, Trial Div) 6, 257–9, 271
Odhavji Estate v Woodhouse [2003] 3 SCR 263 263
Ridley Inc v Bernèche 2006 QCCA 984 (CanLII); 2006 QCCS 3046 (CanLII)
    (sub nom Bernèche v Canada (Procureur Général)) 268–9
Ring v R (2007) 268 Nfld & PEI Rep 204 (Nfld & Labrador Sup Ct, Trial Div) 264
Ryan v Victoria (City) [1999] 1 SCR 201 269–70
Sauer v Canada (Attorney General) (2007) 225 Ontario App Cases 143 (Ont CA);
    (2006) 79 Ontario Rep (3d) 19 (Ont Sup Ct) 268–71, 272

Sauer v Canada (Attorney General) 2008 CanLII 43774 (Ont Sup Ct) 271
Shell v Farrell [1990] 2 SCR 311 262
Western Bank v Alberta [2007] 2 SCR 3 262
Williams v Canada (AG) (2005) 257 DLR (4th) 704 (Ont Sup Ct) 264, 265

EUROPEAN COURT OF JUSTICE AND COURT OF FIRST INSTANCE

Joined Cases T-74/00, T-76/00, T-83-85/00, T-132/00, T-137/00, and
 T-141/00 *Artegodan* [2002] ECR II-4945 97
Case C-6/99 *Association Greenpeace France v Ministère de l'Agriculture et de la Pêche*
 [2000] ECR I-1651 .... 21, 74, 85, 109, 110
Case C-286/02 *Bellio F.lli Srl v Prefettura di Treviso* [2004] ECR I-3465 74–5
Case C-194/94 *CIA Sec Int'l SA v Signalson SA* [1996] ECR I-2201 155
Case C-296/01 *Commission v France* [2003] ECR I-13909 ....103, 166
Case C-429/01 *Commission v France* [2003] ECR I-14439 166
Case C-121/07 *Commission v France* (ECJ, 9 December 2008) 166
Case C-427/07 *Commission v Ireland* (ECJ, 16 July 2009) 21
Case C-165/08 *Commission v Poland* (ECJ, 16 July 2009) 71
Case C-552/07 *Commune de Sausheim v Azelvandre* (ECJ,
 17 February 2009) 22, 99, 156, 172
Case T-257/07 *France v Commission* [2007] ECR II-4153 64
Case C-316/01 *Glawischnig v Bundesminister für soziale Sicherheit und Generationen*
 [2003] ECR I-5995 71
Joined Cases T-366/03 and T-234/04 *Land Oberösterreich v Commission*
 [2005] ECR II-4005 .... 73, 150, 169, 193
Joined Cases C-439/05P and C-454/05P *Land Oberösterreich v Commission*
 [2007] ECR I-7141 .... 73, 117, 119, 150, 169, 193, 195
Case C-236/01 *Monsanto Agricoltura Italia SpA v Prezidenza del Consiglio dei Ministri*
 [2003] ECR I-8105 .... 72, 88, 103, 117
Case T-182/06 *Netherlands v Commission* (CFI, 27 June 2007) 118
Case C-405/07P *Netherlands v Commission* (ECJ, 6 November 2008) 118
Case T-13/99 *Pfizer Animal Health SA v Council of the European Union* [2002]
 ECR II-3305 ....64, 97
Case T-139/07 *Pioneer Hi-Bred v Commission* [2007] OJ C155/28 ....103, 112

FRANCE

Conseil d'État, *Préfecture du Gers c/ Conseil Général du Gers* (30 December 2009) 171
Cour de Cassation, Chambre Criminelle, 19 November 2002, No 02-80788
 [2003] Recueil Dalloz 1315 33
Cour de Cassation, Chambre Criminelle, 18 February 2004, No 03-82951 33
Cour de Cassation, Chambre Criminelle, 28 April 2004, No 03-83783 33
Cour de Cassation, Chambre Criminelle, 7 February 2007, No 06-80.108,
 Darsonville Autrey [2007] Recueil Dalloz 573 ....32, 33, 34

Table of Cases xv

Cour de Cassation Chambre Criminelle, 4 April 2007, No 06-80.512 34
CA Agen, Civ 1, 12 July 2007, No 07/00842 169
CAA Bordeaux, 22 September 2004, No 04BX011452, *Préfet de la Haute Garonne* 170
CAA Bordeaux, 12 October 2004, No 04BX01691, *Commune de Saint-André-sur-Sèvre* 170
CAA Bordeaux, 14 November 2006, No 04BX00265, *Commune d'Ardin* 170
CAA Bordeaux, 15 May 2007, No 06BX01555, *Commune de Londigny* 170
CAA Bordeaux, 15 May 2007, No 04BX02001, *Commune de Mouchan* 170
CAA Bordeaux, 15 May 2007, No 05BX02259, *Département du Gers* 170
CAA Bordeaux, 12 June 2007, No 05BX01360, *Commune de Tonnay-Boutonne* 170
CAA Bordeaux, 26 June 2007, No 05BX00570, *Commune de Montgeard* 170
CAA Lyon, 26 August 2005, No 03LY000696, *Commune de Ménat* (Actualités
    Juridiques – Droit Administratif (AJDA)), 9 January 2006, 38 .... 31, 33, 170
CAA Nantes, Second Chamber, 28 March 2007, No 06NT00627, *Commune de Saran* 31
CAA Nantes, 26 June 2007, No 06NT01032, *Commune de Carhaix-Plouger* 170
TA Nîmes, 5 December 2008, No 0802882, *Préfet de Vaucluze contre Commune de Le Thor* 171
TC Orléans, 9 December 2005, No 2345/S3/2005, *Société Monsanto v Dufour* 30, 34
CA Orléans, 26 February 2008, CT0028, No 07:00472 .... 30–35, 382, 383
TA Pau, 6 April 2005, No 0401315, *Préfet du Gers c/Département du Gers* 170
TA Rennes, 3 October 2005, No 0502631, *Préfet du Morbihan* 170
CAA Versailles, 18 May 2006, No 05VE00098, *Commune de Dourdan* 170
CA Versailles, 22 March 2007, No 06/01902 ....31, 34

GATT/WTO

Australia – Salmon: Measures Affecting Imports of Salmon WT/DS18/R, 12 June 1998;
    WT/DS18/AB/R, 20 October 1998 .... 337, 339, 345
Brazil – Retreaded Tyres: Measures Affecting Imports of Retreaded Tyres
    WT/DS332/R, 12 June 2007 367
EC – Asbestos: Measures Affecting Asbestos and Asbestos-containing Products
    WT/DS135/AB/R, 12 March 2001 .... 66, 75, 372, 373
EC – Biotech: Approval and Marketing of Biotech Products WT/DS291/R,
    WT/DS292/R, and WT/DS293/R, 29 September 2006 .... 7, 12, 45, 53, 64, 237,
                                                   243, 337–44, 349–54, 357, 360, 361, 362–8
EC – Hormones: Measures Concerning Meat and Meat Products WT/DS26/R and
    WT/DS48/R, 18 August 1997; WT/DS26/AB/R and WT/DS48/AB/R,
    16 January 1998 .... 34, 63, 64, 337, 340, 344, 346, 348, 349, 372
*EC – Hormones II: United States – Continued Suspension of Obligations in the EC – Hormones
    Dispute* WT/DS320/AB/R, 16 October 2008; and *Canada – Continued Suspension of
    Obligations in the EC – Hormones Dispute* WT/DS321/AB/R, 16 October 2008 ... 294, 373
Japan – Agricultural Products II WT/DS76/R, 27 October 1998; WT/DS76/AB/R,
    22 February 1999 ....341–2, 346, 348, 351
Japan – Alcoholic Beverages II WT/DS8, 10–11/AB/R, 4 October 1996 362
Japan – Apples: Measures Affecting the Importation of Apples WT/DS245/AB/R,
    26 November 2003 ....345, 348

Korea – Beef: Measures Affecting Imports of Fresh, Chilled and Frozen Beef
  WT/DS161/AB/R and WT/DS169/AB/R, 11 December 2000 373
US – Gasoline: Standards for Reformulated and Conventional Gasoline
  WT/DS2/AB/R, 29 April 1996 362
US – Shrimp: Import Prohibition of Certain Shrimp and Shrimp Products
  WT/DS58/AB/R, 12 October 1998 ....363, 365, 366, 371, 372
US – Upland Cotton WT/DS265/R, 8 September 2004; WT/DS265/AB/R,
  3 March 2005 344

INTERNATIONAL COURT OF JUSTICE

Case Concerning Oil Platforms (Islamic Republic of Iran v United States of America),
  Judgment of 6 November 2003 364

ITALY

Constitutional Court, Judgment 150/2005, issued on 12 April 2005 (*Gazetta Ufficiale*,
  20 April 2005) 175–6
Constitutional Court, Judgment 116/2006, issued on 17 March 2006 (*Gazetta Ufficiale*,
  22 March 2006) 177–8
Council of State, Judgment No 138 (19 January 2010) 178–9

OSPAR TRIBUNAL

Dispute Concerning Article 9 of the OSPAR Convention (2003) 42 ILM 1118, 2 July 2003 ... 363–4

SOUTH AFRICA

Trustees for the time being of the Biowatch Trust v Registrar Genetic Resources [2009]
  ZACC 14 (3 June 2009) 21, 241–2

UNITED KINGDOM

Anns v Merton London Borough Council [1978] AC 728 ....261, 263, 264,
                                                              265, 267, 269, 270, 272
Barker v Corus (UK) Ltd [2006] UKHL 20, [2006] 2 AC 572 ....209, 262
Bard Campaign v Secretary of State for Communities and Local Government [2009]
  EWHC 308 (Admin) 20
Berkeley v Secretary of State for the Environment [2001] 2 AC 603 19–20
Cambridge Water Co v Eastern Counties Leather Ltd [1994] 2 AC 264 ....200, 216
Director of Public Prosecutions v Bayer [2003] EWHC 2567 (Admin), [2004] 1 WLR 2856 ... 28
Director of Public Prosecutions v Tilly [2001] EWHC 821 (Admin), The Times,
  27 November 2001 ....28, 381
Downs v Secretary of State for Environment, Food and Rural Affairs [2009] EWCA
  Civ 664, [2008] EWHC 2666 (Admin) 70

Eriden Properties LLP v Falkirk Council [2007] SLT 966 69
Fairchild v Glenhaven Funeral Services Ltd [2002] UKHL 22, [2002] 3 WLR 89 ... 68, 209, 262
Holtby v Brigham & Cowan (Hull) Ltd [2000] 3 All ER 421 68
Hunter v Canary Wharf [1997] 2 WLR 684 204
Latimer v AEC [1953] AC 643 67
Marquis of Granby v Bakewell UDC (1923) 87 JP 105 204
McGhee v National Coal Board [1973] 1 WLR 1 68
Monsanto plc v Tilly The Times, 30 November 1999 ....27, 380, 381, 383
Morgan and Baker v Hinton Organics (Wessex) Ltd [2009] EWCA Civ 107 21
Murphy v Brentwood DC [1991] AC 398 261
Overseas Tankship (UK) Ltd v Miller Steamship Co Pty Ltd (The Wagon Mound (No 2)
 [1967] 1 AC 617 ....67–8, 200
Poppleton v The Trustees of the Portsmouth Youth Activities Committee [2007]
 EWHC 1567 68
Pride of Derby and Derbyshire Angling Association Ltd v British Celanese [1953] Ch 149 204
R v Colchester Justices, ex p Abbott The Times, 13 March 2001 27
R v HTM Ltd [2006] EWCA 1156 69
R v Secretary of State for the Environment, ex p Watson [1999] Env LR 310 ...... 203, 216, 219
R (Burkett) v London Borough of Hammersmith and Fulham [2004] EWCA Civ 1342 21
R (Edwards) v Environment Agency [2008] UKHL 22; [2006] EWCA Civ 877 69
R (Greenpeace Ltd) v Secretary of State for the Environment [2007] EWHC 311 (Admin) ... 20, 55
R (Horner) v Lancashire County Council [2007] EWCA Civ 784 69
Rylands v Fletcher (1868) LR 3 HL 330 ....200, 261, 330
Southwark Borough Council v Williams [1971] 1 Ch 734 27–8
Transco v Stockport MBC [2003] UKHL 61, [2003] 3 WLR 1467 216

UNITED STATES OF AMERICA

Agra Marke Inc v Aventis CropScience USA LP 2005 WL 327020 (ND Ill) 331
Alliance for Bio-Integrity v Shalala 116 F Supp 2d 166 (DDC 2000) 315, 316
Bates v Dow Agrosciences LLC 544 US 431 (2005) 329
Bennett v Larsen Co 348 NW2d 540, 553 (Wis 1984) 329
Borland v Sanders Lead Co 369 So 2d 523 (Ala 1979) 328
Environmental Defense Fund Inc v Costle 578 F2d 337 (DC Cir 1978) 315
Geertson Seed Farms v Johanns 2007 WL 518624 (ND Cal) 327
Langan v Valicopters Inc 567 P2d 218 (Wash 1977) 329
Martin v Reynolds Metals Co 342 P2d 790 (Or 1959) 328
McDermott v Wisconsin 228 US 115 (1913) 318
Mortellite v Novartis Crop Protection Inc 460 F3d 483 (3d Cir 2006) 329
Nickels v Burnett 798 NE2d 817 (Ill App 2003) 327
Riegel v Medtronic Inc 552 US 312 (2008) 329
Sample v Monsanto 283 F Supp 2d 1088 (ED Mo 2003) 331, 332
Superior Farm Management v Montgomery 513 SE2d 215 (Ga 1999) 327

In re StarLink Corn Products Liability Litigation 212 F Supp 2d 828 (ND Ill 2002) 6, 13, 128, 321, 322, 324, 326, 329, 330, 331–3
Ultramares Corp v Touche (1931) 255 NY 170 268
US v Anderson Seafoods Inc 622 F2d 157 (5th Cir 1980) 314
US v Lexington Mill & Elevator Co 232 US 399 (1914) 314
Village of Wilsonville v SCA Services Inc 426 NE2d 824 (Ill 1981) 326

Table of Legislation

ARGENTINA

Decree 2183/91 .281
Decree-Law 6740/63 281
Law 13636/49 .281
Law 20247/63 . . . . 281, 282
    Art 1 282
    Art 29 282
Law 24375/94 .295
Ministerial Resolution
    No 124 .281
Resolution 39/2003 282
Resolution 57/2003 282
Resolution 26/2004 282

BRAZIL

Decree 2519/98 295
Decree 39314/99 284
Decree 6092/02 285
Decree 4680/03 287
Decree 42618/03 284
Decree 5591/05 285–6
    Art 5 287
Decree 5705/06 295
Federal Constitution 280, 287
    Art 25 287
Law 8974/95 285
Law 9453/99 284
Law 111/05 (Biosafety Law of
    2005) 284, 285–7,
                                                  298
    Art 14 287
State of Acre
    Law 1238/97 285
State of Amapá
    Law 388/97 285
State of Para
    Law 6328/00 285

State of Santa Catarina
    Decree 6092/02 285
    Law 12128/02 285

CANADA

Code of Civil Procedure, RSQ, c C-25
    art 1010 269
Class Actions Act, Statutes of
    Saskatchewan (SS) 2001, c C-12.01,
    s 6 260
Crown Liability and Proceedings Act, Revised
    Statutes of Canada (RSC) 1985,
    c C-50 263
Environmental Assessment Act, SS 1979–80,
    c E-10.1 260
Environmental Management and Protection
    Act, SS 2002, c E-10.21 260
Feeds Act, RSC 1985, c F-9 269
Health of Animals Act, SC 1990, c 21 269

CAMEROON

Law No 2003/006 of 21 April 2003 to Lay
    Down Safety Regulations Governing
    Biotechnology in Cameroon 247–8
    s 6(3) 248
    s 11 248
    s 12 248
    s 18 247
    s 20 247
    s 25 247
    s 35 248
    s 37 248
    s 49(1) and (2) 248
    ss 51–3 247
    ss 60–64 248

EUROPEAN UNION

Decisions

Council Decision (EC) 87/373 of 13
July 1987 laying down the procedures
for the exercise of implementing
powers conferred on the Commission
[1987] OJ L 197/33 110
Council Decision (EC) 99/468 of 28
June 1999 laying down the procedures
for the exercise of implementing powers
conferred on the Commission [1999]
OJ L184/23 ....103, 109
Art 5 110
Commission Decision (EC) 2003/653 of 2
September 2003 relating to national
provisions on banning the use of
genetically modified organisms in the
region of Upper Austria notified by the
Republic of Austria pursuant to Article
95(5) of the EC Treaty [2003] OJ
L230/34.......... 73, 116, 117, 150
Council Decision (EC) 2005/370 of 17
February 2005 on the conclusion, on
behalf of the European Community, of
the Convention on Access to Information,
Public Participation in Decision-making
and Access to Justice in Environmental
Matters [2005] OJ L124/1.........15
Commission Decision (EC) 2006/19 of 3
March 2006 authorising the placing on
the market of food containing, consisting
of, or produced from genetically modified
maize line 1507 (DAS-Ø15Ø7-1)
pursuant to Regulation (EC) No 1829/
2003 of the European Parliament
and of the Council [2006]
OJ L70/82 112
Council Decision (EC) 2006/512 of 17
July 2006 laying down the procedures
for the exercise of implementing powers
conferred on the Commission [2006]
OJ L200/11 113

Commission Decision (EC) 2008/62 of 12
October 2007 relating to Article 111 and
172 of the Polish Draft Act on Genetically
Modified Organisms, notified by the
Republic of Poland pursuant to Article
95(5) of the EC Treaty as derogations
from the provisions of Directive 2001/18/
EC [2008] OJ L16/17....... 116, 117

Directives

Council Directive (EEC) 79/409 on the
conservation of wild birds (Wild
Birds Directive) [1979] OJ
L103/1...........205, 210, 211, 223
Art 4(2) 210
Annex I 210
Council Directive (EEC) 85/337 on the
assessment of the effects of certain public
and private projects on the environment
(EIA Directive) [1985] OJ L175/40 ...20
Art 6 19–20
Art 6(2) 20
Council Directive (EEC) 85/374 on the
approximation of the laws, regulations
and administrative provisions of the
Member States concerning liability for
defective products (Products Liability
Directive) [1985] OJ L210/29, as
amended by Directive (EC) 99/34 of the
European Parliament and of the Council
[1999] OJ L141/20.............152
Council Directive (EEC) 90/219 on the
contained use of genetically modified
micro-organisms [1990]
OJ L117/1 .... 80, 179, 199
Preamble 82
Council Directive (EEC) 90/220 on the
deliberate release into the environment of
genetically modified organisms [1990]
OJ L117/15 ....66, 80, 83, 84,
87, 102, 106, 109,
199, 200, 338, 341, 352
Part B........................200

Part C 200
Arts 12 and 13 106
Art 16 ....149, 343
Council Directive (EEC) 90/313 on the freedom of access to information on the environment [1990] OJ L158/56 71
Council Directive (EEC) 91/414 on the placing of plant protection products on the market [1991] OJ L230/1 70
Council Directive (EEC) 92/43 on the conservation of natural habitats and of wild fauna and flora (Habitats Directive) [1992] OJ L 206/7 ......119–20, 205, 210, 211, 215, 223, 224
Annexes I, II, and IV.............210
Council Directive (EC) 96/61 concerning integrated pollution prevention and control [1996] OJ L257/26 61
Council Directive (EC) 97/11 [1997] OJ L73/5......................20
Directive (EC) 98/34 of the European Parliament and of the Council laying down a procedure for the provision of information in the field of technical standards and regulations [1998] OJ L204/37.....................120
Art 9 155
Council Directive (EC) 98/81 [1998] OJ L330/13 ....82, 179
Directive (EC) 99/2 of the European Parliament and of the Council on the approximation of the laws of the Member States concerning food and food ingredients treated with ionising radiation [1999] OJ L 66/16 95
Directive (EC) 99/3 of the European Parliament and of the Council on the establishment of a Community list of food and food ingredients treated with ionising radiation [1999] OJ L66/24........95
Directive (EC) 2000/60 of the European Parliament and of the Council establishing a framework for Community action in the field of water policy (Water Framework Directive) [2000] OJ L327/1.......................17
Art 14 17
Directive (EC) 2001/18 of the European Parliament and of the Council on the deliberate release into the environment of genetically modified organisms (Deliberate Release Directive) [2001] OJ L106/1........... 2, 3, 17, 31, 33, 34, 56, 64, 66, 73, 80, 83–7, 89, 90, 93, 98, 99, 103, 105–09, 111, 114, 115, 116, 117, 129, 131, 134, 136, 138, 148, 149, 154, 164, 165, 176, 178, 179, 192, 195, 200, 214, 338, 341, 352
Part B..................84, 85, 99
Part C 84, 85, 90, 99, 200
Preamble (4)................. 2, 217
Preamble (10) 17
Preamble (25) 83
Preamble (28) 85
Preamble (46) 17
Preamble (62) 104
Art 1 .... 84, 85, 97–8
Art 2 66
Art 2(2) 79
Art 2(3) 105
Art 2(4) 84
Art 2(8) 104
Art 4 .... 84, 85, 97–8
Art 4(1) 66
Art 6(5) 84
Art 8 84
Art 9 .... 17, 84, 99
Art 10 84
Art 11 84
Art 13 106
Art 13(1) 85, 106
Art 13(2) 85
Art 14(1) and (2).................106
Art 14(2) 85
Art 14(3) 85, 106

Art 15 85
Art 15(1) 106
Art 15(2) 106
Art 15(3) 106
Art 15(4) 65, 200
Art 20 ....86, 114
Art 20(4) 86
Art 21 ....86, 93, 164
Art 21(3) 86, 129
Art 22 ....86–7, 114, 130, 176
Art 23 ....66, 71, 86, 114, 149,
                          176, 196, 343
Arts 24 and 25 171
Art 24 ....99, 104
Art 24(1) 106
Art 25 22
Art 25(4) 99, 172
Art 26a 118, 131, 138, 165,
                          187, 192
Art 26a(1) 131
Art 26a(2) 131
Art 28 ....85, 107
Art 29 104
Art 31 ....86, 156
Art 31(7) 104
Art 31(7)(d) 48, 369
Annex II 48, 66, 104, 200
Annex III 84
Annex IIIA, paras IV and V 84
Directive (EC) 2003/4 of the European Parliament and of the Council on public access to environmental information [2003] OJ L41/26....................71, 80
Directive (EC) 2003/35 of the European Parliament and of the Council relating to the drawing up of certain plans and programmes relating to the environment [2003] OJ L156/17...............20
Directive (EC) 2004/35 of the European Parliament and of the Council on environmental liability with regard to the prevention and remedying of environmental damage (Environmental Liability Directive) [2004] OJ L143/56 ....3, 6, 17, 152, 199, 200, 201, 202, 206–24
Preamble (2) 201
Preamble (14) 152
Preamble (18) 201
Art 2 217
Art 2(1) 210
Art 2(1)(a) 210, 211
Art 2(1)(c) 210
Art 2(2) 211
Art 2(3) 210
Art 2(6) 215
Art 2(7) 215
Art 2(14) 211
Art 3 ....152, 214
Art 3(1) 209
Art 3(1)(b) 215
Art 3(3) 219
Art 4(5) 213
Art 5 213
Art 6 ....213, 216
Art 7 216
Art 8 217
Art 8(4)(a) 220
Art 8(4)(b) 219
Art 12 219
Art 12(1) 17
Art 13(1) 219
Art 14(1) 222
Art 17 ....207, 217, 218
Art 19 207
Annex I 211
Annex II 216, 217
  para 1.1.2..................217
Annex III 209, 213, 214,
                          215, 220
  paras 10 and 11 ....207, 214
  para 11.......................3
Directive (EC) 2006/21 of the European Parliament and of the Council [2006] OJ L102/15 199

Directive (EC) 2009/41 of the European Parliament and of the Council on the contained use of genetically modified micro-organisms [2009] OJ L125/75 . . . . 80, 82, 83, 84, 179
Preamble (17) 82
Art 2 (c) 82
Art 2 (d) 82
Art 4(2) 82
Art 4(3) 82
Art 4(6) 82
Art 6 83
Art 7 83
Art 8(2) 83
Art 8(3) 83
Art 9(2) 83
Art 10(2) 83
Art 10(3) 83
Art 17(1) 83
Art 17(2) and (3). 83
Annex IV 82
Annex IV, Table 1B. 82
Annex V, Part A 83

Recommendations

Commission Recommendation (EC) 2003/556 on guidelines for the development of national strategies and best practices to ensure the coexistence of genetically modified crops with conventional and organic farming (2003 Commission Recommendation) [2003] OJ L189/36 . . . . . .51, 119, 120, 124, 133–8, 139, 152, 154, 155, 156, 164, 165, 175, 178, 180, 183, 186, 187, 191, 192, 194, 195
Preamble (1) 3, 133
Preamble (3) 165, 181
Preamble (4) 178
Preamble (4) and (5) 165
Preamble (4)–(8) 134
Preamble (5) 3, 165, 178, 186
Preamble (6) 134
Annex, para 1.1 . . . 124, 130, 133–4, 165, 178, 194
para 1.2 . . . . 134, 165, 178
para 1.4 165–6
para 1.5 .134
paras 2.1–2.1.12 155
paras 2.1.1–2 134
paras 2.1.3–7 135
para 2.1.4. 156
para 2.1.5. . . . . .135, 136, 187, 194
para 2.1.7. . . . . 135, 139, 185
para 2.1.8. . . . . 134, 191
para 2.1.9. . . . . 3, 135, 152
paras 2.1.10–12. 135
para 2.1.11 155
para 2.2 .135
para 2.2.1. 135
para 2.2.3. . . . . 135, 186, 192
para 2.2.4. 135
para 2.2.5. 135
para 2.2.6. 135
para 2.2.7. 135
para 3 136
para 3.2.1. . . . . .51, 186
paras 3.2.1–3 136
para 3.3 .136
para 3.3.3. 183
paras 3.4–6. 136
paras 3.7–9. 136

Regulations

Council Regulation (EEC) 2092/91 on organic production of agricultural products and indications referring thereto on agricultural products and foodstuffs [1991] OJ L198/1. . . . . 147, 148
Art 5(3)(h) .147
Art 6 147

Regulation (EC) 258/97 of the European
    Parliament and of the Council
    concerning novel foods and novel
    food ingredients [1997]
    OJ L43/1 . . . . .72, 88, 97, 338, 341
    Preamble (8) .88
    Art 3(4) and (5) 88
    Art 12 .72, 343
Regulation (EC) 178/2002 of the European
    Parliament and of the Council laying
    down the general principles and
    requirements of food law, establishing
    the European Food Safety Authority,
    and laying down procedures in
    matters of food safety (Food
    Law Regulation) [2002]
    OJ L31/1 . . . . 81, 85, 89, 94–9
    Preamble (1) .95
    Preamble (1)–(3) 95
    Preamble (2) .95
    Preamble (11) 96
    Preamble (34) 107
    Art 3(13) 99
    Art 7 . . . . .91, 98
    Art 9 99
    Art 10 98
    Art 14(8) 98
    Art 15(5) 98
    Arts 17–20 98
    Art 18 98
    Art 30 108
    Art 36 108
    Arts 53 and 54 . . . . .91, 115
Regulation (EC) 1829/2003 of the
    European Parliament and of the
    Council on genetically modified food
    and feed (Food and Feed Regulation)
    [2003] OJ L268/1 . . . . .51, 72, 80,
                    85, 89–91, 93, 96, 98, 103,
                    105–109, 115, 118, 129,
                    131, 138, 148, 149, 164,
                    165, 181, 192, 221
    Preamble (6) and (7) 89
    Preamble (9) .98
    Preamble (16) 89
    Preamble (21) 89
    Preamble (28) 138
    Preamble (32) 104
    Art 4(1)(a) 90
    Art 4(1)(b) and (c) 90
    Art 5(2)(b) 90
    Art 5(3) 90
    Art 6(1) 90
    Art 6(3) 106
    Art 6(3)(b) and (c) 107
    Art 6(3)(c) .107
    Art 6(4) 90, 107
    Art 6(6) 108
    Art 6(7) 91, 104,
                                                   109
    Art 7(1) 90, 104, 109
    Art 7(5) 91
    Art 9 91
    Art 10 115
    Art 12 93
    Art 12(2) 93, 129,
                                                   164
    Art 12(3) 129
    Art 16(1)(a) .90
    Art 16(1)(b)–(d) 90
    Art 17(2)(b) .90
    Art 17(3) 90
    Art 18(1) 90
    Art 18(4) 90
    Art 18(7) 91, 104
    Art 19(1) 90, 104
    Art 19(5) 91
    Art 21 91
    Art 24 . . . . .93, 129
    Art 24(2) 93, 129, 164
    Art 29 91
    Art 30(1) 91
    Art 30(3) 91
    Art 33 104
    Art 34 . . . . .91, 115, 149
    Art 35 .90, 109

Regulation (EC) 1830/2003 of the European
Parliament and of the Council
concerning the traceability and labelling
of genetically modified organisms and
the traceability of food and feed
products produced from genetically
modified organisms [2003]
OJ L268/24 . . . . 51, 80, 86, 89,
91–3, 97, 99, 129, 164, 183
Preamble (3) 93, 98
Preamble (4) 92
Preamble (11) 92
Art 2 92
Arts 4 and 5 92
Art 4(6) 93
Art 4(7) and (8). 93, 129, 164
Regulation (EC) 1946/2003 of the European
Parliament and of the Council on
transboundary movements of
genetically modified organisms
[2003] OJ L287/1 81
Commission Regulation (EC) 65/2004
establishing a system for the development
and assignment of unique identifiers
for genetically modified organisms
[2004] OJ L10/5 92
Commission Regulation (EC) 641/2004
[2004] OJ L102/14 90
Regulation (EC) 1367/2006 of the European
Parliament and of the Council on the
application of the Aarhus Convention
on Access to Information, Public
Participation in Decision-making and
Access to Justice in Environmental
Matters to Community institutions
and bodies [2006] OJ L264/13. 15
Council Regulation (EC) 834/2007 on
organic production and labelling of
organic products [2007]
OJ L189/1 . . . . 4, 93, 94, 148, 181, 192
Preamble (9) 93
Preamble (9) and (10) 168
Preamble (10) 93

Art 9 168
Art 9(2) 192
Art 23(3) 148
Art 34 148
Regulation (EC) 1333/2008 of the European
Parliament and of the Council on food
additives [2008] OJ L354/16
Art 6 98

EGYPT

Ministerial Decree No 85 of 1995 243
Ministerial Decree No 136 of 1995 243
Ministerial Decree No 242 of 1997 243
Ministerial Decree No 1648 of 1998 243

FRANCE

Charter for the Environment 32, 34
Code de l'Environnement
  Art L.335-1. 170
  Arts L.531-1ff 29, 166
  Art L.531-2-1, paras 1 and 5 166
    para 1 166
    para 5 166
  Art L.533-5. 173
Code Général des Collectivités Territoriales
  Art L.2212-2. 170
Code Pénal
  Art 122-7. 30
  Art 122-5. 30
  Art L.322-1. . . . . 29, 30
Code de Procedure Civile
  Art 808. 169
  Art 809. 169
Code Rural
  Art L.251-18. 173
  Art L.642 173
  Art L.663-1, para 1 171
    para 2 172
    para 3 172
  Art L.663-2. 173
  Art L.663-3. . . . . 173, 174

Art L.663-4-III 32–3
Art L.671-14. 173
Art L.671-15 and 16 174
Art L.671-15. 29
Law No 2008-595 of 25 June 2008 relating
 to GMOs, Journal Officiel de
 la République Française (JORF)
 No 148 of 26 June 2008,
 NOR: DEVX0771876L (codified
 at Articles L.531-1ff of the Code
 de l'Environnement) 29, 30,
   33, 166, 168,171, 173
Ministerial Decree of 7 February 2008
 Suspending the Cropping of
 Genetically Modified Maize Seed
 (Zea Mays L. Line MON810),
 JORF No 34 of 9 February 2008,
 NOR: AGRG0803466A,
 amended 13 February 2008,
 NOR: AGRG0803888A . . . 30, 87, 166

ITALY

Civil Code
 Art 2135. 177
Constitution
 Art 9 175
 Art 114. 175
 Art 117. . . . . 175, 176, 177
 Art 117(3) 178
Government Legislative-Decree
 No 279/2004 . . . . .176, 177,
   178
 Art 1 . . . . 176–7, 178
 Art 2 . . . . 177, 178
 Art 3 . . . . 177, 178
 Art 4 . . . . 177, 178
Statute No 5 of 2005 176
Statute of the Regione Puglia of 4 December
 2003 (No 26). 175
Statute of the Regione Marche of 3
 March 2004 (No 5). 175

KENYA

Biosafety Act 2009 246–7
 Part IV. 247
 s 3 246
 ss 18–23. 247
 s 19 247
 s 28 247
 s 29 247
 s 30 247
 ss 40–42 247
 ss 50–51 247

MALAWI

Biosafety Act (Act No 13
 of 2002). 244
 ss 17 and 18 244
 s 22(a) 244
 s 28 245
 s 30 245
Science and Technology Act of 2003
 (Act No 17 of 2003) 244

PARAGUAY

Decree 18481/97 . . . . 287, 288–9
 Art 2 288
Decree MAG 10661/2000. 288
Decree 12706/08 287
Law 253/93. 295
Law 2309/03 . . . . 289, 295
Law 2459/04 288
Resolutions MAG 62/99, 82/99,
 07/99, 554/99, and 201/00 288
Resolution 10/22/04. 288

SOUTH AFRICA

Consumer Protection Act 2008
 (Act No 68 of 2008) 242–3
 s 24(6) 242

Table of Legislation xxvii

Genetically Modified Organisms Act
(Act No 15 of 1997) 240–42
s 1(m), as amended by the Genetically
Modified Organisms Amendment
Act (Act No 23 of 2006) 241
s 2 240
s 3, as amended 242
s 7, as amended 242
s 17(2) 241
s 18 . . . . .22, 241
s 19, as amended 242
s 19(6), as amended 242
s 20 240
Genetically Modified Organisms
Amendment Act (Act No 23 of
2006) 240, 241, 242
Regulation No 1420 of 26 November
1999 240
reg 2(1) 240
reg 2(2) 240
reg 3(1) 241
reg 3(2) 241
reg 6 242
Annexure, Table 3 240
Regulations relating to the labelling of
foodstuffs obtained through certain
techniques of genetic modifications
(Notice R 25 of 16 January 2004),
made pursuant to the Foodstuffs,
Cosmetics and Disinfectants Act
1972 (Act No 54 of 1972) 242

SPAIN

Constitution
Art 14.1.7a 185
Art 149.1.13.a, 16a, and
23a 185
Law 9/2003 of 25 April 2003, *Boletín
Oficial del Estado* (BOE) No 100 of
26 April 2003 . . . . 179, 180
Law 26/2005 of 30 November 2005,
amending Law 49/2003 of 26
November 2003 on agricultural
tenancies, BOE No 287 of
1 December 2005 183
Law 30/2006 of 26 July 2006, BOE No 178
of 27 July 2006 185
Royal Decree 1697/2003 of 12 December
2003, BOE No 310 of 27 December
2003 185
Royal Decree 178/2004 of 30
January 2004, BOE No 27 of 31
January 2004 . . . . 179, 180

TANZANIA

Environmental Management Act of
2004 249

UNITED KINGDOM

Acts of Parliament
Agricultural Holdings Act 1986,
s 15 183
Criminal Justice and Public Order Act 1995,
s 68(1) 28
Environmental Protection Act 1990
Part VII 205
s 111 221
Magistrates' Courts Act 1980,
s 22(1) and (4) 27
Natural Environment and Rural Communities
Act 2006 215
Part 1 211
s 2 205
s 7 212
Natural Heritage (Scotland)
Act 2003 205
Public Health Act 1875 67
Wildlife and Countryside
Act 1981 . . . . 211, 223
s 28, as amended by Sch 9 to the
Countryside and Rights of Way Act
2000 . . . . 205, 211
s 28E 205

s 28P(6A), as amended by the Natural Environment and Rural Communities Act 2006, s 55(3) 215

Statutory Instruments
Conservation (Natural Habitats etc) Regulations 1994,
SI 1994/2716 . . . . 205, 211
reg 19(2). 205
reg 20. 211
Environmental Damage (Prevention and Remediation) Regulations 2009, SI 2009/153 211
reg 2(1) 211
reg 4. . . . . 212, 215
reg 4(1) 223
reg 4(1)(a). 211
reg 5. . . . . 214, 215
reg 5(1) 215
reg 19(3)(d) 221
reg 19(3)(e). 221
reg 29. 219
reg 33. . . . . 217, 218
Sch 1, para 5. 212
Sch 2, para 9. 214
Sch 3, para 1(g). 221
para 1(h) 221
Environmental Damage (Prevention and Remediation) (Wales) Regulations 2009, SI 2009/995
reg 19(e). 221
Sch 3 221
Genetically Modified Organisms (Deliberate Release) Regulations 2002, SI 2002/2443 221

UNITED STATES OF AMERICA

Cal Food and Agr Code s 52100 319
s 52300–6. 320
s 52305 320
ss 55000–108 320
ss 71000–138 320

Code of Federal Regulations (CFR)
7 CFR 340 302
7 CFR 340.1. 302
7 CFR 340.3(b)(1)–(6) 302
7 CFR 340.3(b)4(iii) 303
7 CFR 340.3(c). 302
7 CFR 340.3(d)2. 302
7 CFR 340.3(d)(5) 306
7 CFR 340.3(d)(6) 302
7 CFR 340.3(e)4 302
7 CFR 340.3(e)5 303
7 CFR 340.4(b). 303
7 CFR 340.4(b)(1)–(14). 303
7 CFR 340.4(f) 303
7 CFR 340.6. . . . . 18, 303
7 CFR 340.6(b). 304
7 CFR 340.6(c). 304
7 CFR 340.6(d) and (f) 304
21 CFR 170.3(i) 314
21 CFR 170.30. 314
21 CFR 170.30(a) and (b) 315
21 CFR 170.30(b). 314
21 CFR 170.35. 315
21 CFR 182, 184, and 186 314
40 CFR 152 307
40 CFR 158 309
40 CFR 167 309
40 CFR 172 308
40 CFR 172.3(a) 308
40 CFR 172.3(b) and (c) 308
40 CFR 172.10(a) 308
40 CFR 174 . . . . 307, 309
40 CFR 174.3. 308
40 CFR 174.501–28 310
Constitution
art I, s 8, cl 3 318
art VI, cl 2 318
amend X. 318
Federal Insecticide, Fungicide, and Rodenticide Act (FIFRA),
7 USC ss 136–136y 307, 308,
318, 332
7 USC s 136a 308

Table of Legislation

7 USC s 136a(c)(5) 308
7 USC s 136(bb) 308
7 USC s 136c 308
7 USC s 136c(b) 308
7 USC s 136c(d) 308
7 USC s 136e 308
7 USC s 136(u) 308
7 USC s 136v(b) 318, 329
7 USC s 136w(b) 308
Federal Seed Act, 7 USC ss 1551–611 . . .320
Fla Stat s 581.083 319
s 604.60 319
Food, Conservation and Energy Act of 2008, Pub L 110–246
s 10204 306
Food, Drug, and Cosmetic Act (FDCA), 21 USC,
ss 301–99 . . . . 307, 308, 309, 310, 314, 315,318, 330
21 USC s 321(n) 316
21 USC s 321(qq) 315
21 USC s 321(s) 314
21 USC s 321(u) 314
21 USC s 342 310
21 USC S 342(a) 314
21 USC s 342(a)(1) 310, 314
21 USC s 342(a)(2)(C) 314
21 USC s 343(a) 315, 316
21 USC s 343(i) 315
21 USC s 343(w) 315, 316
21 USC s 346a 307, 310
21 USC s 346a(a)(1) 309
21 USC s 346a(b)(2)(A)(i) and (ii) . . .310
21 USC s 346a(c)(2)(A)(i) and (ii) . . .310
21 USC s 348 . . . . 310, 315
21 USC s 348(a) 314
21 USC s 348(b) and (c) 314
21 USC s 348(c)(3)(A) 309
21 USC s 360b(d)(1)(H) 309
21 USC s 379e(b)(5)(B) 309
Food Quality Protection Act of 1996, Pub L 104–170, 110 Stat 1489, 1513 (3 August 1996) 309

Hawaii Rev Stat s 321–11.6 318
Idaho Code s 22–2016 319
Illinois 430 ILCS 95/0.01-11 317
430 ILCS 95/1–11 (Illinois Release of Genetically Engineered Organisms Act) 319
Iowa Code
s 15E.209 319
s 199.13A 320
Kan Stat Ann s 2–1450 320
Maine Rev Stat Tit 7
s 1052 320
ss 1053–1054 320
Mich Comp Laws s 211.7ii 319
Mo Rev Stat ss 261.256 and 261.259 . . . 320
National Environmental Policy Act,
42 USC ss 4321–70f 303
42 USC s 4332(2)(C) 303
NJ Stat s 40:8C-2 320
Plant Protection Act (PPA),
7 USC ss 7701–72 . . . . .17–18, 301, 302, 305, 306
7 USC s 7702(2) and (10) 306
7 USC s 7702(14) 302
7 USC s 7711 302
7 USC s 7714 305
7 USC s 7756(b) 319
Vt Stat Ann Tit 6
ss 611 and 644 319
Va Code s 3.2-4008 319

URUGUAY

Decree 249/00 . . . . 290, 291
Decree 37/07 291
Decree 353/08, Official Journal of 28 July 2008 . . . . .279, 292
Law 16408/93 .295
Law 17283/00 .291
Resolution 267/03 291
Seeds Law of 1997 290

Table of Treaties and Conventions

Aarhus Convention on Access to Information, Public Participation in Decision-making and Access to Justice in Environmental Matters 15, 16, 17, 19, 20, 21, 32, 34, 35, 60
Art 2(5) 16
Art 4(4) 22
Art 4(4)(a) and (d) 21–2
Art 6 15
Art 6*bis*. 15–16, 35
Art 7 . . . . 15, 20
Art 8 15
Art 9(2) 16
Art 9(4) 21
Art 15 15
Annex I*bis*
   para 2 16
   para 7 16
African Model Law on Safety in Biotechnology 2, 6, 16, 19, 22, 228, 233–7, 238, 240, 241, 244, 245, 247, 248, 249, 250, 251
Art 2 234–5
Art 4(1) 235
Art 5(2)–(4) 16, 236
Art 5(5)(i) and (ii) 236
Art 6(7) 235
Art 6(9) 236
Art 6(10) 237
Art 8(2) 235
Art 8(5) 235
Art 11 236
Art 12 22
Art 12(1) and (3) 236
Art 12(2) 236
Art 14 236
Art 14(5) 237
Art 14(7) 237
Annex II 236
Basel Convention on the Control of Transboundary Movements of Hazardous Wastes and Their Disposal 367
Cartagena Protocol on Biosafety to the Convention on Biological Diversity 7, 14, 16, 19, 32, 34, 35, 81, 212, 234, 235, 238, 239, 240, 241, 245, 248, 251, 287, 289, 292, 295, 296, 297, 355, 356, 358, 359, 360, 364, 365, 367, 368, 370, 374
Preamble 368, 370
Art 1 289
Art 4 235
Art 5 234
Arts 7–10 235
Art 12 235
Art 15 289
Art 18 . . . . 287, 294
Art 23 236
Art 23(2) 14
Art 26 . . . . 369, 370, 371
Art 26(1) 368
Art 26(2) 368–9
Art 27 212
Charter of the Regions and Local Authorities of Europe on the Subject of Coexistence of Genetically Modified Crops with Traditional and Organic Farming (4 February 2005) (Charter of Florence) 151, 174, 176, 193–4
Codex Alimentarius 296, 346
Convention on Biological Diversity 212, 295, 364

Art 2 212
Art 14(1)(a) 14, 16
(*See also* Cartagena Protocol on Biosafety to the Convention on Biological Diversity)
Declaration of Rennes. 151
Declaration on Biodiversity Strategy for MERCOSUR. 297
EC Treaty (now Treaty on the Functioning of the European Union)
   Art 30 (now Art 36). 116
   Art 32 (now Art 38). 177
   Art 37 (now Art 43). 116
   Art 95 (formerly Art 100a and now Art 114). 66, 86, 114, 115, 117, 118, 120, 148, 150
   Art 95(4) (now Art 114(4)) 116, 150
   Art 95(5) (now Art 114(5)) 73, 112, 115–117, 150
   Art 95(8) (now Art 114(8)) 116
   Arts 152 and 153 (now Arts 168 and 169). 95
   Art 152(4)(b) (now Art 168(4)(b)) . . . 95, 116
   Art 174(2) (now Art 191(2)). 63, 201
   Art 175 (now Art 192). 82
European Convention on Human Rights 32, 34
Framework Agreement on the Environment (MERCOSUR). 294–5
   Art 5 295
   Art 6(j) and (m) 295
GATT/WTO
   Agreement Establishing the World Trade Organization (WTO Agreement) . . . 365
      Preamble 368, 370, 372
   Agreement on Agriculture. 353
      Annex 2 344
   Agreement on the Application of Sanitary and Phytosanitary Measures (SPS Agreement). 7, 53, 337, 338, 339–46, 348–54, 355, 359, 360, 364, 367, 369, 370, 371, 372, 373, 374

      Preamble 337, 348
      Art 1 353
      Art 1.1. . . . . 339, 343
      Arts 2, 3, and 5. 340
      Art 2.1 344
      Art 2.2 . . . . 342, 343, 344, 346, 348, 349
      Art 2.3 . . . . 342, 343
      Art 3.1 . . . . 344, 349
      Art 3.3 . . . . 344, 348, 349
      Art 5 344
      Art 5.1 . . . . 342, 343, 344, 345, 348, 349, 350
      Art 5.2 . . . . 346, 348
      Art 5.3 371
      Art 5.4 372
      Art 5.5 . . . . 342, 343
      Art 5.6 . . . . 341, 342, 346
      Art 5.7 . . . . 45, 348–50
      Art 7 . . . . 343, 350, 351, 352
      Art 8 352
      Art 10(3) 373
      Annex A. 353
      Annex A(1). 339–43, 351
      Annex A(1)(a)–(d) 339, 342
      Annex A(1)(d). 370
      Annex A(4). 345, 350
      Annex B. 341, 343, 350, 351
      Annex B(1). 351, 352
      Annex B(1) and (2). 350–1
      Annex B(3). 351
      Annex C. 352
      Annex C(1)(a). 352, 353
      Annex C(1)(b) 353
      Annex C(1)(c) and (e) 353
   Agreement on Technical Barriers to Trade (TBT Agreement) 340, 341, 369
      Art 1.3 340
      Art 1.4 340
      Art 1.5 340
      Art 2 340

General Agreement on Tariffs and Trade
(GATT)............. 337, 371, 373
Art III 343
Art III:1....................353
Art III:4................340, 353
Art XX.....................337
Art XX(b).......... 337, 372, 373
Art XX(g)............... 366, 371
Lugano Convention on Civil
Liability 6, 199, 206–9, 218
Art 2(1)(b) 208
Art 2(7) and (10)................208
Art 2(7)(c) 209
Art 6(1) 208, 213
Art 8 208
Art 11 209
Art 17(1) 218
Art 17(2) 218
Art 32(3) 208
Protocol of Olivos to the Treaty of
Asunción 277
Protocol of Ouro Preto to the Treaty of
Asunción 276
Art 42 276
Rio Declaration on Environment and
Development 45, 62, 295
Principle 15 62
Statute of the International Court of
Justice 363

Art 38(1)(c)...................363
Technical Regulation on the Labelling
of Packaged Foods
(MERCOSUR)................296
Treaty of Asunción................ 274,
277, 294
Preamble.....................294
Arts 9–18 276
(*See also Protocols of Olivos and of
Ouro Preto to the Treaty of
Asunción*)
Treaty on European Union
New Art 16....................113
Treaty of Montevideo 1980
Art 7 276
Vienna Convention on the Law of
Treaties 1969
Art 30 361
Art 31 .... 363, 365
Art 31(1) 362,
363, 366
Art 31(2) 362
Art 31(3)(a) 362
Art 31(3)(c)............361, 362, 363,
364, 365, 367, 368
Art 31(4) 362
Art 32 362
WTO
(*See GATT/WTO*)

List of Abbreviations

ACP	Advisory Committee on Pesticides
AEBC	Agriculture and Environment Biotechnology Commission
AFSSA	Agence Française de Sécurité Sanitaire des Aliments
AGPM	Association Générale des Producteurs de Maïs
AIA	advanced informed agreement
AJDA	Actualités Juridiques—Droit Administratif
ALADI	Asociación Latinoamericana de Integración
ALALC	Asociación Latinoamericana de Libre Comercio
APB	African Panel on Modern Biotechnology
AU	African Union
BAT	best available techniques
BATNEEC	best available techniques not entailing excessive costs
BOE	*Boletín Oficial del Estado*
BSE	bovine spongiform encephalopathy
CBA	cost-benefit analysis
CBAC	Canadian Biotechnology Advisory Committee
CFIA	Canadian Food Inspection Agency
COMESA	Common Market for Eastern and Southern Africa
CWB	Canadian Wheat Board
DEFRA	Department for Environment, Food and Rural Affairs
ECA	Economic Commission for Africa
ECOSOC	European Economic and Social Committee
ECOWAS	Economic Community of Western African States
EFSA	European Food Safety Authority
ESA	European Seed Association
EU	European Union
FAO	Food and Agriculture Organization
FOEE	Friends of the Earth Europe
FSA	Food Standards Agency
GATT	General Agreement on Tariffs and Trade
GMMs	genetically modified micro-organisms
GMOs	genetically modified organisms
HSE	Health and Safety Executive
ICTSD	International Centre for Trade and Sustainable Development
IFOAM	International Federation of Organic Agricultural Movements
INRA	Institut National de la Recherche Agronomique
ISAAA	International Service for the Acquisition of Agri-biotech Applications
JORF	*Journal Officiel de la République Française*

JRC	Joint Research Centre
MDGs	Millennium Development Goals
MEC	marginal external cost
MWTP	marginal willingness to pay
NASS	National Agricultural Statistics Service
NEPAD	New Economic Partnership for Africa's Development
NGOs	non-governmental organizations
OAU	Organization of African Unity
OECD	Organisation for Economic Co-operation and Development
PNTs	plants with novel traits
RCEP	Royal Commission on Environmental Pollution
RMBM	ruminant meat and bone meal
SAC	special area of conservation
SADC	Southern African Development Community
SCP	Scientific Committee on Plants
SPA	special protection area
SSSI	site of special scientific interest
UNCTAD	United Nations Conference on Trade and Development
UNDP	United Nations Development Program
USDA	United States Department of Agriculture
WFP	World Food Program
WTO	World Trade Organization
WTP	willingness to pay

Introduction

Michael Cardwell

Genetically modified organisms (GMOs) are now an established feature of the agricultural landscape, with 125 million hectares of biotech crops being grown in 2008.[1] Further, in the United States (which accounted for half of that total area), it is estimated that, as at 2009, GM varieties will comprise some 91 per cent of soya and 85 per cent of maize.[2] On the other hand, the picture is far from even. In 2008, only 14 countries grew 50,000 hectares or more of GM crops;[3] and there are notable agricultural 'powerhouses' where the new technology has been less enthusiastically embraced by producers. For example, Spain is the only Member State of the European Union (EU) in which commercialization of GM crops has occurred on a significant scale (but, even there, only some 100,000 hectares were grown in 2008); and in Russia and Ukraine the biotech revolution has barely started.

Despite this somewhat uneven picture, there may be detected a sense that the advance of modern biotechnology is becoming inevitable, with countries finding it increasingly difficult to preserve the integrity of a 'GM-free' food chain. In particular, three major exporters of livestock feed (Argentina, Brazil, and the United States) have all now embraced the GM revolution;[4] and importing countries are faced with very tight supply of conventional maize and soya. As stated by Commissioner Fischer Boel in October 2009, '[e]ven if we can get the soya from elsewhere—which is far from clear—our livestock sector would have to pay higher prices'.[5] In particular, she pointed to the irony of the European livestock sector collapsing and the slack then being taken up by imports of meat produced with GMOs over which the EU had no control.[6] The advancement of modern biotechnology is also advocated as a means of achieving global food security. Thus, the United Kingdom Royal Society saw GM crops as at least part of the solution in its 2009 report, *Reaping the Benefits: Science and the Sustainable*

---

[1] C James, *Global Status of Commercialized Biotech/GM Crops: 2008 (ISAAA Brief No 39)* (Ithaca, NY: International Service for the Acquisition of Agri-biotech Applications (ISAAA), 2009) Executive Summary.
[2] National Agricultural Statistics Service (NASS), *Acreage* (Washington, DC: USDA, 2009); and see further Chapter 12.   [3] C James (n 1 above) Table 1.
[4] See Chapters 11 and 12.
[5] Speech/09/474, 'GMOs: Letting the Voice of Science Speak—Policy Dialogue at "European Policy Centre"', Brussels, 15 October 2009 (speeches of the Commissioner for Agriculture and Rural Development are available at <http://europa.eu/rapid/>).   [6] Ibid.

Intensification of Global Agriculture.[7] That said, there are voices which consistently deny that GMOs could ever be the 'silver bullet' to remove so major a problem.[8]

Importantly, at the same time there is general acceptance that, once commenced, the spread of GMOs may prove irreversible. Indeed, this is expressly countenanced in the Preamble to Directive (EC) 2001/18 of the European Parliament and of the Council on the deliberate release into the environment of genetically modified organisms ('Deliberate Release Directive').[9] Likewise, the United Kingdom GM Science Review Panel has acknowledged that, for the present, there is no method by which GM crops grown on a commercial scale can be genetically isolated from conventional and organic crops.[10] In addition, it is impossible to discount the occurrence of incidents which lead to large-scale escapes of GMOs into the food chain. For example, in the United States there has been the StarLink™ episode (when GM material not authorized for human food was found in taco shells);[11] and in the EU there have been ongoing difficulties with imports of Chinese rice products.[12]

Against this background, the regulatory response would also appear somewhat uneven. There is clear evidence of certain countries, or wider international bodies, adopting a cautious approach. A graphic illustration is the African Model Law on Safety in Biotechnology, which not only accommodates a strong version of the precautionary principle, but also requires that socio-economic criteria be taken into account.[13] By contrast, other countries (such as Argentina and the United States) base their approach on the principle that GM food is substantially equivalent to traditional food, which to date has led to consistent authorization of new varieties.[14] In this context, a matter of some interest is the extent to which individual countries have defined themselves by reference to their approach to modern biotechnology. To take an extreme case, during the 2002 drought which affected Southern Africa, Zambia refused food aid in the form of maize (including milled grains) from the United States, on the basis that there could be contamination from GMOs;[15] and, more generally, several

---

[7] Royal Society, *Reaping the Benefits: Science and the Sustainable Intensification of Global Agriculture* (London: Royal Society, 2009).

[8] See, eg Policy Department of Oxfam (Great Britain), *Genetically Modified Crops, World Trade and Food Security: Position Paper November 1999* (Oxfam, 1999) 3; and see also Friends of the Earth Press Release, *Royal Society Food Report—Friends of the Earth Reaction*, 21 October 2009.

[9] [2001] OJ L106/1, Preamble (4).

[10] GM Science Review Panel, *GM Science Review: First Report—an Open Review of the Science Relevant to GM Crops and Food Based on Interests and Concerns of the Public* (London: 2003) 18.

[11] See further Chapter 12. Another such episode was that involving LibertyLink rice.

[12] See, eg European Commission, IP/08/219, *Commission Requires Certification for Chinese Rice Products to Stop Unauthorised GMO from Entering the EU*, Brussels, 12 February 2008.

[13] Available at <http://www.nepadst.org>, last accessed 3 July 2009; and see further Chapter 9.

[14] See Chapters 11 (Argentina) and 12 (United States).

[15] See, eg R Walters, 'Crime, bio-agriculture and the exploitation of hunger' (2006) 46 British Journal of Criminology 26.

African countries would seem to associate acceptance of GMOs with a loss of sovereignty. Political considerations, and national positioning, may also be detected behind the overtly anti-GM stance adopted by Scotland, in contradistinction to the more equivocal approach of the UK Government.[16] Further, by spring of 2009, the only GM variety grown commercially in the EU (maize MON810) was banned in six Member States (Austria, France, Germany, Greece, Hungary, and Luxembourg). It would be hard to deny that the GM debate resonates far beyond the confines of agriculture and the food chain and continues to act as a lightning rod for wider issues at national and global level.[17]

While national attitudes to GMOs show considerable variation, ambivalence may also be found in individual legislative regimes, as can be illustrated by three examples from that applicable in the EU. First, under the Deliberate Release Directive, great emphasis is laid upon the authorization process (together with its environmental risk assessment) as a means of ensuring that the deliberate release of GMOs will not have adverse effects upon either human health or the environment. Yet, under Directive (EC) 2004/35 of the European Parliament and of the Council on environmental liability with regard to the prevention and remedying of environmental damage ('Environmental Liability Directive'), deliberate releases of GMOs (as defined under the Deliberate Release Directive) fall within the list of activities which may give rise to liability.[18] Secondly, the Community guidelines on measures for the coexistence of GM and non-GM crops encourage the use of insurance against liability for cross-contamination by GMOs.[19] Yet, by 2009 it could be noted by the European Commission that 'insurance products covering risks of GMO admixture seem not to be available on EU markets' (notwithstanding that the coexistence measures of four Member States were already requiring these products, or might do so following a case-by-case assessment).[20] Thirdly, the same Community guidelines recite that no form of agriculture (whether conventional, organic, or using GMOs) should be excluded in the EU, but also that the issue of coexistence addresses 'the most appropriate management measures that can be taken to *minimise* admixture'.[21]

---

[16] See, eg Scottish Government News Release, *Genetic Modification (GM)*, 24 April 2009 (available at <http://www.scotland.gov.uk/News/Releases/2009/04/24150325>, last accessed 29 June 2009).
[17] For discussion of such issues, see S Bonny, 'Factors Explaining Opposition to GMOs in France and the Rest of Europe' in RE Evenson and V Santaniello (eds), *Consumer Acceptance of Genetically Modified Foods* (Wallingford, Oxon: CABI Publishing, 2004) 169.
[18] [2004] OJ L143/56, Annex III, para 11. See further Chapter 8.
[19] Commission Recommendation 2003/556/EC on guidelines for the development of national strategies and best practices to ensure the coexistence of genetically modified crops with conventional and organic farming ('2003 Commission Recommendation') [2003] OJ L189/36, Annex, para 2.1.9.
[20] European Commission, *Report from the Commission to the Council and the European Parliament on the Coexistence of Genetically Modified Crops with Conventional and Organic Farming* COM (2009)153, 5.
[21] 2003 Commission Recommendation, Preamble (1) and (5) (emphasis added).

In consequence, there is acknowledgement that only by permitting cross-contamination up to defined thresholds can the regulatory regime be made to work; and this must raise the question whether, if GM cropping becomes well-established, organic agriculture will be viable throughout the EU. A product may still be 'organic' as decreed by the Community legislation;[22] but, if it is not organic in the mind of the consumer, its commercial position may be jeopardized.

Accordingly, it would seem clear that the regulation of GMOs will increasingly affect the food supply of every person and that different countries will in all likelihood continue to adopt substantially different approaches. It would also seem clear that, even where countries have proceeded to the enactment of regulatory regimes, those regulatory regimes may present internal contradictions and be strongly contested. Significantly, such issues do not seem to dissolve with time. On the contrary, there is some evidence that the passage of years, together with greater information and education, serve only to entrench the views of the public. For example, a 2005 survey found that in France the proportion of the decided public against GMOs rose from 46 to 71 per cent over the period from 1996.[23] This may be a factor behind the possibility of change to what would appear entrenched policies; and a good example is the indication by the President of the European Commission in September 2009 that Member States may be allowed to decide whether or not they would wish to cultivate GM crops in their territory.[24]

For all these reasons, there is always good reason to revisit the regulation of GMOs, and this book will seek to provide new insights into existing issues, while also extending to less-well-trodden ground. It is divided into four parts; and, although there is some emphasis on the position in the EU, there is also a conscious endeavour to reflect the truly international nature of the debate by exploring the legal ramifications of GMOs in representative countries throughout the globe, and also with reference to world trade. That said, it is accepted that no work could hope to be comprehensive where a subject has generated so much controversy and, for example, there is no attempt to cover matters of intellectual property law.

Part I addresses cross-cutting themes in three chapters. In Chapter 1, Luc Bodiguel and Michael Cardwell discuss the impact of the public on the governance of GMOs, considering both formal influence through rights to participation in the legislative process and less formal influence through consumer

---

[22] Under Council Regulation (EC) 834/2007 on organic production and labelling of organic products [2007] OJ L189/1 a product remains 'organic' despite the adventitious presence of authorized GMOs up to a 0.9% threshold.

[23] G Gaskell *et al*, *Europeans and Biotechnology in 2005: Patterns and Trends—Final Report on Eurobarometer 64.3* (Brussels: European Commission, 2006) 21. For a more recent survey, see European Commission, *Attitudes of European Citizens Towards the Environment—Special Eurobarometer 295* (Brussels: European Commission, 2008) 65.

[24] José Manuel Barroso, *Political Guidelines for the Next Commission* (Brussels: European Commission, 2009) 39 (available at <http://ec.europa.eu/commission_barroso/president/pdf/press_20090903_EN.pdf>, last accessed 20 October 2009).

preferences and anti-GM protest. In Chapter 2, Christophe Charlier and Egizio Valceschini examine the importance and limits of cost-benefit analysis (CBA) in relation to GMOs. The chapter demonstrates, not least, the complexity of any CBA in these circumstances, consequent not only upon the novelty of GMOs, but also factors such as the absence of scientific unanimity and consumer fears. On the other hand, it is argued that CBA still renders it possible to make some assessment of the economic legitimacy of biotech products, which can then be fed into the decision-making process. Finally, in Chapter 3, Karen Morrow considers the extent to which traditional risk analysis is appropriate in the case of GMOs. She highlights a greater willingness to contest the views of experts, with the regulation of risk having an increasingly public dimension; and she explores the relationship between risk and the precautionary principle. The chapter concludes with an examination of recent decisions of the courts relating to risk, with particular reference to the United Kingdom and the European Court of Justice.

Part II is concerned with the regulation of GMOs in the EU and consists of five chapters. The first of these, by Marine Friant-Perrot, tracks the development of the Community regulatory regime, with attention directed to the legislation governing deliberate releases into the environment, GM food and feed, and traceability and labelling. It then locates the specific provisions for GMOs within the growing corpus of Community food law and analyses, *inter alia*, the extent to which an integrated approach is being achieved 'from farm to fork'. Chapter 5, by Maria Lee, provides an in-depth analysis of multi-level governance in relation to GMOs, arguably a *locus classicus* for the implementation of legislation through dispersed authority. However, in this context multi-level governance is found to be not straightforward; and there is focus on the role of the European Food Safety Authority and the operation of comitology, both largely immune to the views of 'the public'. The chapter is completed by an evaluation of multi-level governance following the authorization process, together with implications for the internal market. Chapters 6 and 7 both discuss the vexed question of coexistence between GM and non-GM crops. The former, by Peggy Grossman, addresses the question at EU level and provides a detailed treatment of European Commission initiatives, most notably the Community guidelines upon which national measures are based. Emphasis is accorded to the key importance of tolerance thresholds, the special demands of organic agriculture, and attempts to gain 'GM-free' status through bans on GMOs. The latter, by Luc Bodiguel, Michael Cardwell, Ana Carretero García, and Domenico Viti, addresses the same question at Member State level, looking more closely at the development of measures in France, Italy, Spain, and the United Kingdom. Notwithstanding that Community guidelines apply, considerable variation in terms of national implementation is revealed; and the discussion extends to the impact on the regulatory process of the individual constitutional settlements of Members States. For example, in Italy there has been a battle of competences between the regions and the state. Likewise, national imperatives and preferences would seem to be reflected in such legislation as has been enacted, with the regime in France being far more 'top-down' than that in Spain. In Chapter 8, Chris Rodgers assesses the role of environmental liability regimes generally, including the extent to which it is possible to apply a Coasean analysis. He then provides a detailed treatment of the application to GMOs of both the Environmental Liability Directive and the Lugano Convention, exposing several weaknesses.

Part III is more overtly comparative, examining pertinent issues in Africa and North and South America. Chapter 9, by Fikremarkos Merso Birhanu, highlights the unique position of Africa, a continent where agriculture is of great importance to national economies, yet which has low levels of food security. The regulation of GMOs is evaluated at continental, sub-regional, and national level, with the African Model Law on Safety in Biotechnology providing a template which is more or less closely followed. While a developed agricultural sector (as in South Africa) might be an indicator of enthusiasm for GM crops, the position is found to be more complex; and a recurring theme is the need for Africa to own the new technology. In Chapter 10, Jane Matthews Glenn provides a detailed critique of two controversial Canadian cases on GM crops, *Monsanto Canada Ltd v Schmeiser*[25] and *Hoffman and Beaudoin v Monsanto Canada and Bayer Cropscience Inc.*[26] Broader principles of liability are also examined, with particular reference to GMOs. Questions raised are whether it would be possible to recover for pure economic loss; and whether regulatory decisions to authorize GMOs are 'policy' (as opposed to 'operational') decisions, thus becoming immune to challenge before the court. In Chapter 11, Rosario Silva Gilli traces the development of regulatory frameworks for GMOs within the countries of MERCOSUR. While GM crops are cultivated in the territory of all four full members (Argentina, Brazil, Paraguay, and Uruguay), different degrees of enthusiasm for the new technology are observed. The chapter ends with an assessment of whether MERCOSUR might operate as an engine for harmonizing the disparate national regimes. The final chapter in this part is by Peggy Grossman and discusses the governance of GMOs in the United States. She contributes a full study of federal law and policy, including the role of the various regulatory agencies. In addition, there is consideration of the extent to which individual states can, and have, legislated in this field, followed by broader exposition of potential liability under state tort law, with particular reference to nuisance. Notably, this chapter contains an extensive treatment of the *StarLink* litigation.[27]

---

[25] [2004] 1 SCR 902; affirming (2002) 218 DLR (4th) 31 (Federal CA); affirming (2001) 12 Canadian Patent Rep (4th) 204 (Federal Ct, Trial Div).
[26] (2007) 283 DLR (4th) 190 (Sask CA); affirming [2005] 7 Western Weekly Rep 665 (Sask QB).
[27] *In re StarLink Corn Products Liability Litigation* 212 F Supp 2d 828 (ND Ill 2002).

Part IV is avowedly international in focus, with a chapter on the *EC—Biotech* case,[28] and a further chapter locating the regulation of GMOs within the broader context of international law. Chapter 13, by Joe McMahon, investigates the contested role of science within the Agreement on the Application of Sanitary and Phytosanitary Measures ('SPS Agreement'); and, more precisely, it analyses three key aspects of the Panel's decision, namely the concept of a measure, the role of risk assessment, and the provisions on transparency. In addition, he evaluates whether the Panel missed an opportunity to reconcile the demands of science with other factors to be taken into account when making a decision on risk. Chapter 14, by Duncan French, looks both at and beyond the Cartagena Protocol and SPS Agreement, in order to inquire how international law may accommodate new regulatory challenges, such as GMOs. At the same time, it highlights two specific issues: the role of treaty interpretation as an aid to reconciling what at first sight might appear to be conflicting primary rules; and the utility of 'sustainable development' in providing a balanced framework to capture socio-economic aspects within the decision-making process. The conclusion reached is that the regulation of GMOs should only be considered when fully appraised of general international law principles.

---

[28] WT/DS291/R, WT/DS292/R, and WT/DS293/R, 29 September 2006.

PART I

THE PUBLIC, ECONOMICS, AND RISK

1

Genetically Modified Organisms and the Public: Participation, Preferences, and Protest

Luc Bodiguel and Michael Cardwell

1. Introduction

A defining feature of the controversy surrounding genetically modified organisms (GMOs) has been the high level of engagement by the general public exemplified by the GM Nation debate conducted by the UK Government in 2003.[1] Moreover, the informing and educating of the public is widely perceived as central to gaining acceptance of biotechnology,[2] although there is also considerable evidence that increased information and education on the scientific underpinnings of GMOs do not necessarily generate their increased popularity.[3] The heat of this debate has arguably been raised by the high media profile of GMOs. Again in the United Kingdom, the generally negative attitudes of journalists have incurred the criticism of the Government Chief Science Adviser, who had little sympathy with the *Daily Mail* for its use of such charged language as 'Frankenfoods' or with John Humphrys of the BBC *Today* programme for his pro-organic stance.[4] Likewise, in France,

---

[1] See, for the results, Department of Trade and Industry, *GM Nation? The Findings of the Public Debate* (London: Department of Trade and Industry, 2003) ('*GM Nation*').

[2] See, eg MJ Navarro, *Bridging the Knowledge Divide: Experiences in Communicating Crop Biotechnology* (Manila, Philippines: International Service for the Acquisition of Agri-biotech Applications (ISAAA), 2008).

[3] See, eg D Brossard and J Shanahan, 'Perspectives on Communication about Agricultural Biotechnology' in D Brossard, J Shanahan, and TC Nesbitt (eds), *The Public, the Media and Agricultural Biotechnology* (Wallingford, Oxon: CAB International, 2007) 3.

[4] House of Commons Innovation, Universities, Science and Skills Committee, Session 2007–08, HC 115-I, Oral Evidence given by Professor Sir David King (available at <http://www.publications.parliament.uk/pa/cm200708/cmselect/cmdius/115/7120501.htm>, last accessed 1 September 2009).

the activities of the *Faucheurs Volontaires* (and notably José Bové) have been conducted in the full glare of publicity.[5]

That the public should have so strong a 'voice' is arguably not surprising. First and foremost, it is the public as consumers who will eat GMOs; and there is evidence of widespread support for the view that regulation should impose at least a labelling obligation. For example, one UK survey found that only one per cent of respondents did not think it necessary to label all GM products;[6] and, in the United States, where labelling is not required, a clear majority would appear unhappy with Government policy.[7] Further, the concerns of this majority are unlikely to be allayed by the US Food and Drug Administration taking the stance that even cloned food should not be subject to specific labelling rules.[8]

Two other factors liable to engage the public may be highlighted. First, it is generally accepted that consumers are not always convinced by the conclusions of government and industry scientists; and yet it is government and industry scientists who have been accorded so prominent a role in the regulation of GMOs.[9] The bovine spongiform encephalopathy (BSE) crisis may carry much responsibility for this. As concluded in the BSE Inquiry conducted by Lord Phillips, '[e]veryone agreed that the Government had a problem with credibility'; and the recommendation was that, '[w]hen responding to public or media demand for advice, the Government must resist the temptation of attempting to appear to have all the answers in a situation of uncertainty'.[10] Further, in *GM Nation* only seven per cent of respondents were confident that the development of GM crops was being carefully regulated.[11] Such

---

[5] For the activities of the *Faucheurs Volontaires*, see, eg G Hayes, 'Collective action and civil disobedience: the anti-GMO campaign of the *Faucheurs Volontaires*' (2007) 5 French Politics 293.

[6] W Poortinga and N Pidgeon, 'Public Perceptions of Agricultural Biotechnology in the UK: the Case of Genetically Modified Food' in Brossard, Shanahan, and Nesbitt (n 3 above) 21.

[7] See, eg M McGarry Wolf, P Bertolini, and J Parker-Garcia, 'A Comparison of Consumer Attitudes Towards GM Food in Italy and the USA' in RE Evenson and V Santaniello (eds), *Consumer Acceptance of Genetically Modified Foods* (Wallingford, Oxon: CABI Publishing, 2004) 131 (89.6% being in favour of mandatory labelling).

[8] Food and Drug Administration, *Guidance No. 179: Guidance for Industry: Use of Animal Clones and Clone Progeny for Human Food and Animal Feed* (Food and Drug Administration, 15 January 2008).

[9] This 'turn to science' is well-illustrated in the World Trade Organization *EC—Biotech* case, *EC-Approval and Marketing of Biotech Products* WT/DS291/R, WT/DS292/R, and WT/DS293/R, 29 September 2006. See, further, eg J Scott, *The WTO Agreement on Sanitary and Phytosanitary Measures: a Commentary* (Oxford: Oxford University Press, 2007) and Chapter 13.

[10] *The BSE Inquiry: the Report-Vol. 1, Findings and Conclusions*, 14. *Lessons to be Learned* paras 1300–1301 (available at <http://www.bseinquiry.gov.uk.html>, last accessed 17 August 2009). See, generally, eg G Little, 'BSE and the regulation of risk' (2001) 64 Modern Law Review 730.

[11] *GM Nation* (n 1 above) para 122.

scepticism is not confined to the United Kingdom or even Europe.[12] American consumers would generally appear to place more trust in the regulatory authorities;[13] but the StarLink™ incident, where non-authorized GM material was found in taco shells, did shake this greater confidence.[14]

Secondly, there is also evidence that the general public does not see the immediate advantages of GMOs for consumers (as opposed to the immediate advantages for farmers or seed companies). Thus, in *GM Nation* 85 per cent of respondents agreed that GM crops would mainly benefit producers as opposed to ordinary people.[15] That said, there would seem to be considerable support for employing agricultural biotechnology to achieve greater food security in developing countries, as clearly articulated by the UK Environment Minister at the time of the global food crisis in June 2008.[16] Yet, it may also be noted that non-governmental organizations (NGOs) such as Oxfam have long rejected the idea that food security can be remedied by 'technological fixes'.[17] The questionable utility of GM crops (at least from the point of view of the general populace) may also explain findings that, although as a rule they do not give rise to pressing fear, they may nonetheless generate strong resistance;[18] and this paradox is consistent with broader evidence that only a minority of Europeans believe some degree of risk resulting from modern biotechnology to be justified in order to promote economic competitiveness.[19]

Accordingly, against this background, two aspects may be examined: first, the extent to which the public have the ability at a formal level to contribute to the regulation of agricultural biotechnology; and, secondly, the extent to which the public have exerted influence through less-formal channels. In the latter case, there will be the opportunity to examine both the effect of consumer preferences and the impact of public protest against GM crops (including the approach taken by the courts).

---

[12] See, eg K Brooks, 'History, change and policy: factors leading to current opposition to food biotechnology' (2000) 5 Georgetown Public Policy Review 153. On European aspects, see D Chalmers, 'Risk, anxiety and the European mediation of the politics of life' (2005) 30 European Law Review 649.
[13] See, eg M Costa-Font, JM Gil, and WB Traill, 'Consumer acceptance, valuation of and attitudes towards genetically modified food: review and implications for food policy' (2008) 33 Food Policy 99.
[14] For the *StarLink* litigation, see *In re StarLink Corn Products Liability Litigation* 212 F Supp 2d 828 (ND Ill 2002); and see further Chapter 12.   [15] *GM Nation* (n 1 above) para 121.
[16] See, eg 'Genetically Modified Crops "May be Answer to Global Food Crisis"', Telegraph, 19 June 2008 (Phil Woolas).
[17] For an early statement to this effect, see Policy Department of Oxfam (Great Britain), *Genetically Modified Crops, World Trade and Food Security: Position Paper November 1999* (Oxfam, 1999) 3.   [18] Poortinga and Pidgeon (n 6 above).
[19] See, generally, Directorate-General XII, *Eurobarometer 46.1: the Europeans and Modern Biotechnology* (Brussels: European Commission, 1999) (28%) 56.

2. Formal Public Influence on the Regulation of Agricultural Biotechnology

There are several legislative frameworks which provide for public input into the regulation of agricultural biotechnology, these ranging from international agreements, where GMOs are merely a part of their focus, to targeted national regimes.[20]

2.1 International Level

In this context, a leading role is undoubtedly played by the Cartagena Protocol to the Convention on Biological Diversity (concluded on 29 January 2000), since it is both a genuinely international agreement and specifically directed to GMOs.[21] However, it should be noted that the question of public participation had already been addressed by the Convention on Biological Diversity itself (concluded on 5 June 1992).[22] Article 14(1)(a) stipulates that 'as far as possible and as appropriate', the contracting parties were to 'introduce appropriate procedures requiring environmental impact assessment of its proposed projects that are likely to have significant adverse effects on biological diversity with a view to avoiding or minimizing such effects and, where appropriate, allow for public participation in such procedures'. Nonetheless, Article 23(2) of the Cartagena Protocol takes this further, requiring that the parties, in accordance with their respective laws and regulations, shall 'consult the public in the decision-making process regarding living modified organisms and shall make the results of such decisions available to the public'. Accordingly, not only is there express reference to GMOs, but also the obligation seems less qualified, with no proviso that public participation should only be undertaken as appropriate. That said, there would seem to be an obligation only to consult, not an obligation to feed the responses into the decision-making process; and the allowance made for national laws and regulations has apparently resulted in considerable variation, with information being accorded priority over participation.[23]

---

[20] For a useful survey, see, eg L Glowka, *Law and Modern Biotechnology: Selected Issues of Relevance to Food and Agriculture: FAO Legislative Study 78* (Rome: Food and Agriculture Organization, 2003).

[21] Available at <http://www.cbd.int/biosafety/protocol/shtml>. To date, 157 instruments of ratification or accession have been deposited with the United Nations Secretary General: see <http://www.cbd.int/biosafety/signinglist.shtml>, last accessed 6 January 2010.

[22] Available at <http://www.cbd.int/convention/convention.shtml>.

[23] See, eg Institute for Development Studies, *Public Participation and the Cartagena Protocol on Biosafety: a Review for DfID and UNEP-GEF* (Institute for Development Studies, University of Sussex, 2003) 58–9; and see G Jaffe, 'Implementing the Cartagena Biosafety Protocol through national biosafety regulatory systems: an analysis of key unresolved issues' (2005) 5 Journal of Public Affairs 299. For a more general survey, see Secretariat of the Convention on Biological Diversity, *Special Focus: Public Awareness and Participation: Experiences and Lessons Learned from Recent Initiatives*, Biosafety Protocol News, July 2009, Issue 6.

Also featuring prominently in the former category is the Aarhus Convention,[24] as ratified by the Community in 2005.[25] While its coverage extends to environmental matters generally, the recitals expressly recognize 'the concern of the public about the deliberate release of genetically modified organisms into the environment and the need for increased transparency and greater public participation in decision-making in this field'. Further, all its three 'pillars' have the capacity to impact upon the regulation of GMOs (respectively access to information, public participation in decision-making, and access to justice); but it is the second which arguably provides the strongest lever for civil society, since generosity in terms of access to information and even in terms of access to justice would appear materially circumscribed if a restrictive approach is taken to participation in the decision-making process itself.[26]

Under the second 'pillar', provision is made for public participation in respect of: first, decisions on specific environmental activities (Article 6); secondly, plans, programmes, and policies relating to the environment (Article 7); and, thirdly, the preparation of executive regulations and/or generally applicable legally binding normative instruments (Article 8). The scope of Article 8 would seem wide enough to engage participation rights when enacting Community or national legislation on GMOs. Further, and importantly, the original text of Article 6 was not thought sufficiently precise or robust in comparison with the Cartagena Protocol and, by the Almaty amendment of May 2005, it was reconfigured to introduce a specific regime for public participation in decisions on the deliberate release into the environment and placing on the market of GMOs.[27] The new Article 6*bis* requires parties to the Aarhus Convention to

---

[24] Aarhus Convention on Access to Information, Public Participation in Decision-making and Access to Justice in Environmental Matters 1998 (available at <http://www.unece.org/env/pp/documents/cep43e.pdf>, last accessed 26 June 2009). This was concluded under the auspices of the United Nations Economic Commission for Europe. See also, generally, eg M Lee, *EU Environmental Law: Challenges, Change and Decision-making* (Oxford: Hart Publishing, 2005) 113–49; M Lee and C Abbott, 'The usual suspects? Public participation under the Aarhus Convention' (2003) 66 Modern Law Review 80; and C Nadal, 'Pursuing substantive environmental justice: the Aarhus Convention as a "pillar" of empowerment' (2008) 10 Environmental Law Review 28.

[25] Council Decision (EC) 2005/370 of 17 February 2005 on the conclusion, on behalf of the European Community, of the Convention on Access to Information, Public Participation in Decision-making and Access to Justice in Environmental Matters [2005] OJ L124/1; and see also Regulation (EC) 1367/2006 of the European Parliament and of the Council on the application of the Aarhus Convention on Access to Information, Public Participation in Decision-making and Access to Justice in Environmental Matters to Community institutions and bodies [2006] OJ L264/13.

[26] For detailed commentary on the second 'pillar', see, eg Economic Commission for Europe, *The Aarhus Convention: an Implementation Guide* (Geneva: Economic Commission for Europe, 2000) 85–122. It may also be noted that the public may enjoy a role in the enforcement of the Aarhus Convention, since the arrangements to review compliance 'shall allow for appropriate public involvement and may include the option of considering communications from members of the public on matters related to the Convention' (Art 15); and see, generally, eg S Kravchenko, 'The Aarhus Convention and innovations in compliance with multilateral environmental agreements' (2007) 18 Colorado Journal of International Environmental Law and Policy 1.

[27] The Almaty amendment is available at <http://www.unece.org/env/pp/mop2/mop2.decisions.htm>, last accessed 2 September 2009.

'provide for early and effective information and public participation prior to making [such] decisions'. Further, the accompanying modalities stipulate that each party should 'endeavour to ensure that, when decisions are taken on whether to permit the deliberate release of GMOs into the environment, including placing on the market, due account is taken of the outcome of the public participation procedure'.[28] This would seem more prescriptive than the Cartagena Protocol, but it must be recognized that the amendment is not yet in force.[29] Further, a party has the opportunity, if appropriate, to provide for exceptions to the public participation procedure subject to certain conditions: for example, in the case of a deliberate release of a GMO into the environment (other than for its placing on the market), where such a release under comparable bio-geographical conditions has already been approved by the party concerned and sufficient experience has previously been gained with the release of that GMO in comparable ecosystems.[30] On the other hand, it should be emphasized that the Aarhus Convention confers specific rights on environmental NGOs. For the purposes of access to justice (as opposed to public participation under the second 'pillar'), NGOs promoting environmental protection and meeting any requirements under national law are deemed to have sufficient interest, and to have rights capable of being impaired, for the purposes of obtaining access to a review procedure before a court of law and/or other independent and impartial body established by law.[31] An example of national rules would be the twin criteria applicable in France, requiring that the NGO has existed for three years and acts in the field of environmental protection.[32] This provision of the Aarhus Convention may prove particularly relevant in the case of GMOs, where the availability of substantial resources and expertise would seem necessary to mount an effective challenge.

While the Convention on Biological Diversity and the Aarhus Convention represent two important initiatives by the United Nations, participation rights may also be found under the umbrella of other international bodies. By way of illustration, the Organization of African Unity (now the African Union (AU)) encourages states to harmonize their measures in accordance with the African Model Law on Safety in Biotechnology. This provides that the public should participate in decision-making by means of both a notice-and-comment procedure and, in certain cases, public consultations. Importantly, national authorities are to take into account the views and concerns of the public.[33] Likewise, within

---

[28] Annex I*bis*, para 7.
[29] For the ratification status of the Almaty amendment, see <http://www.unece.org/env/pp/ratification.htm>. [30] Annex I*bis*, para 2.
[31] Art 9(2) (by reference to Art 2(5)).
[32] Milieu Ltd, *Summary Report on the Inventory of EU Member States' Measures on Access to Justice in Environmental Matters* (prepared under contract for European Commission DG Environment) (Brussels: Milieu Ltd, 2007) Executive Summary, 10.
[33] African Model Law on Safety in Biotechnology, Art 5(2)–(4) (available at <http://www.nepadst.org>, last accessed 26 June 2009). See further Chapter 9.

the EU it is expressly provided that, in the case of proposed deliberate releases into the environment of GMOs, Member States shall consult the public and, where appropriate, groups.[34] Indeed, the Deliberate Release Directive recites that: '[f]or a comprehensive and transparent legislative framework, it is necessary to ensure that the public is consulted by either the Commission or the Member States during the preparation of measures and that they are informed of the measures taken during the implementation of this Directive'; and that '[c]omments by the public should be taken into consideration in the drafts of measures submitted to the Regulatory Committee'.[35] A matter of some interest is that, as with the Aarhus Convention, the Community has conferred certain privileges on NGOs. These privileges are enjoyed not only for the purposes of consultation in the context of the authorization of deliberate releases into the environment of GMOs, but also for the purposes of enforcement under Directive (EC) 2004/35 of the European Parliament and of the Council on environmental liability with regard to the prevention and remedying of environmental damage ('Environmental Liability Directive').[36] However, the latter provision may prove of limited application in the case of GMOs. Not least environmental damage is defined by reference to protected species and natural habitats; and GMOs are more likely to be grown in areas without conservation status.[37]

2.2 National Level

At national level there would seem to be considerable variation in the degree to which public participation is accommodated within the legislative process. In the United States, such participation is largely confined to comment within a regulatory framework where federal agencies shoulder the greatest responsibilities.[38] To take one example, where a petition is lodged for determination that a GM crop should receive unregulated status under the Plant Protection Act 2000, the Animal and Plant Health Inspection Service must file notice of the

---

[34] Directive (EC) 2001/18 of the European Parliament and of the Council on the deliberate release into the environment of genetically modified organisms ('Deliberate Release Directive') [2001] OJ L106/1, Art 9.

[35] Ibid Preamble (10) and (46). It may also be noted that enhanced participatory rights are conferred on the public under Directive (EC) 2000/60 of the European Parliament and of the Council establishing a framework for Community action in the field of water policy ('Water Framework Directive') [2000] OJ L327/1. The legislation provides that Member States must 'encourage the active involvement of all interested parties': ibid Art 14.

[36] [2004] OJ L143/56, Art 12(1).

[37] See, eg M Cardwell, 'The release of genetically modified organisms into the environment: public concerns and regulatory responses' (2002) 4 Environmental Law Review 156; and see further Chapter 8.

[38] See, eg R Lyster, 'Sustainability, regulatory dilemmas and GMOs: the US and EU compared' (2004) 8 Asia Pacific Journal of Environmental Law 111; and see further Chapter 12.

petition in the Federal Register, inviting public comment.[39] By contrast, in the United Kingdom there has been extensive consultation.[40] As has been seen, *GM Nation* was unusual in the degree to which the public was engaged in the consultation process (it being estimated that nearly 20,000 people attended meetings);[41] and it may be highlighted that more recently (over 2006–7) there has also been specific consultation on coexistence.[42] Besides, an innovative approach has been to employ stakeholder workshops to address key issues such as voluntary 'GM-free' zones and the effects upon the organic sector.[43] That said, the results of such consultation do not appear to have found easy accommodation within the regulatory process. For example, *GM Nation* was but one of three strands in the overall 'GM Dialogue' to inform policy-making, the others being the scientific review and the cost-benefit analysis;[44] and it was openly accepted that the interaction between these three strands had been less effective than had been hoped.[45] Further, in the Government response to the GM Dialogue, commitment was affirmed to 'evidence-based policy-making', with emphasis on science.[46]

Away from Europe, and again at national level, it may be highlighted that a distinctly participatory approach has been adopted in New Zealand.[47] In 1999 the Independent Biotechnology Advisory Council was established to inform and consult the public on matters of biotechnology; and the following year the Royal Commission on Genetic Modification was established to investigate 'the strategic options available to enable New Zealand to address, now and in the future,

---

[39] 7 CFR 340.6.
[40] On consultation in the United Kingdom, see, generally, eg Lee and Abbott (n 24 above); and S Hartley and G Skogstad, 'Regulating genetically modified crops and foods in Canada and the United Kingdom: democratizing risk regulation' (2005) 48 Canadian Public Administration 305.
[41] *GM Nation* (n 1 above) para 82.
[42] Department for Environment, Food and Rural Affairs (DEFRA), *Consultation on Proposals for Managing the Coexistence of GM, Conventional and Organic Crops* (London: DEFRA, 2006); and for the results, see DEFRA, *Summary of Responses to Defra Consultation Paper on Proposals for Managing the Coexistence of GM, Conventional and Organic Crops* (London: DEFRA, 2007).
[43] Details of the workshops, held in 2004, are available at <http://www.defra.gov.uk/environment/gm/crops/index.htm>,last accessed 26 June 2006.
[44] For the scientific review, see GM Science Review Panel, *GM Science Review: First Report—an Open Review of the Science Relevant to GM Crops and Food Based on Interests and Concerns of the Public* (London: 2003); and *GM Science Review: Second Report—an Open Review of the Science Relevant to GM Crops and Food Based on Interests and Concerns of the Public* (London: 2004); and, for the cost-benefit analysis, see Strategy Unit, *Field Work: Weighing Up the Costs and Benefits of GM Crops* (London: Strategy Unit, 2003).
[45] DEFRA, *The GM Public Debate: Learning Lessons from the Process* (London: DEFRA, 2004) para 29. It was also noted that the six-week timescale was not ideal: ibid para 24.
[46] DEFRA, *The GM Dialogue: Government Response* (London: DEFRA, 2004) Executive Summary, para 11.
[47] See, generally, eg Glowka (n 20 above) 27; R Walters, 'Criminology and genetically modified food' (2004) 44 British Journal of Criminology 151; and R Hindmarsh and R Du Plessis, 'GMO regulation and civic participation at the "edge of the world": the case of Australia and New Zealand' (2008) 27 New Genetics and Society 181.

genetic modification, genetically modified organisms, and products', together with 'any changes considered desirable to the current legislative, regulatory, policy or institutional arrangements'.[48] A matter of some significance was the extent of public input into the process. This included scoping meetings, formal hearings, public meetings, Maori consultation workshops, and a youth forum.[49]

2.3 Effectiveness of Formal Public Participation

Nonetheless, despite these initiatives, the effect of formal public participation would seem to be blunted by a combination of factors; and four such factors may be mentioned. First, not all the measures which have been considered are prescriptive in nature. For example, while the AU encourages states to harmonize their measures in accordance with the African Model Law on Safety in Biotechnology, there is no binding obligation to do so. Accordingly, to take two instances, while the legislative framework in South Africa does at least provide a framework for public participation, in Malawi such a requirement is effectively excluded.[50] On the other hand, as has been seen, the Cartagena Protocol imposes an obligation to consult, while the Almaty amendment to the Aarhus Convention would also require that due account is taken of the outcome of its public participation procedure.

Secondly, and more generally, doubts have been raised as to the effectiveness of any such consultation procedure. For example, it may be reiterated that the Cartagena Protocol does not expressly require that responses from the public be fed into the decision-making process. Indeed, as stated by Glowka:

> the extent to which public participation is actually facilitated or exists in a country is difficult to determine from a simple review of the country's biotechnology related legislative instruments. For example, general references to public participation may not translate into actual public participation if additional criteria are not provided on the form that public participation can take. Also the best public participation provisions may not be used if the public cannot have the capacity to effectively participate.[51]

That said, two UK cases may be cited where consultation obligations were taken seriously. In *Berkeley v Secretary of State for the Environment* the House of Lords quashed planning permission to redevelop the Fulham Football Club ground at Craven Cottage, with a reason for the decision being non-compliance with the public participation procedures laid down under the Community regime for environmental impact assessments.[52] In particular, the public had not been

---

[48] Royal Commission on Genetic Modification, *Report of the Royal Commission on Genetic Modification* (2001) 364. The major conclusion was that New Zealand should keep its options open, with a recommendation that coexistence be encouraged between all types of agriculture: ibid Executive Summary, 2.   [49] Ibid Appendix 1, section 3.
[50] See further Chapter 9.   [51] Glowka (n 20 above) 51.
[52] [2001] 2 AC 603. For the Community legislation concerned, see Council Directive (EEC) 85/337 on the assessment of the effects of certain public and private projects on the environment

afforded an opportunity to express an opinion before the project was initiated.[53] Lord Hoffmann stated that '[t]the directly enforceable right of the citizen which is accorded by the [EIA] Directive is not merely a right to a fully informed decision on the substantive issue': rather, what was required was 'the inclusive and democratic procedure prescribed by the [EIA] Directive in which the public, however misguided or wrongheaded its views may be, is given an opportunity to express its opinion on the environmental issues'.[54] Subsequently, a similar approach was adopted in *R (Greenpeace Ltd) v Secretary of State for the Environment*.[55] A 2003 White Paper[56] had indicated that there would be 'the fullest public consultation' before the Government reached any decision to change its policy of non-support for the building of new nuclear power stations. Three years later, in 2006, a consultation exercise on energy policy was initiated, through what appeared to be merely an 'Issues Paper', which would be followed by a more detailed document containing proposals upon which the public could make informed comment. The court found this 'Issues Paper' to be manifestly inadequate where the matter under consideration was of such importance and complexity. In particular, the information on building costs and nuclear waste was insufficient; and the consultation period of 12 weeks was too short.[57] Accordingly, '[t]here could be no proper consultation, let alone "the fullest possible consultation" as promised in the 2003 White Paper';[58] and this amounted to procedural unfairness and a breach of the legitimate expectation of the claimant. Significantly, express reference was made by the court to obligations under the Aarhus Convention, and notably Article 7, requiring each party, to the extent appropriate, to endeavour to provide opportunities for public participation in the preparation of policies relating to the environment.[59]

Thirdly, a potential weakness in regulatory frameworks is the heavy financial cost of challenging non-compliance with environmental measures (whether participation requirements or otherwise).[60] This weakness would, nevertheless,

---

('EIA Directive') [1985] OJ L175/40, Art 6. The original legislation in this regard was amended by Council Directive (EC) 97/11 [1997] OJ L73/5 and, to meet the obligations of the European Community under the Aarhus Convention, by Directive (EC) 2003/35 of the European Parliament and of the Council providing for public participation in respect of the drawing up of certain plans and programmes relating to the environment [2003] OJ L156/17. On environmental impact assessment generally, see, eg J Holder, *Environmental Assessment: the Regulation of Decision Making* (Oxford: Oxford University Press, 2004); and J Holder and D McGillivray (eds), *Taking Stock of Environmental Assessment: Law, Policy and Practice* (London: Routledge-Cavendish, 2007).

[53] The relevant provisions at that date were contained in Art 6(2) of the EIA Directive.
[54] [2001] 2 AC 603, 615.
[55] [2007] EWHC 311 (Admin); [2007] Environmental Law Reports 29. See also P Thompson, 'Consultation and the authorisation of major infrastructure projects' [2009] Journal of Planning and Environmental Law 174. For more successful consultation, see *Bard Campaign v Secretary of State for Communities and Local Government* [2009] EWHC 308 (Admin) (in respect of 'eco-towns'). [56] *Our Energy Future—Creating a Low Carbon Economy*, Cmnd 5761 (2003).
[57] In this regard it may be reiterated that the consultation period for the 2003 GM Public Debate was six weeks. [58] [2007] EWHC 311 (Admin) [117].
[59] Art 7 is, however, subject to the qualification that it is for the relevant public authority to identify the public which may participate, taking into account the objectives of the Aarhus Convention. [60] See, eg Lee and Abbott (n 24 above).

seem to be recognized by the Aarhus Convention, with Article 9(4) providing that review procedures should not be 'prohibitively expensive'; and several national regimes would appear vulnerable (not least, when the 'loser-pays' principle applies).[61] Thus, the UK report *Ensuring Access to Environmental Justice in England and Wales* stated that 'the current principles concerning costs and the potential exposure to costs in judicial review proceedings in England and Wales inhibit compliance with the requirements of Aarhus concerning access to environmental justice';[62] and the same view has been taken by the Court of Appeal in *R (Burkett) v London Borough of Hammersmith and Fulham*, where Brooke LJ declared that 'if the figures revealed by this case were in any sense typical of the costs reasonably incurred in litigating such cases up to the highest level, very serious questions would be raised as to the possibility of ever living up to the Aarhus ideals within our present legal system'.[63] In the specific context of GMOs, such cost implications were thrown into sharp relief by the recent decision of the Saskatchewan Court of Appeal in *Hoffman and Beaudoin v Monsanto Canada and Bayer Cropscience Inc*.[64] Central to the action was whether or not the farmers alleging cross-contamination could establish a class action with all the attendant advantages of sharing costs; and the refusal of the Court of Appeal to so certify effectively brought the case to an end.[65] It is perhaps no coincidence, therefore, that NGOs (as opposed to individual farmers) have played so prominent a role in challenging the regulatory framework for GMOs;[66] and the provisions in the Aarhus Convention granting NGOs privileged status in review procedures may yet prove material.

Fourthly, public participation is inevitably dependent on access to information, yet in many regimes such access is substantially limited for reasons of confidentiality. Thus, the Aarhus Convention exempts from disclosure the proceedings

---

[61] See, eg Milieu Ltd (n 32 above) Executive Summary, 13–16.

[62] Report of the Working Group on Access to Environmental Justice, *Ensuring Access to Environmental Justice in England and Wales* (2008) 34 (available at <http://www.unece.org/env/pp/compliance/C2008-23/AmicusNbrief/AnnexNjusticereport08.pdf>, last accessed 2 July 2009).

[63] [2004] EWCA Civ 1342 [76]. See also *Morgan and Baker v Hinton Organics (Wessex) Ltd* [2009] EWCA Civ 107 (where the Court of Appeal was prepared to proceed on the basis that the Aarhus obligations extended to private nuisance actions while regarding the principles of the Convention as merely something to be taken into account when exercising discretions); and Case C-427/07 *Commission v Ireland* (ECJ 16 July 2009).

[64] (2007) 283 DLR (4th) 190 (Sask CA).

[65] See, eg H McLeod-Kilmurray, '*Hoffman v Monsanto*: courts, class actions, and perceptions of the problem of GM drift' (2007) 27 Bulletin of Science, Technology and Society 188; and see further Chapter 10.

[66] Greenpeace, for example, has frequently been the claimant before the European Court of Justice: see, eg Case C-6/99 *Association Greenpeace France v Ministère de l'Agriculture et de la Pêche* [2000] ECR I-1651. See, generally, eg A Warleigh, '"Europeanizing" civil society: NGOs as agents of political socialization' (2001) 39 Journal of Common Market Studies 619; C Hilson, 'Greening citizenship: boundaries of membership and the environment' (2001) 13 Journal of Environmental Law 335; and J Teel, 'Have NGOs distorted or illuminated the benefits and hazards of genetically modified organisms?' (2002) 13 Colorado Journal of Environmental Law and Policy 137. Also, for successful recovery of costs by a South African NGO in GM litigation, see *Trustees for the time being of the Biowatch Trust v Registrar Genetic Resources* [2009] ZACC 14 (3 June 2009).

of public authorities, where provided by national law, and commercial and industrial information (but on condition that this information is protected by law in order to protect a legitimate economic interest).[67] The exemptions must, nonetheless, be 'interpreted in a restrictive way, taking into account the public interest served in disclosure'.[68] While the African Model Law limits public access to information that may be considered confidential, it generally adopts an open approach and goes so far as to list types of information that may not be considered confidential.[69] That said, its non-binding nature again weakens the force of these measures. In consequence, even in states such as South Africa, where, as has been seen, there is provision for public participation, considerable power is conferred on the Executive Council as to what information may be held back from the public domain.[70] A similar pattern is again adopted in the European Union. In the case of deliberate releases into the environment, the European Commission and the competent authorities are not to divulge to third parties confidential information and are to protect intellectual property rights relating to the data received.[71] That said, there are again circumstances where information may not be held back and this information includes the location of release and intended uses. Significantly, notwithstanding the activities of anti-GM protestors, the European Court of Justice in *Commune de Sausheim v Azelvandre* has expressly held that the location of release cannot be kept confidential on the ground of protecting public order.[72]

3. Informal Public Influence on the Regulation of Agricultural Biotechnology

Against this legislative background, and not least its relatively restrictive approach towards public participation, there may be considered two ways in which wider civil society would yet appear to have shaped the development of GM crops, the first being consumer preferences and the second being public protest.

3.1 Consumer Preferences

As indicated, consumer resistance to GM food in Europe remains substantial. Indeed, a survey conducted in 2005 noted that, while support may have increased between 1999 and 2002, it then decreased between 2002 and 2005.[73] Further,

[67] Art 4(4)(a) and (d). [68] Art 4 (4).
[69] African Model Law on Safety in Biotechnology, Art 12.
[70] Genetically Modified Organisms Act (No 15 of 1997), s 18.
[71] Deliberate Release Directive, Art 25. [72] Case C-552/07 (ECJ 17 February 2009).
[73] G Gaskell et al, *Europeans and Biotechnology in 2005: Patterns and Trends—Final Report on Eurobarometer 64.3* (Brussels: European Commission, 2006) 21. See also, European Commission, *Attitudes of European Citizens Towards the Environment—Special Eurobarometer* 295 (Brussels: European Commission, 2008) 65 (58%, being opposed to the use of GMOs); and see generally, eg S Bonny, 'Factors Explaining Opposition to GMOs in France and the Rest of Europe' in Evenson and Santaniello (n 7 above) 169.

agricultural biotechnology is arguably perceived as less beneficial than other forms of biotechnology. By way of illustration, the same 2005 survey found that, in comparison with nanotechnology, pharmacogenetics, and gene therapy, GM foods were regarded as the least morally acceptable, the least useful, the most risky, and the least to be encouraged.[74] It also sought to ascertain the reasons for buying and not buying GM foods. While a majority were prepared to buy if the product was healthier or contained less pesticide residues, only 44 per cent were influenced to do so by reason of approval from the relevant authorities and only 36 per cent by reason of cheaper price.[75] At the level of the Member State, *GM Nation* in the United Kingdom revealed a very negative attitude towards eating GM food, with only eight per cent of respondents being happy to do so.[76]

Similar views may be found outside Europe. As has again been seen, in the United States public concerns have been exacerbated by the StarLink™ incident; and, while there is evident a generally more pro-GM stance among consumers,[77] it is of interest that the elapse of time has not seen support materially increase. This continued resistance may be detected in the series of polls commissioned by the Pew Initiative on Food and Biotechnology.[78] In the case of the first poll, conducted in 2001, 38 per cent of respondents said that they were likely to eat GM foods, while 54 per cent said that they were unlikely to do so. By 2003 the proportions had changed to respectively 43 and 50 per cent; but by 2006 they had reverted to respectively 38 and 43 per cent.[79] It is also of interest that US consumers would seem to place more trust in regulators than in Europe. Thus, while the same polls found considerable concern at the lack of regulation (in 2006, 41 per cent of those claiming a basic awareness of the regulatory framework said that there was too little regulation, as opposed to 16 per cent who said that there was too much), nearly half were more likely to eat GM foods when supplied with specific information on the role of the Federal Drug Administration.[80] At the same time, research would point to the fact that US farmers are similarly far from oblivious to the broader social and environmental issues raised by GMOs.[81]

---

[74] Gaskell et al (n 73 above) 17. Gene therapy was likewise considered risky, but respondents were clearly of the view that the risk was one worth taking (a view not replicated in the case of GMOs).
[75] Ibid 22. [76] *GM Nation* (n 1 above) para 121. See also, eg Poortinga and Pidgeon (n 6 above).
[77] See, eg McGarry Wolf, Bertolini, and Parker-Garcia (n 7 above); JL Lusk et al, 'Effect of information about benefits of biotechnology on consumer acceptance of genetically modified food: evidence from experimental auctions in the United States, England, and France' (2004) 31 European Journal of Agricultural Economics 179; and International Food Information Council, *2008 Food Biotechnology: a Study of US Consumer Trends* (Washington, DC: International Food Information Council, 2008).
[78] The Pew Initiative on Food and Biotechnology was established in 2001 by The Pew Charitable Trusts, with the aim of serving as an 'honest broker' to the various stakeholders in agricultural biotechnology. Its work was discontinued in March 2007.
[79] Memorandum from the Mellman Group to the Pew Initiative on Food and Biotechnology on Review of Public Opinion Research (available at <http://www.pewtrusts.org/uploadedFiles/wwwpewtrustsorg/Public_Opinion/Food_and_Biotechnology/2006summary.pdf>, last accessed 19 January 2008) 3–4. [80] Ibid 5–6.
[81] See, eg NP Guehlstorf, 'Understanding the scope of farmer perceptions of risk: considering farmer opinions on the use of genetically modified (GM) crops as a stakeholder voice in policy' (2008) 21 Journal of Agricultural and Environmental Ethics 541.

Accordingly, empirical surveys do present a fairly consistent picture of consumer hostility to GMOs. Nevertheless, a note of caution may be sounded. As has been seen, only 36 per cent of the sample in a 2005 European survey stated that they were likely to buy GM crops on the basis that they were cheaper; but the report itself expressed concern that they might be 'responding as citizens rather than as consumers'.[82] Indeed, such discounting of price as a factor is not confined to GMOs. In the case of food quality the European Commission has long been perturbed by the gap between the preference for quality as expressed by consumers and their behaviour when actually shopping.[83] However, in the case of GMOs, particularly compelling evidence has recently been presented by research conducted for the European Commission, *Do European Consumers Buy GM Foods?* (*'Consumerchoice'*).[84] This found that '[w]hatever they may have said in responses to questions, most shoppers did not actively avoid GM-products, suggesting that they are not greatly concerned with the GM issue'. Further, there was evidence that 'the way people respond to prompting via questionnaires and polls is by itself not a reliable guide to what they will buy in a grocery store;[85] and the overall conclusion was 'that a major factor in governing the purchase of GM-products by Europeans is the decision of retailers to make them available to consumers'.[86] It may also be highlighted that the research addressed in some depth the question of labelling. Meetings with focus groups produced data to the effect that, while participants were strongly in favour of labels, few would actually look at them when buying food.[87] Although such a position might at first seem illogical, the prominence and wording of the labels were identified as very relevant criteria, with 'GM-free' in large print on the front of the package more likely to influence shoppers than 'containing GM' in small print on the rear.[88] Indeed, the extent to which consumers look at labels may largely reflect pressure of time.[89]

In light of the overall conclusion of the *Consumerchoice* report, the role of supermarkets and other retailers can only be reinforced. And, notwithstanding such evidence that consumers will in fact buy GM foods if presented with them

---

[82] Gaskell et al (n 73 above) 22. See also, on price aspects, M Burton et al, 'Consumer attitudes to genetically modified organisms in food in the UK' (2001) 28 European Review of Agricultural Economics 479.

[83] See, eg European Commission, *Mid-term Review of the Common Agricultural Policy* COM (2002)394, 6; and A Herrup, 'Eco-labels: benefits uncertain, impacts unclear?' [1999] European Environmental Law Review 144.

[84] King's College London, *Do European Consumers Buy GM Foods?* (*'Consumerchoice'*) (London: Kings College London, 2008). [85] Ibid 1.9–1.10.

[86] Ibid 1–6. [87] Ibid 5–8.

[88] Ibid 1–10. To take an extreme example, it may be questioned whether consumer responses would be as sanguine as that of the focus groups if the presence of GM material was indicated with the same force as health warnings on packets of cigarettes.

[89] See, eg A Warde, E Shove, and D Southerton, *Convenience, Schedules and Sustainability* (Lancaster: Department of Sociology, University of Lancaster, 1998).

on the shelves, retailers across Europe have been reluctant to do so. In fact, they have made a virtue of not doing so. For example, in 2000 Tesco, which had already decided not to stock GM products, even banned suppliers from growing fruit and vegetables on land used for GM crop trials.[90] The strength of this approach was confirmed in *GM Crops? Coexistence and Liability*, the 2003 report prepared by the Agriculture and Environment Biotechnology Commission (AEBC), where it was concluded that 'there appear to [be] no signs at present as far as we can judge of a significant shift on the part of UK retailers from a stated policy of avoiding GM in their own produce'.[91] However, even though consumers might buy GM food if stocked by supermarkets, the approach adopted by supermarkets is not perhaps so illogical. There is the firmest of evidence that at least some potential customers would carefully avoid GM food, while there is little, if any, evidence of customers actually demanding GM food. A blanket ban would have the advantage of retaining the former, while the latter may be so few in number that their alienation would not materially affect profitability. The position might change if GM foods were to be sold at a substantial discount to conventional food, with consumers placing financial savings above perceived risks;[92] but substantial discounts may be difficult to achieve in the case of many food products, owing to the fact that the amounts paid for raw materials (such as soya or maize), whether GM or non-GM, comprise but a small proportion of the price on the shelves.[93]

3.2 Public Protest

A major characteristic of the GM revolution has been the extent to which it has generated public protest. In this context a leading role has been played by well-organized NGOs in Europe, with Friends of the Earth and Greenpeace very much to the fore.[94] For example, in 2005 Friends of the Earth initiated a protest campaign against Sainsburys in the United Kingdom, on the basis that the

---

[90] 'GM Ban is Extended by Tesco', The Guardian, 7 January 2000; but it should be noted that in February 2009 the Chief Executive of Tesco, Sir Terry Leahy, showed support for reopening the GM debate (<http://greenbio.checkbiotech.org/news/leahy_says_supermarkets_too_hasty_judging_gm>, last accessed 11 May 2009).

[91] AEBC, *GM Crops? Coexistence and Liability* (London: AEBC, 2003) 30. The AEBC provided the Government with independent strategic advice on biotechnology issues before being wound up in 2005.

[92] On price implications see, eg Burton et al (n 82 above); and Lusk et al (n 77 above).

[93] See, eg GBC Backus et al, *EU Policy on GMOs: a Quick Scan of the Economic Consequences* (Wageningen: LEI, 2008) 41.

[94] On anti-GM campaigns by NGOs, see, eg Teel (n 66 above). For an interesting survey of the membership of anti-GM and environmental campaign groups (including both Friends of the Earth and Greenpeace), see C Hall and D Moran, 'Investigating GM risk perceptions: a survey of anti-GM and environmental campaign group members' (2006) 22 Journal of Rural Studies 29. This found, *inter alia*, that women and those living near field-trial sites had lower expectations of future benefits flowing from GMOs. See also, generally, S Tromans and C Thomann, 'Environmental protest and the law' [2003] Journal of Planning and Environmental Law 1367.

supermarket was continuing to sell milk and meat produced using GM feed.[95] Less-structured action may also be detected, and in all four corners of the globe. Thus, as early as 1999 there were several incidents in the United States, undertaken by individuals or groups with names such as 'Seeds of Resistance', which did not receive the endorsement of Greenpeace.[96] Indeed, in 2003 members of a local NGO in the Philippines undertook a hunger strike.[97] A distinctive aspect of the campaigns has been the extent to which those participating have been prepared to take actions which might render them criminally liable. This has evinced very different responses from the courts, but a salient fact has been the consistent acquittal of such protestors by juries in the United Kingdom. With juries being the representatives of 'the public', this marks a perhaps unexpected engagement of wider society in the governance of GMOs. Accordingly, anti-GM protest will be examined in two jurisdictions—the United Kingdom and France.

3.2.1 United Kingdom

In 2000, Lord Peter Melchett, together with 27 other Greenpeace volunteers, appeared in court charged with theft and criminal damage for their 'decontamination' of a GM farm-scale trial.[98] They were acquitted of theft by the jury in Norwich Crown Court on 19 April 2000, but a verdict could not be reached on the second charge of criminal damage.[99] Subsequently, on 20 September 2000, they were also acquitted of criminal damage, it being accepted that their actions were justified to protect property and the environment.[100] Five years later Greenpeace volunteers were again before a jury charged with causing a public nuisance.[101] The charge related to their boarding a merchant vessel, the *MV Etoile*, which they claimed was carrying illegal GM animal feed, on the basis that non-GM and GM material had been mixed in the United States. Again they were acquitted. Moreover, the willingness of juries to acquit gave rise to the

---

[95] Friends of the Earth Press Release, *Sainsbury's Targeted in Week of GM Protest*, 10 June 2005 (available at <http://www.foe.co.uk/resources/press_releases/sainsburys_targeted_in_wee_10062005.html>, last accessed 26 June 2009).
[96] JL Fox, 'Anti-GM crop protestors increase activity in the US' (1999) 17 Nature Biotechnology 1053.
[97] 'Anti-GM protest continues in the Philippines' (2003) 3(8) Bridges Trade BioRes 2.
[98] See, generally, eg M Stallworthy, 'Damage to crops—Part 1' (2000) 150 New Law Journal 728; and 'Damage to crops—Part 2' (2000) 150 New Law Journal 801.
[99] Greenpeace Press Release, *Jury Clears Greenpeace Volunteers of Theft and Fails to Convict over Criminal Damage in GM Trial*, 19 April 2000 (available at <http://archive.greenpeace.org/pressreleases/geneng/2000apr19.html>, last accessed 9 January 2010).
[100] Greenpeace Press Release, *28 Greenpeace Volunteers Acquitted in GM Trial*, 20 September 2000 (available at <http://www.greenpeace.org.uk/media/press-releases/28-greenpeace-volunteers-acquitted-in-gm-trial>, last accessed 26 June 2009).
[101] BBC, *GM Protestors Cleared of Charges*, 16 September 2005 (available at <http://news.bbc.co.uk/1/hi/england/bristol/4253212.stm>, last accessed 26 June 2009).

novel case of *R v Colchester Justices ex p Abbott*.[102] Anti-GM protestors were charged with causing criminal damage to a field-trial site, and an issue was whether the case should be tried summarily or whether they could elect for trial by jury. This right to elect for trial by jury might arise only if the value of the criminal damage was clearly over £5,000 or, subject to detailed procedures, if it was unclear whether the value was over £5,000.[103] The scientific report prepared on behalf of the anti-GM protestors estimated the loss to be £13,900, as opposed to loss of £3,250 as advanced by the prosecution; and, with an eye to securing jury trial, it was submitted to the magistrates that it was unclear whether the value of the criminal damage was over £5,000. Although this argument failed before the magistrates, and the Divisional Court dismissed the subsequent challenge to the decision of the magistrates, it remains a pertinent fact that defendants were prepared to place a higher figure on the loss than the prosecution, and consequently risk greater penalties, with a view to appearing before a jury of 12 citizens.

Where juries have not had the final say, anti-GM protestors have generally fared less well. Thus, in *Monsanto plc v Tilly* the Court of Appeal found, *inter alia*, that environmental campaigners did not have an arguable defence to a claim in trespass to land and goods on the basis that their action was necessary to protect third parties or in the public interest.[104] The first five defendants, who were members of GenetiX Snowball, an anti-GM NGO, had only uprooted a relatively small proportion of the crop. Indeed, members of GenetiX Snowball involved in such actions had undertaken not to uproot more than 100 plants.[105] For this reason, Stuart-Smith LJ found the argument that they were averting danger to be inconsistent, since to achieve their stated purpose they would have had to eradicate the whole crop. Importantly, he went on to say that '[t]he defendants are frustrated that they have been unable to change government policy by the strengths of their arguments. It is breaking of the law, with its potential for martyrdom which affords far better publicity than any other'; and this desire for publicity was regarded as the more powerful motive for their actions. The Court of Appeal was also of the view that only in very restricted circumstances could individuals have the right to destroy the property of another in the public interest; and such a right would not be conferred in order 'to attract publicity for what is alleged to be a good cause or to persuade government to legislate against a perceived danger'. Further, the danger must be immediate and obvious; and the defence of justification by necessity was very limited in scope.[106]

---

[102] The Times, 13 March 2001; and [2001] Criminal Law Review 564.
[103] Magistrates' Court Act 1980, s 22(1) and (4).   [104] The Times, 30 November 1999.
[105] For the GenetiX Snowball Handbook see <http://www.gene.ch/pmhp/gs/handbook.htm>, last accessed 10 September 2009 (which refers expressly to non-violent action).
[106] Reference was made to *Southwark Borough Council v Williams* [1971] 1 Ch 734. In that case, the defendants, who were homeless, entered empty houses owned by the council and then pleaded the defence of necessity. This defence was rejected, Lord Denning MR stating: '[i]f homelessness

A similar decision was reached four years later in *Director of Public Prosecutions v Bayer*.[107] The defendants attached themselves to tractors in order to disrupt the lawful drilling of GM maize. When charged with aggravated trespass contrary to s 68(1) of the Criminal Justice and Public Order Act 1994, they initially enjoyed success before the district judge. Although he was of the opinion that they were aggravated trespassers, he held that they could rely on the common law defence of property on the grounds that: they had honest and genuine beliefs about the dangers of GM crops; they had genuine fears for surrounding property; they had reasonable grounds for those beliefs and fears given their scientific knowledge concerning GM crop tests and given their knowledge of the locality; and they acted with all good intentions and had gone no further than was absolutely necessary to try to prevent the sowing of the crops.[108] By contrast, on appeal, the Divisional Court stated that, where a defence of justification was put forward, it was for the court to first determine as a matter of law whether the defence was available to the defendants on the facts of the case and then whether the defence, if available, entitled the defendant to succeed in rebutting the charge. Further, and significantly, it held that a requisite ingredient of the common law defence was that what was being experienced or feared was an unlawful or criminal act; and, notwithstanding that the defendants believed so strongly that the seed represented a danger to neighbouring property, they also knew quite well that the drilling of the GM seed was not unlawful. In consequence, the common law defence was not available.

On the other hand, it may be highlighted that in September 2008 a jury felt able to acquit six Greenpeace activists charged with causing criminal damage to Kingsnorth coal-fired power station; and, perhaps somewhat surprisingly, the jury accepted their defence that they were acting to prevent greater damage caused by climate change.[109] While the case did not concern GMOs as such, the

---

were once admitted as a defence to trespass, no one's house could be safe. Necessity would open a door which no man could shut. ... Each man would say his need was greater than the next man's. The plea would be an excuse for all sorts of wrongdoing.'

[107] [2003] EWHC 2567 (Admin), [2004] 1 WLR 2856; and [2004] Criminal Law Review 663.

[108] Criminal Justice and Public Order Act 1994, s 68(1) provides that '[a] person commits the offence of aggravated trespass if he trespasses on land in the open air and, in relation to any lawful activity which persons are engaging in or are about to engage in on that or adjoining land in the open air, does there anything which is intended by him to have the effect—(a) of intimidating those persons or any of them so as to deter them or any of them from engaging in that activity, (b) of obstructing that activity, or (c) of disrupting that activity'. See also, again in the context of GM crops, *Director of Public Prosecutions v Tilly* [2001] EWHC 821 (Admin), The Times, 27 November 2001; and [2002] Criminal Law Review 128. In that case the High Court held that the offence of aggravated trespass required that there be persons present engaging or about to engage in lawful activity; and it contemplated that they be intimidated or were not able to proceed with what they intended to do. Accordingly, crop destruction in itself was insufficient to make out the offence.

[109] See, eg Greenpeace UK, *Breaking News: Kingsnorth Six Found Not Guilty* (available at <http://www.greenpeace.org.uk/blog/climate/kingsnorth-trial-breaking-news-verdict-20080910>, last accessed 4 September). Evidence for the defendants was given by, among others, Jim Hansen of NASA.

susceptibility of juries towards arguments based upon 'the greater good' environmentally would seem to continue, with the result that anti-GM protestors may still expect a sympathetic hearing if they can secure jury trial. Further, and more broadly, the integrity of the current GMO regime can only be weakened by repeated failure to secure the public endorsement which is conferred by the decision of a jury.[110]

3.2.2 France

3.2.2.1 Introduction

Even if protest in France and the United Kingdom commenced at roughly the same time (1997), it has developed along different lines, not least in terms of its organizational structure.[111] Thus, the main form of protest in France has been that undertaken by the *Faucheurs Volontaires*, a group originally formed around the leader of the peasants' union, José Bové.[112] The destruction of GM crops on both private and public land has bulked large among the activities of this group, although it has had to yield place in terms of publicity to the trashing of a McDonalds in August 1999.

Confronted with these actions, the French courts have in general rejected the arguments of the *Faucheurs* and found them guilty of destroying property belonging to another, contrary to Article L.322-1 of the Code Pénal. Moreover, since enactment of Law No 2008–595 of 25 June 2008 relating to GMOs,[113] a specific offence has been created whereby the destruction or damaging of crops authorized under the GM regime can be punished by two years' imprisonment and a fine of €75,000; and the penalty can increase to three years' imprisonment and a fine of €150,000 if the GMOs destroyed or damaged are for research.[114]

---

[110] On this aspect more generally, see, eg T Brooks, 'A defence of jury nullification' (2004) 10 *Res Publica* 401.

[111] A leading role has been played by established French unions, while in the United Kingdom campaigning NGOs have been most prominent (and, most notably, Greenpeace): see, eg B Doherty and G Hayes, 'A Tale of Two Movements. Manifestations anti OGM en Grande-Bretagne et en France' in D Hiez and B Villalba (eds), *La Désobéissance Civile. Approches Politique et Juridique* (Lille: Septentrion, 2008) 176.

[112] See <http://www.monde-solidaire.org/>, last accessed 25 June 2009. For the Charter of the *Faucheurs Volontaires*, see <http://www.monde-solidaire.org/spip/IMG/pdf/Charte_faucheurs.pdf>, last accessed 25 June 2009. See also, generally, C Heller, 'From scientific risk to *paysan* savoir-faire: peasant expertise in the French and global debate over GM crops' (2002) 11 Science as Culture 5; and Hayes (n 5 above).

[113] Law No 2008-595 of 25 June 2008 relating to GMOs, *Journal Officiel de la République Française* (JORF) No 148 of 26 June 2008, NOR: DEVX0771876L (codified at Arts L.531-1ff of the Code de l'Environnement).

[114] See E Vergès Etienne, 'Loi n°2008-595 relative aux OGM. La recherche d'un équilibre entre respect de l'environnement et respect de la propriété' [2008] *Revue de Science Criminelle* 943: except in the case of research, where the penalties are heavier, the penalties for *fauchage* are the same as those provided for failure to meet the legal rules governing GM cropping, which would indicate that the legislature wishes to attach equal weight to protecting the environment and protecting economic imperatives.

These provisions, contained in the new Article L.671-15 of the Code Rural, therefore raise the level of sanction above that stipulated in general law, since the penalty under Article L.322-1 of the Code Pénal is only two years' imprisonment and a fine of €30,000.

In this context, the decision of the Court of Appeal in Orléans on 26 February 2008 provides a particularly good illustration of the approach adopted by the French judiciary.[115] There was the opportunity to rehearse all the main arguments fully; and, in addition, this was very much regarded as a test case, since the anti-GM movement was hoping for a change of tack by the courts, in line with the recent enactment of legislation to ban MON810.[116] It was also hoping that such a change of tack might impact upon the proposal for a law relating to GMOs (which at that time was being debated in Parliament).[117] In the event, these hopes were dashed, as almost all the defendants received suspended sentences of two, three, or four months and fines ranging from €1,500 to €3,000. A matter of some significance is that in this case Monsanto relied on the lawfulness of its GM crops, authorized in accordance with the procedure then in force. It argued for convictions in respect of the destruction of the crops and for damages of €313,108 to cover its losses. In response, the main argument of the *Faucheurs* was that they should be free of criminal liability on the basis that they had acted through necessity.[118] Reliance was placed on the Code Pénal, which provides that a person is not criminally liable if, confronted by a present or imminent danger ('un danger actuel ou imminent') to himself, another person, or property, he performs an act necessary to protect the person or property, except where the means used are disproportionate to the seriousness of the threat.[119]

3.2.2.2 No danger presented by GMOs?

In the case before the Court of Appeal in Orléans, it was argued that the requisite 'present or imminent danger' was generated by three factors: first, proof of cross-contamination by gene transfer or cross-pollination; secondly, recognition by the public authorities of the need to evaluate health and environmental

---

[115] CA Orléans, 26 February 2008, CT0028, No 07:00472.
[116] Ministerial Decree of 7 February 2008 Suspending the Cropping of Genetically Modified Maize Seed (Zea Mays L Line MON810), JORF No 34 of 9 February 2008, NOR: AGRG0803466A, amended 13 February 2008, NOR: AGRG0803888A.
[117] As indicated, this proposal was subsequently enacted as Law No 2008-595 of 25 June 2008.
[118] They did not dispute that they had carried out acts of destruction: indeed, they celebrated the fact that they had destroyed the crops.
[119] Art 122–7 of the Code Pénal. It should be noted that persons suspected of damaging GM crops sometimes also invoke the defence provided by Art 122–5 of the Code Pénal, which absolves from criminal liability any person who, confronted by an unjustified attack upon himself or another, performs at that moment an action compelled by the necessity of self-defence or the defence of another person, except where the means of defence used are not proportionate to the seriousness of the attack. This line of defence was not advanced in the instant case before the Court of Appeal, but has frequently been invoked and rejected at first instance: see, eg TC Orléans, 9 December 2005, No 2345/S3/2005 *Société Monsanto v Dufour*; and see also J-P Feldman, 'Les "Faucheurs d'OGM" et la Charte de l'Environnement' [2006] *Recueil Dalloz* 814.

aspects of GMOs and to suspend authorizations for commercialization in France pending further assessment (the MON810 moratorium); and, thirdly, the views of national health-regulation bodies, according to which it was impossible to achieve zero risk. Besides, the danger would be the greater in that the damage was irreversible and irremediable.

The appeal judges rejected these arguments. They considered that it was not possible to characterize the danger fully in the absence of a clear answer from science; and they laid emphasis on the fact that the risk of cross-contamination had not been demonstrated either generally (looking to the consequences for other forms of agriculture) or with particular reference to the land concerned (looking to whether in practice it had been contaminated).[120] Further, according to the appeal judges, it was not possible to calculate the risk flowing from the only political choices open to government, namely to invoke the safeguard clause or ban GM crops; and, in any event, such choices formed no part of the criminal action, nor could they remove the criminal liability of the defendants.

The decision therefore entrenched the rule that it is not possible to plead necessity where the danger is only potential or hypothetical. More specifically, the Court of Appeal adopted earlier judicial reasoning that scientific doubt and imminent peril fall into different categories;[121] and that fear based only on a future possibility can never be regarded as present and imminent danger.[122] The Court of Appeal also went on to find that, even if there was risk, this did not justify such action by the *Faucheurs*, since, under the Deliberate Release Directive, risk did not provide a lawful reason for the destruction of the crops grown for research purposes, only for precaution. Thus, the logic adopted was as follows: the Deliberate Release Directive addresses the authorization of GMOs; authorizations are only given after a risk assessment; and, in consequence, the degree of probability of adverse impact after authorization is so small as not to justify collective actions of destruction.[123] Importantly, similar logic may be detected in the decisions of administrative judges in coexistence cases: authorization closes the matter and renders lawful the sowing of the GM crop, with the result that any action intended to avert an imminent risk is precluded, the risk having already been assessed and adjudged innocuous.[124] Against such a background, it is easy to understand how the Deliberate Release Directive has come to be regarded in certain circles, not as a barrier against alleged dangers, but, on

---

[120] See also CAA Nantes, Second Chamber, 28 March 2007, No 06NT00627, *Commune de Saran*.
[121] CAA Lyon, 26 August 2005, No 03LY000696, *Commune de Ménat* (Actualités Juridiques— Droit Administratif (AJDA), 9 January 2006, 38); and see further S Monteillet, 'De la responsabilité pénale des faucheurs jugés à Orléans. Un État des lieux du cadre juridique des OGM' [2007] *Revue Juridique de l'Environnement* 56.
[122] CA Versailles, 22 March 2007, No 06/01902; and see also P Billet, 'Fauchage d'OGM: une relaxe sans nécessité' (2006) 339 *Revue de Droit Rural* 60.
[123] It was also claimed that the actions taken had been only symbolic, but the appeal judges held that such a defence could not be used when the defendants had acted in concert and after careful preparation.   [124] See further Chapter 7.

the contrary, as a dangerous tool ('dangereux outil') for the propagation of GMOs.[125] It might be suggested that such findings by the Court of Appeal should have ended the matter, since, in the absence of any danger, there should have been no need to investigate further the person or property threatened or, more generally, the question of necessity. However, this did not prevent the Court of Appeal moving on to address these further considerations.

3.2.2.3 The defence of necessity more generally

The *Faucheurs* had presented their actions as necessary to safeguard the environment, which is enshrined as the *patrimoine commun* of humanity in the Preamble to the Charter for the Environment (itself incorporated in the Constitution).[126] They saw such protective steps as even more necessary in view of the lack of precautions taken by seed companies against risk of cross-contamination.[127] In addition, they advanced two constitutional principles to counter any argument that they had breached the right to property, namely the protection of public health and the precautionary principle. In the event, the Court of Appeal did not examine these matters in depth, being swift to hold ineffective any arguments drawn from the French Charter for the Environment,[128] the European Convention on Human Rights,[129] the Cartagena Protocol and the Aarhus Convention. In particular, it was not prepared to grant immunity in criminal law to actions taken in defence of the common heritage or in defence of a law, subjective in nature, which was difficult to identify or categorize.

The question of necessity would, nevertheless, seem the more pertinent by reason of the ineffectiveness of damages actions against GM producers. To the extent that these actions can be maintained, they will almost certainly be of little practical benefit, since neither the person suffering the loss nor the person causing the loss are likely to be carrying insurance cover. The fact is that, as of today, insurance companies do not insure risks from GMOs (although, perhaps somewhat optimistically, four Member States, including France, have already provided that insurance or alternative forms of financial guarantee may be legally required).[130] This aspect was not fully addressed by the Court of Appeal,

---

[125] Monteillet (n 121 above) 64.
[126] The French Charter for the Environment is available at <http://www.legifrance.gouv.fr/html/constitution/const03.htm>, last accessed 25 June 2009.
[127] In particular, a buffer of only four rows of conventional maize was considered insufficient to prevent the dispersal of GMOs.
[128] See also Cour de Cassation, Chambre Criminelle, 7 February 2007, No 06–80.108, *Darsonville Autrey* [2007] *Recueil Dalloz* 573; and, generally, J-P Feldman, 'Les "faucheurs" fauchés par la Cour de Cassation' [2007] *Recueil Dalloz* 1310.
[129] Available at <http://www.echr.coe.int/NR/rdonlyres/D5CC24A7-DC13-4318-B457-5C9014916D7A/0/EnglishAnglais.pdf>, last accessed 25 June 2009.
[130] See, eg European Commission, *Report from the Commission to the Council and the European Parliament on the Coexistence of Genetically Modified Crops with Conventional and Organic Farming* COM(2009)153, 5. In the case of France, see Art L.663–4-III of the Code Rural, under which

notwithstanding that it could be argued that the lack of insurance constitutes a material lacuna in the rights of persons who may suffer cross-contamination, so prompting an action based on necessity. On the other hand, a requirement for a successful action would be to demonstrate an insurable risk; and, as has already been seen, this is not something recognized by the courts, so that the need to deal with its consequences evaporates.

According to the *Faucheurs*, a final ground for justifying a defence of necessity was the fact that France had not transposed (or, at least, not transposed correctly) the Deliberate Release Directive, as was the case prior to the enactment of Law No 2008–595 of 25 June 2008. This argument, based on the failure of national authorities to implement Community legislation correctly, raised issues of general application at both European and French level. More precisely, the *Faucheurs* maintained that the seed company did not have the benefit of an authorization which complied with Community law. The reasoning was again rejected by the appeal judges. First, they considered that a defendant could only rely on failure by a Member State to implement Community law correctly if the provision in Community law was sufficiently clear and unconditional to have direct effect; and, on the facts, the defendants did not have in mind any specific provision of the Deliberate Release Directive. Secondly, they also considered that a defendant could not invoke violation of a Community rule when the objectives pursued by that rule had no link with the offence with which he was charged; and the Deliberate Release Directive was not concerned with the destruction of GM crops, let alone potential grounds for escaping criminal liability. The appeal judges went on to say that Community law did the very opposite of banning field trials: the purpose of the Deliberate Release Directive rather was to *regulate* releases. Accordingly, if the Directive had been transposed correctly, it would definitely not have prevented the releases that the defendants sought to bring to an end. Finally, the Court of Appeal affirmed that the non-transposition of the Community legislation engaged only the liability of France as Member State and did not confer on any individuals the right to commit criminal offences.[131]

3.2.2.4 Some conclusions on the decisions of the French courts
The Court of Appeal in Orléans definitely followed precedent.[132] Above all, it held true to the tenet that in such circumstances any resort to a defence of

every farmer growing GM crops authorized for placing on the market must take out a financial guarantee to cover his liability as specified under that article.

[131] On this question, see also CAA Lyon, 26 August 2005, No 03LY000696, *Commune de Ménat* (AJDA, 9 January 2006, 38); and see further Monteillet (n 121 above).
[132] For the earliest authority to this effect, see Cour de Cassation, Chambre Criminelle, 19 November 2002, No 02–80788 [2003] *Recueil Dalloz* 1315; and for cases supporting it see, eg Cour de Cassation, Chambre Criminelle, 28 April 2004, No 03–83783; Cour de Cassation, Chambre Criminelle, 18 February 2004, No 03–82951 (and see also [2004] *Revue Environnement*, July, 22); CA Versailles, 22 March 2007, No 06/01902; Cour de Cassation, Chambre Criminelle,

necessity would be ineffective,[133] since there was no real or present danger, the action was not necessary to protect any person or property, and there was a lack of proportion between the means employed and the seriousness of the threat. Accordingly, three propositions would seem fundamental. First, the courts have consistently held that no danger exists, by reason of the fact that, under Community law, there is no clear proof of danger to the environment or to health (and this proposition resonates loudly within the larger debate on the relationship between the judiciary and science, including the question of the independence and neutrality of scientists).[134] However, there remains the vexed issue of what is to happen if the minority opinion of agro-biotechnologists proves correct.[135]

Secondly, the *Faucheurs* have pleaded a plethora of arguments to justify their actions and this 'chaff' has arguably obscured key points in the eyes of the courts. As indicated, resort has been made 'pell-mell' not just to the Deliberate Release Directive, but also the French Charter for the Environment, the European Convention on Human Rights, the Cartagena Protocol, the Aarhus Convention, and the precautionary principle. That said, the tactics of the *Faucheurs* may simply reflect the limits of the defences available under criminal law, causing them to broaden their arguments before the court. Not least, a hypothetical risk could perhaps be categorized as a present danger to the wider public interest, so legitimating collective and pre-meditated action. Further, there would seem to be inconsistencies with any potential application of the precautionary principle. On the one hand, it could be advocated that the danger is effectively uncertain and that the precautionary principle should be therefore employed. Yet, such reasoning concurrently removes the possibility of employing a defence of necessity, which requires immediate danger and, importantly, a

---

7 February 2007, No 06–80.108, *Darsonville Autrey* [2007] *Recueil Dalloz* 573; and Cour de Cassation, Chambre Criminelle, 4 April 2007, No 06–80.512. It may be noted, nonetheless, that some judges at first instance have sought to adopt a different position. Thus, the Tribunal Correctionnel of Versailles took the view that necessity could be pleaded in light of the fact that it was not possible to take protective actions against GM crops: CA Versailles, 22 March 2007, No 06/01902; and see also TC Orléans, 9 December 2005, No 2345/S3//2005, *Société Monsanto v Dufour* (where 49 *Faucheurs* were acquitted on grounds of necessity). On the latter judgement, generally, see A Gossement, 'Le fauchage des OGM est-il nécessaire?' [2006] *Revue Environnement* (January) 9; Billet (n 122 above); Feldman (n 128 above); and Monteillet (n 121 above).

[133] On this aspect, see also Cour de Cassation, Chambre Criminelle, 7 February 2007, No 06–80.108, *Darsonville Autrey* [2007] *Recueil Dalloz* 573.

[134] For excellent treatment of this larger debate, see, eg S Jasanoff, *The Fifth Branch: Science Advisers as Policymakers* (Cambridge, Mass: Harvard University Press, 1990); S Jasanoff, *Designs on Nature: Science and Democracy in Europe and the United States* (Princeton: Princeton University Press, 2005) 1–93; and, in the context of the World Trade Organization, Scott, (n 9 above) 76–138.

[135] See, eg C Vélot, *OGM. Tout S'explique* (Athée: Edition Goutte de Sable, 2009) 89–191. It may be noted that, in the world trade context, a minority scientific opinion may validate a sanitary or phytosanitary measure: see, eg *EC—Measures Concerning Meat and Meat Products (Hormones)* WT/DS26/AB/R and WT/DS48/AB/R, 16 January 1998, para 194; and Scott (n 9 above) 104–110.

high degree of certainty. On the other hand, it could be advocated that there is clear and present danger, but this would in turn preclude any recourse to the precautionary principle.[136] As a consequence, it is understandable that the Court of Appeal in Orléans did not entertain these wider considerations (short of revolutionizing not only established law on the destruction of GM crops, but also, more generally, the defence of necessity).

Thirdly, the judgment of the Court of Appeal in Orléans confirmed that there was little mileage in seeking to rely on constitutional principles or international law. Instead, there was preference for a positivist view, which rejected recourse to overarching principles of law—principles which have formed the spearhead of civil disobedience.[137]

4. Conclusion

What would seem clear is that the public has engaged extensively in the controversy which surrounds GMOs; and the regulation of GMOs is without doubt an area of law where civil society seeks a 'voice'. It would also seem clear that the legislative framework does grant some latitude for public participation. In particular, the new Article 6bis of the Aarhus Convention, and its accompanying modalities, will require that due account is taken of the outcome of its public participation procedure, while in the United Kingdom the *GM Nation* debate marks a level of public consultation that has rarely been equalled.

Nonetheless, in this context, there are arguably material limitations on effective inclusion within legislative initiatives. Many regimes are less prescriptive than the Aarhus Convention: for example, as has been seen, under the Cartagena Protocol there is only an obligation to consult the public in the decision-making process and make the results of such decisions available to the public. Further, and more generally, there must be real doubt as to the extent the public can engage in authorization procedures which are frequently technocratic and grounded in science.[138] With so much depending upon risk assessments and, with specific reference to the Community, upon the opinions of the European Food Safety Authority,[139] there are relatively high hurdles for the public to cross, both in terms of advancing the necessary expertise and in terms of securing the necessary finance for research. As indicated, for these reasons, the resources of NGOs may prove material.

---

[136] Billet (n 122 above).   [137] See further the concluding Chapter 15.
[138] See, eg Lee (n 24 above) 85–97; DJ Galligan, 'Citizens' Rights and Participation in the Regulation of Biotechnology' in F Francioni (ed), *Biotechnologies and International Human Rights* (Oxford: Hart Publishing, 2007) 335; and M Kritikos, 'Traditional risk analysis and releases of GMOs into the European Union: space for non-scientific factors?' (2009) 34 European Law Review 405.
[139] See Chapters 4 and 5. See also the crucial role of US federal agencies in the authorization of GMOs, as discussed in Chapter 12.

By contrast, in spite of (and maybe even because of) these limitations, sections of the public in several countries would seem to have influenced less formally the pace at which GMOs are being developed, and also their market penetration. For example, in the United Kingdom, GM food is currently not on the shelves of the major supermarkets; and the supermarkets have stated this to be a response to consumer preferences. That said, evidence from surveys such as the *Consumerchoice* report suggests that, if GM products were to be made widely available, they would almost certainly be widely purchased. However, there must be something inherently unsatisfactory in the commercial decisions of supermarkets (as opposed to a regulatory system) playing so great a role in the dissemination of GM food.

Similarly, the consistent destruction of GM trials has had significant impact on research into modern biotechnology and, in the United Kingdom, it has been reported that almost all 54 GM field trials since 2000 have suffered some form of vandalism.[140] A matter of some significance is that the anti-GM protestors undertaking these actions regard themselves as agents of the general public. Thus, the Charter of the *Faucheurs* expressly recites that 70 per cent of French citizens are opposed to GMOs and it claims that their campaign is one to preserve democracy.[141] Civil disobedience of this order will be further considered in Chapter 15, but for the time being it may be highlighted that lack of confidence in the regulatory framework extends well beyond the anti-GM movement, even to governments. This may be illustrated by the fact that by early 2009 six Member States had banned MON810 maize; and it is perhaps unlikely that the governments concerned implemented these bans without an eye to the views of their electorate.[142] In consequence, at some risk of generalization, one conclusion might be that, while members of the public have indeed participated formally in the regulation of GMOs, informally they have exerted greater influence. And, flowing from this, it might also be suggested that greater accommodation of the public will in the legislative process would have produced not only a more widely accepted, but also, importantly, a more structured framework for the introduction of modern biotechnology.[143]

---

[140] See, eg *GM Researcher Despairs as Three Years' Work Destroyed*, Times Higher Education, 7 August 2008; and in 2004 over half of GM maize trials in Metropolitan France were destroyed: Hayes (n 5 above) 296.

[141] For the Charter of the *Faucheurs*, see <http://www.monde-solidaire.org/spip/IMG/pdf/Charte_faucheurs.pdf>, last accessed 25 June 2009: '[d]ans l'état de nécessité actuelle où nous nous trouvons, nous n'avons plus rien à notre disposition pour que la démocratie reste une réalité'.

[142] *Agra Europe Weekly* No 2357, 17 April 2009, EP/1.

[143] For the advantages of including the public within the decision-making process so as to secure better environmental regulation, see, in particular, J Steele, 'Participation and deliberation in environmental law: exploring a problem-solving approach' (2001) 21 Oxford Journal of Legal Studies 415.

2

The Importance and Limits of Cost-Benefit Analysis in the Regulation of Genetically Modified Organisms

Christophe Charlier and Egizio Valceschini

1. Introduction

Economic literature has shown that, for the purposes of securing good 'risk governance' in the agro-food sector, cost-benefit analysis (CBA) should be undertaken, in addition to scientific risk assessment.[1] The aim of CBA is, without doubt, to aid public decision-making and, in that regard, it should be seen as a complementary tool to risk assessment. Using this methodology, policy-makers may evaluate the economic relevance of regulation, comparing this with other possible options (from other types of regulation to the absence of regulation); or, alternatively, information may simply be provided to public authorities as to the costs and benefits of their policies. For such purpose, CBA aims at estimating a monetary value, on the one hand, for environmental or public health degradation and expected benefits from the technological development which would result from the absence of regulation, and, on the other hand, for the expected health and environmental benefits implied by regulation and the implied opportunity costs of constraining the technological development.

Among the many innovations in the agro-food sector, CBA is particularly relevant in the governance of modern biotechnology. Genetically modified organisms (GMOs) can definitely be considered an innovation which has the capacity to present specific problems to both the environment and public health;

---

[1] See, eg J-C Bureau, S Marette, and A Schiavina, 'Non-tariff trade barriers and consumers' information: the case of the EU-US dispute over beef' (1998) 25 European Review of Agricultural Economics 437; and, more recently, CG Turvey and EM Mojduszka, 'The precautionary principle and the law of unintended consequences' (2005) 30 Food Policy 145. For an overview of the economic consequences of GMOs in Europe, see, eg GBC Backus et al, *EU Policy on GMOs: a Quick Scan of the Economic Consequences* (Wageningen: LEI, 2008).

and they also open novel avenues for discussion. It is generally agreed that, with regard to risk, the two main threats are those to biodiversity and human health; and that, with regard to benefit, the first generation of GMOs was engineered to improve productivity for farmers, whereas the second generation presents enhanced nutritional values or environmental characteristics. However, more contentiously, GMOs also provide a 'textbook case' of the complexity which results from any attempt to conduct CBA in relation to innovation, this being a function of their novelty, the ethical concerns that they raise, the danger of potentially irreversible effects on biodiversity, the absence of scientific unanimity in the risk assessment, their economic importance, and consumer fear. These characteristics might indicate that any CBA would inevitably be limited, incomplete, or even impossible: the economic optimum can be very different from the 'ecological optimum',[2] owing to long-term effects on the environment which can be difficult to calculate. That said, CBA can be defended as a useful tool in the development of risk governance for modern biotechnology, adding an economic perspective over and above scientific assessment. This is of particular importance at international level, where divergent national perceptions of GMOs coexist;[3] and where regulation by the World Trade Organization (WTO) favours a science-based approach, leaving little space for economic aspects.

The aim of this chapter is, therefore, to underline the importance of CBA in the regulation of GMOs and to highlight the specific difficulties with which such analysis is confronted. It argues that these difficulties should not be considered as a reason to dispense with economic evaluation. Accordingly, the first section presents the aim and scope of CBA, discussing the originality of the issues generated by GMOs. The second section applies CBA to GMOs, while the third advances a conclusion.

2. The Aim and Scope of CBA as Applied to the Regulation of GMOs

Situations of market failure arise in the presence of externalities, whether positive or negative. In these situations, private operators cannot be expected to take social aspects (either benefits or costs) into account when making decisions. Regulation, in whatever form, is needed to orientate the activity of such operators with the aim of maximizing total well-being or of implementing optimal resource allocation. CBA, in this context, has utility in estimating a monetary value for external variables which have no market and, in

---

[2] See generally, eg DW Pearce and RK Turner, *Economics of Natural Resources and the Environment* (Baltimore: John Hopkins University Press, 1990).
[3] On this aspect, see, generally, eg T Josling, D Roberts, and D Orden, *Food Regulation and Trade: Toward a Safe and Open Global System* (Washington, DC: Institute for International Economics, 2004).

consequence, no monetary price.⁴ This approach can be extended to environmental risk, with CBA aiming to quantify the risk and determine its economic impact;⁵ and, when diverse risks are addressed with diverse methodologies, CBA has the advantage of translating all results into monetary terms. Kleter and Kuiper, for example, when studying the environmental impact created by changes in pesticide use on GM crops, underline the advantage of monetary evaluation over other forms of evaluation which consider physical impact, these being derived from different indicators that cannot easily be compared.⁶

2.1 Presentation of CBA

Before answering the question whether to allow a certain type of pollution and, if so, to what extent, the basic idea of CBA is to produce a demand curve for the environmental good (or bad) concerned.⁷ With 'environmental conservation' considered as an 'environmental good', and 'pollution' as an 'environmental bad' (or 'negative externality'), this curve provides important information: how demand reacts to price change. If the environmental good (or bad) is important to consumers, the reaction should be great, and vice versa. The main difficulty faced by this approach is the absence of a market for environmental goods (or bads), where prices and demand can be observed. CBA has, therefore, to find other ways and use different data in order to construct the demand curve; and, while several approaches exist, they share the basic idea of estimating consumers' willingness to pay (WTP) for the environmental good (or bad).

A consumer's marginal willingness to pay (MWTP) for a unit of good is the maximum price he/she is ready to pay for such unit. To express this in graph form (Figure 1), the points along the demand curve represent the MWTP for the successive units of good. A consumer would not choose to buy a unit for which the market price is higher than his/her MWTP. In consequence, the MWTP of the last unit s/he buys is just equal to the market price, whereas the MWTP of the other units s/he buys is higher than the market price. The difference (measured in monetary unit) between the MWTP and the market price forms, in Figure 1, an area called 'consumer surplus', which is considered a representation (measured in monetary unit) of his/her well-being. The

---

⁴ See, generally, eg R Brent, *Applied Cost-Benefit Analysis* (2nd edn, Cheltenham: Edward Elgar, 2006).
⁵ JA Caswell, 'An evaluation of risk analysis as applied to agricultural biotechnology (with a case study of GMO labelling)' (2000) 16 Agribusiness 115; and E Gozlan and S Marette, 'Analyse des Risques Alimentaires. Aspects Économiques', in M Feinberg et al (eds), *Analyse des Risques Alimentaires* (Paris: Editions TEC and DOC, 2006) 361.
⁶ GA Kleter and HA Kuiper, 'Assessing the Environmental Impact of Change in Pesticide Use on Transgenic Crops' in JH Wesseler (ed), *Environmental Costs and Benefits of Transgenic Crops* (Berlin: Springer, 2005) 33.
⁷ For further discussion of this aspect, see, eg CD Kolstad, *Environmental Economics* (Oxford: Oxford University Press, 2000).

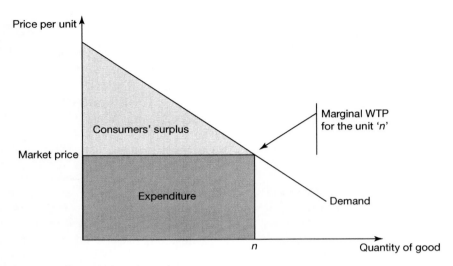

Figure 1 The Consumers' Surplus

rectangular area beneath is total expenditure when n units of goods are bought at the market price.

In case of pollution, no market exists. Nonetheless, the MWTP and the WTP still have a meaning. A consumer suffering from pollution bears the external cost of the economic activity which is responsible. For each unit of pollution, the marginal external cost (MEC) represents the maximum amount the consumer is ready to pay in order to avoid the pollution. In other words, the MEC is the MWTP to reduce the pollution of one unit or to 'buy' one unit of 'environmental' conservation. As a result, the curve representing the MEC (Figure 2) also represents the demand of consumers for pollution reduction or for environmental conservation. The MWTP and WTP thus keep their respective meanings even in the absence of a market for the environmental good. Further, one of the main features of GMOs, as will be seen, is the simultaneous presence of alleged pollution (for example, the creation of 'superweeds') and risks (for example, the risk of transference of allergens into food).

That said, measuring MWTP and WTP cannot be done directly; and two broad approaches have been developed to overcome the absence of markets: the 'revealed preferences' approach and the 'stated preferences' approach. The 'revealed preferences' approach consists in observing consumer choice in markets linked with the pollution problem, so as to deduce information on the way in which they value the environmental good. For example, consumers suffering from noise in urban areas can buy double-glazed windows and/or expend money on trips to relax away from home. The difficulty is to isolate the real weight of

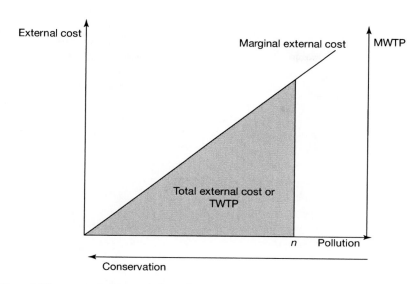

Figure 2 The external costs and the willingness to pay

the environmental motive in such decisions. Both the double-glazed window and the money spent on going to the countryside can be justified by more than just the environmental motive.

The 'stated preferences' approach consists in asking people through surveys how much they valuate an environmental good. This approach is more straightforward, but also more controversial. The absence of real choices introduces the possibility of bias in the responses. Between the revealed preferences approach and the stated preferences approach lies an intermediate area, where experimental markets are artificially constructed. A sample of interviewees are asked to make choices using a fixed amount of money; and the research produces data on how people value environmental goods and react to information on environmental aspects of their consumption (this information being supplied during the course of the research). In consequence, their MWTP is revealed in relation to both particular characteristics of goods and/or health-safety levels in situations where real behaviour is simulated. This form of research has been employed in the case of GMOs, as will be discussed later.

The value estimated for the environmental good, whatever the method used, may be interpreted in more than one way.[8] In particular, a distinction is generally made between 'use value' and 'non-use value'. 'Use value' addresses value derived from current use or expected use. For example, environmental

---

[8] See, eg Pearce and Turner (n 2 above) 120–137.

conservation creates benefits (such as recreation) and thus has a value. This benefit can be either current (if a person is already enjoying the recreation activities) or expected (if he/she only has plans to do so). In the case of GMOs, a person can show that he/she attaches value to consuming non-GM food by spending time and money in finding and purchasing conventional or organic food. Use value can also take account of potential through 'option value' and 'quasi-option value'. 'Option value' represents the WTP for the preservation of an environmental asset in the sense that a person may wish to retain the opportunity to enjoy that asset at some future date. 'Quasi-option value' is very similar in meaning, but takes into account expectation of increased information over time. A positive option value denotes that it is optimal to postpone the destruction of the environmental asset in order to be able to make a better-informed decision at a later date. Quasi-option value is of particular relevance to GMOs, where it is generally understood that environmental damage may be irreversible and where the long-term risks are not firmly established, with the result that it is especially important to be able to obtain further information in the future.

'Non-use value' is the WTP for preservation of an environmental asset expressed independently of any current or future use by a person. Three categories of non-use value are generally distinguished: 'altruistic value', 'bequest value', and 'existence value'. 'Altruistic value' covers situations where a person expresses a WTP for environmental preservation on the basis that others value it. 'Bequest value' is similar, but with focus on the well-being of future generations. 'Existence value' is the value which a person attaches to knowledge that an environmental asset exists, without using it or planning to use it: it is the physical presence of the asset that counts, not its appreciation by other human beings. In the case of GMOs, these non-use values acquire importance when threats to genetic diversity and/or biodiversity more generally are taken into consideration.

While such distinctions may be regarded as somewhat opaque, they are necessary for the purposes of carrying out a comprehensive valuation. Classifications vary from author to author and a single definition of the economic value of an environmental asset consequently does not exist.[9] As indicated, however, whatever the chosen classification, it is difficult, if not impossible, to distinguish the various motives which a consumer expresses by means of a single MWTP. Matters may become yet worse if we consider situations where the environmental risks are not based on clear scientific evidence; or where consumers lack trust in the regulatory system and/or the institutional framework for scientific

---

[9] Ibid; and see also, eg RK Turner, 'The Place of Economic Values in Environmental Valuation' in IJ Bateman and KG Willis (eds), *Valuing Environmental Preferences: Theory and Practice of the Contingent Valuation Method in the US, EU, and Developing Countries* (Oxford: Oxford University Press, 1999) 17.

risk assessment. And such would seem to be the position regarding GMOs.[10] The potential long-term consequences of cultivating GM crops and eating GM food are subject to lively debate; and food-safety crises, such as the bovine spongiform encephalopathy (BSE) crisis, have made consumers suspicious.[11] Indeed, following the BSE crisis, the report by Lord Phillips expressly drew attention to the limitations of CBA, stating that:

> In practice, it was not straightforward to apply the CBA approach. It was difficult to maximise benefits and to evaluate non-monetary factors or uncertainties. Moreover, seeking to attach monetary or other comparative values to human life, health and preferences was controversial and subjective, although it was necessary to attempt this if the costs and benefits of precautionary regulations and procedures were to be assessed.[12]

These words also have resonance with reference to GMOs, but before addressing such issues, it may be reiterated that the use of CBA in the environmental context more generally has not been without criticism.[13] First, the idea of giving a monetary value to environmental conservation is not undisputed. While other valuation units (such as energy units) have been proposed, these are arguably less convenient in a market economy. Secondly, more than one form of value may be in issue, so bringing in train fine distinctions which are not easy to measure. Thirdly, the proposition that the environment enjoys an 'intrinsic value' indicates that ethical concerns are involved (and, with specific regard to GMOs, one could speak of the ethics of preserving biodiversity); and again such concerns are not easy to quantify. Finally, many critics of CBA have denounced its 'consequentialist' approach, since the notion that the right choice is the one that conveys the best consequence can lead to results which are counter-intuitive in ethical terms.

2.2 Novel Issues Generated by GMOs

The use of CBA in the context of GMOs generates many novel issues. As a preliminary point, however, it may be emphasized that the regulatory regime for GMOs in the Community is a complex one, which extends not just to their authorization, but also to their use once authorized (for example, laying down provisions for their traceability and labelling, and also for their coexistence with

---

[10] See, eg JE Hobbs and WA Kerr, 'Will Consumers Lose or Gain from the Environmental Impacts of Transgenic Crops?' in Wesseler (n 6 above) 249.

[11] See, eg *The BSE Inquiry: the Report—Vol. 1, Findings and Conclusions: Executive Summary of the Report of the Enquiry, 1. Key Conclusions* (available at <http://www.bseinquiry.gov.uk.html>, last accessed 17 August 2009); and see, generally, eg G Little, 'BSE and the regulation of risk' (2001) 64 Modern Law Review 730.

[12] *The BSE Inquiry: the Report—Vol. 15, Government and Public Administration, 5. Risk Analysis: an Analytical Approach to Policy-making* para 5.16 (available at <http://www.bseinquiry.gov.uk.html>, last accessed 17 August 2009).

[13] See, eg R Frank, 'Why is cost-benefit analysis so controversial?' (2000) 29 Journal of Legal Studies 913.

conventional and organic crops).[14] While CBA would clearly seem to be important at both these stages, a distinction between authorization and post-authorization measures will be maintained.

Taking first authorization, the decision is, in essence, binary: either to allow or not to allow the GMO in question to be placed on the market. As a consequence, the concept of the 'optimal level of pollution' has no relevance, since there is no logic in a decision to allow the marketing of a GMO 'to a certain extent'. This characteristic is of great relevance for CBA. If the optimal degree of release is sought, an evaluation of the MEC (or benefit) of the GMO should be undertaken and compared to the marginal private benefit. In other words, a CBA would be required. But, as soon as it is recognized that the decision is binary, this imperative disappears;[15] and the monetary valuation of risk can be dispensed with at the authorization stage. The decision whether or not to allow the placing on the market of a GMO is thus to be made simply on the basis of information derived from the scientific risk assessment, with the decision constituting a political choice. For example, as a target for its regulation a government can choose to pursue 'zero-risk', or risk up to a threshold perceived as acceptable; and for these purposes information derived from a scientific risk assessment plays a key role. It is highly relevant whether or not there is risk; but, if there is risk, its monetary valuation is not an issue. Nor are the private benefits and costs associated with the new technologies. This approach is arguably reinforced by the fact that, under the WTO, primacy is given to the evaluation of risk.

Nonetheless, in some circumstances a government may be interested by information on the social value of the new technology, even if it is not considering the optimal level of pollution. These circumstances would seem to arise, in particular, if society is split as to what constitutes a 'reasonable quality of life' or an 'acceptable level of risk'; and a clear illustration would be where there is a very small likelihood of catastrophic consequences (GMOs being frequently cited in this context). Confronted by such a state of affairs, more reference to some ad hoc conception of 'reasonable quality of life' or 'acceptable level of risk' is not a course of action that can comfortably be pursued by government. Rather, the results of CBA can provide information to determine more precisely what constitutes a 'reasonable quality of life' or what qualifies as 'acceptable level of risk'.[16] In other words, a government could then allow the new technology, despite the risks it presents, provided that its net social value is positive.[17]

---

[14] On traceability and labelling, see Chapter 4; and on coexistence, see Chapter 6. The Community regime also addresses the contained use of GMOs (eg for laboratory research) and the experimental release of GMOs into the environment; but such aspects are not considered in this chapter.   [15] Pearce and Turner (n 2 above) 120.
[16] WJ Baumol and WE Oates, 'The use of standards and prices for the protection of the environment' (1971) 74 Swedish Journal of Economics 425.
[17] For example, dangerous herbicides and pesticides can receive marketing approval; and similar reasoning was applied to quarantine policy by James and Anderson: S James and K Anderson, 'On the need for more economic assessment of quarantine policies' (1998) 42 Australian Journal of Agricultural and Resource Economics 425.

A further option would simply be to specify the extent to which economic aspects of the new technology carry weight in the decision-making process and to evaluate if the benefits of a given decision (such as authorizing the placing on the market of a GMO) exceed the cost implied, if there is any cost. This 'economic analysis' should provide more valuable guidance than recourse to the ad hoc criteria mentioned above. Indeed, it goes so far as to question the social value of the innovation that is modern biotechnology. That is to say, when all costs and benefits are taken into account, is modern biotechnology a good innovation from an economic point of view? And, in this regard, GMOs have particular resonance. Because genetic modification concerns 'life' and, moreover, is widely apprehended to be 'risky', society is reluctant to leave it to the market to decide whether the new technology enjoys 'economic validity'. Instead, there are calls for a regulatory decision; and, at this point, the CBA, which would not have appeared necessary at the authorization stage, becomes important. In order to reinforce this line of argument, it may be noted that GMOs form what Lupton has called 'indeterminate goods', that is to say goods in respect of which there are conflicting opinions (in the case of GMOs, the conflicting opinions relating primarily to their health or environmental safety).[18] Such uncertainty is shared by all agents (including producers and consumers) and cannot be considered asymmetric information. Accordingly, the lack of scientific knowledge (and consequent divergence in opinions) plays a major role in this context, in that it can be used to reject the goods from the market.

Surprisingly, some formulations of the precautionary principle provide an additional ground for elevating the importance of CBA so far as GMOs are concerned (and the relevance of the precautionary principle to GMOs is well-illustrated by the use of Article 5.7 of the Sanitary and Phytosanitary Agreement by the European Union in the *EC-Biotech* dispute).[19] The precautionary principle is generally interpreted as giving primacy to uncertainty rather than to the balance of costs and benefits in the decision-making process. Nonetheless, economic evaluation is not irrelevant, as is evident in the Rio Declaration itself, which states that '[w]here there are threats of serious or irreversible damage, lack of full scientific certainty shall not be used as a reason for postponing cost-effective measures to prevent environmental degradation'.[20] Thus, CBA can ensure that the chosen measures are not more costly than necessary; and its application may extend both to authorization and post-authorization measures. However, as proposed by Aslaksen and Myhr, it should be complemented by

---

[18] S Lupton, 'Shared quality uncertainty and the introduction of indeterminate goods' (2005) 29 Cambridge Journal of Economics 399.
[19] *EC—Approval and Marketing of Biotech Products* WT/DS291/R, WT/DS/292/R, and WT/DS293/R, 29 September 2006; and see further Chapter 13.
[20] Rio Declaration on Environment and Development, Principle 15 (available at <http://www.unep.org/Documents.multilingual/Default.asp?DocumentID=78&ArticleID=1163>, last accessed 10 August 2009).

other approaches, stressing, for example, the interests of stakeholders and disagreements between scientific experts.[21]

Turning further to post-authorization measures, it has already been noted that the Community regime lays down provisions for the traceability and labelling of GMOs, and also for their coexistence with conventional and organic crops (examples of the latter provisions being the rules governing isolation distances). Here, again, any decision by the authorities could be regarded as binary: to trace and label or not; to have coexistence measures or not. In consequence, it could be argued that CBA is not a necessity. That said, measures governing traceability, labelling, and coexistence can be more or less stringent, which resurrects the question of the costs and benefits. By way of illustration, Golan et al have defined three characteristics of traceability rules that can be adjusted for their strength, namely: their breadth (the amount of data); their depth (the length of time that the data may be traced); and their precision (the means of tracking the GM material, whether by reference, for example, to a day's production or the truck/container in which it was transported).[22] Yet, Charlier and Valceschini have shown that in practice the Community traceability regime is not very demanding, it being quite possible to impose greater rigour on batching.[23] Such a regulatory choice would inevitably entail higher costs, with the result that it becomes clear that the question of costs and benefits cannot easily be dispensed with when enacting post-authorization measures. Interestingly, the question of the 'optimal level of pollution' implicitly reappears as a corollary. More stringent coexistence measures aim at lowering 'GM pollution', as does more stringent labelling. A reduction of the threshold for presence of GM material from 0.9 per cent to 0.1 per cent would indeed impose a tight definition of what constitutes (from the point of view of the regulator) such 'GM pollution'.[24] And herein lies a novelty/paradox of the regulatory regime. At the authorization stage, no mention is made of the optimal level of pollution: only risks matter and, in consequence, primacy is accorded to risk assessment. Nevertheless, once a GMO is authorized, implicit reference to the optimal level of pollution is introduced via the thresholds for mandatory labelling and for coexistence measures; and this brings into question the need for CBA.

---

[21] I Aslaksen and AI Myhr, '"The worth of a wildflower": precautionary perspectives on the environmental risk of GMOs' (2007) 60 Ecological Economics 489.

[22] E Golan et al, *Traceability in the US Food Supply: Economic Theory and Industry Studies: Agricultural Economic Report 830* (United States Department of Agriculture, Economic Research Service, 2004).

[23] C Charlier and E Valceschini, 'Coordination for traceability in the food chain. A critical appraisal of European regulation' (2008) 25 European Journal of Law and Economics 1. Not least, in order to facilitate any required withdrawal of products, there could be more emphasis on content (the homogeneity of the batches).

[24] For the cost impacts of trying to achieve a 0.1 per cent threshold, see, eg A-K Bock et al, *Scenarios for Co-Existence of Genetically Modified Conventional and Organic Crops in European Agriculture: Tech. Rep. EUR 20394* (Joint Research Centre, European Commission, 2002) 127; and see further below.

3. The Application of CBA to GMOs

3.1 The Authorization Stage

Any assessment of the social value of modern biotechnology implicitly involves two sets of variables: private variables (such as productivity and yields) and public variables (such as the external cost on the environment). Considering both sets is unavoidable if a complete picture is to be obtained of the social consequences of the innovation. That said, CBA for the purpose of GMOs should arguably begin with their potential effect on the environment and human health. Only then can there be an evaluation of whether social benefits (in other words, private benefits plus external benefits) cover social costs (in other words, private costs plus external costs). Such an approach would thus take the form of a list of the consequences of GMOs for farmers and society, with each consequence being evaluated as to whether it is good or bad and whether it is private or public.[25] If only private financial consequences are at issue, then it would seem that no regulation is required, and any CBA is effectively useless, for the reason that the markets would provide all the information needed to orientate the decisions of individual operators and the 'social adoption' of the new technology. If both private financial and external consequences are at issue, then regulation is required, for the reason that the markets do not provide information on the external consequences; and, as already seen, for this purpose CBA can be a useful tool. Such external consequences will receive greatest emphasis in this discussion, since they create real complexity in any CBA. However, this does not mean that private variables (such as enhanced productivity, yields, and trade foregone) should not also be considered, as they materially assist in providing a complete picture of the consequences of adopting GMOs.

In this analysis, two classes of external consequences can be distinguished. The first are direct external consequences, namely consequences beyond the farmer. The second are private consequences for farmers, which nonetheless present public aspects that should be evaluated. The main direct external consequence is public health and, in particular, food safety concerns over the as yet unknown, long-term effects of eating GM food (of which potential allergens and the use of antibiotic-resistant marker genes would be just two examples). Once a GMO has been approved for marketing, such considerations fall totally beyond the farmer; and this creates an imperative that the evaluation of the risks and their translation into monetary terms should be undertaken prior to authorization. The second class of external consequences may be categorized as

---

[25] RD Weaver, 'Ex post Evidence on Adoption of Transgenic Crops: US Soybeans' in Wesseler (n 6 above) 125.

environmental; and an example would be adverse effects on non-target organisms, such as beneficial or neutral insects.[26]

In this context, it is important to note that GMOs very often present both private and public characteristics. The first generation, 'Ht' and 'Bt' crops, were engineered in order to improve productivity (Ht crops are tolerant to the herbicide glyphosate and allow, *inter alia*, for better weed control, while Bt crops are insect resistant and allow, *inter alia*, for reduced pesticide use). The opportunity to reduce the application of chemical herbicides and pesticides directly concerns farmers; and it might be thought that, in consequence, no CBA would be required, since no public dimension is present. However, to take such a view would be to ignore important social benefits and costs,[27] as closer consideration of Ht crops illustrates.[28] With less need for the application of herbicides as compared to conventional crops, farmers may adopt 'conservation tillage methods', which offer economies in terms of fuel use, labour, and machine time and which also improve carbon sequestration.[29] To this may be added better weed control and easier weed management, with the result that there are understood to be a bundle of private benefits, which may be set against the greater cost of the seed. *A priori* it may be contended that no decision by public authorities is required to orientate private decisions by farmers whether or not to grow Ht crops. On the other hand, scientific studies suggest that the intensive use of the glyphosate-tolerant herbicide has led to the appearance of superweeds, which also enjoy a measure of resistance;[30] and this resistance could be considered an externality, since the phenomenon cannot be limited to a particular field. Furthermore, to control these hybrid plants, farmers might respond by increasing the amount or toxicity of the herbicides they apply, a cost that arguably does not arise directly from genetic engineering, but which needs to be taken into account in any overall CBA. In any event, both the appearance of superweeds and increased herbicide application to combat them constitute long-term effects which are unlikely to have been in the contemplation of most farmers when making their original choice to purchase Ht seed. Besides, ecological studies indicate that conservation tillage methods can themselves create a negative environmental effect on the soil structure, although these must in turn

---

[26] Directive (EC) 2001/18 of the European Parliament and of the Council on the deliberate release into the environment of genetically modified organisms ('Deliberate Release Directive') [2001] OJ L106/1, Annex II requires that interactions between GMOs and non-target organisms be evaluated in the environmental risk assessment.

[27] It may be noted that the Deliberate Release Directive also requires that the report be issued every three years by the European Commission should address 'the socioeconomic implications of deliberate releases and placing on the market of GMOs': ibid Art 31(7)(d).

[28] See further, eg DE Ervin et al, *Transgenic Crops: an Environmental Assessment* (Henry A Wallace Center for Agricultural and Environmental Policy at Winrock International, 2000); and S Bonny, 'Genetically modified glyphosate-tolerant soybean in the USA: adoption factors, impacts and prospects. A review' (2007) 28 Agronomy for Sustainable Development 21.

[29] For the claimed benefits of Ht crops, see Monsanto at <http://www.monsanto.com/products/benefits/conservation_till.asp>, last accessed 22 July 2009.   [30] Bonny (n 28 above).

be set against reduced wind and water erosion, together with the improved carbon sequestration already mentioned.[31] Accordingly, there would seem to be sufficient positive and negative external effects to justify use of CBA: without such 'quantitative' monetary valuation, they are purely qualitative and not easy to balance.

By contrast, GMOs of the second generation are designed specifically to improve the environmental or health characteristics of the products concerned. As a consequence, their 'external effects' are emphasized. An example of this second generation would be 'Golden Rice', engineered for higher Vitamin A content;[32] and certain varieties of apple, already commercially well-established (such as Gala or Elstar), have been engineered so as to render them scab-resistant.[33] Although these varieties of apple have not yet been authorized for placing on the market, they should permit lower application of chemical fungicide; and this reduction in chemical input should be taken into account by apple-growers in assessing their profit margins. At the same time, a positive external effect on the environment should be created and fed into the equation.

Despite strong opinions on the use of modern biotechnology in agriculture (both positive and negative), the extent to which CBA has been undertaken is not great. Such studies as have been undertaken broadly adopt the same approach, proposing that there should be an evaluation of consumers' WTP for GM foods. The rationale is that people reveal their preferences for GM foods through their WTP. If GM foods are considered as an environmental and/or health gain, the WTP of consumers should be positive. Alternatively, if GM foods are considered as an environmental and/or health loss, consumers should be willing to pay to prevent/escape that loss. Within this framework, Marks, Kalaitzandonakes, and Vickner provide an overview of evaluations of WTP for GM foods, setting out their varying predictions, as well as highlighting their technical advantages and limitations.[34] In particular, they distinguished between 'opinion surveys', 'choice experiment', and 'experimental auction market methods', with the last-mentioned being considered the most reliable. Further, as shall be seen later, a matter of some interest is that studies which use an experimental economics approach generally reveal positive WTP in respect of both US and European respondents.[35] These results contradict the generally accepted idea, especially in the case of Europe, that consumers would shun GM

---

[31] Weaver (n 25 above).
[32] See, eg D Dawe and L Unnevehr, 'Crop case study: GMO Golden Rice in Asia with enhanced Vitamin A benefits for consumers' (2007) 10 AgBioForum 154.
[33] G Vanloqueren and PV Baret, 'Les pommiers transgéniques résistants à la tavelure. Analyse systémique d'une plante transgénique de "seconde génération"' (2004) 52 *Le Courrier de l'Environnement de l'INRA* 5.
[34] LA Marks, N Kalaitzandonakes, and SS Vickner, 'Evaluating consumer response to GM foods: some methodological considerations' (2003) 4 Current Agriculture, Food & Resource Issues 80.
[35] See also, eg King's College London, *Do European Consumers Buy GM Foods?* ('*Consumerchoice*') (London: Kings College London, 2008).

products if they were introduced, appropriately labelled, onto the market. But in any CBA, questions must remain as to the exact meaning of WTP expressed in a laboratory setting. And how should such WTP be interpreted in terms of the environment or health?

Hobbs and Kerr note that consumers have two concerns in relation to transgenic crops: direct (tangible) consumption effects and indirect (intangible) existence-value effects.[36] From a theoretical point of view, the WTP can thus cover many elements already considered, such as existence value and option value. Besides, in any empirical studies the resultant WTP will depend on the knowledge of respondents prior to the survey of health or environmental implications of GMOs, on the information that they receive from the researchers, and on the way it which the questions are phrased. Different protocols will produce different results, rendering comparisons difficult.[37] In particular, much will depend on the extent of their knowledge of potential externalities which may flow from GMOs.

To take one example, Noussair, Robin, and Ruffieux have used experimental economics to produce findings that the consumers surveyed did not refuse to buy GM food.[38] However, they also noted that a survey would 'accord greater weight to public dimensions such as negative externalities that result from widespread use of the product, than a bid in auction market'.[39] As a consequence, much may depend on the methodology employed; and it would be possible to argue that, if the respondents did not know the environmental implications of their consumption of GM products, the research protocol (and, in particular, the amount of information provided) could materially affect responses. In their research protocol, Noussair, Robin, and Ruffieux gave only general information to those surveyed, namely: the definition of a GMO; the criteria used for classifying a product as containing GM material; the list of GMOs authorized in France; food products sold in France which contain GMOs; and, finally, French law governing GMOs. Importantly, no information was provided on the scientific debate over the safety of GMOs, which precluded full data on how those surveyed understood potential external effects. The fact that the GMOs in question were authorized effectively closed the door on that debate. Thus, the WTP so evaluated reflects consumers' preferences based on the information supplied for the survey, and not necessarily broader social benefits/costs.

That said, such an approach should not be rejected, since it forms a useful and relatively precise tool for evaluating WTP where what is sought is a

---

[36] Hobbs and Kerr (n 10 above).
[37] Comparisons may be rendered even more difficult by the development of parallel methodologies in ecological sciences.
[38] C Noussair, S Robin, and B Ruffieux, 'Do consumers really refuse to buy genetically modified food?' (2004) 114 Economic Journal 102.   [39] Ibid 106.

Application of CBA to GMOs 51

monetary evaluation of the environmental impact of the GMO. Furthermore, Shrogren et al, when describing food safety choices in the context of three markets (retail, survey, and laboratory auction), showed that the responses of consumers to prices in all three contexts were comparable.[40] Accordingly, there is authority for the fact that consumers' health-safety choices as observed under experimental conditions may approximate to those in a 'real world' retail environment.

3.2 The Post-authorization Stage: Traceability and Labelling, and Coexistence

The Community regime for traceability and labelling implies specific and additional costs for suppliers;[41] and these costs have been widely studied.[42] Such measures are also inextricably linked with the Community coexistence regime, which seeks to maintain any cross-contamination of conventional and organic crops within the labelling thresholds. While the details of the coexistence regime are considered elsewhere in this book,[43] it may be highlighted that the Community framework is currently provided by Commission Recommendation (EC) 2003/556 on guidelines for the development of national strategies and best practices to ensure the coexistence of genetically modified crops with conventional and organic farming ('2003 Commission Recommendation').[44] Importantly, at farm level, it includes within an indicative catalogue of measures the establishment of isolation distances/buffer zones.[45] If the land within the isolation distance is not planted at all, that is a segregation cost; and production from buffer zones planted with non-GM crops is to be sold as 'GM', which may again entail a cost. Moreover, the larger the isolation distance/buffer zone, the higher the cost (and their extent will be crop-specific).[46] However, cross-contamination can also occur by reason of, for example, seed impurity and during the course of handling, storage, and transportation to end-users, with the result that coexistence measures must be broadly construed. Indeed, at farm level alone, the

---

[40] JF Shrogen et al, 'Observed choices for food safety in retail, survey, and auction markets' (1999) 81 American Journal of Agricultural Economics 1192 (with regard to irradiated food).
[41] Regulation (EC) 1829/2003 of the European Parliament and of the Council on genetically modified food and feed [2003] OJ L268/1; and Regulation (EC) 1830/2003 of the European Parliament and of the Council concerning the traceability and labelling of genetically modified organisms and the traceability of food and feed products produced from genetically modified organisms [2003] OJ L268/24.
[42] For a survey of these studies, see WW Wilson, X Henry, and BL Dahl, *Costs and Risks of Conforming to EU Traceability Requirements: the Case of Hard Red Spring Wheat* (Agrobusiness and Applied Economics Report No 564, 2005).
[43] See further Chapters 6 (at Community level) and 7 (at Member State level).
[44] [2003] OJ L189/36.   [45] Ibid Annex, para 3.2.1.
[46] See, eg A Messean et al, *New Case Studies on the Coexistence of GM and non-GM Crops in European Agriculture: Tech. Rep. EUR 22012* (Joint Research Centre, European Commission, 2006).

2003 Commission Recommendation identifies four main areas for consideration: preparation for sowing, planting, and soil cultivation; harvest and post-harvest field treatment; transportation and storage; and field monitoring.[47] In consequence, there may be additional costs to cover, for example, the expense of cleaning combines and storage equipment if successively used for GM and non-GM crops. Alternatively, farmers may dedicate equipment and storage units to one of the two forms of production, but this might be at the price of under-utilization. However, a further strategy might simply be to specialize in either GM or non-GM crops.[48] Indeed, Bullock and Desquilbet have argued that the major component of segregation costs comes from 'flexibility loss', due to the necessity of diverting equipment down either the GM or non-GM channel, as opposed to any necessity to clean sowing and harvesting equipment.[49]

To be added to the equation is the cost of GM-certified seed, which attracts a premium. For example, Wilson, Henry, and Dahl have estimated this to be 1.45 dollars per bushel in the case of hard red spring wheat.[50] A third party can also be employed to certify production as non-GM, so as to provide confidence for consumers; but again this is a cost that flows from the introduction of GM crops. Finally, the specific and detailed requirements of the traceability and labelling regime also imply financial burdens (such as those incurred in respect of lot number and inspection and control measures).

More generally, the cost of compliance with post-authorization measures tends to rise as a function of two factors: the complexity of the supply chain and the strength of tolerance thresholds. With regard to the first factor, if the demand for non-GM food is sufficiently strong, a genuinely twin-track supply chain can emerge in response, so lowering the cost of segregation. With regard to the second factor, Huygen, Veeman, and Lerohl have studied the segregation costs of tolerance thresholds varying from 0.1 per cent to five per cent.[51] The subject of their study was wheat produced in Alberta, and they confirmed that segregation costs increased with the severity of the threshold. More precisely, they estimated that, per tonne of wheat at farm level, costs climbed from 1.04 dollars applying a tolerance threshold of five per cent to 6.45 dollars applying a tolerance threshold of 0.1 per cent. Further, the impact of the traceability and labelling regime on the well-being of consumers should be evaluated. Loureiro and Hine, for example, have used contingent valuation to estimate the premium that US consumers are willing to pay in the case of both mandatory and voluntary labelling.[52] Their results show that the willingness to

---

[47] Annex, para 3.2.   [48] Wilson, Henry, and Dahl (n 42 above).
[49] DS Bullock and M Desquilbet, 'The economics of non-GMO segregation and identity preservation' (2002) 27 Food Policy 81.   [50] Wilson, Henry, and Dahl (n 42 above).
[51] I Huygen, M Veeman, and M Lerohl, 'Cost implications of alternative GM tolerance levels: non-genetically modified wheat in Western Canada' (2004) 6 AgBioForum 169.
[52] ML Loureiro and S Hine, 'Preferences and willingness to pay for GM labeling policies' (2004) 29 Food Policy 467.

pay is low compared to the total cost of mandatory labelling; and, on that ground, they claim that voluntary labelling should be preferred.

4. Conclusion

If only private consequences (whether costs or benefits) flowed from the introduction of GMOs, CBA and regulation would arguably not be required, since, unless modern biotechnology offered the prospect of profit for producers, it would not be adopted. But when external benefits and external costs are at stake, there arises a need to enter the public arena, since private decisions alone cannot provide sufficient direction. In the context of GMOs, which engage sanitary and phytosanitary measures, risk assessment is currently presented as the cornerstone of regulation. Nonetheless, CBA can give monetary value to any risks detected, so as to complete the risk assessment. Feeding the results of such economic analysis into the decision-making process introduces the possibility of accepting risk on 'economic grounds' or of quantifying from an economic standpoint what can be considered an 'acceptable level of risk'.

CBA in the case of GMOs is no easy task, since it includes health and environmental aspects, and even cultural and ethical aspects. These difficulties are exacerbated at international level, where different nations adopt different perceptions of modern biotechnology. That said, what may be regarded as limitations of CBA also emphasize its importance. National regulations which aim to correct the externalities generated by GMOs can, in part at least, be characterized as protectionist, as is well-illustrated by the *EC—Biotech* dispute. Balancing the costs and benefits of such regulations can test their 'economic legitimacy'; and CBA is ideally suited for this task,[53] even though the WTO Sanitary and Phytosanitary Agreement leaves little scope for economic considerations. Such an exercise could, accordingly, permit determination of the extent to which protection is motivated by ethical or political considerations.

---

[53] JC Beghin and J-C Bureau, 'Quantitative policy analysis of sanitary, phytosanitary and technical barriers to trade' (2001) 87 *Economie Internationale* 107.

3
Genetically Modified Organisms and Risk

Karen Morrow

1. Introduction

... as we know, there are known knowns; there are things we know we know. We also know there are known unknowns; that is to say we know there are some things we do not know. But there are also unknown unknowns—the ones we don't know we don't know.[1]

While Donald Rumsfeld may have garnered both ridicule and a 'Foot in Mouth' award from the Plain English Campaign[2] in respect of the remark quoted above, it actually serves as an excellent starting point for the consideration of risk in the modern world. Risk, both natural and man-made, is necessarily part and parcel of human endeavour and experience, and some risks, the 'known knowns', are apparent and can, in principle, be engaged with in terms of prevention and/or cure. However, as we are increasingly coming to appreciate, other risks, in particular those associated with cutting-edge science and technology, are not immediately obvious in terms of their nature and/or extent; these fall into the category of 'known unknowns' or even 'unknown unknowns', and addressing them presents enormous legal and practical challenges. Threats of these types have necessitated a fundamental re-examination of the very conception of risk and its nature, resulting in a comparatively rapid change in its characterization and status in recent years, as it has increasingly come to the fore in society, policy, and law.

In a modern context, it is virtually impossible to frame any discussion of risk without discussing Ulrich Beck's seminal work on the risk society,[3] which contextualizes current debate in this area in the broader social milieu. Very briefly, Beck's argument is founded on a conception of post-industrial risk, which is more complex and far-reaching than previous views of the issue. Whether or not one subscribes to Beck's vision of risk as the key determinant in

---

[1] Donald Rumsfeld: quoted in *Rum Remark wins Rumsfeld an Award*, 2 December 2003 (available at <http://news.bbc.co.uk/1/hi/world/americas/3254852.stm>, last accessed 2 July 2009).
[2] Ibid.   [3] U Beck, *Risk Society: Towards a New Modernity* (London: Sage, 1992).

modern social and political praxis, it is certainly arguable that deeper and wider societal awareness of it now ensures that, far from being regarded as a technical issue and the exclusive province of expert opinion, risk is increasingly viewed as a more broadly contested concept. One result of this emerging understanding of risk, and of the enhanced status that it now enjoys, is that it has become incumbent on the law to engage with it in new ways, extending beyond well-established (if often problematic) cost-benefit analysis as applied to known risk, into the more controversial application of risk and the precautionary principle in the face of uncertainty, of which more below.

2. Engaging with Risk

A variety of decision-makers are called upon to act as the arbiters of acceptable risk at a number of levels: legislators determine the range and statutory basis of applicable regimes; policy-makers provide guidance on their operation; and regulators administer permits and the day-to-day running of regulatory systems. With public engagement being increasingly pursued through mandated information and consultation procedures, questions of risk are also ever more likely to form the subject of profound public concern at all stages of policy and decision-making processes, ultimately allowing the contested nature of risk to crystallize into litigation, where the courts are called upon to determine resultant disputes and resolve uncertainties.[4] Historically, societal responses to risk have been founded on the practice of risk analysis which is commonly understood as being based on an initial scientific evaluation in the form of a risk assessment followed by a political risk management decision.[5] The former is a two-stage process involving: first, the identification of hazard in terms of specific adverse effect(s); and, secondly, the assessment of the likelihood/probability that harm will occur and the consequences or damage that may result. The latter is defined by Guruswamy as 'a process of decision-making that integrates the scientific findings of [risk assessment] within the broader structure of policy and law'.[6]

Human endeavour necessarily impacts on the environment and, in consequence, poses a range of challenging risk questions. The introduction of

---

[4] *R (Greenpeace Ltd) v Secretary of State for Trade and Industry* [2007] EWHC 311 (Admin) provides an interesting example of the courts being called upon to rule on the adequacy of the Government's consultation on reviewing energy policy, the challenge being founded at least in part on Greenpeace's objection to a reconsideration of nuclear power options founded on notions of risk.

[5] N de Sadeleer, 'The Precautionary Principle in European Community Health and Environmental Law: Sword or Shield for the Nordic Countries' in N de Sadeleer (ed), *Implementing the Precautionary Principle: Approaches from the Nordic Countries, the EU and USA* (London: Earthscan, 2007) 10, 18.

[6] LD Guruswamy, 'Sustainable agriculture: do GMOs imperil biosafety?' (2001–2002) 9 Indiana Journal of Global Legal Studies 461.

genetically modified organisms (GMOs) raises particular concerns in this regard by introducing new and unpredictable features into an already complex environment. The challenges are multi-faceted, being at once problematic in their own right and also in interaction with each other and the existing environmental context.[7] The unforeseen and potentially dynamic synergistic, temporally and spatially unlimited impacts that they may generate mean that they give rise to equal or perhaps greater cause for concern than all but the most serious anthropogenic impacts on the environment to date.

For current purposes, the key role ascribed to 'environmental' risk assessment in Directive (EC) 2001/18 of the European Parliament and of the Council on the deliberate release into the environment of genetically modified organisms ('Deliberate Release Directive')[8] serves to underline the continuing central importance of risk analysis in regulatory decisions in this area. In principle, this hybrid scientific and political process should deliver an optimum approach to questions of risk rather than involving any necessary dichotomy. In theory, risk analysis features a clear distinction between the role of science in *informing* decisions and the political exercise of actually taking decisions as to what constitutes acceptable risk. Thus, as one commentator puts it: '[r]isk assessment, therefore, should be viewed as a policy tool based in part on the work of scientists, not as science'.[9] The final part of this quotation is extremely revealing. In past practice, such was the authority of science that, under the Cartesian model, it came to dominate risk assessment, if not to the exclusion of all else, then certainly as the pre-eminent basis of decision-making. In light of this, it is arguable that risk analysis came to be popularly understood as a technocratic process rather than a democratic one. At the same time, this provided decision-makers with the opportunity to abdicate political responsibility for unpopular or controversial regulatory decisions involving new technologies; and, by allowing science to assume primacy in decision-making, rather than acting as an important element within a broader process, the tendency was to devalue other points of view. This approach arguably lends (an at least partially spurious)[10] air of objectivity to decision-making processes, with alternative views of risk being regarded as subjective and therefore less valuable. It is also perhaps inevitable that societal evaluations are likely to be even more contested than scientific material, lacking, as they often do, the veneer of objectivity: what is regarded as acceptable risk after all tends to depend on who is being asked to weigh the issues. At the very least, in operational terms, risk assessment historically divorced facts (the province of science) from (social and political) values; and this was ultimately neither particularly helpful nor completely plausible.

---

[7] Natural and human-crafted.   [8] [2001] OJ L106/1.
[9] A Babich, 'Too much science in Environmental Law' (2003) 28 Columbia Journal of Environmental Law (2003) 119.
[10] Science is, after all, not entirely objective: see Beck (n 3 above).

However, from the mid-1960s onwards the strength of the hold of science on society has waned considerably. Public faith in the ability of science to deliver solutions to manifest risks has been significantly eroded by experience, and science is now viewed not so much as the solution to, as the cause of, risk.[11] On the other hand, this development has not yet changed the fact that risk assessment, with science playing the lead role, remains the primary mechanism underpinning action in this area. Babich succinctly explains that the reason for this is primarily pragmatic: '[i]n terms of the standard set by the scientific method, most risk assessment is neither good nor reliable science—it is merely the best science we have on the subject'.[12]

As things stand then, in scientific terms, risk assessment is relied on at present in the absence of something better; but it is also true to say that, in many cases, risk assessment continues to be presented as a prescriptive quasi-science, masking or even minimizing the political element of decision-making in risk analysis more generally. In any event, the dominant position of risk analysis and the way in which it is approached mean that risk assessment must form the starting point in considering current responses to questions of risk. It should of course, by its very nature, be a cross-cutting and multi-disciplinary[13] undertaking, but in reality it is open to criticism as narrow and compartmentalized.[14]

Increasing and complex problems of uncertainty within science itself[15] also create difficulties, as decision-makers can no longer rely on science that frames itself in terms of certainty as a basis for regulatory governance. In consequence, the resulting relatively easy dialogue that had prevailed between the law and science for decades has been replaced by a more open, but also more contentious debate. Science in the modern world is a very different (and indeed still emerging) construct, framed in terms of and responsive to uncertainty. Having said this, science as a discipline is, in many ways, more comfortable with risk and uncertainty than law, and one result of this is that the status of science as a co-traveller with law has become ever more strained, and a new paradigm for interdisciplinary interaction is urgently required.

In principle, while science may still be viewed as of foundational importance, it cannot be considered adequate in and of itself as the basis for decision-making in risk contexts. There is also arguably a contagion of distrust of science where the public is concerned—having been sold the benefits of science without due regard to costs on numerous occasions in the past,[16] people are now suspicious rather than credulous where 'new miracles' are concerned. This is particularly

---

[11] N de Sadeleer, *Environmental Principles: From Political Slogans to Legal Rules* (Oxford: Oxford University Press, 2002) 152.   [12] Babich (n 9 above) 142.
[13] de Sadeleer (n 11 above) 193–4.   [14] Ibid 186.
[15] AI Myhr, 'Uncertainty and Precaution: Challenges for Science and Policy of Genetically Modified Organisms' in de Sadeleer (n 5 above) 185.
[16] See, generally, eg R Carson, *Silent Spring* (London: Readers Union, 1964).

remarkable given that technological optimism has been so thoroughly ingrained in Western culture and consciousness since the Enlightenment. Nonetheless, in Europe, suspicion is now a marked feature of public reaction to new technologies, genetic engineering notable among them.

In examining risk, while we may be able to achieve a certain amount by extrapolation from past experience, the fact remains that, in the case of new technologies, uncertainty and even ignorance must be acknowledged and engaged with in decision-making processes. This, coupled with the fact that risk is now openly acknowledged as having prominent social dimensions, has ensured that other factors, particularly those relating to political aspects of the decision-making process, are increasingly coming to the fore. An important corollary is that public participation and acceptance has become more important in decisions relating to risk. Public input is, however, problematic in a number of ways, not least in that, on a very fundamental level, it challenges orthodox political and regulatory decision-making processes. In a representative democracy, these have long involved elected representatives making these choices (or delegating them to experts) on behalf of the electorate. In the current context, however, there is increasing public distrust of politicians, regulators, and even of science itself, in light of past episodes where the public is of the opinion that it has been let down by all three.[17] It is, therefore, arguable that setting the parameters for acceptable risk is now coming to be viewed as 'more of a philosophical—or at least political—question than a technical one'.[18]

It is without doubt not only the decision-making process, but also the science upon which it is founded, that is subject to public criticism, ranging from the facile to the sophisticated. Thus, where risk is concerned, the expert view alone is no longer likely to have things all its own way; technical expertise now increasingly finds itself embattled in an often hotly contested arena rather than forging a direct route to a foregone conclusion. The genetic modification of food, which thus far constitutes a relatively untried and immature field of science, raises particularly emotive concerns[19] (including the speed and scale of change[20] and the likelihood/inevitability of 'genetic pollution') in the already controversial realm of industrial agriculture. Thus, the GM debate in many respects represents an excellent crucible in which to test emerging methods of engaging with risk.

---

[17] A good illustration is the bovine spongiform encephalopathy (BSE) crisis: see, eg A Nucara, 'Precautionary principle and GMOs: protection or protectionism' (2003) 9 International Trade Law and Regulation 47.   [18] Babich (n 9 above) 122.
[19] Guruswamy (n 6 above) 482.
[20] JS Applegate, 'The Prometheus Principle: using the precautionary principle to harmonise the regulation of genetically modified organisms' (2001) 9 Indiana Journal of Global Legal Studies 207, 216.

3. Participation and Risk

Established practice features political decision-making processes that have been so dominated by conventional (and, in particular, economic) priorities and interests for so long that, while other stakeholders are now being heard, the actual impact of this development remains to be seen. Given the emerging importance of consultation and dialogue in this area, the way in which risk is presented and, indeed, to whom it is presented, will assume particular importance. The identity and characteristics of the stakeholders involved is therefore crucial: industry bodies, for example, are likely to be well-resourced and regarded as enjoying considerable expertise, while the same may not necessarily be said for non-governmental organizations (NGOs) and members of the public at large. In addition, it must be remembered that for NGOs and the public, participation brings burdens as well as benefits.[21] As a result, the dispersed inputs provided by the latter are weighed in the process with the much more concentrated sectoral inputs of industry bodies—like is not compared with like. However, as de Sadeleer states:

> Because risks are co-determined by sets of social values which affect their public acceptability, it might be more productive to accept those values as an integral part of the procedure, by making them explicit and co-ordinating them with regulatory goals instead of excluding them.[22]

While this would be the case in an ideal world, it may not be readily achievable in practice, as demonstrated, for example, by aspects of the GM Nation debate in the United Kingdom;[23] and, as indicated in Chapter 1, participation can generate considerable frustration and even ultimately litigation, when it is viewed as amounting to mere tokenism. It has also been seen that such frustration and litigation are the more likely where participation is (or is perceived to be) limited to informing[24] or consulting[25] citizens; and, in this regard, there is less currency nowadays in the long-cherished view of scientists that, if the public is 'educated' on the facts relating to new technologies, these will be rendered

---

[21] H Cullen and K Morrow, 'International civil society in International Law: the growth of NGO participation' (2001) 1 Non-State Actors and International Law 7.
[22] de Sadeleer (n 11 above) 191.
[23] Department of Trade and Industry, *GM Nation? The Findings of the Public Debate* (London: Department of Trade and Industry, 2003) ('*GM Nation*'); and, for a brief overview of key aspects of this debate in the United Kingdom, see, eg M Grekos, 'Conduct of the GM public debate, Eighteenth Report of Session 2002–03 (November 20, 2003)' [2004] Journal of Planning and Environmental Law 579.
[24] SR Arnstein, 'A ladder of citizen participation' (1969) 35 Journal of the American Institute of Planners 216, 3.3 (reproduced at <http://lithgow-schmidt.dk/sherry-arnstein/ladder-of-citizen-participation.html>, last accessed 2 July 2009).
[25] Ibid 3.4.

acceptable.[26] Further, even where considerable efforts are made to foster effective participation, there may still be considerable dissatisfaction with the outcome of regulatory decision-making processes, and opponents will use such means (primarily procedural) as are available to them to call the decisions into question in the courts (which will be discussed below).

Without doubt, public participation in what may be broadly termed environmental decision-making has been significantly enhanced over recent years, not least at international level by the Aarhus Convention;[27] and it has also been a focus of more general transparency-based initiatives, such as the UK Government's Cabinet Office Code of Practice on Consultation.[28] Indeed, participation is a core concern of the increasingly prominent environmental justice movement.[29] Having said this, it remains to be seen whether such recent and thus far primarily procedural developments will be sufficient to address the particular issues that arise with respect to public participation in matters of environmental risk, such as GMOs. In this context, public input is subject to difficult questions regarding the appropriate provision, presentation, and reliability of information, and, perhaps most importantly, its subsequent evaluation by the public (though any difficulties are not necessarily insurmountable).[30] These thorny questions are further complicated by the fact that, where risk is concerned, it is identified by scientific experts, and science plays a prime role in decision-making, standard-setting, and the application of regulatory law in practice—an uneasy state of affairs, as distrust of the scientific foundation of the system tends to compound public suspicion. While these matters may remain unresolved, recent developments in the United Kingdom serve to show that public participation in environmental processes can be viewed not so much as a dethroning of technocracy by popularism, or of expert by lay opinion, as a broadening of the range of factors that require consideration in risk evaluation;

---

[26] See, eg AgBioView Special, 'Waiter ... There is a gene in my soup! ... communicating with the public, media and policymakers on AgBiotech issues' (available at <http:///www.agbioworld.org/biotech-info/articles/biotech-art/communicating.html>, last accessed 2 July 2009).

[27] Aarhus Convention on Access to Information, Public Participation in Decision-making and Access to Justice in Environmental Matters 1998 (available at <http://www.unece.org/env/pp/documents/cep43e.pdf>, last accessed 2 July 2009); and, for a seminal overview, see Arnstein (n 24 above).

[28] Cabinet Office Better Regulation Executive, January 2004 (available at <http://www.berr.gov.uk/files/file44364.pdf>, last accessed 2 July 2009).

[29] See, eg C Stephens, S Bullock, and A Scott (eds), *Environmental Justice: Rights and Means to A Healthy Environment for All: ESRC Global Environmental Change Programme* (2001) (available at <http://www.foe.co.uk/resource/reports/environmental_justice.pdf>, last accessed 2 July 2009); and the report of the Working Group on Access to Environmental Justice, *Ensuring Access to Environmental Justice in England and Wales* (2008) (available at <http://www.unece.org/env/pp/compliance/C2008-23/Amicus%20brief/AnnexNjusticereport08.pdf>, last accessed 2 July 2009).

[30] See, eg W Kenyon and G Edwards-Jones, 'What level of information enables the public to act like experts when evaluating ecological goods?' (1998) 41 Journal of Environmental Planning and Management 463 (with reference to commonly used contingent valuation approaches applied in a land use planning context).

and it may be highlighted that parallel developments have taken place in the generally less technical, but equally controversial, area of public concern in the planning system.[31]

4. The Limits of Risk-based Approaches

In addition to the limiting factors discussed above relating to public participation and risk analysis, further weaknesses are apparent in the fact that the process is generally undertaken on a case-by-case basis;[32] and the very ad hocism[33] that results can see risk compartmentalized in a way that fails adequately to take into account its interconnected and cumulative nature,[34] a particularly acute concern in an environmental context. Even at their best, risk-based approaches do not answer all of the questions posed by the need to regulate emerging technologies: for example, some uncertainty cannot be eliminated, and it may not be possible to identify the full extent of risk, with the result that other avenues are worthy of consideration. Babich, for example, favours an alternative approach to regulation based on what he terms 'the best level of protection based on technological and economic feasibility'.[35] This is an appealing solution in some respects, and an example of a concept that bears some resemblance to it can be found in the best available techniques not entailing excessive costs (BATNEEC) approach adopted to integrated pollution control under Part I of the Environmental Protection Act 1990, and (less explicitly) in the concept of best available techniques (BAT) in Council Directive (EC) 96/61 concerning integrated pollution prevention and control.[36] Such approaches may, however, be problematic in their own right;[37] and there do appear to be limits to their utility, in particular where unknown risk is concerned and the cost/benefit calculation is compromised by lack of knowledge.

In light of the above discussion, our understanding of risk in an environmental context, and in particular where GMOs are concerned, must surely benefit from the recognition that it is intimately linked to some of the key principles of environmental law, notably those of prevention, the polluter pays, and precaution. In theory, the polluter-pays principle can ensure that those who

---

[31] See, eg C Hilson, 'Planning law and public perceptions of risk: evidence of concern or concern based on evidence?' [2004] Journal of Planning and Environmental Law 1638.

[32] This is the basic approach adopted to regulation of GMOs in the Deliberate Release Directive. See also, eg Guruswamy (n 6 above) 481.

[33] This is of course a common feature of environmental law, although more usually a product of the crisis-driven nature of much activity in this area: see, eg ME Kahn, 'Environmental disasters as risk regulation catalysts? The role of Bhopal, Chernobyl, Exxon Valdez, Love Canal, and Three Mile Island in shaping U.S. environmental law' (2007) 35 Journal of Risk and Uncertainty 17.

[34] de Sadeleer (n 11 above) 186–191.   [35] Babich (n 9 above) 126.

[36] [1996] OJ L257/26.

[37] See, eg N Haigh, 'Integrated Pollution Prevention and Control: UK and EC approaches and possible next steps' (1996) 8 Journal of Environmental Law 301.

benefit from environmental risks bear (at least some of) the associated costs: thus, current regulatory regimes place the burden of proof where risk is concerned on those promoting GM products. However, the relatively supine approach of regulators in this regard means that material on risk provided by the promoters of GM products is arguably not properly counterbalanced by, first, taking alternative views sufficiently seriously and, secondly, regulators casting a truly critical eye over all the material made available to them. The preventative principle may be employed to address known risks, as for example with the imposition of isolation distances to tackle cross-pollination between GM and non- GM crops. It is, nonetheless, in relation to the precautionary principle that risk plays its most pronounced and contentious role with respect to GMOs, as it is here that the law must attempt to engage with uncertainty. In cases where there is incomplete knowledge, determining what degree of risk is acceptable becomes virtually impossible, and the workability of a risk-based approach, as traditionally understood, is at issue. It is in situations like this that an approach based on, or more accurately contextualized by, the application of the precautionary principle could prove beneficial. Such an approach, however, is fraught with difficulty as, if expansively interpreted, it would stand to fundamentally challenge prevailing social, political, and legal values, arguably extending well beyond the 'modest overlay on the liberal institutions of private property and consumer sovereignty'[38] presented by environmental regulation as currently practised.

5. Risk and the Precautionary Principle

The precautionary principle,[39] as expressed, for example, in Principle 15 of the Rio Declaration,[40] represents, at the very least, an attempt to forge a new discourse by endeavouring (amongst other things) to accommodate uncertainty in the law through justifying legal action in the face of incomplete scientific

---

[38] AD Tarlock, 'Ideas without institutions: the paradox of sustainable development' (2001–2002) 9 Indiana Journal of Global Legal Studies 35, 42.

[39] There is a wealth of valuable literature on the precautionary principle, taking in many aspects of its development. This includes: A Deville and R Harding, *Applying the Precautionary Principle* (New South Wales: Federation Press, 1997) (looking at the Australian experience); T O'Riordan, J Cameron, and A Jordan (eds) *Reinterpreting the Precautionary Principle* (London: Cameron May, 2001) (taking a policy- and management-based approach); and P Harremoes (ed), *The Precautionary Principle in the 20th Century: Late Lessons from Early Warnings* (London: Earthscan, 2002) (which takes an historical approach).

[40] Rio Declaration on Environment and Development, Principle 15: '[i]n order to protect the environment, the precautionary approach shall be widely applied by States according to their capabilities. Where there are threats of serious or irreversible damage, lack of full scientific certainty shall not be used as a reason for postponing cost-effective measures to prevent environmental degradation' (available at <http://www.unep.org/Documents.multilingual/Default.asp?DocumentID=78&ArticleID=1163>, last accessed 10 August 2009).

knowledge. According to de Sadeleer, 'precaution aims to bridge the gap between scientists working on the frontiers of scientific knowledge and decision-makers willing to act to determine how safe is safe enough'.[41]

In general terms, the precautionary principle offers an intuitively appealing way of engaging difficult issues, and some commentators[42] believe that it has now crystallized into a recognized principle of international environmental law. This view is, however, by no means universal: for example, the World Trade Organization (WTO) Appellate Body in *EC—Measures Concerning Meat and Meat Products (Hormones)* questioned such status.[43] Nonetheless, it has certainly gained mainstream status in Community law. Not least, it features in (though is not defined by) Article 174(2) of the EC Treaty (now Article 191(2) of the Treaty on the Functioning of the European Union) and applies outside the environmental context.[44] That said, the full legal implications of the precautionary principle have not perhaps always been fully appreciated in practice and it remains, in many ways, a profoundly problematic and contested concept, despite featuring widely in the law at all levels. Guruswamy, for example, argues that:

> ...it represents a major leap backward from a focus on risk to a focus on hazard. ... 'risk' and 'hazard' have distinct meanings in the risk analysis literature. 'Hazard' is the intrinsic potential of an agent to cause an adverse effect, whereas 'risk' is the likelihood and magnitude of the adverse effect occurring under real world exposure scenarios.[45]

Having said this, it may be argued that the precautionary principle, if applied in its most extreme form, would effectively paralyse societal development, as nothing can ever be proved to be totally safe, but in practice, as seen below, a qualified approach is taken. In real terms, where the precautionary principle is invoked, it is not the elimination of risk that is sought; rather it is a question of the risk being identified, insofar as this is possible, and a decision then being taken as to what constitutes an acceptable level of risk.

Acting on even a qualified precautionary basis does, nevertheless, represent a challenge to current orthodoxy in legal decision-making, which is based so profoundly on the concept of proof. Thus, successfully grafting the precautionary principle on to existing systems and values is necessarily extremely challenging and ultimately depends on a high degree of political backing in order to be successfully pursued.

The relationship between the precautionary principle and risk may be viewed in a variety of ways, ranging from regarding it as principally part of risk assessment[46] to considering it predominantly as a tool for risk management.[47]

---

[41] de Sadeleer (n 5 above) 6.   [42] de Sadeleer (n 5 above).
[43] WT/DS26/AB/R and WT/DS48/AB/R, 16 January 1998.
[44] For example, public health law and the regulation of dangerous substances: de Sadeleer (n 11 above) 111ff.   [45] Guruswamy (n 6 above) 484.
[46] See, eg de Sadeleer (n 5 above) 18.
[47] See, eg European Commission, *Communication from the Commission on the Precautionary Principle* COM(2000)1.

However, in reality, identifying a strict division between risk assessment and risk management is highly artificial and not terribly convincing. In any event, where unknown risk is in play, the role to be undertaken by the precautionary principle is perhaps best expressed as being to contextualize both risk assessment and management elements of risk analysis in a new way. This has implications for both scientific praxis and political decision-making in terms of the breadth of the interdisciplinary approach, the need for iteration, and the scope of their temporal application.[48] As a result, invoking the precautionary principle also necessitates a number of significant changes in how we view regulatory decision-making, not least in the fact that the longevity and predictability of regulatory decisions are potentially reduced by the need to revisit them in response to new scientific developments.[49] In any event, it ultimately falls to the courts to deal with the most profound and intractable challenges posed by making the precautionary principle operational.[50] Principles are particularly significant in this regard, since in what one commentator terms a 'post-modern context',[51] they aid the courts in filling the gaps left by rapid development of the law.

The association between the precautionary principle and risk analysis is, accordingly, a complex one; and Applegate has identified the precautionary principle as comprising four distinct elements:[52] trigger (variously worded as anticipated serious/significant risk of irreversible harm, based on available scientific evidence); timing (it is, by its very nature, anticipatory); response (which must be proportionate); and iteration (decisions must be revisited in response to new scientific knowledge). The trigger and response elements provide the most obvious control mechanisms for decision-makers. The former, in particular, is hugely significant, as it keeps science centre stage in decision-making processes and underlines the fact that 'speculative' risk will not suffice to invoke the precautionary principle.[53] Thus, much depends on how one interprets the relationship between the precautionary principle and existing science. As stated by de Sadeleer, what is involved at this stage is a 'threshold' approach: '[t]he precautionary measure must ... be linked to a minimum of knowledge: that is to say, to scientific grounds with a demonstrated degree of consistency';[54] and, as

---

[48] See, eg Myhr (n 15 above).
[49] These issues are partially reflected in the GM regulatory regime under the Deliberate Release Directive (notably the use of the safeguard clause and time-limited authorizations).
[50] For Community cases, see, eg Case T-13/99 *Pfizer Animal Health SA v Council of the European Union* [2002] ECR II-3305; and Case T-257/07 *France v Commission* [2007] ECR II-4153; and, for WTO cases, see, *eg EC—Measures Concerning Meat and Meat Products (Hormones)* WT/DS26/AB/R and WT/DS48/AB/R, 16 January 1998; and *EC—Approval and Marketing of Biotech Products* WT/DS291/R, WT/DS292/R, and WT/DS293/R, 29 September 2006. On the *EC—Biotech* case, see further Chapter 13.
[51] N de Sadeleer, 'Environmental Principle, Modern and Post-modern Law' in R Macrory et al (eds), *Principles of European Environmental Law: Proceedings of the Avosetta Group of European Environmental Lawyers* (Gronigen: Europa Publishing, 2004) 235.
[52] See Applegate (n 20 above) 249.   [53] Nucara (n 17 above) 48–9.
[54] de Sadeleer (n 11 above) 159.

indicated, this is borne out in the approach adopted by both the European Court of Justice and WTO dispute-settlement body.[55]

It is clear then that the precautionary principle does not by any means represent the end of technocracy, though it does require a realignment of the technical/scientific and political elements of decision-making. At the very least, in this context, it ensures that science cannot be viewed as solely determinative. This may be seen as re-casting risk analysis in something more akin to the form that it was originally envisaged as taking. It may also be viewed as promoting transparency by laying bare the fact that political choices (based on scientific knowledge and its limits) are of central importance. Precaution also requires that regulatory decisions themselves must be subject to re-examination in light of the changing state of knowledge, with ongoing decision-making processes replacing open and closed decisions, at least in some contexts. While this type of approach may create problems in terms of legal certainty, these are not insurmountable: for example, under the Deliberate Release Directive, consents for placings on the market are subject to a maximum period of 10 years.[56] Having said this, any time periods adopted also need to be precautionary in nature.

6. The Community GMO Regime

The development and exploitation of GMOs promises a variety of potential benefits, but also poses possible serious social, economic, and environmental concerns (both established and unknown) at national, international, and global levels.[57] Whether one takes the view that genetic engineering is different in kind or only in degree from genetic manipulation as traditionally practised in agriculture, at the very least, the time periods involved are shorter and the scale of immediate application larger. In consequence, potential impacts may be made manifest more swiftly and widely. It is also true to say that the degree of manipulation involved is of a different magnitude, being more profound and using material from a much wider range of genetic sources than would be viable in traditional practice. Further, trade issues tend to be particularly contentious where GMOs are concerned, as national perspectives on risk vary considerably,[58] so that '[d]ifferences in the definition of what can be felt as a risk to human health are often culturally determined and therefore lead to differences concerning the need, the form and the duration of the precautionary principle'.[59]

These factors, amongst others, have pushed the development of regulatory law in this area to the limits and beyond. For example, the Community's first

---

[55] See n 50 above.   [56] [2001] OJ L106/1, Art 15(4).
[57] See, eg Applegate (n 20 above); and M Rosso Grossman, 'Genetically modified crops in the United States: Federal Regulation and State tort liability' (2003) 5 Environmental Law Review 86.
[58] Rosso Grossman (n 57 above).   [59] Nucara (n 17 above) 50.

tranche of GM legislation, especially the process-focussed Council Directive (EEC) 90/220 on the deliberate release into the environment of genetically modified organisms,[60] proved limited and profoundly flawed. In particular, Directive 90/220 was found lacking in terms of both transparency and opportunities for public input; and it was further weakened by a failure to impose demanding monitoring requirements.[61] Moreover, the initial regulatory regime failed to convince Member States and the public at large, leading to the adoption of the moratoria on the authorization of GMOs, which only ended on 19 May 2004 with the authorization of BT11 sweet maize.[62]

When the new regime was put in place under the Deliberate Release Directive, there was a clear attempt to respond to the conflicting issues of national/public concern and the demands of international trade law. Since it was based on Article 95 of the EC Treaty, the market remains firmly centre stage in the regulatory regime;[63] and there also continues to be focus on environmental risk assessment (in Article 2 and Annexe II). However, under Article 4(1), it is expressly stipulated that regulatory action in this area is to be based on the precautionary principle, Member States being required to 'take all appropriate measures to avoid adverse effects on the human health and the environment which might arise from the deliberate release or placing on the market of GMOs'. Further, the legislation is designed to ensure that risk is addressed on an inclusive basis between Community institutions and Member States (including provision for public participation). This may help to counter the widely held perception that the regime is both highly politicized (conflict is to be resolved primarily through the comitology process) and centralized (the Community institutions represent the prime locus of decision-making power, this being especially so in the case of placings on the market).[64] Experience has nonetheless shown that tensions will remain in the regulation of GMOs and that both Member States and NGOs, dissatisfied with a permissive approach (arguably, at least in part, imposed upon the Community by the WTO)[65] will not shrink from using whatever avenues are open to them in order to litigate (and otherwise contest decisions).[66]

---

[60] [1990] OJ L117/15. [61] For the early Community legislation, see further Chapter 4.
[62] As discussed at <http://www.gmo-safety.eu/en/archive/2004/289.docu.html>, last accessed 2 July 2009. [63] Under the earlier Directive 90/220, the EC Treaty basis had been Art 100a.
[64] See, eg T Christoforu, 'The regulation of genetically modified organisms in the European Union: the interplay of science, law and politics' (2004) 41 Common Market Law Review 651; and T Christoforu, 'Genetically Modified Organisms in the EU' in N de Sadeleer (ed), *Implementing the Precautionary Principle: Approaches from the Nordic Countries, the EU and USA* (London: Earthscan, 2007) 197. Not least, in the case of placings on the market, the main avenue open to Member States is resort to the safeguard clause: Deliberate Release Directive, Art 23.
[65] See, eg de Sadeleer (n 11 above) 98; and see further Chapter 13.
[66] In the context of health, see, eg the protests surrounding *EC—Measures Affecting Asbestos and Asbestos-containing Products* WT/DS135/AB/R, 12 March 2001.

7. Risk and the Courts

While it may be true to say that risk assessment is a predominantly scientific matter, and risk management a primarily political concern, neither of these is exclusively so, and the courts can play a role in these areas in their own right through their role in policing regulatory law and in the development of the common law. In addition to long-recognized risks to human health,[67] the law has more recently come to address what may be termed 'environmental risk'; and, in areas where new technologies are concerned, science plays a key role in law-making, since it pre-figures[68] and subsequently plays a role in shaping regulatory regimes, grounding standard-setting, and guiding the application of the law in licensing systems. It also plays a key substantive role when regulatory decisions are challenged. This section will consider the way in which the courts have addressed risk from two significant perspectives. It will look first at the evolution of broad engagement with risk in the domestic courts in the United Kingdom, reflecting on progress and limitations in this regard. It will then consider developments, primarily in the European Court of Justice, with specific respect to risk and GMOs.

The UK courts have long experience in dealing with known risks at common law, most frequently in the context of the tort of negligence. In this area, cases are founded upon damage allegedly caused to the claimant by the defendant's acts or omissions. The courts will undertake an objective *ex post facto* examination of the magnitude and likelihood of that damage in an (admittedly often fairly rudimentary) form of cost-benefit analysis when examining whether or not a defendant has been in breach of the duty of care that he/she owes to the claimant. This really amounts to a basic type of retrospective risk assessment, as is evident in many well-known cases including *Latimer v AEC*.[69] This case concerned injury to the claimant as the result of slipping on oil spread on the defendant's factory floor by floodwater. The defendant recognized that there was a risk of injury to workers and spread sawdust on the floor to deal with it. However, there was not sufficient sawdust to cover the entire shop floor and the claimant was injured crossing an untreated area. The House of Lords found that the defendant was not negligent as its response was deemed proportionate to the risk involved. The Privy Council also dealt with risk in *Overseas Tankship (UK) Ltd v Miller Steamship Co Pty Ltd, (The Wagon Mound (No 2))*, where the defendants' negligence led to a quantity of bunker oil escaping into Sydney harbour.[70] Although the risk of this particular type of oil igniting in the open air

---

[67] State intervention to ensure sanitation has been a priority since the nineteenth century: see, eg the UK Public Health Act 1875.
[68] See, eg Royal Commission on Environmental Pollution Thirteenth Report, *The Release of Genetically Modified Organisms to the Environment*, Cm 720 (1989). [69] [1953] AC 643.
[70] [1967] 1 AC 617.

was extremely small, the consequences if it did so (which in fact happened) were very grave indeed. All that would have been required to avert the risk in question was to turn off a tap that had been left on. The Judicial Committee took the view that a reasonable person would not ignore even a small risk if it could be addressed without investing any substantial effort or resources.

Recently, case law has demonstrated a much more developed and explicit form of discussion of risk assessment as in *Poppleton v The Trustees of the Portsmouth Youth Activities Committee*;[71] but, despite such ongoing developments, the common law continues to experience greater difficulty in engaging with what may be described as partially unknown risks (in particular, where causation is uncertain), as in *McGhee v National Coal Board*.[72] This case concerned a claimant who contracted dermatitis through exposure to brick dust in the workplace. Part of the exposure was 'innocent' (in other words, it was an inevitable consequence of the work environment in question). On the other hand, part of the exposure was 'guilty', in that it was due to the defendant's negligent failure to provide staff with washing facilities, thus prolonging their exposure to the dust. Medical science at the time could not determine whether the innocent or guilty exposure caused the claimant's condition. This caused problems for the claimant in terms of causation (the normal rule in negligence being that the claimant must prove on the balance of probabilities that the defendant's negligence caused or materially contributed to his/her damage). The House of Lords, recognizing that the claimant could not meet the relevant standard of proof in the circumstances, found the defendant liable on the basis of conduct that created a material increase in risk of the claimant's damage.

The question of unknown risk re-emerged in *Fairchild v Glenhaven Funeral Services Ltd*, which involved claimants who developed mesothelioma as a result of exposure to asbestos during their working lives.[73] Mesothelioma can be caused by exposure to a single asbestos fibre, but current science cannot pinpoint the exact incident that causes the disease. In this case, the claimants had worked for multiple employers for a variety of periods. Each employer had been negligent, but it was not possible to prove which defendant's negligence had caused the damage. The House of Lords decided that all of the employers should be liable, as each of them had increased the risk of the claimants' damage. At the same time, the Law Lords were very anxious to underline that the approach that they were adopting in *Fairchild* was very much the exception to the general rule on causation and would only be applicable where lack of scientific evidence precluded an application of the usual approach.[74] Nonetheless, it is obvious that the special approach to causation adopted in *McGhee* and *Fairchild*, in placing

---

[71] [2007] EWHC 1567 (relating to personal injury which took place when undertaking indoor climbing). [72] [1973] 1 WLR 1.
[73] [2002] UKHL 22, [2002] 3 WLR 89.
[74] This was confirmed in the asbestosis case of *Holtby v Brigham & Cowan (Hull) Ltd* [2000] 3 All ER 421 (asbestosis being caused by cumulative exposure).

risk centre stage where there is scientific uncertainty, has the potential to exert a much wider influence in the development of the law. As stated by de Sadeleer, '[t]he uncertainty that attaches to causation of post-industrial risks affects the calculation of the probability as well as the nature and scope of the damages they may entail'.[75] One result of this is that, while the decisions in the cases considered may fit well on the facts, they sit uneasily with the general tenor of the law. The cases are admittedly very difficult and ultimately controversial, involving what may be termed multi-dimensional uncertainty, with the law pressed somewhat uncomfortably into service despite gaps in scientific understanding; but such judicial creativity could provide valuable insights on how to tackle risk elsewhere in the legal system.

Risk is also a key feature of much regulatory law, and perhaps more obviously so than at common law. This has long been the case in, for example, health and safety law, where known risk and risk evaluation are typically at issue (as, for example, in *R v HTM Ltd*[76]). However, the courts are also gaining experience of dealing with risk in regulatory contexts of more direct concern to the environment. In particular, they have had occasion to examine the role of the Health and Safety Executive (HSE) and local planning authorities when dealing with hazardous substances consents;[77] and in such cases they are called upon to engage with risk assessment in a very explicit fashion, as in *Eriden Properties LLP v Falkirk Council* (albeit *obiter*).[78] This decision ultimately turned on the view of the court that the local planning authority, in carrying out what was termed a difficult balancing exercise, had acted reasonably in determining the risk-management approach to be adopted toward the risks in question. In this area, the delineation of science and political aspects of the risk-evaluation process, and the correct relationship between them, is both clearly understood and articulated. More commonly, however, risk is an implicit consideration for the courts, as in the environmental impact assessment (EIA) case of *R (Edwards) v Environment Agency*, where there were strong differences of opinion between the Environment Agency and the public on the risks associated with proposed changes to the fuel mix applied to an industrial kiln.[79] Similar concerns arose in *R (Horner) v Lancashire County Council*, a case involving judicial review of a grant of planning permission (without an EIA) for an incinerator to burn animal-waste-derived fuel, in the wake of the BSE crisis.[80] However, a particularly clear example of litigation following upon dissatisfaction with the treatment of

---

[75] de Sadeleer (n 11 above) 153.   [76] [2006] EWCA 1156.
[77] According to para 14 of Circular 5/93, *Planning Controls for Hazardous Substances*, the role of the HSE is to undertake risk assessments and advise on the nature and severity of risk in hazardous substances consents; and, although it is for the local planning authorities to make the final decision, they will not disregard the advice of the HSE without good reason.   [78] [2007] SLT 966.
[79] [2006] EWCA Civ 877. The case subsequently went to the House of Lords, which affirmed the Court of Appeal decision and did not consider these matters in detail: [2008] UKHL 22.
[80] [2007] EWCA Civ 784. In the event, the case was not ultimately determined on this basis, with the court focussing instead on technical questions of statutory interpretation.

risk in consultation processes can be found in *Downs v Secretary of State for Environment, Food and Rural Affairs*.[81] Ms Downs, a well-known anti-pesticides campaigner, brought a claim for judicial review in respect of the Secretary of State's decision not to impose a mandatory 'no-spray buffer zone'. The decision of the Secretary of State followed public consultation, detailed research by the Royal Commission on Environmental Pollution (RCEP), and the taking of advice from the Advisory Committee on Pesticides (ACP). The RCEP and the ACP differed in their views, with the latter (broadly favouring the status quo) prevailing with government. Ms Downs preferred the more precautionary approach of the RCEP and challenged the decision for failure to comply with Council Directive (EEC) 91/414 on the placing of plant protection products on the market,[82] both in respect of risk assessment and the obligation to protect public health. She was successful at first instance, where, importantly, Collins J held that she had scientifically justified her argument.[83] The Secretary of State was, however, successful on appeal, where it was found that he had complied with the requirements of Directive 91/414, which allowed a certain amount of discretion to Member States. In particular, the Court of Appeal determined that Collins J had substituted his own view of the evidence for that of the Secretary of State. More significantly, the judgment also made clear that it was necessary to prove a 'manifest error' by a decision-maker in such a technical area in the absence of scientific consensus, a task that was acknowledged to present formidable difficulties; and the claimant had failed to do so.

In such environmental contexts, the close relationship between science and law is, of course, particularly evident. Science can drive progress as certainty increases, a good example being the case of ozone-depleting substances; but, where science cannot be so swiftly consolidated, action can be retarded, a good example being global warming. To summarize de Sadeleer, the normal relationship between science and environmental law may be described in terms of a fourfold process: first, science identifies the societal problems to be tackled; secondly, it mediates the physical world to decision-makers; thirdly, it dominates the setting of environmental standards (especially through risk assessment); and, fourthly, it is often decisive in judicial decisions.[84] However, while this type of dynamic may have been adequate in the past, it may not serve us so well in the future, as the precedence of science can potentially hamper the development of other mechanisms that may be better suited to dealing with our limited understanding of the impacts of our actions, such as the precautionary principle. Having said this, the precautionary principle, while much referred to, has not yet provided an equivalent motive force to challenge the status quo.

---

[81] [2009] EWCA Civ 664.   [82] [1991] OJ L230/1.
[83] [2008] EWHC 2666 (Admin); and [2009] Journal of Planning and Environmental Law 885.
[84] de Sadeleer (n 11 above) 175.

Accordingly, the UK courts have a track record, although admittedly not an unchequered one, of dealing with risk. This task is complicated in the environmental context in general, and where GMOs are concerned in particular, where courts (and notably the European Court of Justice) are called upon to deal with often novel and sensitive issues that, despite carefully crafted decision-making regimes, are still the focus of discontent. Here, on the one hand, the courts are often being called upon to adjudicate on nice legal distinctions, but, on the other hand, they are actually being asked to deal with profound social, political, and sometimes even conceptual disagreements about risk between the Community institutions, Member States, and NGOs. And the concerns that they have been required to address may be illustrated by the following cases.

First, a reluctance on the part of the European Court of Justice to engage with the broader social concerns relating to GMOs may be detected in *Glawischnig v Bundesminister für soziale Sicherheit und Generationen*.[85] An Austrian politician sought a preliminary ruling on whether a variety of matters relating to the labelling of GMOs fell under the provisions of Council Directive (EEC) 90/313 on the freedom of access to information on the environment.[86] The European Court of Justice, agreeing with the European Commission, held that information on GMOs did not fall within any of the definitions of 'environmental information' contained in the Directive. This determination rested on the view that GM labelling provisions were deemed not to be geared to protect the environment, but to facilitate the market and inform consumers. The approach of the European Court of Justice shows how difficult it is to adopt a comprehensive approach towards GMOs, which, while they undoubtedly have social and economic impacts, may obviously also have environmental ramifications. Nonetheless, the European Court of Justice seemed to resort to a technical approach that, while justified on narrow legal grounds, in many ways seems to miss the point.[87] It is therefore significant that such information would fall more easily within the subsequent wider and more detailed information requirements in the new regime provided for in Directive (EC) 2003/4 of the European Parliament and of the Council on public access to environmental information.[88]

Secondly, a particular pressure point in the regulation of GMOs has been attempts by Member States to invoke 'safeguard clauses' in order to avoid the introduction of Community-approved GM crops in their own jurisdictions.[89] Such safeguard clauses provide Member States with a rare opportunity to wrest back some control from the Community institutions; and litigating on

---

[85] Case C-316/01 [2003] ECR I-5995. See also, more recently, eg Case C-165/08 *Commission v Poland* (ECJ 16 July 2009). [86] [1990] OJ L158/56.
[87] The European Court of Justice is not alone in adopting this approach. For a similar approach adopted in the context of the WTO, see J Scott, *The WTO Agreement on Sanitary and Phytosanitary Measures: a Commentary* (Oxford: Oxford University Press, 2007). [88] [2003] OJ L41/26.
[89] For the safeguard clause in the current legislation governing deliberate releases into the environment of GMOs, see the Deliberate Release Directive, Art 23.

differences of opinion as to their use also allows Member States the opportunity to raise public concern within the system. The precautionary principle, what is required to trigger it, and its relationship with risk assessment, provides the focus for much of the litigation in this area, as, for example, in *Monsanto Agricoltura Italia SpA v Prezidenza del Consiglio dei Ministri*.[90] This case concerned disagreement as to whether foods containing residues of transgenic proteins could be subjected to simplified regulatory procedures under Regulation (EC) 258/97 of the European Parliament and of the Council concerning novel foods and novel food ingredients. In particular, the question was whether such foods were 'substantially equivalent' to existing foods and could thus avoid a detailed risk assessment.[91] The European Commission, relying on the advice of the Advisory Committee on Novel Foods and Processes, favoured the adoption of the simplified procedure, but the Italian Government objected (on grounds of a potential threat to human health) and legislated under Article 12 of the Regulation, a safeguard clause, to restrict trade in goods containing the GM maize in question. The claimant sought an annulment of the Italian instrument and compensation. The European Court of Justice was evidently uncomfortable at being asked to adjudicate in an area of such complexity and reiterated that, in such a context, its judicial review function was limited to determining whether there had been a 'manifest error of assessment or a misuse of powers or whether the legislature [had] manifestly exceeded the limits of its discretion'.[92] Nonetheless, it agreed with the national court that Article 12 constituted an expression of the precautionary principle, but added that the provision could not be invoked on a speculative basis. There had to be as complete a risk assessment as possible under the circumstances, using the most reliable scientific evidence available. Such an assessment would be required, in light of the precautionary principle, to identify a detailed/specific risk of dangerous effects to human health which justified the adoption of restrictive measures. Further, the European Court of Justice took the view that it was for national courts to decide whether such dangerous effects existed; and it found that the simplified procedure could apply if no risk of potentially dangerous effects on human health had been identified at the time of the initial assessment. However, it also found that, where a Member State had conducted a proper risk assessment, Article 12 could be validly invoked. Once again, it is highly significant that, following considerable scientific progress, the simplified procedure had already been effectively abandoned as from January 1998, by agreement between the European Commission and Member States; and this informal agreement was subsequently carried into effect by Regulation (EC) 1829/2003 of the European Parliament and of the Council on genetically modified food and feed.[93]

---

[90] Case C-236/01 [2003] ECR I-8105.   [91] [1997] OJ L43/1.
[92] Case C-236/01 [2003] ECR I-8105, para 135.   [93] [2003] OJ L268/1.

A similar issue arose in *Land Oberösterreich v Commission*.[94] Austria had taken the view that invoking the safeguard clause in the Deliberate Release Directive would be insufficient to protect conventional and organic farming in Land Oberösterreich, so instead requested a derogation from the Directive to allow the imposition of a total ban on GMOs. The request was rejected by the European Commission.[95] The reason given was that no new scientific evidence had been produced establishing a specific (uniquely local) need to protect the environment or the working environment. Accordingly, the requirements of Article 95(5) of the EC Treaty had not been satisfied;[96] and emphasis was placed on the fact that the European Commission had taken into consideration the opinion of the European Food Safety Authority (to the effect that there was no scientific evidence demonstrating the existence of a specific problem).

Austria failed both in the Court of First Instance[97] and on appeal.[98] The case is complex;[99] and among the key issues that it discusses is the precautionary principle, with the Opinion of the Advocate-General raising some most interesting observations on risk. In particular, the Opinion states that:

> ... relevant though the precautionary principle may undoubtedly be when assessing new evidence concerning a new situation, no amount of precaution can actually render that evidence or situation new. The novelty of both situation and evidence is a dual criterion which must be satisfied before the precautionary principle can come into play.[100]

The Advocate-General was well aware of the social implications of the decision, but ultimately made an extremely important observation and one that goes to the heart of much of the litigation in this area: 'the concerns in question are policy concerns which must be dealt with in political fora. ... the concerns in question are not in themselves directly relevant to the legal issues raised in this case ....'[101]

---

[94] Joined Cases C-439/05P and C-454/05P [2007] ECR I-7141; and see generally FM Fleurke, 'What use for Article 95(5) EC? An analysis of *Land Oberösterreich and Republic of Austria v Commission*' (2008) 20 Journal of Environmental Law 267.

[95] Commission Decision (EC) 2003/653 of 2 September 2003 relating to national provisions on banning the use of genetically modified organisms in the region of Upper Austria notified by the Republic of Austria pursuant to Article 95(5) of the EC Treaty [2003] OJ L230/34.

[96] '[I]f, after the adoption by the Council or by the Commission of a harmonisation measure, a Member State deems it necessary to introduce national provisions based on new scientific evidence relating to the protection of the environment or the working environment on grounds of a problem specific to that Member State arising after the adoption of the harmonisation measure, it shall notify the Commission of the envisaged problems as well as the grounds for introducing them'.

[97] Joined Cases T-366/03 and T-234/04 *Land Oberösterreich v Commission* [2005] ECR II-4005.

[98] Joined Cases C-439/05P and C-454/05P *Land Oberösterreich v Commission* [2007] ECR I-7141.

[99] The complexity of the case is compounded by the fact that in several respects the Advocate-General and ECJ adopted different approaches.

[100] Joined Cases C-439/05P and C-454/05P [2007] ECR I-7141, para 134.

[101] Ibid para 145.

Thus, while in such situations there may be a technical case to take to the European Court of Justice, in reality the issues are much more profound.

In decision-making involving GMOs, on balance it appears that the European Court of Justice has been developing a sort of synthesis, invoking a hybrid view of the precautionary principle and traditional science. Employing conventional risk assessment, insofar as this is feasible, and using it as a trigger to invoke the precautionary principle, makes for an appealing starting point from a judicial point of view—though importantly this approach fails to recognize that not all risks may be identifiable on the basis of current knowledge. Nonetheless, the European Court of Justice would appear anxious not to be used as a forum for the continued airing of what in substance amount to political disagreements about GMOs between the Community institutions and Members States/NGOs; and, accordingly, it has effectively curtailed grounds of objection in such cases to new scientific issues, in order to ensure that it is not simply being asked to go back over old ground. This had earlier been evident in *Association Greenpeace France v Ministère de l'Agriculture et de la Pêche*.[102] The case centred on the technical issue of whether a Member State (in this case, France), having transmitted a favourable opinion to the European Commission during the authorization procedure for placing a GMO on the market, was then obliged to issue written consent allowing the product to be placed on the market. Greenpeace challenged the French action, arguing that it was procedurally flawed through lack of complete information (regarding the impact of ampicillin-resistant genes on public health) and breached the precautionary principle. The European Court of Justice ruled that a Member State could withhold consent in writing provided that it had received new information in the interim that indicated a possible risk to human health and the environment and provided that the European Commission and other Member States had been informed, so allowing a decision to be taken under the Community legislation.

In addition, in its treatment of risk more generally, the approach adopted by the European Court of Justice tends to depend very much upon the statutory context. For example, it happily upheld a zero-tolerance approach to risk in the context of contamination of animal feedstuffs in *Bellio F.lli Srl v Prefettura di Treviso*.[103] The view taken was that the seizure and destruction of a consignment of animal feed contaminated with mammalian bone tissue satisfied both the precautionary principle and the requirements of proportionality, as that state of scientific knowledge made it impossible to identify what degree of contamination created a risk of BSE in humans. There was discussion *obiter* of the acceptable contamination provisions applicable to GMOs, but the European

---

[102] Case C-6/99 [2000] ECR I-1651. See also, eg A Mastromatteo, 'A lost opportunity for European regulation of genetically modified organisms' (2000) 25 European Law Review 425.
[103] Case C-286/02 [2004] ECR I-3465.

Court of Justice was very clear that such an approach was not applicable in the instant case.

8. Conclusion: Can the Law Provide Answers in Respect of Risk and GMOs?

The law alone cannot deliver a definitive solution to the problems posed by risk where GMOs are concerned. Much of the decision-making in this area is ultimately political and strongly led by developments in scientific knowledge, which, in turn, feed back into regulatory regimes. Law, however, can and must contribute a great deal in this area, giving, as it does, expression to both a variety of perspectives on GMOs and furnishing opportunities for dissent through litigation. At best, as arguably in *EC—Asbestos*,[104] it can provide opportunities creatively to re-examine even very dominant free trade norms—or at least acknowledge the need for a degree of flexibility in their application in cases involving particularly pronounced risk.

Having said this, regulatory regimes do not, in and of themselves, deliver acceptability. Nonetheless, suitable stakeholder inclusion may serve to diffuse some public hostility; and, in this regard, the provision of information is a key element, though whether or not it is regarded as trustworthy rather depends on its point of origin.[105] Public inclusion in decision-making processes themselves, and the subsequent provision of choice through the adoption of labelling and other information strategies, may be 'explained as a way of devolving decisions on risk to the public'.[106] As an informed public is free to choose whether or not to accept and purchase GM products, and, ultimately, where regulatory regimes ignore or are perceived as sidelining public concerns, the end result may (where GM products are labelled) be detrimental to the market.[107] Consumer choice is predicated by a number of complex and interacting factors, not least the psychology of risk; and, where products are distrusted, there will not be the demand for them, thus making further development in the area less attractive.

The fact remains, nevertheless, that there is a vast spectrum of opinion on GMOs, running from its commercial proponents, on the one hand, to its public/NGO opponents on the other; and, however inclusive decision-making processes are, it is impossible to satisfy everyone with a stake in the issues. In consequence, it will always be the case that, whatever efforts are made to regulate GMOs, profound disagreements about the very principle, let alone the practice, of genetic modification will remain. One result will inevitably be a role for the

---

[104] WT/DS135/AB/R, 12 March 2001.
[105] C Hilson, 'Information disclosure and the regulation of traded product risks' (2005) 17 Journal of Environmental Law 305. [106] Ibid 310.
[107] Nucara (n 17 above) 52.

courts, adjudicating disputes that, while superficially based on narrow questions of procedure, are actually attempts to call broader issues of risk back into question in a judicial forum; and such resort to litigation reflects the fact that political and legislative debate has not delivered results that are palatable to opponents of genetic engineering. This demonstrates one of the key limitations in attempts by regulatory law to address risk. Another limitation lies in the fact that case law to date on GMOs makes it abundantly clear that the courts in their adjudicatory capacity are themselves not as yet in a much better position to deal satisfactorily with uncertain risks. Having said this, the evolution of judicial approaches to risk in other areas of law offers another lens through which to view the profound societal problems generated by the lack of certainty that surrounds GMOs. It is worth examining the possibility of translating into the GM context both the expertise shown in areas such as the regulation of hazardous substances and the creativity shown in areas such as tort law. These other areas of legislative activity also demonstrate a willingness to attempt to deal with issues that pose particular problems to existing legal culture. While the solutions available may themselves be flawed or incomplete, they do provide further avenues for discussion and may perhaps offer an opportunity to move forward the jurisprudential debate on risk and GMOs. Otherwise, judging by the litigation to date in this area, there is danger of stagnation if no escape is made from the comparatively narrow confines that currently apply.

PART II
THE EUROPEAN UNION

4

The European Union Regulatory Regime for Genetically Modified Organisms and its Integration into Community Food Law and Policy

Marine Friant-Perrot

1. Introduction

In its 2002 Communication, *Life Sciences and Biotechnology—a Strategy for Europe*, the European Commission clearly stated the approach towards genetically modified organisms (GMOs) that has since been followed by the European Union: rather than passively accepting the advent of the new technologies, it was preferable to 'develop proactive policies to exploit them in a responsible manner, consistent with European values and standards'.[1] There were great aspirations for GMOs[2] in the spheres of health, agriculture, food production, and environmental protection. Consequently the promotion of biotechnology appeared a necessary policy objective to reinforce competitiveness in Europe. Nevertheless, anti-GM protestors and some scientists continue to cite risks to the environment (let alone health); and a 2007 European Parliament Report highlighted that '[t]he long-term consequences of GMO technology are still unknown. There are contradictory scientific statements, and many people

---

[1] European Commission, *Communication from the Commission to the European Parliament, the Council, the Economic and Social Committee and the Committee of the Regions: Life Sciences and Biotechnology—a Strategy for Europe* COM(2002)27 [2002] OJ C55/3, para 1.
[2] For a French definition of GMOs, see *Assemblée Nationale Française, Rapport de M. Christian Ménard sur les Enjeux des Essais et de l'Utilisation des Organismes Généfiquement Modifiés: No 2254* (2005) Vol 1, 14: the wording 'utilisé pour désigner les organismes obtenus par transgénèse'. Community law has adopted a more narrow definition: see, eg Directive (EC) 2001/18 of the European Parliament and of the Council on the deliberate release into the environment of genetically modified organisms ('Deliberate Release Directive') [2001] OJ L106/1, Art 2(2): '"genetically modified organism (GMO)" means an organism, with the exception of human beings, in which the genetic material has been altered in a way that does not occur naturally by mating and/or natural recombination'.

are afraid of the possible dangers and risks.'[3] Moreover, the pro-GMO policy adopted by the European Commission has not found general support among civil society, with a clear majority being against the use of agricultural biotechnology at any point in the food chain 'from farm to fork'.[4] In addition to health and the environment, these fears are an expression of ethical and cultural attitudes, which lends further weight to arguments that they should be accommodated in the legislative framework: not least, this would accord with the right to information in environmental matters, as already enshrined in Community law.[5] Thus, with its advocacy of managed use of GMOs, the European Union has chosen a 'third way' between risk control and economic development;[6] and, according to Sophie Mahieu, the objective ascribed to the legislation is to ensure the protection of health, the provision of information to consumers, and the protection of the environment, all without shackling the development of a modern biotechnology sector which can benefit society.[7]

The specific regime which governs GMOs has undergone a long gestation and comprises measures which address different stages in their exploitation, at all times distinguishing between their contained use and their deliberate release into the environment. Thus, the former is regulated by Directive (EC) 2009/41 of the European Parliament and of the Council on the contained use of genetically modified micro-organisms,[8] which repealed the earlier Council Directive (EEC) 90/219 on the contained use of genetically modified micro-organisms,[9] and the latter is regulated by the Deliberate Release Directive,[10] which repealed the earlier Council Directive (EEC) 90/220 on the deliberate release into the environment of genetically modified organisms.[11] In the case of food and feed, consumer protection and information are addressed by Regulation (EC) 1829/2003 of the European Parliament and of the Council on genetically modified food and feed ('Food and Feed Regulation')[12] and also by Regulation (EC) 1830/2003 of the European Parliament and of the Council concerning the traceability and labelling of genetically modified organisms and the traceability of food and feed products produced from genetically modified organisms.[13] The specific regime for GMOs is then completed by Council

---

[3] European Parliament, *Report on the Proposal for a Directive of the European Parliament and of the Council Amending Directive 2001/18/EC Concerning the Deliberate Release into the Environment of Genetically Modified Organisms, as Regards the Implementing Powers Conferred on the Commission*, A6-0292/2007, 11.
[4] See, eg European Commission, *Attitudes of European Citizens Towards the Environment— Special Eurobarometer 295* (Brussels: European Commission, 2008) 65 (58% of the those polled were against the use of GMOs, while only 21% were in favour); and see further Chapter 1.
[5] See Directive (EC) 2003/4 of the European Parliament and of the Council on public access to environmental information [2003] OJ L41/26.
[6] See, eg S Mahieu, 'Le nouveau cadre juridique européen applicable aux O.G.M. ou le paradoxe d'une réforme inachevée' (2003) 4 *Revue Européenne de Droit de la Consommation* 295.
[7] Ibid 296.   [8] [2009] OJ L125/75.   [9] [1990] OJ L117/1.
[10] [2001] OJ L106/1.   [11] [1990] OJ L117/15.   [12] [2003] OJ L268/1.
[13] [2003] OJ L268/24.

Regulation (EC) 1946/2003 of the European Parliament and of the Council on transboundary movements of genetically modified organisms,[14] which implements the obligations of the European Community under the Cartagena Protocol.[15] Governance of GMOs 'from farm to fork' must, however, also take into account legislation that is broader in ambit; and in this context may be highlighted the umbrella Regulation (EC) 178/2002 of the European Parliament and of the Council laying down the general principles and requirements of food law, establishing the European Food Safety Authority, and laying down procedures in matters of food safety ('Food Law Regulation').[16]

Accordingly, this chapter will be divided into three parts. First, a more detailed overview will be provided of the specific measures relating to contained use, deliberate releases into the environment, and placings on the market (other than in food and feed); secondly, focus will then be directed more fully to the regulation of GMOs in food and feed; and thirdly, there will be discussion of the extent to which these measures are integrated into wider Community policy on food law.

2. Contained Use and Deliberate Release

As indicated, Community legislation distinguishes between the nature of the risks that may arise; and, for that reason, provides separately for the contained use of GMOs and for their deliberate release into the environment.[17]

2.1 The Contained Use of GMOs

Before any GMOs are placed on the market, or indeed before any form of their deliberate release into the environment, preliminary research and development is conducted within the confines of laboratories or greenhouses. During this stage

---

[14] [2003] OJ L287/1.
[15] Available at <http://www.cbd.int/biosafety/protocol/shtml>. See, generally, eg AH Qureshi, 'The Cartagena Protocol on Biosafety and the WTO—co-existence or incoherence?' (2000) 49 International and Comparative Law Quarterly 835; TJ Schoenbaum, 'International trade in living modified organisms: the new regimes' (2000) 49 International and Comparative Law Quarterly 856; J Bourrinet and S Maljean-Dubois, *Le Commerce International des Organismes Génétiquement Modifiés* (Aix-Marseille/Paris: CERIC/La Documentation Française, 2002); JP Beurier, 'Les OGM et l'Évolution du Droit International' in *Mélanges Hélin* (Paris: Litec, 2004) 90; and P Deboyser and S Mahieu, 'La Régulation Internationale des OGM: une Nouvelle Tour de Babel' in P Nihoul and S Mahieu (eds), *La Sécurité Alimentaire et la Réglementation des OGM -Perspectives Nationale, Européenne et Internationale* (Brussels: Larcier, 2005) 241. See further Chapter 14.
[16] [2002] OJ L31/1.
[17] On the European Community regime, see, eg C Macmaoláin, *EU Food Law: Protecting Consumers and Health in a Common Market* (Oxford: Hart Publishing, 2007) 241–54; and see, generally, M Lee, *EU Regulation of GMOs: Law and Decision Making for a New Technology* (Cheltenham: Edward Elgar, 2008).

there is a slim, yet genuine, risk that the '[m]icro-organisms, if released into the environment in one Member State in the course of their contained use, may reproduce and spread, crossing national frontiers and thereby affecting other Member States'.[18] For this reason Directive 90/219 was originally adopted,[19] so as to control emissions and to prevent accidents.[20] Contained use is now defined as:

any activity in which micro-organisms are genetically modified or in which such GMMs are cultured, stored, used, transported, destroyed, disposed of or used in any other way, and for which specific containment measures are used to limit their contact with, and to provide a high level of safety for, the general population and the environment.[21]

Interestingly, even though the scope of Directive 2009/41 is limited to microorganisms (such as bacteria and enzymes), Member States have extended national implementation to contained use of all GMOs.[22] As an environmental measure, it was based on Article 175 of the EC Treaty (now Article 192 of the Treaty on the Functioning of the European Union);[23] and the preventive steps required are graduated in accordance with the perceived level of hazard.

As a matter of procedure, the user must carry out a prior assessment of the risks that the contained uses could present to human health and the environment.[24] A record of this assessment is to be kept by the user and made available in an appropriate form to the competent authority as part of the notification procedure or upon request.[25] Under Directive 2009/41, the assessment is to result in a final classification of contained uses into four classes: Class 1 (activities of no or negligible risk); Class 2 (activities of low risk); Class 3 (activities of moderate risk); and Class 4 (activities of high risk).[26] Consequent upon this classification, respective containment levels (Levels 1–4) are imposed to protect human health and the environment. The details of the containment and other protective measures are set out in Annex IV (which also includes general principles); and, to give one example, glasshouses must be permanent structures for Levels 2–4, but not for Level 1.[27]

In terms of procedure, where premises are to be employed for the first time for contained uses, the user must submit prior notification to the competent

---

[18] Directive 2009/41, Preamble (17).
[19] Directive 90/219, Preamble (the Directive being subsequently amended by Council Directive (EC) 98/81 [1998] OJ L330/13).
[20] For these purposes, in the current legislation an 'accident' means 'any incident involving a significant and unintended release of GMMs in the course of their contained use which could present an immediate or delayed hazard to human health or the environment': Directive 2009/41, Art 2(d).   [21] Ibid Art 2(c).
[22] S Mahieu, 'Le Contrôle des Risques dans la Réglementation Européenne Relative aux OGM: Vers un Système Conciliateur et Participatif' in Nihoul and Mahieu (n 15 above) 153, 170. See, eg in the case of England and Wales, the Genetically Modified Organisms (Contained Use) Regulations 2000, SI 2000/2831.
[23] See also R Romi (with G Bossis and S Rousseaux), *Droit International et Européen de l'Environnement* (Paris: Montchestien 2005) 125.   [24] Directive 2009/41, Art 4(2).
[25] Ibid Art 4(6).   [26] Ibid Art 4(3).   [27] Ibid Annex IV, Table 1B.

authorities.[28] The information required for the notification is to include the class of the contained uses.[29] However, in the case of Class 1, there need only be a summary of the assessment, and then the contained use may proceed without further notification.[30] In the case of Classes 2–4, the procedure is more complex. Thus, in the case of Class 3 or 4, the prior consent of the competent authority is required, its decision to be communicated in writing.[31] Nonetheless, it is important to emphasize that, apart from any ability to grant or withhold consent (which is largely restricted to Classes 3 and 4),[32] the role of the competent authorities is not extensive. For example, they are to examine conformity of the notifications with the requirements of the Directive, the accuracy and completeness of the information given, and the correctness of the assessment and the class of contained use, together with (where appropriate) the suitability of the containment and other protective measures, waste management, and emergency measures.[33] Finally, Member States must submit an annual summary report to the European Commission on Class 3 and class 4 contained uses of genetically modified micro-organisms;[34] and they must also submit every three years a summary report on their experience with the Directive, these in turn forming the basis of a summary by the European Commission.[35] It may be emphasized, however, that this simplified procedure is specific to contained use, on the ground that such use does not give rise to serious risk to the environment or health: the position is different where GMOs are cultivated in the open.

2.2 The Deliberate Release of GMOs into the Environment

After contained use of a GMO for research purposes and before its placing on the market, field trials may be necessary; and the Deliberate Release Directive (replacing Directive 90/220) addresses the deliberate release of GMOs into the environment (with particular reference to such field trials).[36] As stated at Preamble (25): '[n]o GMOs, as or in products, intended for deliberate release are to be considered for placing on the market without first having been subjected to satisfactory field testing at the research and development stage in ecosystems which could be affected by their use'.

---

[28] Ibid Art 6.   [29] Ibid Annex V, Part A.   [30] Ibid; and Art 7.   [31] Ibid Art 9(2).
[32] A decision can be specifically requested by the applicant in the case of Class 2 contained use: ibid Art 8(2); and, again in the case of Class 2, the competent authority can submit indication to the contrary if the premises have not been the subject of a previous notification: ibid Art 8(3).
[33] Ibid Art 10(2). In addition, if necessary, the competent authority may ask the user to provide further information, modify the conditions of the proposed contained use, or amend the class assigned. Further, it may also time-limit consents or impose specific conditions: ibid Art 10(3).
[34] Ibid Art 17(1).   [35] Ibid Art 17(2) and (3).
[36] See, generally, eg E Brosset, 'The prior authorisation procedure adopted for the deliberate release into the environment of genetically modified organisms: the complexities of balancing Community and national competences' (2004) 10 European Law Journal 555.

These controversial field trials have generally concerned transgenic varieties of maize, oilseed rape, and potatoes and they are covered by a specific authorization procedure, as set out in Part B of the Deliberate Release Directive. It is for the notifier to submit a notification containing the technical information indicated in Annex III and an environmental risk assessment. The technical information should cover, *inter alia*, interactions between the GMOs and the environment, together with monitoring, control, waste treatment, and emergency plans.[37] The competent authorities of the Member State must examine the application and consider also any observations from other Member States (through the exchange of information procedure established under Article 11). They must then either authorize or reject the application within 90 days of the notification.[38] Subsequently, the notifier must send to the competent authority the result of the release, in relation to any risk to human health or the environment.[39] By virtue of the precautionary principle (which is expressly mentioned in both Articles 1 and 4), if modifications or unintended changes take place or new information emerges which could have consequences in terms of risk to human health and the environment, then measures must be taken to address those risks and appropriate revisions implemented.[40] Further, under Article 9, Member States must make available to the public information on all Part B releases in their territory and, in addition, they must consult with the public and, where appropriate, groups, consistent with the transparency principle. Thus, in many respects, the Deliberate Release Directive provides stronger protection for the environment than Directive 2009/41 on contained use. Further, its procedure confers considerable powers on the Member States, whereas, in the case of placings on the market, the emphasis is on regulation at Community level.[41]

2.3 Placings on the Market

Placings on the market[42] constitute the procedural stage that is most tightly controlled, with precautionary measures bolstered and the highest degree of public participation. As stated by Macmaoláin, '[i]t is clear from the outset that Directive 2001/18 is largely moulded by a vastly increased consideration for consumer concerns, an input that was largely missing from the provisions of Directive 1990/220'.[43] The regulatory regime is set out in Part C of the Deliberate Release Directive, a clear distinction being made between placings on the market of

---

[37] See, in particular, Deliberate Release Directive, Annex IIIA, paras IV and V.
[38] Ibid Art 6(5).   [39] Ibid Art 10.   [40] Ibid Art 8.
[41] See, eg Romi (with Bossis and Rousseaux) (n 23 above) 129 (who refers in this latter context to 'forte communautarisation').
[42] Defined as, subject to exceptions, 'making available to third parties, whether in return for payment or free of charge': Deliberate Release Directive, Art 2(4).
[43] Macmaoláin (n 17 above) 246. See also Mahieu (n 22 above) 189 (who affirms that, since this stage is regarded as potentially the most hazardous for society and the environment, it is correspondingly 'la phase la plus délicate et nécessairement conciliatrice').

GMOs as or in products within the Community and the carrying-out of deliberate releases into the environment for any other purposes (which, as has been seen, are regulated by Part B).[44] Above all, Part C is characterized by a procedure which provides for authorization at Community level.[45] However, as from 2004 the scope of the Deliberate Release Directive has been narrowed, in that where GMOs are destined for human food or animal feed, the single authorization procedure under the Food and Feed Regulation is that applicable. Nonetheless, in this context the Deliberate Release Directive does remain relevant where the GM product is not destined for human food or animal feed; and it may be noted that a double application may be possible: one for cropping and one for the food or feed use.[46]

First, the Deliberate Release Directive strengthens the requirements relating to the potentially harmful effects of GMOs. Thus, any person (for example, manufacturers or importers) wishing to place a GMO on the market must submit a notification to the competent authority of the Member State where the GMO is to be placed on the market for the first time.[47] Within 90 days of the notification, the competent authority must prepare an assessment report,[48] this task to be undertaken in conformity with the general obligation to respect the precautionary principle as set out in Article 4. The report must indicate whether or not the GMO(s) in question should be placed on the market (together with any conditions if the indication is favourable);[49] and it must be sent to the European Commission, to be forwarded to the competent authorities in the other Member States. Where the competent authority which prepared the report decides in favour of placing the GMO on the market, then the other Member States and the European Commission have 60 days in which to make reasoned objections; otherwise, consent is to be given by that competent authority, for a maximum duration of 10 years.[50] However, as a rule objections are made; and this sets in motion a second phase, which unfolds before the Community authorities, commencing with consultation of the relevant scientific committee, the European Food Safety Authority (EFSA).[51] The European Commission then submits a draft decision to a Committee composed of the representatives of the Member States, under the comitology procedure.[52] This Committee

---

[44] Deliberate Release Directive, Art 1.   [45] Ibid, Preamble (28).
[46] See further below.
[47] Deliberate Release Directive, Art 13(1). The information required for the notification dossier is set out in Art 13(2).   [48] Ibid Art 14(2).
[49] Ibid Art 14(3).
[50] Ibid Art 15. The Member State is bound to issue consent, unless it has received new information which leads it to consider that the product may constitute a risk to human health and the environment, and immediately informs the European Commission and the other Member States: Case C-6/99 *Association Greenpeace France v Ministère de l'Agriculture et de la Pêche* [2000] ECR I-1651; and see also A Mastromatteo, 'A lost opportunity for European regulation of genetically modified organisms' (2000) 25 European Law Review 425.
[51] Deliberate Release Directive, Art 28; and, for the establishment of the EFSA, see the Food Law Regulation.
[52] On comitology in this context, see Lee (n 17 above) 66 and 70–1; and see further Chapter 5.

operates on the basis of qualified majority voting and, in the absence of agreement, the Council must come to a decision within three months. If again no agreement can be reached, the European Commission is to adopt the proposal. It may be observed that, in practice, entrenched positions adopted by Member States respectively in favour of and against commercialization of GMOs has resulted in the Council abrogating its political responsibility in favour of the European Commission.[53]

Secondly, the Deliberate Release Directive lays down rules for the monitoring and handling of new information on risks.[54] The notifier shall ensure that monitoring reports are submitted to the European Commission and the competent authorities of the Member States, where this is specified in the consent; and the results of monitoring must be made available to the public, so as to ensure transparency.[55] Again in accordance with the precautionary principle, if new information emerges which could have consequences in terms of risk to human health and the environment, the notifier must immediately take the necessary protective measures and inform the competent authority, with further provision for onward transmission of this information to the European Commission. A procedure is then set in train which provides for amendment to the conditions of the consent.

Thirdly, the Deliberate Release Directive required that GMOs placed on the market as or in products must be labelled, but subject to a *de minimis* threshold in the case of adventitious or technically unavoidable traces of authorized GMOs.[56] The threshold was subsequently fixed at 0.9 per cent, under Regulation 1830/2003;[57] but it may be highlighted that, for the exemption to apply, the GMO must be authorized and its presence must be adventitious or technically unavoidable.

Finally, provision is also made for the European Commission to publish every three years a summary based upon reports submitted by Member States on their experience with GMOs placed on the market.[58]

It is immediately apparent that the logic behind the Deliberate Release Directive and the emphasis on procedure at Community level indicate that the legislature was operating under constraints imposed by World Trade Organization (WTO) agreements. There is no coincidence in the Directive being based on Article 95 of the EC Treaty, as a harmonization measure, nor in Article 22 stipulating that, without prejudice to the safeguard clause in Article 23, 'Member

---

[53] See, eg *Assemblée Nationale Française, Rapport de M. Christian Ménard sur les Enjeux des Essais et de l'Utilisation des Organismes Généfiquement Modifiés: No 2254* (2005) Vol. 1, 100: such inability to reach a decision by qualified majority voting 'conduit le Conseil des ministres à se défausser de sa responsibilité politique sur la Commission européenne'.

[54] Deliberate Release Directive, Art 20. [55] Ibid Art 20(4). [56] Ibid Art 21.

[57] Ibid Art 21(3), as amended by Regulation 1830/2003.

[58] Deliberate Release Directive, Art 31. See, eg European Commission, *Second Report from the Commission to the Council and the European Parliament on the Experience of Member States with GMOs placed on the market under Directive 2001/18/EC on the Deliberate Release into the Environment of Genetically Modified Organisms* COM(2007)81.

States may not prohibit, restrict or impede the placing on the market of GMOs, as or in products, which comply with the requirements of this Directive'. The margin of appreciation enjoyed by Member States is, accordingly, narrow. They do have the benefit of the safeguard clause provisionally to restrict or prohibit the use and/or sale of GMOs as or in products within their territory; but this can only be invoked where new or additional information is made available since the date of the consent and affects the environmental risk assessment or where existing information is reassessed on the basis of new or additional scientific knowledge. In consequence, the Member State must have detailed grounds for considering that a duly authorized GMO constitutes a risk to human health or the environment. However, as is common knowledge, the safeguard clause has been invoked many times. For example, it was authority for the provisional ban on the growing and commercialization of Bt maize (MON810) as implemented in France by a Decree of 7 February 2008,[59] and it may be noted that MON810 has also been provisionally banned by Austria, Germany, Greece, Hungary, and Luxembourg.[60] On the adoption of measures under the safeguard clause, Member States must immediately inform the European Commission and the other Member States. The advice of the EFSA may be sought to evaluate the information supporting the ban; and then it is for the Council to make a decision by qualified majority as to whether the safeguard measure should be upheld or overturned. It may be highlighted that, under this procedure, environment ministers in March 2009 rejected a European Commission proposal to overturn three safeguard measures, Austrian and Hungarian bans of MON810 and an Austrian ban of another GM maize variety, T25.[61] Indeed, only four Member States supported the European Commission proposal on all three counts.

The legislative framework as augmented by the Deliberate Release Directive did not, however, specifically address the use of GMOs in food and feed; and, to facilitate consumer choice, these aspects were regulated by a legislative package enacted in 2003.

3. Genetically Modified Food and Feed

3.1 General

The first GMOs used in human food and animal feed were originally authorized on the basis of Directive 90/220. These included transgenic plants (such as

---

[59] Ministerial Decree of 7 February 2008 Suspending the Cropping of Genetically Modified Maize Seed (Zea Mays L Line MON810), *Journal Officiel de la République Française* (JORF) No 34 of 9 February 2008, NOR: AGRG0803466A, amended 13 February 2008 NOR: AGRG0803888A.
[60] *Agra Europe Weekly* No 2357, 17 April 2009, EP/1.
[61] 2928th Council Meeting: Environment, Brussels, 2 March 2009 (Presse 53); and see, eg *Agra Europe Weekly* No 2351, 6 March 2009, EP/1. The French and Greek bans also remained in place at that date.

maize and soya), food produced from these plants, and enzymes produced from bacteria and yeasts used as processing aids.[62] Subsequently, Regulation (EC) 258/97 of the European Parliament and of the Council concerning novel foods and novel food ingredients became the specific measure to govern GMOs destined for food use.[63] This provided for two distinct procedures: first, an authorization procedure based on a unified, Community system for risk assessment, to ensure both environmental and food safety; and, secondly, a simplified procedure, consisting merely of notification requirements. The simplified procedure applied where the novel food and novel food ingredients were considered substantially equivalent to existing food or food ingredients, as regards their composition, nutritional value, metabolism, intended use, and the level of undesirable substances which they contained.[64] An example would be sugar from GM sugar beet.[65] The Regulation had the merit of providing specific labelling rules which went beyond safety for human health. Thus, it expressly recited that 'defined population groups associated with well established practices regarding food should be informed when the presence in a novel food of material which is not present in the existing equivalent foodstuff gives rise to ethical concerns as regards those groups'.[66] That said, it did not adequately assure that foodstuffs are free from risk, or guarantee to the consumer either a sufficiently robust regime in terms of traceability or a sufficient level of information. In response to such deficiencies, a moratorium was adopted by six Member States; and, in the case of France, this commenced on 26 June 1999.

The same year, following the bovine spongiform encephalopathy (BSE) crisis, the European Commission, in its *White Paper on Food Safety*, urgently called for the adoption of legislation to cover assessment, authorization, and labelling in the case of novel foods for animals, with particular reference to GMOs and feed derived from GMOs; and, in addition, it proposed improvement of the authorization procedure under Regulation 258/97, by adopting a single assessment for all ingredients in foodstuffs and harmonizing the measures governing the labelling of food, additives, and flavourings containing GMOs or derived

---

[62] See, eg *Avis No 17 du Conseil National de l'Alimentation Relatif à l'Étiquetage des Nouveaux Aliments et des Nouveaux Ingrédients Constitués d'OGM ou Issus d'OGM* 2.

[63] [1997] OJ L43/1. It may be noted that there are in train proposals to replace Regulation 258/97: European Commission, *Proposal for a Regulation of the European Parliament and of the Council on Novel Foods* COM(2007)872; and political agreement on the draft regulation was reached by the Council on 22 June 2009 (Presse 188) (available at <http://www.consilium.europa.eu/uedocs/cms_data/docs/pressdata/en/misc/108678.pdf>, last accessed 10 January 2010).

[64] For the simplified procedure, see ibid Art 3(4) and 5; and see also Case C-236/01 *Monsanto Agricoltura Italia SpA v Prezidenza del Consiglio dei Ministri* [2003] ECR I-8105 (considered more fully in Chapter 3).

[65] For discussion of substantial equivalence more generally, see, eg *Avis No 17 du Conseil National de l'Alimentation Relatif à l'Étiquetage des Nouveaux Aliments et des Nouveaux Ingrédients Constitués d'OGM ou Issus d'OGM* 2. [66] Regulation 258/97, Preamble (8).

from GMOs.[67] As indicated, these proposals were enacted by the Food and Feed Regulation and Regulation 1830/2003 (on traceability and labelling); and the two Regulations fitted into the legislative framework already being developed by the Deliberate Release Directive and the Food Law Regulation. In particular, they put in place, first, a single authorization procedure and, secondly, traceability and labelling obligations to allow consumers to be informed of the history of the food concerned (such as the ingredients and additives used) and, thereby, to follow their preferences when choosing what to eat. These two aspects will be considered in turn.

3.2 The Single Authorization Procedure

The Food and Feed Regulation establishes a single authorization procedure for all food and feed consisting of, containing, or produced from GMOs; and the labelling provisions enjoy the same broad ambit. Importantly, these labelling provisions apply whether or not GM material can be found in the end product, the Preamble reciting that:

> labelling should include objective information to the effect that a food or feed consists of, contains or is produced from GMOs. Clear labelling, irrespective of the detectability of DNA or protein resulting from the genetic modification in the final product, meets the demands expressed in numerous surveys by a large majority of consumers, facilitates informed choice and precludes potential misleading of consumers as regards methods of manufacture or production.[68]

Moreover, the Food and Feed Regulation further tightens the earlier regime in two material respects: first, it abandons the notification procedure for novel foods considered substantially equivalent to existing foods; and secondly, it extends to feed produced from GMOs, which had not previously been subject to an authorization procedure.[69] That said, it does not extend to food and feed produced 'with' a GMO, although as indicated, it does cover both food and feed produced 'from' a GMO'. The key criterion to be met for the former category is that no material derived from the GM source should be present in the food or feed;[70] and this would be satisfied by food or feed manufactured with the help of a GMO (for example, cheese produced with a GM enzyme that does not remain in the end product), or by products obtained from animals fed with GM feed or treated with GM medicinal products.[71]

---

[67] European Commission, *White Paper on Food Safety* COM(1999)719, Annex—Action Plan on Food Safety: Actions 6, 50, 51, and 52.
[68] Food and Feed Regulation, Preamble (21). The detailed labelling provisions will be considered later. [69] Ibid Preamble (6) and (7).
[70] Ibid Preamble (16).
[71] Ibid; and for fuller discussion of this distinction, see European Commission, *Proposal for a Regulation of the European Parliament and of the Council on Genetically Modified Food and Feed* COM(2001)425, 4; MR Grossman, 'European Community Legislation for Traceability and

The single authorization procedure is conducted entirely at Community level. The application is to be submitted to the competent authority of the Member State where the product is to be placed on the market for the first time, but it acts only as a 'reception desk'.[72] The procedure can cover both food and industrial uses. Further, the applicant can also choose to make one application on the basis of Part C of the Deliberate Release Directive for GMOs and GM products that are not destined to enter the food chain, and another on the basis of the Food and Feed Regulation for those that are destined to do so.[73] As soon as the application has been submitted, the national competent authority is to inform the EFSA and make available to the EFSA the application and any supplementary information supplied by the applicant (the required information being set out in Article 5(3) of the Food and Feed Regulation in the case of food and Article 17(3) in the case of feed).[74] In the case of both food and feed, it is required that they must not have adverse effects on human health, animal health, or the environment.[75] However, these are not the only requirements: in the case of food, there must also be no misleading of the consumer and the food must not 'differ from the food which it is intended to replace to such an extent that its normal consumption would be nutritionally disadvantageous for the consumer';[76] and similar provisions are imposed in the case of feed.[77]

The EFSA must at once make available the application and any supplementary information to the other Member States and the European Commission, and also make available to the public a summary of the dossier as supplied by the applicant.[78] It then has six months from receipt of the application to deliver its opinion.[79] With regard to environmental risk, the assessment regime established under the Deliberate Release Directive must be employed.[80] On the basis of this opinion, as well as any relevant provisions of Community law and other legitimate factors, the European Commission must submit a draft decision to the Standing Committee on the Food Chain and Animal Health, which operates by qualified majority.[81] If no agreement can be

Labeling of Genetically Modified Crops, Food, and Feed' in P Weirich (ed), *Labeling Genetically Modified Food: The Philosophical and Legal Debate* (New York: Oxford University Press, 2007) 32; and C Macmaoláin, 'The new genetically modified food labelling requirements: finally a lasting solution?' (2003) 28 European Law Review 865. For a definition of 'processing aid' in the context of the Codex Alimentarius, see General Standards for the Labelling of Food Additives: 'a substance or material not including apparatus or utensils and not consumed as a food ingredient by itself, intentionally used in the processing of raw materials, foods or its ingredients to fulfil a certain technological purpose during treatment or processing and which may result in the non-intentional but unavoidable presence of residues or derivatives in the final product'.

[72] See, eg S Mahieu and P Nihoul, 'La réglementation applicable aux OGM dans l'Union européenne' (2004) 111 *Journal des Tribunaux—Droit Européen* 193, 198.
[73] Mahieu (n 22 above) 191.
[74] For the detailed rules, see Commission Regulation (EC) 641/2004 [2004] OJ L102/14.
[75] Food and Feed Regulation, Arts 4(1)(a) (food) and 16(1)(a) (feed).
[76] Ibid Art 4(1)(b) and (c).   [77] Ibid Art 16(1)(b)–(d).
[78] Ibid Arts 5(2)(b) (food) and 17(2)(b) (feed).   [79] Ibid Arts 6(1) (food) and 18(1) (feed).
[80] Ibid Arts 6(4) (food) and 18(4) (feed).   [81] Ibid Arts 7(1) (food), 19(1) (feed), and 35.

reached, the matter falls to the Council; and, if no qualified majority can be obtained in the Council, the European Commission is to adopt the proposal. Any authorization so granted is valid for 10 years, subject to renewal.[82]

After any placing on the market, the authorization-holder must submit monitoring reports to the European Commission where this is a condition of the authorization; and he is also obliged to alert the European Commission of any new scientific or technical information which might influence the evaluation of the safety of use in food or feed.[83] By virtue of the precautionary principle, which receives express reference in Article 7 of the Food Law Regulation, emergency measures may be adopted by the European Commission under Article 34 of the Food and Feed Regulation where it is evident that products which have been authorized are likely to constitute a serious risk to human health, animal health, or the environment. If a Member State has officially informed the European Commission of the need for such measures and no action has been taken, the Member State may adopt interim protective measures.[84] In this respect, the legislation differs from the Deliberate Release Directive, since the Member States must engage the European Commission prior to any emergency measure and can only act in the event of its inactivity.[85]

Finally, there are obligations to inform and consult the public, subject to exceptions so as to assure confidentiality. Thus, in principle, the application for authorization, supplementary information from the applicant, EFSA opinions, monitoring reports, and information from the authorization-holder are to be made accessible to the public.[86] Where the applicant wishes for information to be treated as confidential on the ground that its disclosure might significantly harm its competitive position, there must be verifiable justification.[87] Moreover, certain information, such as the general description of the GMO and the address of the authorization-holder cannot be considered confidential.[88] Nevertheless, labelling proves the most effective tool in terms of providing information to civil society and consumers.

3.3 Traceability and Labelling

Without doubt, a key objective of Regulation 1830/2003 was to guarantee reliable information to the consumer by the establishment of a full traceability regime for GMOs. This is not based upon being able to detect proteins or DNA resulting from GM modification, but upon being able to show, by working back through the traceability regime, whether products derived from GMOs were in

---

[82] Ibid Arts 7(5) (food) and 19(5) (feed).   [83] Ibid Arts 9 (food) and 21 (feed).
[84] For the procedures concerned, see the Food Law Regulation, Arts 53 and 54.
[85] See, eg M Le Prat and L Verdier, 'Le Conseil d'État justifie l'interdiction de la mise en culture du maïs MON 810' (2008) 363 *Revue de Droit Rural* 56.
[86] See, eg Food and Feed Regulation, Arts 6(7), 18(7), and 29.   [87] Ibid Art 30(1).
[88] Ibid Art 30(3).

fact used in the production process.⁸⁹ Community law thus creates a legal distinction between foods in accordance with their method of production (whether they be GM, conventional, or organic).⁹⁰ Significantly, Regulation 1830/2003 played a major role in bringing to an end in 2004 the de facto moratorium on GMOs;⁹¹ and its scope is extensive, applying to all stages of the placing on the market of:

(a) products consisting of, or containing, GMOs, placed on the market in accordance with Community legislation;

(b) food produced from GMOs, placed on the market in accordance with Community legislation; and

(c) feed produced from GMOs, placed on the market in accordance with Community legislation.⁹²

As recited in the Preamble, '[t]raceability requirements for food and feed produced from GMOs should be established to facilitate accurate labelling of such products';⁹³ and the obligations which are imposed seek to preserve any history of use of GMOs at any stage in the food chain, so as to guarantee for consumers full and reliable information.⁹⁴ Accordingly, any operator who places on the market products consisting of or containing GMOs or products destined for food or feed produced from GMOs must transmit in writing the specified information to the operators receiving the products.⁹⁵ To improve the identification of the GMO and to ensure effective controls, GMOs are assigned a unique identifier. This has been comprised of a combination of nine letters and numbers since the entry into force of Commission Regulation (EC) 65/2004 establishing a system for the development and assignment of unique identifiers for genetically modified organisms, which lays down the detailed rules.⁹⁶ The traceability regime has the multiple advantages of securing food production as far as reasonably possible in accordance with the precautionary principle, of permitting the swift withdrawal of products considered dangerous, and of tracking back the chain of causation to establish liability on producers and distributors.⁹⁷ Indeed, it is expressly stated that '[t]raceability should also facilitate

---

⁸⁹ See, eg *Avis No 17 du Conseil National de l'Alimentation Relatif à l'Étiquetage des Nouveaux Aliments et des Nouveaux Ingrédients Constitués d'OGM ou Issus d'OGM* 2; and, on labelling generally, see, eg P Weirich (ed), *Labeling Genetically Modified Food: The Philosophical and Legal Debate* (New York: Oxford University Press, 2007).

⁹⁰ F Collart Dutilleul, 'Conclusion Générale' in *La Régulation du Commerce Communautaire et International des Aliments* (Les Dossiers de la RIDE, Dossier No 1, *Revue Internationale de Droit Économique*, 2007) 111.

⁹¹ See, eg <http://www.gmo-safety.eu/en/archive/2004/289.docu.html>, last accessed 2 July 2009. ⁹² Regulation 1830/2003, Art 2.
⁹³ Ibid Preamble (4). ⁹⁴ Ibid Preamble (11). ⁹⁵ Ibid Arts 4 and 5.
⁹⁶ [2004] OJ L10/5.
⁹⁷ L Boy, 'Précaution, Traçabilité et Droits du Consommateur. L'Exemple de la Directive UE du 12 Mars 2001 Relative à la Dissémination Volontaire d'OGM dans l'Environnement' in *Liber Amicorum Jean Calais-Auloy* (Paris: Dalloz, 2004) 131.

the implementation of risk management measures in accordance with the precautionary principle'.[98]

Importantly, in addition to the rules on traceability, Regulation 1830/2003 stipulates that, in the case of pre-packaged products consisting of or containing GMOs, there must appear on the label the words 'This product contains genetically modified organisms' or 'This product contains genetically modified [followed by the name of the organism(s)]'; and identical wording is used in the case of non-pre-packaged products.[99] Similarly, the Food and Feed Regulation provides for labelling in the case of GM food and feed.[100] Both Regulations address the thorny question of the threshold level for adventitious or technically unavoidable presence of unauthorized GM material, conferring exemption from labelling requirements provided that the traces of such GM material do not exceed 0.9 per cent.[101] That said, it must again be emphasized that any presence of GMOs must be adventitious or technically unavoidable, so encouraging operators in the agrifood sector to reduce risk by adopting hazard analysis and critical control points (HACCPs) methods. More recently, Council Regulation (EC) 834/2007 on organic production and labelling of organic products has retained for organic crops the same 0.9 per cent threshold for adventitious or technically unavoidable presence of unauthorized GMOs.[102] This has led Professor Boy to describe the legislation as 'déceptif', in the sense that there is a risk of misleading consumers.[103] Certain NGOs and the European Parliament were in favour of adopting a threshold of 0.1 per cent in the case of organic production.[104] However, the eventual compromise illuminates the economic and technical obstacles to coexistence between GM and non-GM crops, since the threshold of 0.9 per cent amounts to a confession that it is impossible to prevent cross-contamination by GMOs.[105]

In conclusion, it would seem that this aspect of the Community legislative framework is based upon a division of roles between the public authorities and private operators: the former adopt the rules, the latter carry out the research,

---

[98] Regulation 1830/2003, Preamble (3).   [99] Ibid Art 4(6).
[100] Food and Feed Regulation, Art 12 (GM food) and Art 24 (GM feed).
[101] Ibid, Arts 12(2) (GM food) and 24(2) (GM feed); and Regulation 1830/2003, Art 4(7) (by reference to the Deliberate Release Directive) and (8) (by reference to the Food and Feed Regulation). As has been seen, Regulation 1830/2003 also fixes the threshold for the purposes of the Deliberate Release Directive, by amendment to Art 21.   [102] [2007] OJ L189/1.
[103] L Boy, 'Production et étiquetage des produits "bio" en droit communautaire' [2007] *Droit de l'Environnement* 294. See also L Bodiguel and M Cardwell, 'La Coexistence des Cultures GM et Non GM: Approche Comparative Entre l'Union Européenne, le Royaume-Uni et la France' in G Parent (ed), *Production et Consommation Durables: de la Gouvernance au Consommateur Citoyen* (Québec, Canada: Éditions Yvon Blais, 2008) 325.
[104] See, eg the vote of the European Parliament on Text A6-0061/2007, 29 March 2007.
[105] This tension is also found in the Preamble to Regulation 834/2007. Preamble (9) states that GMOs and products produced from or by GMOs should be 'incompatible with the concept of organic production and consumers' perceptions of organic products'; and that they should 'not be used in organic farming or in the processing of organic products'. Yet Preamble (10) then accepts that the aim is simply to have the lowest possible presence of GMOs in organic products. Thus, even though they cannot be used in the production process itself, they may be present in the end product.

and then the consumer is the final arbiter.¹⁰⁶ But the role of final arbiter lacks credibility unless the consumer has a genuine choice between GM and non-GM food; and, as indicated, Regulation 834/2007 is evidence enough of the internal contradictions present in the current regime. Indeed, a central ambiguity resides in the fact that the consumer does not realistically have access to food that is guaranteed free from GMOs. At best, he is offered food which is 'without GMOs', in the sense that any cross-contamination is light, falling within a threshold that is fixed by reference to the ability to detect GM material using methods that are economically viable.¹⁰⁷ The sectoral regulation of GMOs may thus be considered a legal test-bed for the close intersection of health law, environmental law, and consumer law.¹⁰⁸ In this regard, the inclusion of the regulation of GM food within the broader legislative framework for food law provides an excellent opportunity to address such reciprocal considerations.

4. GMOs and Food Law

4.1 Introduction

The adoption of the Food Law Regulation marked, according to Professor Collart Dutilleul, the advent of food law in the sense that the food sector as an area of economic activity had not previously been brought under a single legal umbrella (with all the benefits in terms of coherence that such entailed).¹⁰⁹ The new legislation is characterized by, first, an integrated approach towards policies and products; and, secondly, by affirmation that the general principles and general rules do very much apply to GM food production. Again, these two aspects will be considered in turn.

4.2 The Integrated Approach Towards Policies and Products

4.2.1 The integrated approach towards policies

Following the BSE crisis, the European Union was determined to restore the confidence of consumers in the safety and quality of food;¹¹⁰ and the construction

---

¹⁰⁶ Mahieu and Nihoul (n 72 above) 199.
¹⁰⁷ P Billet, 'Index raisonné de la loi OGM du 25 juin 2008' (2008) 368 *Revue de Droit Rural* 9, 23. In practice, it is possible to detect GM material down to a level of approximately 0.1 per cent: see Department for Environment, Food and Rural Affairs (DEFRA), *Consultation on Proposals for Managing the Coexistence of GM, Conventional and Organic Crops* (London: DEFRA, 2006) 40 (where 0.1% was regarded as effectively the practical limit for detecting GM material).
¹⁰⁸ See Boy (n 97 above) 132.
¹⁰⁹ F Collart Dutilleul, 'Eléments pour une Introduction au Droit Agro-alimentaire' in *Mélanges en l'Honneur du Professeur Yves Serra* (Paris: Dalloz, 2006) 91.
¹¹⁰ See, eg Macmaoláin (n 17 above) 175–95; D Chalmers, '"Food for thought": reconciling European risks and traditional ways of life' (2003) 66 *Modern Law Review* 532; and G Berends

of a dedicated policy allows better co-ordination of economics with considerations of public health and environmental and consumer protection. In the first place, food safety issues have led the Community legislature to reinforce the protection of public health and of the consumer, in conformity with Articles 152 and 153 of the EC Treaty (now Articles 168 and 169 of the Treaty on the Functioning of the European Union): not least, as recited in the Food Law Regulation, '[a] high level of protection of human life and health should be assured in the pursuit of Community policies'.[111] Such 'sanitarization' of the food chain finds expression in the mantra that '[t]he free movement of safe and wholesome food is an essential aspect of the internal market';[112] and this necessary reconciliation of interests can also be found in the regime governing GMOs, where, in the case of GM food and feed, placing on the market is subordinated to safety assessment.[113] Further, the Food and Feed Regulation was expressly stated to be based upon Article 152(4)(b) of the EC Treaty (now Article 168(4)(b) of the Treaty on the Functioning of the European Union), and is not just an agricultural or harmonizing measure.

Secondly, food law is an arena which allows a degree of reconciliation between the consumer and the environment, this being achieved through, in particular, the notion of 'sustainable consumerism', now at the heart of Community policies.[114] Agrifood law is, in this respect, the precursor of a more general evolution towards sustainable consumerism,[115] based upon the fact that food is not bought and sold like other products and qualifies as what Marcel Mauss would categorize as a '*fait social total*'.[116] In this sense, it combines social and environmental values more than any other product. Taking account of this overarching imperative, the Community legislature has imposed on professionals in the agrifood sector a series of specific obligations to provide information on the history of production. In consequence, the consumer is enlightened as to the origin of meat or the possibility that food or food ingredients have been treated with ionizing radiation.[117] This movement in favour of labelling to describe the methods and process of production may definitely be explained by the various health scares which have dented the confidence of

---

and I Carreño, 'Safeguards in food law—ensuring food scares are scarce' (2005) 30 European Law Review 386.

[111] Food Law Regulation, Preamble (2). [112] Ibid Preamble (1).
[113] Food and Feed Regulation, Preamble (1)–(3). It may be noted that the first two recitals are identical to those in the Food Law Regulation. [114] Boy (n 97 above) 132.
[115] M Friant-Perrot, 'La Consommation Durable et la Protection des Consommateurs: Réflexions sur les Nouveaux Rapports entre le Droit de la Consommation et le Concept de Développement Durable' in G Parent (ed), *Production et Consommation Durables: de la Gouvernance au Consommateur Citoyen* (Québec, Canada: Éditions Yvon Blais, 2008) 567.
[116] M Mauss, *Essai sur le Don. Forme et Raison de l'Échange dans les Sociétés Archaïques (1925): Introduction de Florence Weber* (Paris: Quadrige/Presses Universitaires de France, 2007).
[117] See, eg Directive (EC) 99/2 of the European Parliament and of the Council on the approximation of the laws of the Member States concerning food and food ingredients treated with ionizing radiation [1999] OJ L66/16; and Directive (EC) 99/3 of the European Parliament and of the Council on the establishment of a Community list of food and food ingredients treated with ionizing radiation [1999] OJ L66/24.

consumers: it may also, however, be explained by the need to delimit what constitutes an 'acceptable risk', on the basis of empirical evidence derived from civil society as well as scientific considerations. It is against this background that Community regulation on traceability and labelling of GMOs has been enacted; and GMOs provide a classic illustration that consumers wish for more than food that is safe and wholesome: rather, they wish to be able to choose a production process that is acceptable in environmental terms and, at the end of the day, to be able to reject the use of more controversial means of production, such as GMOs.[118]

4.2.2 The integrated approach towards products

Food law must be seen in the round, beyond just health or environmental considerations; and the Food Law Regulation has adopted a fully integrated approach 'from farm to fork'.[119] Drawing on the lessons learnt from the BSE crisis, it recites that:

[i]n order to take a sufficiently comprehensive and integrated approach to food safety, there should be a broad definition of food law covering a wide range of provisions with a direct indirect effect on the safety of food and feed, including provisions on materials and articles in contact with food, animal feed and other agricultural inputs at the level of primary production.[120]

With regard to GMOs, this approach is essential if safety is to be assured and freedom of choice to be effective for consumers. As has been seen, the Food and Feed Regulation is broad in its ambit, with the single authorization procedure extending to all food and feed containing GMOs, consisting of GMOs, or produced from GMOs (but not to food or feed produced 'with' GMOs). That said, as again noted, there has been heavy criticism by certain consumer organizations that exclusion of food and feed produced 'with' GMOs means that the Food and Feed Regulation does not cover food or feed manufactured with the help of a GM processing aid, or products obtained from animals fed with GM feed or treated with GM medicinal products; and this is a material gap in the concept of a comprehensive and integrated approach.[121] Nevertheless,

---

[118] See, eg P Borghi, 'La Diversité des Définitions de la Qualité des Aliments' in *La Régulation du Commerce Communautaire et International des Aliments* (Les Dossiers de la RIDE, Dossier No 1, *Revue Internationale de Droit Économique*, 2007) 61.

[119] On this aspect see, eg European Commission, *Communication from the Commission to the Council and the European Parliament: Integrated Product Policy—Building on Environmental Life-Cycle Thinking* COM(2003)302. [120] Food Law Regulation, Preamble (11).

[121] There is some indication of concern on the part of the European Commission that, if the full force of the GMO legislation were to be applied to products from animals raised on GM feed, there might be difficulties in sourcing sufficient non-GM feed to meet consumer demands: see, eg Commissioner Fischer Boel, Speech/07/665, 'Reflections on the Future of the CAP', Budapest, 26 October 2007 (speeches of the Commissioner for Agriculture and Rural Development are available at <http://europa.eu/rapid>). See also, eg Friends of the Earth Press Release, *British Retailers Call for Tough Stand on GM Soya*, 26 August 2005 (available at <http://www.foe.co.uk/resource/press_releases/british_retailers_call_for_26082005.html>, last accessed 29 June 2009).

Regulation 1830/2003 does ensure reliable information for the consumer, by the establishment of a comprehensive traceability regime for GMOs and also by the introduction of systematic labelling, which is based not on the possibility of detecting proteins or DNA resulting from GM modification in the end product, but on the possibility of tracing back to establish whether or not products derived from GMOs have been used at any point in the food chain. In such context, therefore, the legislation does indeed adopt a more integrated approach, which follows the full history of the product (whether it be food, feed, a flavouring, or an additive).

This broad coverage of products and activities linked to production, of manufacture, and of distribution constitutes a point of convergence between general food law and the specific regime attaching to GMOs. Without doubt, the idea was to put in place a system to govern risks, articulated round common principles and, in particular, the precautionary principle and the right to information.[122]

4.3 Convergence of Principles in the Regulation of GMOs and Community Food Law

4.3.1 *The precautionary principle*

The precautionary principle is expressly mentioned as a general principle in the Food Law Regulation;[123] and, derived from environmental law, it has now become a general principle of Community law, extending notably to the governance of food.[124] Significantly, in this respect the Food Law Regulation was, at least partially, prefigured by earlier legislation. In particular, Professor Collart Dutilleul has regarded Regulation 258/97 as instituting a procedure (as with medicines) which is precautionary in nature in that it allows the imposition of assessment whenever there is uncertainty as to whether or not the novel food is safe.[125] Further, the principle is expressly affirmed in Articles 1

---

[122] C Noiville, *Du Bon Gouvernement des Risques* (Paris: Presses Universitaires de France, 2003) 86.

[123] Food Law Regulation, Art 7. See also *Avis No 48 du Conseil National de l'Alimentation sur la Préparation de l'Entrée en Vigueur au 1ᵉʳ Janvier 2005, de Certaines Dispositions du Règlement CE n° 178/2002 du Parlement Européen et du Conseil qui Concernent les Entreprises, sur le Rapport de F. Collart Dutilleul*, 9 November 2004.

[124] For the development of the precautionary principle generally see, eg C Joerges, 'Law, science and the management of risks to health at the national, European and international level—stories on baby dummies, mad cows and hormones in beef' (2001) 7 Columbia Journal of European Law 1; N Salmon, 'A European perspective on the precautionary principle, food safety and the free trade imperative of the WTO' (2002) 27 European Law Review 138; and N de Sadeleer, 'The precautionary principle in EC health and environmental law' (2006) 12 European Law Journal 139. For consideration by the European Court of Justice in the health context, see, eg Case T-13/99 *Pfizer Animal Health SA v Council* [2002] ECR II-3305; and Joined Cases T-74/00, T-76/00, T-83–85/00, T-132/00, T-137/00, and T-141/00 *Artegodan* [2002] ECR II-4945.

[125] F Collart Dutilleul, 'Le Consommateur Face au Risque Alimentaire. Pour une Mise en Oeuvre Raisonnable du Principe de Précaution' in *Liber Amicorum Jean Calais-Auloy* (Paris: Dalloz, 2004) 314.

and 4 of the Deliberate Release Directive, which goes beyond mere expression of principle to give body to precautionary measures through detailed, and technical, rules ('procéduralisation').[126] Importantly, there is clear admission of the need to accommodate uncertainty in risk management,[127] as may be evidenced by the institution of monitoring of the effects of releases of GMOs on human health and the environment and by the putting in place of a traceability regime.

The Food Law Regulation adopts a similar approach, but doubt remains as to its scope in relation to risk management. It applies, as a rule, to public authorities.[128] Equally, however, it does impose certain responsibilities on business operators.[129] In particular, the general traceability obligation introduced by Article 18 fully involves private operators in the management of risks, both proven and emerging, pursuing the logic of precaution and prevention; and this pattern has been followed in subsequent legislation. Thus, as indicated, Regulation 1830/2003 recites that '[t]raceability should also facilitate the implementation of risk management measures in accordance with the precautionary principle';[130] and the Food and Feed Regulation expressly includes within its authorization procedures the new principles (including the precautionary principle) introduced in the Deliberate Release Directive and the new framework for risk assessment laid down by the Food Law Regulation.[131]

This convergence of general food law and specific GMO legislation in giving effect to the precautionary principle can, nonetheless, be nuanced. For example, the decision whether or not to authorize the placing on the market of a GMO is based entirely on scientific evaluation of risk, an approach which fails to secure social acceptance, since consumers manifestly do not think that the only issue is whether or not GMOs are safe. And yet in food law other criteria are often employed as the basis for authorizations: in particular, the utility of the product is a criterion in the case of additives or processing aids.[132] But no evaluation of this sort is required in the case of GMOs or GM products.[133] Accordingly, 'acceptable risk' becomes essentially a scientific and technical notion, with little importance being attached to factors such as the benefit (if any) to consumers or the social and economic impact on agriculture.[134] Besides, even if a criterion of utility was added to that of safety, it is important that consumers retain the

---

[126] Boy (n 97 above) 153.   [127] Romi (with Bossis and Rousseaux) (n 23 above) 39.
[128] See, eg Food Law Regulation, Arts 7, 10, 14(8), and 15(5); and see also *Avis No 48 du Conseil National de l'Alimentation sur la Préparation de l'Entrée en Vigueur au 1er Janvier 2005, de Certaines Dispositions du Règlement CE n°178/2002 du Parlement Européen et du Conseil qui Concernent les Entreprises*, sur le Rapport de F. Collart Dutilleul, 9 November 2004, 10.
[129] Food Law Regulation, Arts 17–20.   [130] Regulation 1830/2003, Preamble (3).
[131] Food and Feed Regulation, Preamble (9).
[132] See, eg in the case of additives, Regulation (EC) 1333/2008 of the European Parliament and of the Council on food additives [2008] OJ L354/16, Art 6.
[133] M-A Hermitte, *Relire l'Ordre Juridique à la Lumière du Principe de Précaution* (Paris: Dalloz, 2007) 1518.
[134] M Lee, *EU Environmental Law: Challenges, Change and Decision-making* (Oxford: Hart Publishing, 2005) 250–2; and Noiville (n 122 above) 86.

ability to guard against risks, which must be regarded as an ongoing source of controversy, through a system of product labelling.

4.3.2 Information for consumers and civil society

In its resolution in 2000 on the precautionary principle, the Council considered 'that civil society must be involved and special attention must be paid to consulting all interested parties as early as possible'.[135] This affirmation, linking access to information with a precautionary approach to the enactment of legislation, demonstrates the importance of ensuring freedom of choice for consumers where science is uncertain. According to Hermitte, the need to involve civil society may be explained by the fact that, in this context, it is difficult for the executive simply to follow its own preferences, on the basis that, at the end of the day, any risk is collective.[136] The principle that consumers should be informed of risks is enshrined in the Food Law Regulation;[137] and it forms part of a broader principle of transparency, which also encompasses consultation of civil society on matters of food law.[138] The principle of transparency may likewise be found in the Deliberate Release Directive. As has been seen, in the case of deliberate releases into the environment,[139] Article 9 provides for both consultation of and information to the public on the proposed deliberate release, while information is likewise to be made available where GMOs are placed on the market.[140] In particular, this allows citizens access to data, so that they can make their own evaluation of risks and also find out the distribution of field trials in their area. Importantly, the location of such field trials may not be kept confidential;[141] and the right to know this information has been defended by the European Court of Justice in *Commune de Sausheim v Azelvandre*.[142] Moreover, as has again been seen, Regulation 1830/2003 guarantees reliable information to consumers by establishing a comprehensive traceability and labelling regime.

Although the right to information may be enshrined in the legislation, there remains debate as to what exactly the content of that information should be, with the focus of this debate being on GM food. Indeed, in the case of not just GMOs, but also animal cloning, or other novel production processes or

---

[135] Presidency Conclusions: Nice European Council, 7–9 December 2000, Annex III, para 15 (available at <http://www.consilium.europa.eu/ueDocs/cms_Data/docs/pressData/en/ec/00400-r1.%20ann.en0.htm>, last accessed 29 June 2009).
[136] Hermitte (n 133 above) 1518; and see also J Black, 'Regulation as facilitation: negotiating the genetic revolution' (1998) 61 Modern Law Review 621.
[137] For a definition of 'risk communication', see Food Law Regulation, Art 3(13).
[138] Ibid Art 9: '[t]here shall be open and transparent public consultation, directly or through representative bodies, during the preparation, evaluation and revision of food law, except where the urgency of the matter does not allow it'.
[139] Ie under the Deliberate Release Directive, Part B, for any other purpose than placing on the market.
[140] Ie under ibid Part C: Art 24.   [141] Ibid Art 25(4).
[142] Case C-552/07 (ECJ 17 February 2009); and see further Chapter 1.

methods, there is the problem of dealing with a form of food that is different in kind. Yet, since 'you are what you eat', there is good reason to ban the consumption of 'foreign bodies', such as GMOs.[143] Accordingly, the mere provision of objective information on the risks attached to their consumption is not sufficient.[144] Rather (to adopt the language of sociologists), it is necessary for the consumer to be confident that traditional products, which have constructed the identity of their eater, should not be sullied by novel foods, which are still not understood. In other words, consumers must be informed of every last trace of GM cross-contamination, however small, so as to be able to regain mastery of their choice of food. But will such interests be respected? At an individual level, perhaps yes, since niche areas of consumption can be preserved free of GMOs at economically acceptable cost. Yet at a general level, the question becomes more thorny, since under Community law as it now stands the general public cannot force through a ban of GM food on their plates. Nevertheless, rather surprisingly, GMOs are not yet on the shelves of supermarkets in some countries of the European Union;[145] and it is a fact that economic operators are keen to respond to the demands of their customers by requiring from their suppliers both food and food ingredients that are non-GM. Further, the general emphasis accorded to environmental protection finds echo in use of the safeguard clause, with, as already indicated, considerable success.[146]

5. Conclusion

The Community regulation of GMOs allows evaluation, tracing, and monitoring of GM production from 'farm to fork'. Originally a part of environmental policy, the legislation has become suffused with consumer considerations, this being particularly prevalent where GMOs are to enter the food chain. Moreover, such intersection between the protection of the environment and the protection of public health is a characteristic found more broadly under the new European food law. The market actor—the consumer—must be informed of the history of the product; and this would seem to give him a measure of choice in respect of both his health and economic interest. However, if his choice is to be preserved, there must be genuine ability to select food that is GM *and* non-GM.

---

[143] For such terminology, see, eg C Fischler, *L'Homnivore* (Paris: Odile Jacob, 1990).
[144] See, eg Chalmers (n 110 above) 546.
[145] See, eg in the case of the United Kingdom, Agriculture and Environment Biotechnology Commission (AEBC), *GM Crops? Coexistence and Liability* (London: AEBC, 2003) 30; and see Chapters 6 and 7.
[146] See, eg the decision of the Environment Ministers on 2 March 2009 to retain the bans on MON810 and T25: 2928th Council Meeting: Environment, Brussels, 2 March 2009 (Presse 53); and see also, eg *Agra Europe Weekly* No 2351, 6 March 2009, EP/1.

5
Multi-level Governance of Genetically Modified Organisms in the European Union: Ambiguity and Hierarchy

Maria Lee

1. Introduction

It is increasingly difficult to identify clear lines dividing national from European Union (EU) competences. The EU has always been characterized by its multi-level governance, in which authority is not allocated in a straightforward way to either the EU or the Member States. Instead, authority is dispersed along a spectrum of more or less national/centralized control. That is certainly the case in respect of the regulation of genetically modified organisms (GMOs) in the EU, where legislators have made an apparently self-conscious effort to share authority and avoid hierarchy. The regulation of GMOs provides plenty of opportunities for different levels of governance to intervene in decision-making, and the legislative framework generally avoids recourse to single levels of authority, as well as building in opportunities for early discussion. For example, the centralization implied by the role of the European Food Safety Authority (EFSA) in assessing risk is mediated by obligations of networking and consultation with national authorities, and special obligations of transparency in respect of disagreement. This fits well with multi-level governance, whose heritage of governance or even 'new governance'[1] takes seriously the complexity of governing in a society with widely dispersed resources of information or knowledge and (especially pertinent for GMOs) of authority and legitimacy. Multi-level governance calls on institutions that enable collaboration, learning,

---

[1] There is now a large amount of literature on 'new' governance in a range of fields, including the environment: see, eg the publications on the CONNEX network (available at <http://www.connex-network.org/>, last accessed 1 July 2009); G de Búrca and J Scott (eds), *Law and New Governance in the EU and the US* (Oxford: Hart Publishing, 2006); and G de Búrca and Joanne Scott (eds), 'Narrowing the gap? Law and new approaches to governance in the European Union' (2007) 13(3) Columbia Journal of European Law, Special Issue.

and discussion in order to escape the limited solutions or deadlocks that occur when using harder-edged techniques of law and government.[2]

The classic mechanisms of multi-level governance are present in the regulatory framework for GMOs. However, multi-level governance is not straightforward in respect of GMOs. GMOs are politically highly visible, and the different actors are so divided that collaboration and compromise are frequently not proving possible. The dispersal of authority associated with multi-level governance is generally unable to provide a decision on applications for authorization of GMOs. Crucially, however, there is a perception that, after so many years of inoperative law, a decision cannot wait. This means that an authoritative decision-maker is sought. But as the literature on governance might have warned us, the single site of uncontested authority is also elusive: the shadow of hierarchy only falls over soft governance techniques if the traditional mechanisms of 'government' are willing and able to take action.[3] And, in the case of GMOs, this reversion to hierarchy is not straightforward. Decisions are slow and contested, and disagreement rumbles on after decisions are supposed to have been settled. There is a danger that the language of multi-level governance is simply euphemistic in this context of significant conflict over the exercise of power.

The difficulties of sharing authority in this area force us to face squarely the dilemma of authority. The compromise in the legislation, of sharing and mediating authority, is probably not sustainable, and the resulting battle for authority has the potential to spill over into broader debates about the place of GMOs in EU agriculture, and even indeed about the legitimacy of the EU. This is not the place for detail, but rather than the determined centralization of authority described below, an alternative to the inability to share authority would be to permit Member States to exercise their authority in a way that allows them to respond in diverse ways to democratically expressed concerns about GMOs.

2. The Regulatory Context

The broad regulatory context for GMOs is by now relatively familiar. Since 1990, the marketing or deliberate release of GMOs has required authorization.[4]

---

[2] Governance also generally implies the involvement of civil society beyond 'government', although the focus here is particularly on the legal arrangement of EU and national contributions to decisions. Multi-level governance should also encourage consideration of the role and influence of sub-national regions and international bodies such as the World Trade Organization (WTO).

[3] See, eg the discussion of hierarchy in CF Sabel and J Zeitlin, 'Learning from Difference: The New Architecture of Experimentalist Governance in the European Union' *European Governance Papers*, C-97-02.

[4] Council Directive (EEC) 90/220 on the deliberate release into the environment of genetically modified organisms [1990] OJ L117/15.

The Regulatory Context 103

This 1990 legislation quickly came to be seen as deeply inadequate, and eventually fell spectacularly apart during the infamous moratorium on new authorizations. Following declarations from 12 (of then 15) Member States that they were opposed to further authorizations of GMOs,[5] the European Commission stopped putting GMOs through the authorization process. Accordingly, between 1998 and 2004 no applications reached the end of the decision-making process, and a number of Member States introduced measures barring national market access to GMOs that had already been authorized.

From 1998, the efforts of the EU institutions and the Member States were directed towards the negotiation of new legislation that could accommodate national concerns, whilst allowing the commercialization of GMOs. The main results of that process are Directive (EC) 2001/18 of the European Parliament and of the Council on the deliberate release into the environment of genetically modified organisms ('Deliberate Release Directive') and Regulation (EC) 1829/2003 of the European Parliament and of the Council on genetically modified food and feed ('Food and Feed Regulation').[6] Only after the deadline for implementation of the new rules did the European Commission begin to step up its formal pressure on Member States.[7] Others, such as the affected commercial entities, also hesitated, apparently sharing a sense that formal action would be counterproductive.[8]

---

[5] Two declarations were made by different groups of Member States in the 2194th Council Meeting of 24–25 June 1999. These Member States had the majority they needed in the Council to deny authorization on applications in respect of individual GMOs: Council Decision (EC) 99/468 of 28 June 1999 laying down the procedures for the exercise of implementing powers conferred on the Commission [1999] OJ L184/23, allowing European Commission proposals on authorization to be rejected by qualified majority voting, entered into force on 18 July following the declarations. Ireland, Portugal, and the United Kingdom did not join either declaration.

[6] Respectively, [2001] OJ L106/1 and [2003] OJ L268/1. There is now a large amount of literature on this legislation: see, eg E Brosset, 'The prior authorisation procedure adopted for the deliberate release into the environment of genetically modified organisms: the complexities of balancing Community and national competences' (2004) 10 European Law Journal 555; T Christoforou, 'The regulation of genetically modified organisms in the European Union: the interplay of science, law and politics' (2004) 41 Common Market Law Review 637; and S Poli, 'The overhaul of the European legislation on GMOs, genetically modified food and feed: mission accomplished. What now?' (2004) 11 Maastricht Journal of European and Comparative Law 13.

[7] By the time of the mid-term review of the *Life Sciences Strategy* (European Commission, *Commission Staff Working Document—Document Accompanying the Communication on the Mid Term Review of the Strategy on Life Sciences and Biotechnology* SEC(2007) 441), two cases were pending: SEC(2007) 441, 44. The European Commission had earlier brought a partially successful action against France in respect of failure to implement the legislation: Case C-296/01 *Commission v France* [2003] ECR I-13909.

[8] In 2001, Monsanto challenged Italy's safeguard measures (Case C-236/01 *Monsanto Agricoltura Italia SpA v Presidenza del Consiglio dei Ministri* [2003] ECR I-8105). Pioneer Hi-Bred has challenged the European Commission's failure to submit its application in respect of insect-resistant GM maize 1507 to Committee under the Deliberate Release Directive (see Case-T139/07 *Pioneer Hi-Bred v Commission* [2007] OJ C155/28). Pioneer Hi-Bred brought its challenge in May 2007, having applied for authorization to the Spanish competent authority in 2001. The European Food Safety Authority (EFSA) on 19 January 2005 concluded that its GM maize is as safe as conventional maize: see *Opinion of the Scientific Panel on Genetically Modified Organisms on a*

The new legislation strengthens the risk assessment of GMOs. So, for example, the meaning of environmental risk assessment is clarified and broadened: 'the evaluation of risks to human health and the environment, whether direct or indirect, immediate or delayed, which the deliberate release or the placing on the market of GMOs may pose', including 'cumulative long-term effects', defined as 'the accumulated effects of consents on human health and the environment, including inter alia flora and fauna, soil fertility, soil degradation of organic material, the feed/food chain, biological diversity, animal health and resistance problems in relation to antibiotics'.[9] This, in turn, contributes to the clarification of which 'adverse effects' may be addressed, the ambiguities on which had previously included the relevance of insecticide resistance and antibiotic resistance, now both clearly within the legislation. The processes and criteria for risk assessment, whilst still controversial, are also more settled in the new legislation.

Most interesting, albeit still ambiguous, is the new legislative framework's effort to respond to concerns as to whether risk assessment really captures what is interesting about this technology. Protection of human health and the environment are enormously important and very complex, but so may be the way agricultural biotechnology distributes risk, benefit, and power, globally and locally, and the profound uncertainties about its physical and socio-economic impacts. The new regulation of GMOs allows for the importance of capturing factors beyond risk, apparently accepting that the authorization of GMOs is not simply a technical decision. There are opportunities for public participation,[10] and for the consultation of an ethical committee;[11] there is even reference (somewhat oblique) to the socio-economic impacts of GMOs.[12] And in respect of food and feed GMOs, there is an explicit recognition that 'scientific risk assessment alone' may not provide all the necessary information, and that a decision can be based on 'other legitimate factors'.[13] The recognition that decisions on GMOs are complex and multi-faceted is enormously significant. It is a response to decades of empirical and theoretical work in the social sciences, absorbing very important insights on the complexity of decision-making in areas of high technological content. But whilst the reference to 'other legitimate

---

Request from the Commission Related to the Notification (Reference C/ES/01/01) for the Placing on the Market of Insect-tolerant Genetically Modified Maize 1507 for Import, Feed and Industrial Processing and Cultivation, under Part C of Directive 2001/18/EC from Pioneer Hi-Bred International/Mycogen Seeds.

[9] Deliberate Release Directive, Art 2(8) and Annex II.
[10] Ibid Art 24; and Food and Feed Regulation, Arts 6(7)(food) and 18(7) (feed).
[11] Deliberate Release Directive, Art 29; Food and Feed Regulation, Art 33.
[12] The European Commission is to include, in its periodic reports on the legislation, a chapter on the socioeconomic advantages and disadvantages of each category of GMO, 'which will take due account of the interests of farmers and consumers': Deliberate Release Directive, Preamble (62) and Art 31(7).
[13] Food and Feed Regulation, Preamble (32) and Arts 7(1) (food) and 19(1) (feed).

factors' is more than mere rhetoric, it is likely to be extremely difficult to make this formula meaningful. The legal and policy context of decision-making is deeply entrenched in a technical risk framework. The legal and political incentives to justify all decisions on a scientific basis are not likely to be overcome by simply adding on 'other legitimate factors'.[14]

This move beyond risk is an important aspect of multi-level governance, specifically of the effort to unblock the relationship between centre and Member State, and between different Member States. Without wishing to oversimplify, it emphasizes the difficulties of any straightforward centralization of authority, because political concerns about GMOs tend to be most intense at the national level. It is more difficult for 'the public' to be heard at the European level, but it is public involvement that most readily captures the non-scientific aspects of decisions, and is most alert to the uncertainty that pervades regulation of GMOs. Equally importantly, the response to GMOs varies considerably around the Member States. Not only are agriculture and food production differently organized, with different economic structures, ecologies, and geographies, but Member States may also have competing scientific information and even competing ideas of what counts as science. The new legislation attempts to share authority between the centre and the Member State, rather than allocate it to one or the other.

3. The Decision-making Process: the Dispersal of Authority

The Deliberate Release Directive imposes an obligation to seek authorization for the marketing or deliberate release[15] of any GMOs. It was almost immediately amended by the Food and Feed Regulation, which imposes an obligation to seek authorization for any GMO destined for use in food or feed. The regulatory framework thus varies in its detail according to whether we are concerned with a GMO used in food or (animal) feed, or a non-food GMO, such as cotton or flowers. The two key pieces of legislation overlap, so that food or feed that is also a GMO (eg a tomato rather than tomato ketchup) is also subject to environmental risk assessment provisions of the Deliberate Release Directive. There is still considerable lack of clarity over the correct relationship between the legislation in respect of seeds, but the pattern of the legislation does suggest a role for

---

[14] The ambiguity of the move beyond 'risk' is discussed in detail in M Lee, *EU Regulation of GMOs: Law and Decision Making for a New Technology* (Cheltenham: Edward Elgar, 2008).

[15] Meaning 'any intentional introduction into the environment of a GMO or a combination of GMOs for which no specific containment measures are used to limit their contact with and to provide a high level of safety for the general population and the environment': Art 2(3). A separate process applies to deliberate releases for purposes other than marketing (experimental releases) under Part B. Authorization is granted by the single Member State (other Member States are informed) and release is limited to that Member State. Experimental authorization does not, however, imply free movement. See further Chapter 4.

the Food and Feed Regulation.[16] Heavily processed food or feed is addressed under the Food and Feed Regulation alone, whilst GMOs such as flowers or cotton are authorized under the Deliberate Release Directive only.

For non-food and non-feed GMOs, the applicant notifies the competent authority of the Member State where the GMO is to be placed on the market for the first time.[17] The notification must contain a range of information, including the environmental risk assessment carried out by the applicant, and a plan for monitoring the GMO following its release into the environment.[18] The notification is forwarded by the competent authority to the European Commission and thence to the competent authorities of the other Member States. This is basically the same process as under Council Directive (EEC) 90/220 on the deliberate release into the environment of genetically modified organisms.[19] However, in an effort to use governance mechanisms to stave off disagreement, information is shared earlier in the process, providing an opportunity for discussion before positions are entrenched.[20] The summary of the notification dossier is also made available to the public when received by the European Commission.[21]

The initial competent authority examines the notification for compliance with the Deliberate Release Directive, and prepares an assessment report, which is circulated to other Member States via the European Commission.[22] The assessment report indicates whether the GMO should (and under what conditions) or should not be placed on the market.[23] In the latter case, the application is rejected, with reasons,[24] but the applicant can make the same application to any other competent authority. In the former case, in the absence of a reasoned objection from another Member State or the European Commission, the competent authority gives its consent, valid throughout the EU.[25] A period is built into the legislation for Member States and the European Commission to 'discuss any outstanding issues with the aim of arriving at an agreement'.[26]

Clarifying and strengthening the scope of the risk assessment under the Deliberate Release Directive (as discussed above) was one way of trying to enhance the acceptability of other Member States' risk assessments. These efforts to build in time for discussion and agreement are an additional legislative

---

[16] If nothing else, the reference to 'seeds or other plant propagating materials' in Art 6(3) suggests that authorization can be sought for seeds under the Food and Feed Regulation. The idea had been that the Regulation would provide a 'one-stop shop' process for all uses of a GMO, but applicants seem to be cautious, applying separately for different uses (eg feed use first, then cultivation).   [17] Deliberate Release Directive, Art 13(1).
[18] Ibid Art 13.   [19] [1990] OJ L117/15.
[20] Under the 1990 legislation, Member States were just informed of the original Member State's decision on the application: Directive 90/220, Arts 12 and 13; under the Deliberate Release Directive, Art 13(1), the summary of the applicant's dossier is forwarded 'immediately': see Brosset (n 6 above) for discussion.   [21] Deliberate Release Directive, Art 24(1).
[22] Ibid Art 14(1) and (2).   [23] Ibid Art 14(3).   [24] Ibid Art 15(2).
[25] Ibid Art 15(3).   [26] Ibid Art 15(1).

response to the same dilemma. In the event of continued disagreement, however, we turn from mutual recognition of national risk assessments to a more laborious 'Community procedure'. Notwithstanding the opportunities for discussion and the extra detail on risk assessment in the post-moratorium legislation, disagreement between Member States, and, hence, the 'Community procedure', continues to be the norm for GMOs. At this stage the process is very similar to the process under the Food and Feed Regulation. The European Commission calls on the expertise of the EFSA, specifically its GMO Panel,[27] and then turns to political decision-making at EU level through comitology.

The Food and Feed Regulation on its face considerably reduces the role of the national competent authorities. The initial application is sent to a national competent authority, but that authority simply passes the application to the EFSA. The EFSA then passes the application to the European Commission and thence to other Member States. The EFSA is 'an independent scientific point of reference in risk assessment',[28] and its primary tool and responsibility is risk assessment, with risk management in principle for the political institutions.[29]

This looks like a simple centralization of scientific authority. Members of the different bodies of the EFSA are not appointed by the Member States, and do not provide for representation of national interests, making this a firmly central body. But when it is drawing up an opinion under the Food and Feed Regulation, the EFSA *may* ask a national competent body to carry out the risk assessments.[30] In the case of food and feed GMOs that also fall under the Deliberate Release Directive,[31] an environmental risk assessment as in that Directive is a mandatory part of the application, and the EFSA is obliged to *consult* all national competent authorities.[32] If the application is for authorization of 'seeds or other plant propagating material', the EFSA '*shall* ask a national competent authority to *carry out*' the environmental risk assessment.[33] The role of national authorities provides an important opportunity for the incorporation of national perspectives on risk assessment.[34] More generally, supplementing the explicit obligations on GM food and feed, and so applying 'Community

---

[27] Ibid Art 28 sets out the obligation to consult the 'relevant scientific committee'.
[28] Regulation (EC) 178/2002 laying down the general principles and requirements of food law, establishing the European Food Safety Authority and laying down procedures in matters of food safety ('Food Law Regulation') [2002] OJ L31/1, Preamble (34).
[29] The Food Law Regulation formalizes the division of 'risk analysis' into separate stages of 'risk assessment', 'risk management', and 'risk communication'. The EFSA does have responsibilities in risk communication. For discussion, see D Chalmers, '"Food for thought": reconciling European risks and traditional ways of life' (2003) 66 Modern Law Review 532.
[30] Food and Feed Regulation, Art 6(3)(b) and (c).
[31] GMOs, or food containing or consisting of GMOs.
[32] Food and Feed Regulation, Art 6(4).   [33] Ibid Art 6(3)(c) (emphasis added).
[34] But note that 'EFSA experiences difficulties to find [competent authorities] willing to carry out the initial risk assessment': European Commission, *Report on the Implementation of Regulation 1829/2003 on Genetically Modified Food and Feed* COM(2006)626, 10.

procedure' under both the Deliberate Release Directive and the Food and Feed Regulation, the EFSA is required by its own constitution to promote 'the European networking of organisations' operating in its area: to facilitate a scientific co-operation framework by the co-ordination of activities, the exchange of information, the development and implementation of joint projects, the exchange of expertise, and best practices.[35]

'Networking' of risk assessors through agency structures invites national risk regulators and national perspectives on risk into the EU system. This attempts to compromise between the easier and more coherent decision-making that goes with increased centralization, and the danger that too much centralization damages the legitimacy of decisions, blurring the boundaries between 'central' and 'national' institutions.[36] In the event of disagreement between national and central risk assessors, networks provide an opportunity for that disagreement to be addressed. More concretely, the EFSA is required to 'exercise vigilance' in respect of 'any potential source of divergence' between its scientific opinions and those of other bodies.[37] There is an obligation to contact the body in question to ensure the sharing of scientific information in order to identify 'potentially contentious scientific issues', followed by an obligation to co-operate in order either to resolve the divergence, or to prepare and publish a joint document 'clarifying the contentious scientific issues and identifying the relevant uncertainties in the data'. The opportunity for consensus is enhanced, and remaining disagreement is brought into the open: if the diverging opinion comes from a Community body, the joint document is 'presented to the Commission'; if it is a national body, it is simply prepared and made public. The EFSA does not arbitrate between different approaches, and there is no hierarchy of scientific information in the legislation, for example according to whether it is nationally or centrally produced. But persistent disagreement and associated uncertainty should be transparent in the political decision-making process. These general obligations of transparency on divergent opinions are reinforced by the obligation on the EFSA in the Food and Feed Regulation to state the reasons for its opinion and the information on which the opinion is based, including the responses of consulted competent authorities.[38]

The European Commission seems to have hoped that the existence of the EFSA would mean that Member States would no longer produce their own opinions on GMOs, entirely sidestepping this issue of divergent opinions.[39]

---

[35] Food Law Regulation, Art 36.

[36] See, for discussion, R Dehousse, 'Regulation by networks in the European Community: the role of European Agencies' (1997) 4 Journal of European Public Policy 246; E Chiti, 'The emergence of a Community administration: the case of European Agencies' (2000) 37 Common Market Law Review 309; and G Majone, 'The credibility crisis of Community regulation' (2000) 38 Journal of Common Market Studies 273.   [37] Food Law Regulation, Art 30.

[38] Food and Feed Regulation, Art 6(6).

[39] D Chalmers, 'Risk, anxiety and the European mediation of the politics of life' (2005) 30 European Law Review 649, 654.

That did not happen, and the European Commission is now clearly anxious to improve the liaison between the EFSA and national competent authorities. In similar terms to the legislation, it has 'invited' the EFSA to 'liaise more fully with national scientific bodies, with a view to resolving possible diverging scientific opinions with Member States'. This places hope in the informal techniques of governance that have been productive in less contentious areas. Anticipating continued disagreement, where the 'divergence' is not capable of resolution, the EFSA is to 'provide more detailed justification, in its opinions on individual applications, for not accepting scientific objections raised by the national competent authorities'.[40] European Commission decisions on GMOs have to date followed EFSA opinions very closely. This demand for greater justification of scientific opinions suggests that the European Commission is very aware of the contentiousness of these decisions, and arguably also of how elusive uncontroversial facts are in this area.

The EFSA's opinion is sent to the European Commission and the Member States for political decision-making. The opinion is published and there is an opportunity for the public to 'make comments'.[41] This marks the beginning of the comitology stage of decision-making, which applies to all food and feed applications, and to those deliberate release applications that are subject to the 'Community procedure' (so, to date, all applications). Comitology is formally a mechanism through which Member States can supervise the European Commission's exercise of implementing powers. In the case of the Deliberate Release Directive, the decision is formally that of the Member State to whom the application was initially sent, but if authorization is sanctioned at EU level, that Member State has no further discretion (in the absence of new information),[42] and so cannot choose to respond to the sensitivity of the discussions in Council by declining authorization. Under the Food and Feed Regulation, the European Commission formally takes the decision, with the 'assistance' of the regulatory committee.[43]

The European Commission submits a draft of measures to a regulatory committee, consisting of Member State representatives and chaired by the European Commission.[44] The committee delivers its decision by qualified majority, and if the committee gives a positive opinion, the European Commission adopts the decision. If not, the decision goes to Council. In the vast majority of cases in EU law, comitology committees agree with the European Commission, and so the decision never reaches the Council.

---

[40] European Commission, *Report on the Implementation of Regulation 1829/2003 on Genetically Modified Food and Feed* COM(2006)626, 11. For detailed consideration of EFSA Opinions and their handling of disagreement, see Chalmers (n 39 above).
[41] Food and Feed Regulation, Art 6(7).
[42] Case C-6/99 *Association Greenpeace France v Ministère de l'Agriculture et de la Pêche* [2000] ECR I-1651 (discussing the same wording under Directive 90/220).
[43] Food and Feed Regulation, Art 35.   [44] Ibid Arts 7(1) and 35; and Decision 99/468.

GMOs are, however, an anomaly in this respect, and every decision so far has had to go to Council. Council consideration slows the process down, but at least ensures that authorization of GMOs receives high-level political consideration. If the Council fails to act (including a failure to reach a qualified majority decision in either direction) the European Commission 'shall' adopt its decision.[45] So far (end of 2008), the Council has not managed to reach a qualified majority in either direction in respect of any application for authorization of a GMO.

Comitology is a classic institution of multi-level governance in the EU, characterized by Joanne Scott and David Trubek as 'new old governance'.[46] GMOs have been something of a *cause célèbre* for comitology. Under the 'old' comitology decision, the Council could only reject a European Commission proposal if it acted unanimously.[47] That was obviously unlikely given that at least one Member State had wished to approve and at least one had wished to deny the application in order to end up in the 'Community procedure' at all. Notoriously in *Association Greenpeace France v Ministère de l'Agriculture et de la Pêche*, the European Commission had adopted its decision on a GMO in spite of angry objections from a number of Member States, a European Parliament resolution against authorization, and the positive approval of only one Member State in Council.[48] The ability of the Council to reject the European Commission's proposal by qualified majority under the current legislation obviously removes this extreme scenario. It demands more negotiation between the centre and the Member States, and possibly a more subtle approach to national perspectives. However, Member State disagreement has so far meant that the Council has been unable to reach a qualified majority either for or against any European Commission proposal on authorization of a GMO. This leaves considerable power in the hands of the European Commission, which is the decision-maker of last resort in a process that is supposed to be unusual, but is now the norm in this area. When there is conflict between the Member States, the default position is crucial, and here decisions revert to the centre (the European Commission), rather than back to the national level.

Even if the arrangement of national and central responsibilities in the regulation of GMOs satisfied the Member States and the central institutions that their needs were properly heard, one of the oddities of multi-level governance is

---

[45] Decision 99/468, Art 5.
[46] J Scott and D Trubek, 'Mind the gap: law and new approaches to governance in the European Union' (2002) 8 European Law Journal 1, 2.
[47] Council Decision (EC) 87/373 of 13 July 1987 laying down the procedures for the exercise of implementing powers conferred on the Commission [1987] OJ L197/33.
[48] Case C-6/99 [2000] ECR I-1651. See the discussion in T Hervey, 'Regulation of genetically modified products in a multi-level system of governance: science or citizens?' (2001) 10 Review of European Community and International Environmental Law 321.

its closed nature. 'New governance' is generally considered to be more open and participative than 'old government'. But neither comitology nor EFSA committees are transparent to national publics, elite views are better represented than popular views, and peers can exercise more pressure than publics. Even with the best will of national governments, it is striking how difficult it would be for the Member States to represent the diverse views of European publics through the various committees or in Council.

4. Multi-level Governance and Reversion to Hierarchy

The authorization of GMOs (especially food and feed) is in one respect heavily centralized. The EFSA's role in the authorization process, for example, is a visible centralization of authority; as is the decision-making (albeit *in extremis*) of the European Commission in the years following the end of the moratorium. However, these are not simple allocations of power. Networking is an intrinsic part of the EFSA's legal constitution, bringing the Member States into the process and implying the openness of scientific views to critique from national technical experts. And whilst the final political decision is taken by the most 'European' of EU institutions, all Member States are involved through comitology. The legislation makes clear efforts to compromise between national and central authority. It attempts to mediate between national and central authority, providing both national and EU 'levels' of governance with opportunities to persuade, inform, and educate, rather than command. Multi-level governance is an appropriate description.

But these mechanisms of multi-level governance are not leading to an outbreak of peace. The opportunities for discussion built into the Deliberate Release Directive have not prevented the routine use of the more complex 'Community procedure'. And EFSA decisions are surprisingly self-contained: it seems not to be willing to engage with competing science from the Member States.[49] Even the reiteration by the European Commission of the need to explain divergent opinions (discussed above), whilst providing an opportunity for the sharing of scientific authority, arguably enhances the Europeanization of scientific authority where there is no consensus, since the EFSA then provides the European Commission with more reasons for preferring EFSA opinions over national approaches. Nor does the provision of a scientific 'voice' to Member States seem to render acceptable those decisions with which any particular Member State and its citizens disagree, convincing too few Member States for a qualified majority to be found in favour of authorization. And finally, the decisions of the European Commission,

---

[49] See Chalmers (n 39 above).

notwithstanding a national voice in comitology, are highly contentious, meeting with a stream of criticism, but importantly also with little commercialization of GMOs in the EU[50] and a considerable number of national measures limiting access of GMOs to national markets.[51]

National involvement in authorization decisions is weak because the Member States disagree too profoundly to act collectively. The perception (especially on the part of the European Commission) that decisions *must* be taken, that the law cannot any longer be left in abeyance, compounds the difficulties of collaboration and co-operation. In 2007 Pioneer Hi-Bred brought an action before the Court of First Instance against the European Commission's failure to put to comitology an application for authorization of cultivation of its maize 1507.[52] This legal action suggests that non-decisions and failure to progress applications will no longer be tolerated by the industry. It also suggests, however, that the European Commission does not just plough on regardless of the sensitivities. In this case, notwithstanding a positive opinion from the EFSA, the European Commission simply failed to draw up a draft decision for consideration in comitology. The maize has been authorized for other uses by European Commission action,[53] but apparently cultivation was a step too far. No GMO has yet been authorized for cultivation under the new legislation.[54] The European Commission has even held off authorizing cultivation of a GM potato that was put to comitology, but where, consistently with other applications, the Council was unable to reach a qualified majority decision in either direction.[55]

---

[50] So, eg, the very first authorization at the end of the moratorium was for Syngenta's sweetcorn in 2004, but it was not commercialized because of consumer pressure. European Commission, *Report on the Implementation of Regulation 1829/2003 on Genetically Modified Food and Feed* COM (2006)626 states that 'few food products labelled as genetically modified are at the present time on the Community market', because of 'factors that are not related to the legislative framework, such as consumer demand and the policies of food producers and retailers' (p 13). GM fruit and vegetables are simply not being marketed in the EU at present, and maize is still the only GM crop being commercially cultivated.

[51] So, eg in 2008, France notified its application of the safeguard clause to MON810 maize on 11 February 2008. See also the safeguard clauses discussed below (nn 63 and n 65), and the Austrian and Polish notifications under Art 95(5) of the EC Treaty (nn 69 and 70). Member States are also using coexistence measures to restrict access to national markets (see n 86 below).

[52] Case T-139/07 *Pioneer Hi-Bred v Commission* [2007] OJ C155/28. We await a decision.

[53] Commission Decision (EC) 2006/197 of 3 March 2006 authorizing the placing on the market of food containing, consisting of, or produced from genetically modified maize line 1507 (DAS-Ø15Ø7-1) pursuant to Regulation (EC) 1829/2003 of the European Parliament and of the Council [2006] OJ L70/82.

[54] A number of GMOs were authorized for cultivation under the 1990 legislation, and need re-approval.

[55] The European Commission drafted a proposal for the authorization of cultivation of a GM potato, on which the Council failed to reach a qualified majority on 16 July 2007: see *Draft Commission Decision of [. . .] concerning the placing on the market, in accordance with Directive 2001/18/EC of the European Parliament and of the Council, of a potato product (Solanum tuberosum L. line EH92-527-1) genetically modified for enhanced content of the amylopectin component of starch*. European Commission delay in approving meant that the variety of potato could not be grown in

The institutional framework within which the Council will make its decisions on GMOs is changing. A new comitology decision has introduced the 'regulatory committee with scrutiny', which allows the European Parliament, like the Council, to reject European Commission comitology proposals.[56] It applies, however, in respect of 'measures of general scope designed to amend non-essential elements of a basic instrument' and, although it reminds us that understanding of European Parliament exclusion continues to evolve, it is not likely to apply to the authorization of GMOs.[57] More significantly, the Lisbon Treaty introduces a new 'double majority' approach to qualified majority decision-making, which is supposed to make the majorities that are so elusive in the respect of authorization of GMOs easier to find in the enlarged EU. Double majorities require at least 55 per cent of Member States, representing at least 65 per cent of the EU's population. A blocking minority needs to be formed of enough Member States to make up 35 per cent, plus one more Member State—preventing a small number of large Member States dominating procedures.[58] Council is more likely to agree with the European Commission now, albeit by overriding rather than convincing minorities.[59] Given that many Member States are already unwilling to accept either force of numbers or force of argument in the Community processes, this is not likely to resolve the debate over GMOs, in the short term at least.

For now, collaboration and opportunities for debate and sharing authority are not leading to swift (or indeed any) decisions. But because there is a perception that final decisions *have* to be taken, recourse is made to the hierarchy—in this case the European Commission. In turn, the European Commission relies very heavily on EFSA opinions, rather than competing scientific information or indeed 'other legitimate factors'. But as the new governance literature has warned us, this reversion to hierarchy is problematic. The legislation turned to collaborative governance because an uncontentious single site of authority on GMOs was elusive; the European Commission's decisions remain slow, contentious, and poorly implemented.

2008. There is no sign that approval is on the horizon. Indeed, there seems to be disagreement in the College of Commissioners: see <http://www.gmo-compass.org/eng/news/349.docu.html>, last accessed 1 July 2009.

[56] Council Decision (EC) 2006/512 of 17 July 2006 laying down the procedures for the exercise of implementing powers conferred on the Commission [2006] OJ L200/11.

[57] It does apply to other aspects of the GMO legislation (for example, adapting the labelling thresholds). Note that the European Parliament Committee on Environment, Public Health and Food Safety would have applied the regulatory procedure with scrutiny to authorization of GM food: see Report, 14 July 2007. This was not picked up in the text adopted by the European Parliament on 29 November 2007.

[58] New Art 16 Treaty on European Union. Transitional measures, introduced at the insistence of Poland, mean that for a period a 'significant' opposition to a measure falling short of this (one-third of Member States and 25% of the population) will lead to efforts to seek a compromise solution. [59] It is possible that GMOs will initially be deemed too sensitive to move to a vote.

5. Multi-level Governance after Authorization

To have a minimally complete story on the multi-level governance of GMOs, some discussion of the post-authorization framework is necessary. Once a GMO has been authorized in the EU system, it enjoys the protection of internal market law, and free movement of goods throughout the EU.[60] Once authorized, and subject to conditions in that authorization, a seed can in principle be grown anywhere in the EU, and food can in principle be sold anywhere in the EU. The possibility for autonomous Member State action after Community authorization is found in the 'safeguard clauses' contained in the GMO legislation and in Article 95 of the EC Treaty itself. Further space for Member States might be found in the approach to coexistence between GM and other forms of farming.

The safeguard clause in the Deliberate Release Directive allows a Member State provisionally to restrict use and/or sale of a GMO where there is 'new or additional information ... affecting the environmental risk assessment or reassessment of existing information on the basis of new or additional scientific knowledge' and it has 'detailed grounds for considering that a GMO ... constitutes a risk to human health or the environment'.[61] This is a narrow, science-based provision, focused on human health and the environment to the exclusion of any other concerns. The Member States collectively are involved in each other's safeguard decisions: a Member State decision to take safeguard measures is notified to the European Commission and other Member States, and a European Commission decision is taken on the matter, with Council involvement through comitology.[62] If the Member States fail to agree, the European Commission acts.

The European Commission was reluctant to take action against national measures at the height of the controversy over GMOs. And apparently rightly so, because in 2005, the Council actually defeated, by qualified majority, a European Commission proposal to request the withdrawal of Austrian safeguard measures in respect of GM maize. In response, the European Commission consulted the EFSA on these measures for a second time. Following the EFSA's conclusion that it could still see no reason to believe that the continued placing on the market of this GMO would have adverse effects, the European Commission resubmitted its proposals to the Council. Again, a qualified majority rejected the European Commission's proposals, challenging the failure to

---

[60] This is explicit in the Deliberate Release Directive, Art 22.
[61] Ibid Art 23. Art 20 provides for Community action if new information becomes available that applies to the whole EU.
[62] Note that the European Parliament Committee on Environment, Public Health and Food Safety would have applied the regulatory procedure with scrutiny to safeguard clauses: see Report, 23 July 2007. This was not picked up in the text adopted by the European Parliament on 14 November 2007.

reassess the GMOs in question under the Deliberate Release Directive, and arguing that 'the different agricultural structures and regional ecological characteristics in the European Union need to be taken into account in a more systematic manner in the environmental risk assessment of GMOs'.[63] Responding to the concerns expressed in the Council, the European Commission's next proposal on the Austrian safeguard measures targeted only the food and feed aspects (and so not the cultivation) of the GMO. The Council was this time unable to reach a qualified majority in either direction, leaving the European Commission free to adopt its proposal.[64] From a multi-level governance perspective we see here the ability of the Member States to assert their authority in a case that is particularly sensitive to one of their peers, and possibly even a fairly constructive compromise. The Council has also rejected a European Commission proposal seeking to request Hungary to withdraw safeguard measures in respect of the cultivation of GM maize MON810.[65]

Under the Food and Feed Regulation, 'emergency measures' may be triggered either by an opinion from the EFSA, or where 'it is evident that products ... are likely to constitute a serious risk to human health, animal health or the environment'.[66] Not only is this a high standard ('serious risk', 'evident'), but emergency measures are centralized, addressed by European Commission action through comitology.[67] Autonomous Member State action is possible only where the Member State has informed the European Commission of the need to take emergency measures, and the European Commission has not acted in accordance with the Regulation. Again, the measures go to comitology.

Whilst they are restrictive, the safeguard provisions are more generous than the general internal market provisions of Article 95(5) of the EC Treaty. And by contrast with the safeguard provisions, Article 95 gives to the European Commission alone (subject to appeal to the Court of First Instance and European Court of Justice) the authority to pass judgment on national measures, without Member State involvement through comitology. Article 95(5) allows

---

[63] 2773rd Council Meeting, Brussels, 18 December 2006, rejecting European Commission, *Proposal for a Council Decision concerning the provisional prohibition of the use and sale in Austria of genetically modified maize (Zea mays L. line T25) pursuant to Directive 2001/18/EC* COM(2006) 510.

[64] European Commission, *Proposal for a Council Decision concerning the provisional prohibition of the use and sale in Austria of genetically modified maize (Zea mays L. line T25) pursuant to Directive 2001/18/EC* COM(2007)589; European Commission, *Proposal for a Council Decision concerning the provisional prohibition of the use and sale in Austria of genetically modified maize (Zea mays L. line MON810) pursuant to Directive 2001/18* COM(2007)586; and 2826th Council Meeting, Brussels, 14 November 2007.

[65] 2785th Council Meeting, Brussels, 20 February 2007, rejecting European Commission, *Proposal for a Council Decision concerning the provisional prohibition of the use and sale in Hungary of genetically modified maize (Zea mays L. line MON810) expressing the Bt cryIA(b) gene, pursuant to Directive 2001/18/EC* COM(2006)713.

[66] Food and Feed Regulation, Art 34. See also Art 10 on new information affecting the authorization. [67] Food Law Regulation, Arts 53 and 54.

derogations from measures adopted (like the GMO legislation[68]) under the internal market provisions of the Treaty. Whilst safeguard clauses are really practical only in respect of individual GMOs, on a case-by-case basis, Article 95(5) might apply more generally. Both Austria and Poland have notified the European Commission of measures banning or restricting the use of GMOs. The Austrian measures were concerned with the protection of organic agriculture, as well as protection of nature and the environment.[69] Poland argued that its measures aimed to protect conventional farming and likely moves to organic farming, as well as biodiversity.[70] Article 95(5) allows for the introduction of new national measures for the protection (only) of 'the environment or the working environment'.[71] This is a very narrow basis for divergence, and Austria fell foul of this narrow framework. Austria's efforts to protect organic agriculture were rejected as relating 'more to a socio-economic problem than to the protection of the environment or the working environment'.[72] The European Commission did not even consider Poland's detailed concerns about its farming future, because of the failure to provide new scientific evidence, discussed below.

New measures under Article 95(5) must be 'based on new scientific evidence relating to the protection of the environment or the working environment'. The requirement for 'new scientific evidence' is already slightly narrower than the possibility for a 'reassessment of existing information on the basis of new or additional scientific knowledge', as in the Deliberate Release Directive's safeguard clause. The question of the 'novelty' of scientific information is central to the operation of Article 95(5).[73] The Polish application under Article 95(5) was rejected by the European Commission on the basis of a failure to provide new

---

[68] The Food and Feed Regulation is also based on Arts 37 and 152(4)(b) of the EC Treaty.

[69] Commission Decision (EC) 2003/653 of 2 September 2003 relating to national provisions on banning the use of genetically modified organisms in the Region of Upper Austria notified by the Republic of Austria pursuant to Article 95(5) of the EC Treaty [2003] OJ L230/34.

[70] Commission Decision (EC) 2008/62 of 12 October 2007 relating to Article 111 and 172 of the Polish Draft Act on Genetically modified Organisms, notified by the Republic of Poland pursuant to Article 95(5) of the EC Treaty as derogations from the provisions of Directive (EC) 2001/18 [2008] OJ L16/17.

[71] Even public health is not included in Art 95(5) of the EC Treaty: a 'specific problem on public health in a field which has been the subject of prior harmonisation measures' can be brought to the attention of the European Commission under Art 95(8), and the European Commission must 'immediately examine' whether to propose an 'adaptation' of those measures. Art 95(4) allows for the *maintenance* of *existing* national measures on the 'grounds of major needs referred to in Article 30' (public morality, public policy, or public security; the protection of health and life of humans, animals, or plants; the protection of national treasures possessing artistic, historic, or archaeological value; or the protection of industrial and commercial property) or 'relating to the protection of the environment or the working environment'.

[72] Commission Decision 2003/653, para 67

[73] For an excellent analysis of the developing case law, see M Doherty, 'The application of Article 95(4)–(6) of the EC Treaty: is the emperor still unclothed?' (2008) (8) Yearbook of European Environmental Law 48. For early reviews, see H Sevenster, 'The environmental guarantee after Amsterdam: does the emperor have new clothes?' (2000) 1 Yearbook of European

(or any) scientific evidence.[74] Austria relied on a report published after adoption of the legislation, but the data in the report 'were for a large part available prior to the adoption of Directive 2001/18/EC' and 'the vast majority of the sources were published prior to the adoption of Directive 2001/18/EC'.[75] Demanding brand new *data* places a very heavy burden on the Member State. Importantly, when the Austrian GMO case was appealed to the European Court of Justice, Advocate-General Sharpston concluded that 'new conclusions drawn from existing data *may* constitute new scientific evidence within the meaning of Article 95(5) EC'.[76] Austria could not provide even this type of evidence, but this more generous approach is a realistic understanding of what might provoke a change in national policy. But in the case of GMOs, the disagreement is really over the appropriate role of agricultural biotechnology, and over the risks that might be worth taking in this area, rather than new scientific information. Article 95 does not address this concern.

Article 95(5) applies only if a Member State can establish 'grounds of a problem specific to that Member State, arising after the adoption of the harmonisation measure'. The meaning of 'specific' is not wholly clear: in the words of Advocate-General Sharpston, 'a specific problem clearly lies somewhere between one which is unique and one which is common, generalised or widespread'.[77] Both Austria and Poland in their Article 95(5) applications referred to their particular farming contexts. Citing the EFSA, the European Commission held that Austria had failed to establish either that its 'small structured farming systems' are specific to this region, or that Austria has 'unusual or unique ecosystems' requiring separate risk assessments.[78] Poland argued that 'Polish agriculture is fragmented to a very high degree', with almost two million farms, averaging less than eight hectares in size, and that '[g]iven this high level of fragmentation, it is not possible to isolate GM crops from conventional and organic crops, and this may also pose a serious threat to Poland's developing organic farming'.[79] Given the absence of new scientific information within the terms of Article 95(5), the European Commission did not even discuss this. Whilst the meaning of 'specific' remains uncertain, it will

Environmental Law 281; and N de Sadeleer, 'Procedures for derogations from the principle of approximation of laws under Article 95 EC' (2003) 40 Common Market Law Review 889.

[74] Commission Decision 2008/62.
[75] Commission Decision 2003/653, para 65. The European Commission relied heavily on the EFSA's opinion.
[76] Joined Cases C-439/05P and C-454/05P *Land Oberösterreich v Commission* [2007] ECR I-7141, para 124. This point was not pursued by the European Court of Justice. Note that in Case C-236/01 *Monsanto Agricoltura Italia SpA v Presidenza del Consiglio dei Ministri* [2003] ECR I-8105, the Advocate-General took the position that the moratorium falls within the scope of new information or a reassessment of existing information (explicitly allowed in the safeguard clause at issue in that case): para 151.
[77] Joined Cases C-439/05P and C-454/05P *Land Oberösterreich v Commission* [2007] ECR I-7141, para 110. [78] Commission Decision 2003/653, para 71.
[79] Commission Decision 2008/62, para 26.

always be challenging in an environmental context.[80] To rely on Article 95, a Member State will need to establish something distinctive about the impact of GMOs in its territory rather than others, and not just a different understanding of acceptable risks. In short, the strict approach taken to Article 95 means that it does not play the role that it might in developing a flexible regime of multi-level governance.

The ability of Member States to block the commercialization or cultivation of authorized GM agriculture or food in their territory is focused on the science of risk to the environment or (under the safeguard clauses only) public health. Agricultural biotechnology raises a host of other difficult social questions, which are very likely to manifest themselves differently around the Member States. The sensitivity of national preferences is clearly recognized in the Council's political support for the Austrian and Hungarian safeguard measures. But on the face of the law, the scope for engagement with citizen values or preferences in the Member States after authorization is virtually non-existent. To return to the multi-level governance theme, there is very little scope for redistributing authority on GMOs after the grant of an authorization. The regulation of 'coexistence', however, provides a further opportunity for the negotiation of authority. But from almost the opposite starting point, we again have a picture of the centre exerting its control. Coexistence is explicitly a Member State competence under the legislation: 'Member States may take appropriate measures to avoid the unintended presence of GMOs in other products'.[81] This language is open, and in theory leaves room for distinctive national approaches to the balance between different forms of farming. The apparent de-centralization of this area is rationalized on the basis of diversity around the EU, diversity of the farming industry (small family farms or huge agri-business, different degrees of attachment to organic farming), as well as of ecologies, geography, and climate. Just as at the authorization stage, however, the division of national and central authority is more complex than it appears. The internal market is again a very powerful centralizing force, although alongside it run softer approaches to collaboration and learning.

The European Commission has established a 'coexistence network' in which national and Commission-appointed experts discuss coexistence. In the authorization context, these informal networks of experts are used to mitigate centralization. With coexistence, this appears to work in reverse, with the

---

[80] See also Case C-405/07P *Netherlands v Commission* (ECJ 6 November 2008). Although the European Court of Justice overturned the Court of First Instance decision (Case T-182/06) on the basis that the European Commission had failed to take account of relevant information, the discussion by the Court of First Instance and the European Commission of the importance of 'specificity' stands. This case concerned particulate air pollution, and comparisons were drawn with other Member States, 'such as Belgium, Austria, Greece, the Czech Republic, Lithuania, Slovenia and Slovakia': para 67 of the Court of First Instance decision.

[81] Deliberate Release Directive, Art 26a, as amended by the Food and Feed Regulation.

network establishing a forum for continued central and mutual (Member State/Member State) influence in a context of de-centralization, again avoiding a single site of authority. The role of this network is familiar, focused on sharing resources, including information and expertise, and providing for the spread of good practice and for peer review. But whilst the networks look as if they are mediating de-centralization, it is at least arguable that they reinforce centralization. Although they are national representatives, the coexistence experts are distant from national publics, and their meetings are chaired by the European Commission, in the framework of a narrow European Commission understanding of coexistence.[82] And these soft approaches are more than matched in hard law that focuses on a European Commission-defined policy, not on the open-ended national autonomy that appears in the legislation.

The same internal market considerations as arise at authorization restrict national autonomy on coexistence. Internal market law is, moreover, placed in a very narrow context by the European Commission, in its non-binding Recommendation (EC) 2003/556 on guidelines for the development of national strategies and best practices to ensure the coexistence of genetically modified crops with conventional and organic farming ('2003 Commission Recommendation').[83] The non-binding 2003 Commission Recommendation is another soft tool of governance, relying on persuasion and information, but, although non-binding, it is potentially very influential. The European Commission recommends specific farm management measures such as buffer zones, cleaning machinery, or co-ordinating planting. The overarching themes of the 2003 Commission Recommendation are more interesting for current purposes, and point to the European Commission's efforts to control national autonomy. It takes a minimalist approach to coexistence, advocating a crop-by-crop, farm-level approach. Most significantly, and repeated again and again in subsequent policy,[84] the European Commission presents coexistence as a *purely* economic issue. This is based on the assumption that all environmental or health issues are dealt with in the authorization process, or will be dealt with by safeguard clauses. This economic approach is very well-entrenched, and has indeed been confirmed by Advocate-General Sharpston.[85] Nevertheless, the European Commission's failure to acknowledge the ongoing environmental and socio-economic questions implicit in coexistence is a significant misunderstanding of what is at stake in the regulation of GMOs, if anything emphasized by its own acceptance that Council Directive (EEC) 92/43 on the conservation of natural

---

[82] So, eg in the very first meeting, the European Commission emphasized the wholly economic nature of coexistence regulation: see *Minutes of the Meeting of COEX-NET of 22 September2005*, para 2.   [83] [2003] OJ L189/36. See further Chapters 6 and 7.
[84] For example, European Commission, *Report on the Implementation of National Measures on the Coexistence of Genetically Modified Crops with Conventional and Organic Farming* COM(2006)104.
[85] Joined Cases C-439/05P and C-454/05P *Land Oberösterreich v Commission* [2007] ECR I-7141.

habitats and of wild fauna and flora ('Habitats Directive')[86] applies to the cultivation of GMOs. This must be an implicit recognition of the continued relevance of ecological concerns post-authorization.

For current purposes, it is important to note that an economic understanding of coexistence dramatically restricts national freedom of action. Most fundamentally, basic internal market law in the EU prevents the Member States using economic arguments to justify an interference with the free movement of goods. Whilst strict requirements would still apply, greater leeway could be available if coexistence measures were also about social regulation; nor are Article 95 or the safeguard clauses apt to address coexistence measures if coexistence is wholly economic. Directive (EC) 98/34 of the European Parliament and of the Council laying down a procedure for the provision of information in the field of technical standards and regulations, which requires Member States to notify draft standards and technical regulations to the European Commission, provides for a dialogue around possible internal market impact of proposed coexistence measures, allowing more of a multi-level governance approach.[87] A number of Member States have notified coexistence measures. There is some give and take,[88] but, on the whole, the European Commission is using the process to keep a very tight rein on the subject. It consistently reiterates its own 2003 Commission Recommendation, especially on the purely economic nature of coexistence. An economic approach limits the scope of the debate; the concern is not that environmental arguments (say) will lose, but that they cannot officially even be heard.

At the time of the moratorium, the European Commission had resisted legislating on coexistence, preventing harmonization. But now, paradoxically, it seems to be busy 're-Europeanizing' the whole area, asserting its own authority in a heavy-handed manner. The indirect control that the European Commission

---

[86] [1992] OJ L206/7. This is discussed in a number of detailed opinions issued by the European Commission in respect of coexistence measures notified under Directive (EC) 98/34 of the European Parliament and of the Council laying down a procedure for the provision of information in the field of technical standards and regulations [1998] OJ L204/37. Detailed opinions are not publicly available, but I am grateful to the European Commission for making detailed opinions available to me in respect of notifications 2003/475/A, 2004/459/A, 2006/73/P, 2005/634-637/HU, 2005/271/P, 2004/133/D, 2006/455/SK, 2004/538/A, 2005/005/A, 2003/200/A, and 2005/610/A. See also *Commission Staff Working Document, Annex to the Report on the Implementation of National Measures on the Coexistence Of Genetically Modified Crops with Conventional and Organic Farming* SEC(2006)313.   [87] [1998] OJ L204/37.

[88] The movement seems to have been mainly from the Member States, but the European Commission has, for example, accepted the relevance of the Habitats Directive. See the detailed opinions in n 86 above; I am also grateful to the European Commission for making the following responses available: SG(2006) D/53080 on Notification 2006/455/SK; SG(2006) D/51855 on Notification 2005/634/HU, 2006/637/HU; SG(2004) D/51849 on Notification 2004/133/D; SG(2006) D/51039 on Notification 2005/610/A; SG(2005) D/52124 on Notification 2005/5/A; SG(2005) D/51017 on Notification 2004/538/A; SG(2005) D/51284 on Notification 2004/459/A; SG(2004) D/51007 on Notification 2003/475/A; SG(2003) D/52110 on Notification 2003/200/A; and SG(2005) D/50052 on Notification 2004/311/A.

exercises now, however, is more autonomous than anything that it might have achieved through legislation. Unfortunately, it has left us with a framework that is obscure, both in its forum and in its technical, highly legalistic approach to a complex and subtle area. More generally, the role of the internal market tends to centralize authority on GMOs after authorization.

To conclude, the authorization process for GMOs provides for a careful arrangement of authority between the Member States and the centre. In its operation, nonetheless, it is currently looking increasingly centralized. That centralization of authority, the emphasis on uniformity and harmonization, persists after authorization.

6. Multi-level Governance and the Role of the Internal Market

Authority for GMOs seems to be settling at the EU level. The most pressing impetus to this centralization are the demands of the internal market. Because they are products, GMOs demand more sensitivity to internal market law than many areas of environmental governance. As the capacity and ambitions of the EU expand, we perhaps need to remind ourselves how central the internal market is to the purposes of the EU. The free movement of goods is important as an instrument of economic prosperity, but since the birth of the European Economic Community, economic integration has also been an important symbol and precursor of increasing political integration. So if different national approaches to the regulatory challenge of GMOs raise barriers to the completion of the internal market, they may be (mis)understood as a threat to the core of the EU's ambitions. External pressures from trade partners, particularly through the WTO, amplify the importance of maintaining internal market disciplines, and presumably increase the concern of the European Commission (which answers to the WTO) to hold its line.

There is an enormous tension in the regulation of GMOs, between the demands of the internal market and the reluctance of Member States to reduce their control over such a politically sensitive issue. In these circumstances, a turn to ideas of multi-level governance, avoiding single sites of authority and emphasizing space for collaboration, discussion, influence, and negotiation was entirely to be expected. But because of the sensitivity of the topic, these governance techniques are not at the moment overcoming disagreement. To speak of the failure of multi-level governance would be to overstate the case, not least because it assumes that what law and governance is for is entirely uncontroversial; it may be that perpetual postponement of decisions was exactly the purpose of the legislation. But there is a perception that decisions are needed, and in these circumstances, we see reversion to hierarchy and formal direction from the centre.

However, as might have been predicted, this reversion to hierarchy is not a straightforward alternative to new governance, which was precisely an effort to overcome the limitations of a strictly hierarchical response. So the Europeanization of GMOs remains controversial. Not only are some decisions (on cultivation, discussed above) conspicuously avoided by the European Commission, but all decisions are slow and painful, and Member States continue to apply legally questionable barriers to the use of GMOs. The controversial nature of the European Commission's role in authorization risks spilling over into broader debates on the legitimacy of the European Commission and even of the EU. Greater autonomy and discretion for the Member States, allowing diversity in national approaches to GMOs, would be a preferable solution. It is, of course, extremely difficult to increase autonomy and flexibility whilst maintaining internal market disciplines. The principles of free movement that characterize the EU will be insisted upon. But perhaps a greater challenge may be insisting upon uniform agricultural systems in the EU. Allowing Member States to take a different view on whether the risks (socio-economic as well as environmental or health risks, uncertainty as well as scientifically established risk) of GMOs should be borne, need not lead to a free for all—evidence would still be demanded, and challenge would still be possible.[89]

7. Conclusions

GMOs provide an anomalous tale of multi-level governance in a number of ways: GMOs are products, and so the internal market context is emphasized; comitology rarely moves past committee, and then usually the Council is able to make a decision; and most studies of EU governance observe diversity and flexibility rather than the overreaching centralization that we see here. But we can learn a lot from anomalies, and the case of GMOs forces us to think about the location of authority in the EU. The GMO legislation builds in opportunities for collaboration and compromise, but perhaps this was just wishful thinking. The basic absence of agreement over the most fundamental questions of what agricultural biotechnology is for, how it should be regulated, and by whom, means that consensus and compromise are hard to find. A simultaneous perception that non-decisions are no longer acceptable means that decision-making is now reverting to the centre, but in a very uncomfortable way. The legislative compromise of dispersed and shared authority does not look likely to survive, and has simply postponed (or disguised) the inevitable decision over where authority will ultimately lie.

---

[89] This is not the place for a detailed discussion, but, for more detail on my position, see Lee (n 14 above).

6

Coexistence of Genetically Modified, Conventional, and Organic Crops in the European Union: The Community Framework

Margaret Rosso Grossman[*]

1. Introduction

Genetically modified crops make up an ever-larger proportion of the world's maize and soybeans, and the coexistence of these crops with other, more traditional varieties is a matter of increasing concern, especially in the European Union (EU). Agricultural production systems can be described in three categories: conventional, conventional using genetically modified (GM) crops, and organic. Conventional systems often use fertilizers and pesticides. Genetically modified systems use crops that have been genetically engineered to resist pests or disease or to tolerate herbicides, and some view GM systems as a 'continuum' of conventional systems. Organic systems, by contrast, do not usually use chemical fertilizers or pesticides, nor do they use GM materials in the production process.[1] These three types of agricultural production may occur in the same geographic region. When they do, producers who grow organic or conventional crops often want to avoid the 'adventitious presence' of GM material in their crops. In the EU, where GM crops have been viewed with suspicion, the goal of avoiding adventitious presence is particularly important. Of course, farmers in other nations often share the same concern.

---

[*] This chapter is based on work supported by the National Institute of Food and Agriculture, United States Department of Agriculture (USDA), under Hatch Project No ILLU-470-309. It is derived in large part from Grossman's article, 'The coexistence of GM and other crops in the European Union' (2007) 24 Kansas Journal of Law and Public Policy 324, used here with permission.
[1] Pew Initiative on Food and Biotechnology, *Peaceful Coexistence Among Growers of Genetically Engineered, Conventional, and Organic Crops* (Washington, DC: Pew Initiative on Food and Biotechnology, 2006) 7.

For the European Commission, the term 'coexistence' 'refers to the ability of farmers to make a practical choice between conventional, organic and GM crop production, in compliance with the legal obligations for labelling and/or purity standards'.[2] The European Association for Bioindustries (including 'green' or agricultural biotechnology) offered a more expansive definition:

[c]oexistence is about how crops intended for different markets can be grown in the same vicinity without becoming commingled and thereby possibly compromising the economic value of each other. Coexistence is based on the premise that farmers should be free to cultivate the crops of their choice using the production system they prefer, whether they are GM, conventional or organic.[3]

In the last few years, coexistence has engendered significant controversy in Europe.[4] As the Dutch Minister of Agriculture noted at a 2006 conference, coexistence is both complex and highly emotional: '[i]t is a subject that ... can give rise to heated discussions. The debate is often dominated by polarized views, which sometimes makes a dialogue hard to achieve'.[5]

This chapter focuses on coexistence of GM and other crops in the EU. It begins with explanations of the concepts of adventitious presence and coexistence and emphasizes the importance of Community regulatory labelling thresholds for coexistence.[6] The chapter then analyses Community policy on coexistence, with a special focus on the guidelines provided in Commission Recommendation (EC) 2003/556 on guidelines for the development of national strategies and best practices to ensure the coexistence of genetically modified crops with conventional and organic farming ('2003 Commission Recommendation')[7], and explores several critical issues, including purity of seeds, organic production, and efforts to ban genetically modified organisms (GMOs). The chapter refers only briefly to the important question of liability, which is discussed further in Chapter 8. Finally, the chapter looks generally at Member State approaches to coexistence.

---

[2] Commission Recommendation (EC) 2003/556 on guidelines for the development of national strategies and best practices to ensure the coexistence of genetically modified crops with conventional and organic farming ('2003 Commission Recommendation') [2003] OJ L189/36, Annex, para 1.1.

[3] ABE/EuropaBio, *Understanding Coexistence: Science, Principles and Practical Experiences* (2006) (available at <http://www.europa-bio.be>, last accessed 24 May 2008).

[4] Coexistence is also an issue in the United States: '[i]t is a basic principle in the U.S. that farmers should be able to produce commodities by any method they prefer and to market them in any market available, assuming they meet all safety and marketing standards': Pew Initiative on Food and Biotechnology (n 1 above) 3.

[5] C Veerman, 'The Dutch Approach to Co-existence—Remarks at the EU Conference on Co-existence of Genetically Modified, Conventional and Organic Crops: Freedom of Choice Conference', Vienna, 4–6 April 2006 (papers presented at the Freedom of Choice Conference are available at <http://ec.europa.eu/agriculture/events/vienna2006/index_en.htm>, last accessed 6 October 2009).

[6] Chapter 4 discusses Community regulatory measures that govern authorization, traceability, and labelling of GMOs.   [7] [2003] OJ L189/36.

2. GM Crops and Coexistence in the EU

Coexistence is a critical issue because global cultivation of GM crops has increased rapidly since their introduction in 1996. Though producers in EU Member States now grow only a small proportion of the world's GM crops, as GM cultivation increases in the EU, coexistence will raise practical concerns for producers and consumers. Coexistence is also significant internationally. Asynchronous approvals of GMOs by the EU and its main trading partners (eg the United States, where more varieties have been approved) interfere with trade, because the EU does not permit imports of unauthorized GM material in food and feed, even at low levels.[8]

Indeed, in 2006, the EU Commissioner for Agriculture and Rural Development noted that the complex issue of coexistence often reflected conflicting goals among producers and consumers. Some farmers focus on enhanced competitiveness and see an important role for GMOs, while others prefer traditional practices. Some European consumers value 'quality production' and products 'linked to traditional practices and geographic origin'.[9] These values are consistent with the European model of agriculture that balances socio-economic, as well as environmental and territorial considerations. Other consumers value food produced by organic farmers or with traditional methods; some believe that GMs are not compatible with traditional production. Many producers and consumers prefer to avoid commingling GM and other crops.

In some instances, these attitudes go beyond mere preferences. For example, Friends of the Earth Europe (FOEE) asserted that:

GMO contamination is a new type of pollution created by industry. It involves living and replicating organisms, and because it involves the building blocks of life (genes), is irreversible as well as increasing over time. It can occur at any stage along the food chain as a result of natural processes and human intervention: from seed production, to crop growing, to harvesting, to storage, to transport, to processing and packaging.[10]

Perhaps reflecting this attitude, another non-governmental organization (NGO), GMO-free Europe, calculated that 289 regions in Europe have declared themselves GM-free or would like to restrict GM crops, and 4,567 localities would also like to restrict GMOs in their territories.[11]

---

[8] See, eg DG-Agri, European Commission, *Economic Impact of Unapproved GMOs on EU Feed Imports and Livestock Production* (2007).

[9] Commissioner Fischer Boel and J Pröll, 'Concluding Remarks at the EU Conference on Co-existence of Genetically Modified, Conventional and Organic Crops: Freedom of Choice Conference' (Freedom of Choice Conference, n 5 above).

[10] FOEE, *Contaminate or Legislate? European Commission Policy on 'Coexistence'* (FOEE, 2006).

[11] GMO-free Europe, *GMO-free Regions and Areas in Europe* (2009) (available at <http://www.gmo-free-regions.org/gmo-free-regions.html>, last accessed 26 April 2009). Regions include provinces, prefectures, and departments. In addition, 30,370 individuals are GMO-free. On agri-food paradigms and coexistence as a 'new battlefield for contending political

As with GMOs themselves, coexistence is controversial and has engendered years of vigorous debate. The European Council, European Parliament, and various committees have considered the issue, and the European Commission has provided advice. While some argue that Community-level legislation should govern coexistence, others believe that Member States should develop measures and implement them, building experience and knowledge to facilitate Community measures, should they be necessary.[12] In the end, as the discussion below indicates, policy-makers believe that it is in the best interest of the various Member States to regulate coexistence to ensure that critical 'geographical, ecological and climatic conditions' that affect crop production can be taken into account.[13]

2.1 Agriculture and Coexistence

2.1.1 Adventitious presence

Before the development of GM technology, adventitious presence—the 'unavoidable variability in seed, grain, and food'—was considered natural.[14] Seeds may be mixed at some stage of the production process (at planting, harvest, or in storage) or during a later stage of the distribution process (transportation or handling).[15] Even when seed purity is high, various substances, including 'weed seeds, seeds from other crops, dirt, insects, or foreign material', infiltrate at various steps in the chain of production from seed to product.[16] Despite the best efforts of farmers, crops will include the adventitious presence of other material, sometimes resulting from cross-pollination with crops from neighboring fields.[17] Indeed, some level of cross-pollination cannot be avoided, especially when crops flower at the same time. Farming is practised in the open air, and farmers cannot control the wind and insects that play a role in

agendas', see L Levidow and K Boschert, 'Coexistence or contradiction? GM crops versus alternative agricultures in Europe' (2008) 39 Geoforum 174, 188.

[12] Commissioner Fischer Boel and J Pröll (n 9 above).
[13] Commissioner Dimas, Speech/06/229, 'Co-existence of Genetically Modified Crops with Conventional and Organic Crops—Freedom of Choice Conference on GMO Co-existence', Vienna, 5 April 2006 (speeches of the Commissioner for Environment are available at <http://europa.eu/rapid/>). For Commissioner Dimas, the key issue for coexistence is natural cross-pollination.
[14] S Vandegrift and C Gould, 'Issues surrounding the international regulation of adventitious presence and biotechnology' (2003) 44 Jurimetrics Journal 81, 83. See also American Seed Trade Association, *What is Adventitious Presence in Seed?* (available at <http://www.amseed.com/qaDetail.asp?id=52>, last accessed 24 May 2008). [15] ABE/EuropaBio (n 3 above).
[16] DL Kershen and A McHughen, *Adventitious Presence: Inadvertent Commingling and Coexistence among Farming Methods* (Cast Commentary QTA2005-1, 2005). The authors note that the Association of Official Seed Certifying Agencies 'tolerates 0.5% of seed of other varieties or off-types and 2% of inert material in certified hybrid corn' and that hybrid maize is considered pure if it has 98% purity: ibid 1.
[17] K Schumacher, 'Segregation—Challenges for the Trade in Agricultural Commodities: Freedom of Choice Conference' (n 5 above).

pollination.[18] Recognizing that '100% purity is impossible', industry standards for adventitious presence in conventional varieties depend on the species, the importance of purity, and the difficulty of avoiding contamination. In general, however, the accepted level of variability has been one to two per cent.[19]

Adventitious presence has attracted attention in recent years, due to the development and widespread cultivation of GM crops, the reluctance of some European governments, producers, and consumers to grow or consume those crops, and the desire to avoid the 'contamination' of non-GM crops with GM material. The Community policy of zero tolerance for unauthorized GMOs is also significant. Moreover, as GM crops with pharmaceutical and industrial traits are developed and planted in field trials, concern about adventitious presence will become more compelling, because pharmaceutical or industrial compounds must be kept clear of the food supply.[20]

2.1.2 Coexistence

Coexistence focuses on the practices used to minimize adventitious presence. In the context of agricultural production systems using GM crops, like in other production systems, coexistence has both horizontal and vertical elements. Horizontal coexistence focuses on the relationships between farmers who cultivate GM crops and those who cultivate conventional or organic crops. Horizontal management requires growers of GM crops to implement measures such as notice to neighbours, isolation distances, and segregation. Vertical coexistence focuses on seeds and the on-farm production process; it requires those who plant traditional and organic crops to ensure that their seeds and fields are free of GM material, insofar as possible, and to segregate products along the chain of production. Both the level of purity of seeds and the transfer of GM materials on the farm (for example, through admixture) affect the level of presence of GM in food and feed.[21]

In fact, farmers and others have long managed horizontal (production) and vertical (pre- and post-production) activities to achieve coexistence. For example, in the United States, production of certified seeds or specialty crops has required practices like separation distances and time intervals between crops to ensure seed and product purity. US producers have also channelled some GM and non-GM crops into different markets—for example, domestic or export.[22]

---

[18] Veerman (n 5 above). When the EU imports grains and oilseeds in bulk, often in large vessels, varieties cannot be separated completely, and some adventitious presence will occur: Schumacher (n 17 above). [19] Vandegrift and Gould (n 14 above) 85.
[20] This chapter does not focus on disputes between GM seed companies and producers whose crops are discovered to have GM material that violates property rights of the seed company or on pharmaceutical or industrial crops.
[21] L Girsh, 'The Precautionary Principle in Coexistence Management: Freedom of Choice Conference' (Freedom of Choice Conference, n 5 above).
[22] G Brookes and P Barfoot, *Co-Existence in North American Agriculture: Can GM Crops be Grown With Conventional and Organic Crops?* (Dorchester: PG Economics, 2004) 21.

Similarly, in the United Kingdom and Germany, production of high erucic acid oilseed rape, which cannot be consumed, has been governed by contracts that require separation distances, clean equipment, and post-harvest segregation.[23]

Although farmers have long experience with coexistence, not all efforts at coexistence have been successful. Unhappy sagas of failed stewardship that resulted in admixture of unapproved GM crops, followed by significant financial losses, illustrate the importance of effective management practices for coexistence. The case of StarLink™ maize in the United States is a well-known example. StarLink™, approved for feed and industrial uses, but not for human consumption, was found in taco shells. Allegations that StarLink™ had contaminated the US maize supply resulted in recalls of maize products, disruptions in trade, and litigation. The developers of the maize incurred significant losses, including compensation to farmers and the costs of monitoring so that the protein in StarLink™ could be contained and removed from the food supply.[24]

Moreover, as more acres are planted with GM crops, the likelihood and rate of adventitious presence will increase.[25] Indeed, a report published in 2004 referred to research suggesting that many conventional seeds in the United States have been 'contaminated', albeit at low levels (0.05 to 1 per cent), by DNA from GM crop varieties; about 50 per cent of traditional maize and soybean varieties and over 80 per cent of canola seed samples in the reported research had GM material.[26]

Researchers have identified two main economic consequences of coexistence: 'the costs involved in meeting tolerances for the adventitious presence of unwanted material (eg, by having to change farming practices, initiating on-farm segregation of crops)' and 'the economic consequences of not meeting tolerances (eg, possible loss of non GM or organic price premia)'.[27] The latter can be characterized as 'market access problems'; that is, adventitious presence of GM plant material may prevent crops of conventional and organic growers from meeting market specifications.[28] Although much of the discussion of coexistence in recent years has focused on the economic effects of adventitious presence of GM crops in conventional and organic products, it is useful to note that adventitious presence of non-GM genetic material in GM crops or even in other non-GM crops may also have economic consequences.[29]

---

[23] G Brookes, *Co-existence of GM and Non GM Crops: Current Experiences and Key Principles* (Dorchester: PG Economics, 2004) 9–10.
[24] *In re Starlink Corn Products Liability Litigation* 212 F Supp 2d 828 (ND Ill 2002). See further Chapter 12. [25] Vandegrift and Gould (n 14 above) 85.
[26] M Mellon and J Rissler, *Gone to Seed: Transgenic Contaminants in the Traditional Seed Supply* (Cambridge MA: Union of Concerned Scientists, 2004) 1–2 (noting that seeds 'reproduce and carry genes into future generations' and are the 'wellspring of our food system').
[27] G Brookes, *Co-existence of GM and Non GM Crops: Economic and Market Perspectives* (Canterbury: Brookes West, 2003) 4.
[28] Growers of GM crops with 'high-value output traits' may eventually face similar problems: Pew Initiative on Food and Biotechnology (n 1 above) 3. [29] Brookes (n 27 above) 2.

2.2 A Key to Coexistence in the EU: Thresholds for Labelling and Traceability

2.2.1 Thresholds

Although this chapter does not focus on the details of the Community regulatory regime for GM crops and their products, an understanding of the system of labelling and traceability—and especially thresholds for its application—is critical for the concept of coexistence.[30] In brief, Directive (EC) 2001/18 of the European Parliament and of the Council on the deliberate release into the environment of genetically modified organisms ('Deliberate Release Directive') emphasizes the importance of traceability and requires labels for GMOs placed on the market.[31] Regulation (EC) 1830/2003 of the European Parliament and of the Council concerning the traceability and labelling of genetically modified organisms and the traceability of food and feed products produced from genetically modified organisms ('Regulation 1830/2003') requires traceability for products consisting of or containing GMOs, as well as food and feed produced from GMOs, placed on the market; it also requires labels for products consisting of or containing GMOs.[32] In addition, Regulation (EC) 1829/2003 of the European Parliament and of the Council on genetically modified food and feed ('Food and Feed Regulation') requires labels for GM food and feed that contain or consist of GMOs or are produced from or contain ingredients produced from GMOs.[33]

Regulatory thresholds, however, mean that some products that contain 'adventitious or technically unavoidable' traces of *authorized* GMOs may not have to be labelled, and some traceability requirements will not apply.[34] That is, traces of GMOs do not trigger traceability and labelling requirements if the traces do not exceed the thresholds, which have been set at 0.9 per cent.[35] A threshold for presence of authorized GM material in seeds, however, has not yet been established.[36]

2.2.2 Coexistence in the EU

Coexistence in the EU often focuses on these thresholds. That is, producers prefer to keep the adventitious presence of GM material in non-GM crops and products below the regulatory threshold to avoid the obligation to label products as GMOs.[37] Because of the limited area in the EU devoted to GM crops

---

[30] See further Chapter 4. [31] [2001] OJ L106/1 (as amended).
[32] [2003] OJ L268/24. [33] [2003] OJ L268/1.
[34] To prove that the presence of GM material is adventitious or technically avoidable, operators must be able to show that they have taken appropriate steps to avoid the presence of GM material: ibid Arts 12(3) and 24.
[35] Deliberate Release Directive, Art 21(3), as amended by Regulation 1830/2003; Food and Feed Regulation, Arts 12(2) (GM food) and 24(2) (GM feed); and Regulation 1830/2003, Art 4 (7) and (8). [36] See below.
[37] In the EU, as in other nations, thresholds for labelling GM content 'are not safety-based standards, but rather market specifications that trigger labeling requirements if not met': Pew

(about 108,000 hectares in 2008) and the authorization of only MON810 maize for commercial cultivation, European producers have little experience with coexistence. In Spain, which claims successful coexistence, GM maize (almost 80,000 hectares in 2008) still represents a small percentage of the total maize crop.[38]

Perhaps because of the technical problems in achieving coexistence, some would claim coexistence as a uniquely European concept. For example, a European Commission report referred to the 'novel concept of coexistence between GM and non-GM agriculture developed by the EU'.[39] Though farmers in other countries have similar concerns about protecting the purity of crops, the report indicated that the coexistence rules being developed in Member States are unique, noting that '[a] similar framework does not exist currently in other areas of the world where GM crops are cultivated'.[40] Indeed, some aspects of the Community principles of coexistence described below may be unique, especially insofar as the burden of coexistence falls primarily on growers of GM crops. As economists have noted, under most systems, producers who will benefit (that is, receive premium prices) from production of a special crop (for example, organic or seed) normally bear the burden of costs for achieving purity and meeting tolerances for integrity and identity.[41]

In a sense, the Community focus on coexistence might be seen in the context of the precautionary principle, which plays a significant role in the regulatory scheme for GMOs and their products.[42] Having effective plans for coexistence in place before the cultivation of GM crops on a wide scale implements a precautionary approach and may help to avoid some of the economic losses that have followed failures of coexistence elsewhere.

3. Community Policy on Coexistence

3.1 Background

Coexistence allows producers to choose between conventional, organic, and GM crops and to comply with legal requirements for labelling and purity.[43]

Initiative on Food and Biotechnology (n 1 above) 8. Japan's threshold is 5%, while Australia, New Zealand, and Brazil use 1%.

[38] Brookes (n 27 above) 4. Current statistics on GM hectares are available at GMO Compass, <http://www.gmo-compass.org/>, last accessed 1 July 2009.
[39] M Gómez-Barbero and E Rodríguez-Cerezo, *Economic Impact of Dominant GM Crops Worldwide: Tech. Rep. EUR 22547* (Joint Research Centre (JRC), European Commission, 2006) 37.
[40] Ibid.   [41] Brookes (n 27 above) 3 and 11.
[42] On the precautionary principle and GMOs in the EU, see, eg HT Anker and MR Grossman, 'Authorization of genetically modified organisms: precaution in US and EC Law' (2009) 4 European Food and Feed Law Review 1. See also MR Grossman, 'Traceability and labeling of genetically modified crops, food, and feed in the European Union' (2005) 1 Journal of Food Law and Policy 43.   [43] 2003 Commission Recommendation, Annex, para 1.1.

The issue of coexistence focuses on 'economic consequences of adventitious presence' of GM crops in non-GM crops, which can result from 'seed impurities, cross-pollination, volunteers ... harvesting-storage practices and transport'.[44] As already noted, the farmer's practical choice and its economic consequences are affected by Community requirements that crops and their products be labelled as GMOs if they contain more than the prescribed minimal threshold of GMO content, 0.9 per cent. As also indicated, producers of conventional and organic crops, of course, would prefer to avoid the required labeling triggered by the adventitious presence of GM material above the regulatory threshold.

Two provisions in the Deliberate Release Directive are critical for coexistence. Article 22 indicates clearly that, subject to narrow safeguards, Member States must permit cultivation and sale of authorized GMOs and their products: '[w]ithout prejudice to [the safeguard clause], Member States may not prohibit, restrict or impede the placing on the market of GMOs, as or in products, which comply with the requirements of this Directive'.[45] Article 26a, added in 2003, states that 'Member States may take appropriate measures to avoid the unintended presence of GMOs in other products'.[46] The Deliberate Release Directive also requires the European Commission to study the issue, 'observe the developments regarding coexistence in the Member States', and 'develop guidelines on the coexistence of genetically modified, conventional and organic crops'.[47] Though Community policy-makers have rejected, at least for now, the possibility of setting Community-wide standards for coexistence, the European Commission has provided policy guidelines to Member States, and Member States are implementing those guidelines, which will now be discussed.

3.2 The European Commission and Coexistence

Even by 2002 the European Commission had indicated that it would 'take the initiative to develop, in partnership with Member States, farmers and other private operators, research and pilot projects to clarify the need, and possible options, for agronomic and other measures to ensure the viability of conventional and organic farming and their sustainable coexistence with genetically modified crops'.[48] For this purpose, it has sponsored research into scientific and

---

[44] *Communication from Mr. Fischler to the Commission, 'Co-existence of Genetically Modified, Conventional and Organic Crops'* SEC(2003)258/4 ('Fischler Communication') 2.
[45] Deliberate Release Directive, Art 22.
[46] Deliberate Release Directive, Art 26a(1), as amended by the Food and Feed Regulation.
[47] Deliberate Release Directive, Art 26a(2).
[48] European Commission, *Communication from the Commission to the European Parliament, the Council, the Economic and Social Committee and the Committee of the Regions: Life Sciences and Biotechnology—a Strategy for Europe* COM(2002)27 [2002] OJ C55/3, Action 17. See also European Commission, *Communication on the Mid Term review of the Strategy on Life Sciences and Biotechnology* COM(2007)177.

practical aspects of coexistence,[49] with important studies appearing in 2002[50] and 2006.[51]

February 2003 saw issue of the Fischler Communication; in a comment a few weeks later, the Commissioner for Agriculture emphasized that '[c]o-existence is about economic and legal questions, not about risks or food safety, because only authorised GMOs can be cultivated in the EU'.[52] The Fischler Communication itself reminded the European Commission that authorization of GMOs and their products requires a comprehensive assessment of health and environmental effects. Therefore, in Fischler's opinion, coexistence must focus on *economic* issues triggered by adventitious presence of GMOs. For conventional crops, adventitious presence may trigger labelling, which may lead to loss of income or difficulty selling the crops. For organic crops, which cannot be produced from GMOs under Community law, contamination has potentially more severe economic consequences. The probability of admixture of GM and non-GM crops and the practices to avoid it depend on the type of crop and on geographic factors like natural conditions, farm structure, and field sizes. Coexistence measures must therefore address the feasibility and cost of management practices to avoid admixture and adventitious presence.[53]

The document accepted that some Member States wanted to ban GM crops in all or part of their territories, but concluded that mandatory GM-free zones should not be permitted in the EU: the protection of mere economic interests by strong limits on fundamental liberties cannot be justified legally, and the approach 'runs counter to the very principle of co-existence'.[54] Fischler also reflected on the Community's possible role in allocating liability. He noted that economic loss from admixture raises difficult questions of causation. If mandatory management rules established a standard of behaviour, then the producer who did not comply could be presumed liable; alternatively, collective funds could compensate for economic losses.[55]

In considering the role of the Community institutions in helping to achieve coexistence, Fischler indicated that they could act as co-ordinator and advisor, or they could legislate. In the role of co-ordinator and advisor, they would gather information, provide advice, and issue guidelines, leaving Member States to take

---

[49] See, eg Scientific Committee on Plants, Opinion Concerning the adventitious presence of GM seeds in conventional seeds (SCP/GMO-Seeds-Cont/002-Final, 13 March 2001) ('SCP Opinion').

[50] A-K Bock et al, *Scenarios for Co-Existence of Genetically Modified Conventional and Organic Crops in European Agriculture: Tech. Rep. EUR 20394* (JRC, European Commission, 2002).

[51] A Messean et al, *New Case Studies on the Coexistence of GM and non-GM Crops in European Agriculture: Tech. Rep. EUR 22012* ('*New Case Studies*') (JRC, European Commission, 2006).

[52] European Commission Press Release IP/03/314, *GMOs: Commission Addresses GM Crop Coexistence*, Brussels, 5 March 2003.   [53] Fischler Communication (n 44 above) 2–4.

[54] Ibid 6. See also Written Question E-1891/02, Eluned Morgan to the Commission [2003] OJ C28E/149 (rejecting GM-free zones).   [55] Fischler Communication (n 44 above) 7.

measures consistent with Community law and suited to their own geographical, environmental, and agricultural situations. In contrast, a legislative approach would involve adoption of a framework measure to establish basic principles, with details to be provided by Member States; alternatively, the European Commission could impose specific and detailed requirements for risk management at Community level. Fischler seemed to prefer the role of co-ordinator and advisor (based on the principle of subsidiarity), which would allow Member States to choose 'efficient and cost-effective measures', but he did not exclude Community legislation, should it prove necessary.[56]

Some in the agriculture sector disagreed with Fischler's preference. For example, COPA-COGECA, an umbrella organization that represents National Farmers Unions and National Agriculture Cooperative Federations (15 million farm workers and 40,000 co-operatives), agreed that coexistence should focus on economic aspects and preferred that the GM sector develop in a way that safeguards organic and conventional agriculture, but without imposing additional costs on producers. COPA-COGECA insisted that producers who follow rules for production should not be held liable for adventitious presence, and that management and other measures be developed to ensure identity preservation. The organization did not support Fischler's emphasis on subsidiarity: '[t]he GMO-issue is transnational and, therefore, a Community issue', so a flexible Community framework should be enacted.[57]

3.3 The 2003 Commission Recommendation

A few months later, in July 2003, the European Commission issued its influential and controversial Recommendation on guidelines for the development of national strategies and best practices to ensure the coexistence of genetically modified crops with conventional and organic farming.[58] This agreed with Fischler that '[n]o form of agriculture, be it conventional, organic, or agriculture using [GMOs] should be excluded in the European Union'.[59] Farmers should be allowed to choose the type of crops they grow, but consumer choice should also be protected through traceability, labelling, and an agricultural sector that can maintain the required production systems.

As already noted, agriculture has often been required to preserve the identity of crops, and the 2003 Commission Recommendation could observe that

[t]he coexistence of different production types is not a new issue in agriculture. Seed producers, for example, have a great deal of experience of implementing farm

---

[56] Ibid 7–9.
[57] COPA-COGECA, *Comments on the Communication from Mr. Fischler to the Commission on the Co-existence of Genetically Modified, Conventional and Organic Crops* (PR(03)46F1, 19 May 2003) 2–3 and 5.   [58] [2003] OJ L189/36.
[59] Ibid Preamble (1).

management practices to ensure seed purity standards. Other examples of segregated agricultural production lines include yellow dent field maize for animal feed, which successfully coexists in European agriculture with several types of 'speciality maize' grown for human consumption, and waxy maize grown for the starch industry.[60]

Because the protection of health and the environment are already addressed as part of GM authorization under the Deliberate Release Directive, the 2003 Commission Recommendation (like Fischler) addressed other issues: economic loss from the admixture of GM and non-GM crops and the management measures designed to minimize admixture.[61] 'Farm structure and farming systems, and the economic and natural conditions ... are extremely diverse' in the Member States,[62] and for this reason the European Commission echoed Fischler's preference that Member States should develop their own measures for coexistence, with information, support, and advice from itself.[63]

3.3.1 Guidelines for Member State strategies

The 2003 Commission Recommendation provides guidelines for Member States as they develop their national strategies and best practices. The European Commission did not insist on any particular form of policy instrument for coexistence, but instead welcomed voluntary agreements, legislation, soft law, or a combination of these instruments, as long as they would 'achieve effective implementation, monitoring, evaluation, and control'.[64] It should also be mentioned that 2003 Commission Recommendation itself sets out only non-binding guidelines. They are not measures to be adopted, but instead general principles to apply and factors to consider in designing Member State-specific measures. In terms of scope, the guidelines apply 'from the seed to the silo' and address commercial seed and crop production, but not experimental releases.[65]

Principles to support development of Member State coexistence strategies include transparency, stakeholder involvement, and management measures based on the best available scientific evidence. That is, management measures should allow cultivation of both GM and non-GM crops and ensure that 'non-GM crops remain below the legal thresholds for labelling and purity standards with respect to genetically modified food and feed and seeds, as defined by Community legislation'.[66] Those measures should be proportional, however, and not stricter than necessary to meet legislative tolerance thresholds. Member States should focus on authorized GM varieties that are appropriate for their territories and use a coexistence system built on existing means of crop

---

[60] Ibid Annex, para 1.1.
[61] Ibid Annex, para 1.2. See M Lee, 'The governance of coexistence between GMOs and other forms of agriculture: a purely economic issue?' (2008) 20 Journal of Environmental Law 193 (asserting that the European Commission's narrow understanding of coexistence as an economic issue restricts Member States and ignores relevant environmental concerns).
[62] 2003 Commission Recommendation, Preamble (6).   [63] Ibid Preamble (4)–(8).
[64] Ibid Annex, para 2.1.8.   [65] Ibid Annex, para 1.5.   [66] Ibid Annex, paras 2.1.1–2.

segregation. Measures should emphasize farm-scale management and encourage co-operation and voluntary arrangements between farmers.[67] Two recommendations that have been applied in Member States are that operators who introduce a new production type should be responsible for limiting gene flow and that farmers who plan to grow GM varieties should inform neighbours.[68]

With regard to liability for GM admixture, the 2003 Commission Recommendation states that

[t]he type of instruments adopted may have an impact on the application of national liability rules in the event of economic damage resulting from admixture. Member States are advised to examine their civil liability laws to find out whether the existing national laws offer sufficient and equal possibilities in this regard. Farmers, seed suppliers and other operators should be fully informed about the liability criteria that apply in their country in the case of damage caused by admixture.[69]

Importantly, it is then suggested that an existing or new insurance scheme might help to compensate damage from admixture of GM and other crops.[70]

In addition, the 2003 Commission Recommendation provides a list of factors for Member States to consider in developing national strategies and best practices. These factors focus on more practical concerns, such as the level of coexistence to be achieved, sources of admixture, threshold values for labels, characteristics of specific crops, and region-specific considerations.[71] Different levels of coexistence include crops on a single farm (during the same or successive years), on neighbouring farms (in the same year), or in the region.[72] Sources of admixture also vary—pollen transfer over long or short distances, mixing of crops at harvest or post-harvest, transfer of seeds in the production chain or by animals, volunteers (perhaps from seed left from prior years), and seed impurities. These may have cumulative effect, which Member State measures should recognize. Moreover, some crops are more liable to generate volunteers or outcrossing, factors that must also be considered.[73] As indicated, measures should take account of the legal thresholds for labelling crops with GM presence, both for conventional and organic crops, with the European Commission noting that regulatory thresholds for crops and for seeds may eventually differ.[74]

---

[67] Ibid Annex, paras 2.1.3–7. In particular, '[r]egion-wide measures should only be considered if sufficient levels of purity cannot be achieved by other means. They will need to be justified for each crop and product type': ibid Annex, para 2.1.5.

[68] Ibid Annex, para 2.1.7. In addition, farmers should be able to choose their production type 'without imposing the necessity to change already established production patterns in the neighbourhood'.

[69] Ibid Annex, para 2.1.9.

[70] Ibid. Other principles focus on monitoring and evaluation, EU-level exchange of information, and shared research results: ibid Annex, paras 2.1.10–12.

[71] Ibid Annex, para 2.2.

[72] Ibid Annex, para 2.2.1. The share and distribution of GM crops in a region and geographic factors are significant: ibid Annex, para 2.2.6.

[73] Ibid Annex, paras 2.2.2. and 2.2.4. The European Commission looked forward to '[b]iological methods to reduce gene flow': Ibid Annex, para 2.2.7.

[74] Ibid Annex, paras 2.2.3 and 2.2.5. The 2003 Commission Recommendation noted that 'for seed production, specific legislation currently being prepared by the Commission will be adopted'.

The European Commission also provided an 'indicative catalogue' of measures, including on-farm measures, that might 'in varying degrees and in various combinations' be part of a Member State's coexistence strategy.[75] Some of these are technical on-farm measures for crop production (for example, isolation distances, choice of planting dates, and cleaning seed drills), harvest and post-harvest measures (for example, cleaning harvesting equipment), and transportation and storage measures (for example, physical segregation).[76] Monitoring schemes, on-farm record keeping (including that in respect of traceability and labelling), and the land register required by the Deliberate Release Directive will help Member States to monitor GM developments.[77]

Although the European Commission rejected the idea of mandatory prohibitions of GM crops (except, perhaps, exceptional regional measures for specific crops and production types),[78] the guidelines indicated that voluntary co-operation of farmers would be appropriate. Notice to neighbouring farms (preferably before seed is ordered) is an important step. Farmers could agree to establish zones of a single production type, which would reduce costs of crop segregation. Alternatively, producers could cluster fields with similar crop varieties, plant varieties with different flowering times, use different sowing dates, or co-ordinate crop rotations.[79]

3.3.2 Responses to the 2003 Commission Recommendation

Commissioner Fischler presented the 2003 Commission Recommendation to members of the European Agriculture Council in September 2003. Although the Council welcomed it as a step toward coexistence, Member State representatives expressed important concerns. Some wanted to consider GM-free zones, if necessary, in some regions. Others called for regulatory adoption of thresholds for GM presence in seeds and for discussion of a threshold for GM content in the case of organic agriculture. Still others, concerned about liability and possible distortions of competition, wanted to consider Community legislation to govern liability in case of 'contamination' of conventional or organic crops.[80]

Subsequently, the European Parliament issued a report on coexistence in December 2003,[81] followed by a Resolution, published in April 2004.[82] It emphasized the importance of information about GM presence in seeds

---

[75] Ibid Annex, para 3.   [76] Ibid Annex, paras 3.2.1–3.
[77] Ibid Annex, paras 3.4–6. Other measures might include training for farmers, exchange of information, and dispute-settlement procedures: Ibid Annex, paras 3.7–9.
[78] Ibid Annex, para 2.1.5.   [79] Ibid Annex, para 3.3.
[80] Presidency Conclusions, 2528th Council Meeting, Agriculture and Fisheries, Brussels, 29 September 2003 (Presse 270) 5–6.
[81] European Parliament, *Report on Coexistence Between Genetically Modified Crops and Conventional and Organic Crops* A5-0465/2003, 4 December 2003.
[82] European Parliament Resolution, Coexistence of GM crops with conventional and organic crops (2003/2098(INI)) [2004] OJ C91E/680.

and demanded a labelling threshold at the detection level. In addition, there was call for a number of strict measures to foster coexistence, such as formal enactment of binding Community rules on coexistence, as well as Member State measures to safeguard coexistence. The European Parliament further wished for legally binding definitions of 'adventitious' and 'technically unavoidable' and demanded Community legislation to allocate liability for economic loss and to require insurance as a condition for authorization of GMOs. Indeed, it believed that no further GMOs should be approved until binding rules for coexistence and a system of liability, based on the polluter pays principle, were in place.[83]

The European Economic and Social Committee (ECOSOC), eager to join the debate, issued its own-initiative opinion on coexistence, both to comment on important substantive issues and to propose areas to be governed by Community or Member State legislation.[84] Influenced by the scepticism or even hostility of EU citizens and the fact that coexistence affects agriculture, the food sector, land use, consumer protection, and environmental protection, the ECOSOC called for strict measures.[85] Coexistence, the ECOSOC insisted, 'cannot be limited to the economic aspects of cultivation alone, but is an integral part of the risk management and prevention laid down by law'.[86] Like the European Parliament, the ECOSOC emphasized the importance of seed and demanded Community legislation to govern both labelling of GM presence in seeds (at the detection threshold) and purity requirements for non-GM seeds.[87] The ECOSOC also wanted Community legislation to regulate 'targets, results, legal provisions and minimum standards of good practice', and to establish principles of civil liability for users and suppliers of GMOs. Member States and regions should govern other aspects, including on-farm measures and regional provisions to allow or ban certain GMOs.[88]

NGOs, too, reacted to the 2003 Commission Recommendation. As might be expected, reactions were critical. For example, a position paper from the FOEE called for strict liability for GMOs, as well as a legal framework for coexistence to ensure farmer choice and to protect those who grow organic or conventional crops.[89] FOEE has continued to engage with the issue and responds regularly to

---

[83] Ibid 681–2. On the polluter pays principle in the Community, see, eg MR Grossman, 'Agriculture and the polluter pays principle: an introduction' (2005) 59 Oklahoma Law Review 1, 11–25; and, more generally, N de Sadeleer, *Environmental Principles: from Political Slogans to Legal Rules* (Oxford: Oxford University Press, 2002) 21–60.

[84] Opinion of the European Economic and Social Committee on the 'Co-existence between genetically modified crops, and conventional and organic crops' [2005] OJ C157/155.

[85] Ibid paras 2.7–8.   [86] Ibid para 3.2.2.

[87] Ibid paras 3.5, 4.6, and 4.9.1. The ECOSOC believed that the extra costs of coexistence should be 'shared out and compensated for according to the polluter pays principle', and that those who supply and use GM crops should bear the cost: ibid paras 4.1.2 and 4.8.3.

[88] Ibid paras 4.9.1–2.

[89] FOEE, *The Coexistence of GMO Crops with Conventional and Organic Agriculture* (24 April 2003).

legislative and other developments.⁹⁰ Greenpeace reacted to the 2003 Commission Recommendation and in 2005 joined other NGOs to recommend more freedom to regions in banning GMOs, as well as strict liability provisions.⁹¹ Moreover, FOEE, Greenpeace, and other organizations sought professional advice. The resulting document heavily criticized the 2003 Commission Recommendation, in particular for its focus on labelling thresholds in implementing coexistence, its restriction of coexistence to economic issues, and the possible harm to organic production. The advice (which FOEE cites as a legal opinion) even went so far as to call the 2003 Commission Recommendation 'fundamentally flawed' and asserted that its approach has 'no basis in Community legislation and [is] wrong in law'.⁹²

Nevertheless, one issue that it raised may be significant: the use of the legislative threshold for labelling as the goal for coexistence. The advice noted that references to coexistence in both the Deliberate Release Directive and the Food and Feed Regulation prescribe that 'the unintended presence of GMOs in other products' should be avoided,⁹³ and the authors believed that coexistence measures are not 'as a matter of law constrained by the labelling thresholds'.⁹⁴ Indeed, coexistence measures that permit a threshold level of GM content do not attempt to avoid the unintended presence of GMOs; as a result, GM presence may no longer be adventitious or technically unavoidable, as laid down in the labelling exemptions in Community legislation. As the advice argues, coexistence measures 'are directed towards preventing the avoidable contamination of non-GM produce and not to merely minimising such contamination to (acceptable) tolerance levels'.⁹⁵

3.4 European Commission Research on Coexistence

The JRC of the European Commission has sponsored several studies of coexistence.⁹⁶ A synthesis report published in 2002 focused on agronomic

---

⁹⁰ See, eg FOEE, *Co-existence—Looking for a Real Solution?* (no date) (link from <http://www.foeeurope.org/>, last accessed 1 July 2009); and FOEE, Media Briefing, *FOEE Commentary on the European Commission Communication* (10 March 2006).
⁹¹ Letter from European Environmental Bureau et al to the Commissioner, 'Time to change European policy on GMOs in agriculture' (17 March 2005).
⁹² Advice of Paul Lasok QC and Rebecca Haynes (2005) (available at <http://www.greenpeace.org.uk/media/press-releases/european-gm-crop-co-existence-recommendations-legally-flawed>, last accessed 26 June 2009) paras 20 and 55.
⁹³ Respectively, [2001] OJ L106/1, Art 26a; and [2003] OJ L268/1, Preamble (28).
⁹⁴ Advice of Lasok and Haynes (n 92 above) para 26.   ⁹⁵ Ibid para 32.
⁹⁶ The EU continues to contribute to the study of coexistence. A project called 'Co-Extra' focuses on 'GM and non-GM supply chains: their CO-EXistence and TRAceability'. Co-Extra works as a consortium with international partners to enable coexistence and improve traceability: see <http://www.coextra.eu/>, last accessed 1 July 2009. The European Commission's network group, COEX-NET, assists in the field of coexistence: Commission Decision (EC) 2005/463 establishing a network for the exchange and coordination of information concerning coexistence of genetically modified, conventional and organic crops [2005] OJ L164/50. The European

and economic issues connected with GM crop production in the EU.[97] This report looked in part at the role of conventional and organic farmers in avoiding the adventitious presence of GM material in their crops, with an analysis of coexistence measures and some associated costs. It assumed that the burden of these measures would fall on non-GM farmers.[98] The 2003 Commission Recommendation, however, indicated that 'during the phase of introduction of a new production type in a region, operators (farmers) who introduce the new production type should bear the responsibility of implementing the farm management measures necessary to limit gene flow'.[99] Because guidelines in the 2003 Commission Recommendation led Member States to enact measures to be implemented by GM farmers, the JRC initiated a new series of studies and, in 2006, published a second synthesis report, *New Case Studies*.

With a focus on seed and crop production (especially for maize—a major crop in the EU and the only GM crop approved for commercial cultivation) and a scope limited to 'agricultural production up to the farm gate', the JRC case studies were intended to identify and analyse agronomic measures for coexistence, estimate gene flow and adventitious presence at a landscape scale, evaluate specific measures for meeting various thresholds of purity for seed production, and study the level of adventitious presence over long time periods.[100] For example, for maize crop production, *New Case Studies* identified the main sources of adventitious presence: 'traces of GM seeds in non-GM seed lots, cross-pollination from neighbouring GM fields, and the sharing of harvesting machinery between GM and non-GM fields'.[101] Both the geographic position of GM and non-GM fields (with respect to wind direction) and the size of adjacent fields affect gene flow. Measures to minimize adventitious presence include isolation distances, non-GM buffers around GM crops, and (although less reliable) planting varieties with different flowering dates. The economic impact of these practices varies, depending on location and other factors.

Commission also established the European Coexistence Bureau, to provide technical help with coexistence. In co-operation with Member States, the Bureau is developing crop-specific guidelines and best agricultural management practices for coexistence. A best practice document for maize production is expected by 2010: European Coexistence Bureau, JRC, Home Page <http://ecob.jrc.ec.europe.eu/>, last accessed 1 July 2009.

[97] Bock et al (n 50 above).   [98] *New Case Studies* (n 51 above) 17.
[99] 2003 Commission Recommendation, Annex, para 2.1.7. The 2003 Commission Recommendation continues, '[f]armers should be able to choose the production type they prefer, without imposing the necessity to change already established production patterns in the neighbourhood'.
[100] *New Case Studies* (n 51 above) 10. Some seeds have a long life and dormancy time and can build up in the soil. For each case study, the report '(i) identifies key sources of adventitious GM presence ..., (ii) estimates the levels of adventitious presence ... with current and adapted farming practices ..., (iii) proposes adapted agronomic practices and technical measures ... and (iv) evaluates the techno-economic feasibility of such proposals'.
[101] Ibid 11. Cleaning harvesters costs about €50–60 per operation: ibid 13.

The JRC report also considered seed maize production, as well as sugar beet, cotton, and oilseed rape.[102]

New Case Studies reached several conclusions about coexistence. First, on the basis of expert opinions and simulations, the report concluded that coexistence in seed production is 'technically feasible for a threshold of 0.5%, with few or no changes in current practices', at least for non-GM and GM seed production; for GM maize crops and non-GM seed production, larger separation distances would be required.[103] Secondly, GM presence in seed of 0.5 per cent or lower can facilitate coexistence in crop production at the regulatory threshold of 0.9 per cent, though some GM maize growers will need additional measures. Thirdly, *New Case Studies* concluded that model simulations are valuable for estimating adventitious presence in various landscapes and using various agricultural practices, but cannot replace more time-consuming and costly field research.[104]

Although *New Case Studies* was heralded as a science-based reference that concludes that coexistence at regulatory thresholds is feasible,[105] it has engendered criticism. FOEE, for example, deduced from the report that successful coexistence would require the designation of certain areas for clustered fields of GM crops or case-by-case assessments of individual fields, using a 'flexible decision-support system' that would analyse isolation distances and other geographic factors. In light of the labelling requirement that excepts adventitious or technically unavoidable presence of authorized GMOs up to the 0.9 per cent threshold, FOEE wondered if acceptance of 'built-in' contamination, as both the JRC and the European Commission seem to do, might mean that such GM contamination can no longer be considered adventitious or technically unavoidable. Coexistence, according to FOEE, 'must aim to avoid GM contamination of non-GM crops, not aim to allow a certain level of contamination'.[106]

---

[102] Ibid 11–13. For production of seed maize, seed purity and contamination from farm machinery are controlled carefully, so cross-pollination is the main source of adventitious GM presence. Depending on the threshold required, coexistence between GM and non-GM seed production is feasible, perhaps with increased separation. Coexistence between GM crops and non-GM seed is more difficult, because large separation requirements (400–600 metres) might preclude GM crop production: ibid 13–14.

[103] Ibid 15. Current isolation distances are 100–200 metres, which would increase to 400–600 metres for coexistence of non-GM maize seed and GM maize crops. German researchers found that 'GM DNA in harvested grain resulting from outcrossings can be managed to levels below 0.9% by simply planting 20 m of conventional maize as a pollen barrier between adjacent fields': WE Weber et al, 'Coexistence between GM and NON-GM maize crops—tested in 2004 at the field scale level (Erprobungsanbau 2004)' (2007) 193 Journal of Agronomy and Crop Science 79, 79.

[104] *New Case Studies* (n 51 above) 15–16.

[105] European Commission Press Release IP/06/230, *New Report Considers Co-existence of GM and Non-GM Crops and Seeds*, Brussels, 24 February 2006.

[106] FOEE (n 10 above) 8. Built-in GM presence is not adventitious, FOEE argues, and technically unavoidable should not apply to contamination that could be avoided with stronger measures. See Advice of Lasok and Haynes (n 92 above).

4. Coexistence Issues

Concerns about coexistence often focus on several recurring issues. GM presence in seeds influences the GM content of crops, so a brief discussion of Community attempts to establish seed thresholds is important here. The organic sector often expresses concern about the effect of GM crops on organic products, and a brief overview of Community restrictions on GM content in organic products follows. In addition, this section will raise the question of GM-free areas and draw brief attention to the vexing issue of liability.

4.1 Seeds

Although seed purity, in the EU and worldwide, is critical for coexistence, 100 per cent purity in seeds is impossible, a reality reflected in seed tolerances.[107] Because much seed is 'increased' in countries where GM varieties are grown, adventitious presence (including presence of varieties not yet authorized in Europe) in non-GM varieties cannot be avoided completely. Asynchronous approvals raise particular challenges; that is, GM varieties have been authorized in some countries (for example, the country of export), but not in others (for example, the country of import). This situation 'will continue to disrupt trade, pose high liability risks to the entire food and feed chain, erode confidence among consumers and increase the burden on regulatory agencies of importing countries'.[108] The European Seed Association (ESA) insisted that GM thresholds should have been fixed first for seeds and only thereafter for products derived from seeds. Without reasonable thresholds for adventitious presence, the supply of seed for European farmers is in jeopardy. Moreover, the ESA noted that, absent 'clear and practical regulations' that recognize the inevitability of adventitious presence, the seed industry is subject to liability and risk.[109]

Thus, in the context of rules for coexistence, seed production deserves special attention. As Commissioner Fischer Boel made clear, seed purity is critical, but controversial:

[a]n important issue in relation to coexistence [is] the purity standards that apply to seeds. There was broad agreement that common labelling thresholds for seeds are

---

[107] ABE/EuropaBio (n 3 above) 4.

[108] Schumacher (n 17 above). See also EuropaBio, *Green Biotechnology Manifesto* (2007) 10 (referring to delays, caused by the 'continued asynchronous approval speed' in the EU, that have caused trade disruptions). EuropaBio is the political voice of the European biotech industry; its members are national biotechnology associations (representing 1,500 companies), corporations, regions, and associates.

[109] ESA, *ESA Position on Thresholds for Adventitious Presence of GM Seeds in Conventional Seeds* (ESA_01.0010, 7 November 2001) 2–3 and 6. Because seed production and trade are international, the ESA preferred an international harmonization of thresholds, perhaps through the Organization for Economic Co-operation and Development (OECD) or the *Codex Alimentarius*.

necessary. In our view, these should be set in such a way that it is in any case possible to respect the labelling threshold for the final product at the end of the food production chain.[110]

4.1.1 Seed thresholds?

In 2001, the Scientific Committee on Plants (SCP), in the Health and Consumer Protection Directorate General, issued the SCP Opinion on the adventitious presence of GM seeds in conventional seeds.[111] The SCP Opinion answered several questions raised by the European Commission as it considered legislation to govern adventitious presence in seeds. The questions focused on the appropriate standards for thresholds for adventitious presence, possible restrictions on production of conventional seeds in fields used for GM production in prior years, and minimum separation distances for specific crops.

On the issue of thresholds, the European Commission queried the wisdom of proposing 0.3 per cent for cross-pollinating crops and 0.5 per cent for self-pollinating and vegetatively propagated crops. The Scientific Committee advised that those thresholds could be achieved only under ideal conditions for seed production. Even that would be more difficult as more GM crops are grown in the EU. Segregation, on the farm and at all stages of harvest, transport, and storage, will play an important role in achieving those thresholds.[112] On the issue of planting sequence, the European Commission had considered proposing that fields used for conventional seeds be free of GM production for five years in the case of fodder, oil, and fibre plants and for two years in the case of other plants. The Scientific Committee advised that seed longevity be considered. That is, seeds with short persistence (soya and maize) would require a one-year break of rotation; seeds with long persistence (oilseed rape, potatoes, and others) would require five years; and seeds with medium persistence (wheat and barley) would require two to three years. The Scientific Committee recommended specific best practices for different crops.[113] On the issue of isolation distances, the European Commission suggested that to avoid GM presence it might double the isolation distances for cross-pollinating species already required under Community seed legislation. The Scientific Committee advised that current isolation distances are sufficient for some oilseed rape (swede rape) and for maize (with additional measures, like barriers), while for hybrid oilseed rape, even

---

[110] Commissioner Fischer Boel and Pröll (n 9 above) 1.
[111] SCP Opinion (n 49 above). This was reviewed and confirmed twice, on 24 April 2002 and 30 January 2003: European Commission, MEMO/03/186, *Questions and Answers About GMOs in Seeds*, Brussels, 24 October 2003, 2.
[112] SCP Opinion (n 49 above) 7 and 9. With some crops (eg hybrid seed), meeting the thresholds will be difficult: ibid 17.
[113] Ibid 11–13. For example, one year is sufficient for soya and maize; oilseed rape may require seven years.

doubled isolation distances would be insufficient.[114] Of course, local practices and environmental conditions play a role.

In addition to these questions, the Scientific Committee reacted to the European Commission's comment that zero tolerance should apply to unauthorized GM seed. On that issue, the Scientific Committee indicated that 'a zero level of unauthorised GM seed is unobtainable in practice', because field crops 'are always subject to unintended pollen and seeds from various sources'.[115]

4.1.2 Draft Community legislation

In light of the SCP Opinion, the European Commission issued a working document on seed thresholds in early 2002.[116] That draft, designed to amend a number of seed marketing measures, would have established break of rotation periods, isolation distances, and thresholds for a number of crops. The recommended thresholds for adventitious presence of approved GM seed in conventional seed were 0.3 per cent for swede rape and cotton; 0.5 per cent for maize, potato, tomato, beet, and chicory; and 0.7 per cent for soya.[117]

The ESA welcomed the regulatory approach of setting thresholds for seed labelling, but believed that the thresholds were unrealistically low. It preferred a threshold level of at least 1.0 per cent and also wanted the seed industry to be responsible for meeting thresholds without other regulatory production requirements (isolation distances and rotation breaks).[118] COPA-COGECA commented that thresholds that are too low may make coexistence impossible, while thresholds that are too high, in comparison to thresholds for food and feed, may raise difficulties for conventional seed producers.[119]

In 2004, the European Commission published the Draft Commission Decision. This document, prepared by the Agriculture and Environment Directorate Generals, specified thresholds values for several crops: 0.3 per cent for oilseed rape and maize (reduced from the 2002 proposal) and 0.5 per cent for sugar beet, fodder beet, potatoes, and cotton. In addition, it provided that Member States may allow seed producers to label conventional crops either with

---

[114] Ibid 13–14.   [115] Ibid 16.
[116] See Draft Commission Directive SANCO/1542/02 (29 January 2002).
[117] Ibid Preamble (3). COPA-COGECA, which had demanded seed thresholds since the 1999 publication of the European Commission *White Paper on Food Safety* COM(1999)719, supported these recommended thresholds: TG Kofoed, 'The EU Legal Framework on Genetically Modified Organisms is Still Missing the Seed Side to be Complete! Freedom of Choice Conference' (Freedom of Choice Conference, n 5 above).
[118] See ESA, *Position Paper on Draft Commission Directive* (ESA_02.0001, 1 March 2002). The ESA advised that thresholds also be set for varieties authorized in OECD countries and approved commercially: ESA (n 109 above) 3–4.
[119] R Volanen, *Comments of COPA/COGECA* (SEM(02)099P3-jf, 11 March 2002) 1. Other comments focused on the omission of unauthorized GMOs, isolation distances, and rotation breaks; it was advocated that thresholds should apply to imported seed.

the exact value of GM admixture or with a guaranteed content of GM seeds below a specified threshold.[120]

In response, the ESA urged the European Commission to provide a threshold of 0.5 per cent for maize, rather than 0.3 per cent, which the ESA believed to be economically infeasible and unnecessarily burdensome. In addition, the ESA urged the European Commission not to allow voluntary labelling, which would interfere with the internal market and be inconsistent with Community traceability and labelling measures.[121] Like the ESA, EuropaBio criticized the lower maize threshold and provision for voluntary labelling.[122]

The European Commission subsequently withdrew its proposal to establish thresholds for the adventitious presence of approved GM seeds in conventional seed.[123] Seed industry professionals, however, believe that legislation to set labelling thresholds for adventitious presence of GMOs remains critical.[124] Moreover, while various seed associations have called for a reasonable approach to thresholds, opponents of GMOs have demanded stricter limits. The Network of the GMO-Free European Regions and Local Authorities, for example, demanded a threshold equivalent to the limit of detection.[125] Similarly, FOEE and the Assembly of European Regions demanded that the European Commission 'guarantee a GMO free seed supply and EU wide standards for seed labelling at the practical detection limit'.[126]

4.1.3 Still no thresholds?

The threshold level for seeds is still at issue. In 2006, the European Commission noted that '[n]o labelling thresholds [for seed lots] have been defined yet, which means that seed lots containing detectable traces of GMOs have to be labelled as containing GMOs and the unique identifiers of the GMOs have to be

---

[120] Draft Commission Decision establishing minimum thresholds for adventitious or technically unavoidable traces of genetically modified seeds in other products (July 2004) 4, Annex (available at <http://www.saveourseeds.org/downloads/com_draft_seeds_04_2004.pdf>, last accessed 20 May 2009). Soya was omitted and the threshold values are the percentage modified DNA/total genomic DNA of the species. Commissioner Fischer Boel once called for a threshold of 0.1%, the detection level: Presidency Conclusions, 2578th Council Meeting: Agriculture and Fisheries, Luxembourg, 26 April 2004 (Presse 180) 8.

[121] ESA, *Contribution to Draft Commission Decision* (ESA_04.0099, 22 April 2004) 1–2.

[122] EuropaBio, *Comments on Draft Commission Decision* (29 April 2004) 1–2.

[123] EuropaBio, Press Release, *Contradictory Signals by the European Commission* (8 September 2004) (alleging that 'the Commission has chosen to ignore its responsibility').

[124] See ESA, *Resolution on Labelling Thresholds for the Adventitious Presence of GMOs in Seed of Conventional Plant Varieties* (ESA_06.0474, 17 October 2006).

[125] S Cenni, 'Agriculture, Biodiversity and Rural Development at Regional Level: Which Coexistence? Freedom of Choice Conference' (Freedom of Choice Conference, n 5 above). The European Parliament had suggested that 'the limit value for the labelling of GMO-impurities in seed should be set at the ... detection threshold': European Parliament Resolution, Coexistence of GM crops with conventional and organic crops (2003/2098(INI)) [2004] OJ C91E/680, 681.

[126] FOEE and the Assembly of European Regions, *Safeguarding Sustainable European Agriculture* (17 May 2005).

mentioned on the label'.[127] In 2007, it reported that Member States and industry both expressed the need for thresholds for adventitious presence in seeds,[128] and it planned to reconsider the issue of labelling thresholds for seeds, in light of further studies carried out in 2007.[129]

At its December 2008 meeting, the Environment Council emphasized the importance of seed thresholds. The Council noted that the European Commission will complete a study on the establishment of seed thresholds and reaffirmed the need for labelling thresholds for the adventitious presence of GMOs in conventional seeds. The Council insisted that thresholds be 'at the lowest practicable, proportionate and functional levels for all economic operators, [and] must contribute to ensuring freedom of choice to producers and consumers of conventional, organic and GM products alike'.[130] The Council invited the European Commission to set those thresholds as soon as possible, using current scientific information and, in its April 2009 report, the European Commission indicated that an impact assessment would be an important basis for the legislative text on seed thresholds.[131]

Meanwhile, seed professionals have become ever more insistent. In October 2006, the ESA had again demanded 'clear and legally certain rules for the adventitious presence of trace amounts of GM material in conventional seed'.[132] In the absence of thresholds, the requirement that seed with detectable traces of GMOs must be labelled imposes a costly burden and raises legal uncertainties.[133] In 2007, the ESA and EuropaBio highlighted the importance of seed thresholds to plant breeders and seed producers. Such thresholds would avoid trade disputes and ensure respect for freedom of choice both in Europe and in other countries.[134] More recently, in March 2009, after 10 years of industry demand for seed thresholds, the ESA 'considers it incomprehensible and

---

[127] European Commission, *Report on the Implementation of National Measures on the Coexistence of Genetically Modified Crops with Conventional and Organic Farming* COM(2006)104 ('2006 Implementation Report').

[128] European Commission, *Second Report on the Experience of Member States with GMOs Placed on the Market Under Directive 2001/18/EC on the Deliberate Release into the Environment of Genetically Modified Organisms* COM(2007)81, 5.

[129] European Commission, *Commission Staff Working Document Accompanying the Communication on the Mid Term Review of the Strategy on Life Sciences and Biotechnology* SEC(2007)441, 41.

[130] Council Conclusions on Genetically Modified Organisms, 2912th Council Meeting: Environment, Brussels, 4 December 2008, paras 13 and 14.

[131] European Commission, *Report from the Commission to the Council and the European Parliament on the Coexistence of Genetically Modified Crops with Conventional and Organic Farming* COM(2009)153 ('2009 Implementation Report'), 3.

[132] See ESA (n 124 above). In March 2007, EuropaBio insisted that the European Commission propose a labelling threshold for seed and that the measure receive 'utmost priority': n 108 above 8–9.

[133] The European Parliament noted that if GMOs were present in seed, farmers cannot claim that its presence is 'adventitious and technically unavoidable' and must therefore label the crops, with potential loss of income: European Parliament Resolution (n 125 above) 681.

[134] ESA and EuropaBio, *Adventitious Presence—Bringing Clarity to Confusion* (2007) 13.

unacceptable that the Commission still is not answering this unanimous request'.[135] In a strongly worded call for action, the ESA demanded 'a legislative proposal on thresholds for adventitious presence of GMOs in conventional seed ... without further delay'.[136] The 10 years of inaction is 'financially and legally intolerable and threatens the future of the EU's plant breeding and variety development and with that its sustainable agriculture'.[137]

4.2 Organic Production

Consumer demand for organic products has increased in recent years, and some of that growth, in the EU and in the United States, may be attributed to consumers 'who seek to avoid products derived from [GM] crops'.[138] In 2007, 32.2 million hectares, on over 1.2 million farms worldwide, were managed organically. In the EU, about four per cent of the agricultural land area (7.2 million hectares on about 180,000 farms) used organic management; in contrast, North America had only about 2.2 million hectares under organic management, just 0.6 per cent of the agricultural area.[139] Researchers have indicated, however, that the organic share of maize, oilseed rape, and sugar beet in the EU is extremely low, less than 0.5 per cent. Besides, organic expansion in the EU may be expected to focus on fruits and vegetables, rather than arable crops.[140] In Spain, with the most GM maize in the EU, GM and traditional varieties of maize have coexisted 'without economic and commercial problems' and with only a few incidents of adventitious GM presence in organic crops.[141]

Nonetheless, much of the objection to GM crops comes from the organic sector. For example, the International Federation of Organic Agricultural Movements (IFOAM) would prefer to ban GMOs and genetic engineering.[142] Part of the objection from organic growers and consumers might be philosophical. Analogizing to the pesticide spray drift that moves from conventional farms to nearby organic farms, some argue that GM farming cannot coexist with organic farms. By contrast, to those who promote GM crops, coexistence

---

[135] ESA, *European Seed Industry Call for Action* (ESA_09.0133.1, 2 March 2009).
[136] Ibid.   [137] Ibid.   [138] Pew Initiative on Food and Biotechnology (n 1 above) 3.
[139] H Willer, 'The World of Organic Agriculture 2009: Summary' in H Willer and L Klicher (eds), *The World of Organic Agriculture: Statistics and Emerging Trends 2009* (Bonn: International Federation of Organic Agricultural Movements, 2009) 19, 22, and 23. Italy, Spain, and Germany lead the EU in land under organic management.
[140] G Brookes and P Barfoot, *Co-existence of GM and Non GM Arable Crops: The Non GM and Organic Context in the EU* (Dorchester: PG Economics Ltd, 2004) 4 and 11.
[141] G Brookes and P Barfoot, *Co-existence of GM and Non GM Crops: Case Study of Maize Grown in Spain* (Dorchester: PG Economics Ltd, 2003) 1, 3, and 7–9.
[142] See IFOAM, *Position on Genetic Engineering and Genetically Modified Organisms* (P01, 2002) (available at <http://www.ifoam.org/press/positions/pdfs/IFOAM-GE-Position.pdf>, last accessed 24 May 2008). IFOAM recently advocated a 0.1% GM threshold for all seed: G Azeez, 'The Position of the European Organic Farming Movement and Experience of GM Contamination—Freedom of Choice Conference' (Freedom of Choice Conference, n 5 above).

involves managing conflicting values and also requires certain technical issues to be resolved such as ... minimizing the physical transfer of material between GM and non-GM systems. ... But the problems are much deeper than that because the differences between biotechnology-based and organic agriculture are so fundamental that both systems are based on totally different ecological rationales.[143]

Indeed, it may be argued, 'GM crops are totally incompatible with agroecologically-based approaches'.[144] Unlike the large-scale monoculture often used with GM production, organic agriculture relies on diversification, local varieties, low input management, and small or medium farms.[145]

For organic producers, certification is based on production standards rather than product testing.[146] Accordingly, regulatory measures in both the EU and the United States preclude the deliberate use of GM materials in organic production. Council Regulation (EEC) 2092/91 on organic production of agricultural products and indications referring thereto on agricultural products and foodstuffs ('Regulation 2092/91'), effective only until the end of 2008, provided that organic production implies that GMOs and/or their products may not be used.[147] Moreover, product labels for food could refer to organic production only if 'the product has been produced without the use of genetically modified organisms and/or any products derived from such organisms'.[148] This Regulation did not specifically address gene flow of GMOs into organic crops; though it allowed a threshold for adventitious presence, no threshold was established.[149] In 2004, the *European Action Plan for Organic Food and Farming* recommended new regulatory provisions to address the issues of labelling and thresholds for adventitious presence of GMOs.[150]

In 2005, the European Commission proposed a new Regulation to govern organic production.[151] In the proposal, it considered the place of GMOs in organic farming:

[t]o maintain consumer confidence, the use of GMOs and of products produced from or by GMOs should continue to be prohibited in organic farming, as it is the case in the current Regulation. Despite this, in cases where products have been accidentally

---

[143] See MA Altieri, 'The myth of coexistence: why transgenic crops are not compatible with agroecologically based systems of production' (2005) 25 Bulletin of Science, Technology & Society 361, 363.   [144] Ibid 364.
[145] Ibid 364–7.   [146] See Brookes (n 27 above) 8.   [147] [1991] OJ L198/1, Art 6.
[148] Ibid Art 5(3)(h).
[149] The 0.9% labelling threshold for GM content in non-GM food or feed was more stringent than organic standards: ABE/EuropaBio (n 3 above) 5. Foods produced under organic standards had a 5% tolerance for non-organic material: see Regulation 2092/91, Art 5, which allowed organic labelling if at least 95% of agricultural ingredients were produced with organic production methods.
[150] European Commission, *European Action Plan for Organic Food and Farming* COM(2004) 415, 5.
[151] European Commission, *Proposal for a Council Regulation Amending Regulation No 2092/91 on Organic Production of Agricultural Products and Indications Referring Thereto in Agricultural Products and Foodstuffs* COM(2005)671.

contaminated by GMOs, the current organic rules do not prohibit the simultaneous labelling as organic and GMO. As announced ..., *the proposal prohibits the use of the term 'organic' for GMO labelled products*. Finally, the labelling thresholds for organic and non-organic products should be identical, unless detailed rules foresee specific thresholds for example possibly for organic seeds.[152]

In June 2007, after consultation with the European Parliament, the Council enacted a new Council Regulation (EC) 834/2007 on organic production and labelling of organic products ('Regulation 834/2007').[153] This Regulation repealed Regulation 2092/91, as from 1 January 2009.[154] Article 23(3) of the new Regulation indicates that terms used to designate that a product complies with the rules for organic production 'shall not be used for a product for which it has to be indicated in the labelling or advertising that it contains GMOs, consists of GMOs or is produced from GMOs according to Community provisions'.[155] Thus, under Regulation 834/2007, products that exceed the labelling threshold for adventitious presence of GMOs are no longer considered organic. In consequence, whether from cross-pollination or commingling, the presence of GM material in organic products may trigger marketing problems, and the crop may be rejected by the buyer, causing loss of income.[156]

4.3 GM-free?

The 2003 Commission Recommendation and other Community policy statements make it clear that coexistence should not exclude GM, conventional, or organic farming. Nonetheless, Member State and regional governments, as well as NGOs, have attempted to prevent GM cultivation, either in a whole Member State or regionally.[157]

4.3.1 Member State GM bans

Some Member States have attempted to ban GMOs, using the safeguard clauses in the Deliberate Release Directive and the Food and Feed Regulation or invoking Article 95 of the EC Treaty.

The safeguard clause in the Deliberate Release Directive allows a Member State provisionally to restrict or prohibit use or sale of a GMO or a GM product

---

[152] Ibid 6, para 23 (emphasis added). COPA-COGECA recommend that the same thresholds apply to organic as to other seeds: see Kofoed (n 117 above) 3.
[153] [2007] OJ L189/1.   [154] Ibid Art 39.
[155] Ibid Art 23(3). Some objected to this provision; though the Regulation does not permit GMOs in organic food, products with adventitious presence of GMOs below the 0.9% labelling threshold can still be labelled organic: EurActive, *Accidental GMO Content Permitted in Organic Food* (13 June 2007) (available at <http://www.euractiv.com>, last accessed 27 May 2008).
[156] Pew Initiative on Food and Biotechnology (n 1 above) 8.
[157] Center for Food Safety, *Genetically Engineered Crops and Foods: Regional Regulation and Prohibition* (June 2006) (part of an ongoing project to identify GM-free zones throughout the world).

in its territory if the Member State has new information to demonstrate that the GMO poses a risk to human health or the environment. The new information must be made available since the date of consent or a reassessment of existing information using new scientific information.[158] Similarly, the safeguard clause in the Food and Feed Regulation allows a Member State to take emergency measures if an authorized product constitutes a 'serious risk' to health or the environment.[159]

A number of Member States have invoked these safeguard clauses in attempts to ban GMOs in their territories. In each instance, scientific committees and the European Food Safety Agency (EFSA) found no scientific justification for Member State bans, but the Council has not acted to lift them.[160] Most notably, Austria has continued to press its ban against several GM varieties, including MON810 maize, under the Directive's safeguard clause. In April 2008, the European Commission asked the EFSA for a scientific opinion on additional information submitted by Austrian authorities, and in December 2008 the EFSA again concluded that no scientific evidence of risk to health and the environment justified the ban.[161] MON810 maize has also proved controversial in other Member States. Hungary invoked the safeguard clause in January 2005, and the EFSA again found the ban unjustified, but the Council has failed to lift it.[162] Similarly, in February 2008, France notified its ban of MON810 to the European Commission both under the safeguard clause in the Deliberate Release Directive and as an emergency measure under the Food and Feed Regulation. Once again, the European Commission requested an opinion from the EFSA, which concluded that no new scientific evidence justified invocation of the safeguard clause or an emergency measure.[163]

A matter of some importance is that, despite such lack of scientific evidence, Member States have not been forced to overturn their bans. In March 2009, for

---

[158] Deliberate Release Directive, Art 23.   [159] Food and Feed Regulation, Art 34.
[160] Six Member States (Austria, France, Germany, Luxembourg, Greece, and the United Kingdom) had invoked the safeguard clause, in nine applications, under Council Directive (EEC) 90/220 on the deliberate release into the environment of genetically modified organisms [1990] OJ L117/15, Art 16. Eight were re-notified under the Deliberate Release Directive: European Commission, MEMO/07/117, *Questions and Answers on the Regulation of GMOs in the European Union*, Brussels, 6 March 2007) 6 and Annex 5. Several of the banned products are no longer available for sale.
[161] EFSA, 'Scientific Opinion of the panel on Genetically Modified Organisms on a request from the European Commission related to the safeguard clause invoked by Austria on maize MON810 and T25 according to Article 23 of Directive 2001/18/EC' (2008) EFSA Journal 891, 1.
[162] European Commission (n 160 above) 7. On other bans, see Foreign Agricultural Service (FAS), United States Department of Agriculture, *EU-27: Biotechnology Annual 2008* (GAIN Rep E48137, November 2008).
[163] EFSA, 'Scientific Opinion of the Panel on Genetically Modified Organisms on a request from the European Commission related to the safeguard clause invoked by France on maize MON810 according to Article 23 of Directive 2001/18/EC and the emergency measure according to Article 34 of Regulation (EC) No 1829/2003' (2008) EFSA Journal 850, 1.

example, the Council again rejected European Commission proposals to lift national safeguards in Hungary and Austria.[164] Controversy continues, and in April 2009, the German agriculture minister banned sale and cultivation of MON810 maize.[165]

Article 95 of the EC Treaty offers an additional safeguard, which permits Member States to introduce national provisions after adoption of a Council or European Commission harmonization measure 'based on new scientific evidence relating to the protection of the environment ... on grounds of a problem specific to that Member State arising after the adoption of the harmonisation measure'.[166] Relying on this Treaty provision, Austria notified the European Commission of a proposed ban on the use of GMOs in Upper Austria, a measure justified as a means to protect traditional and organic production systems, nature, the environment, and biodiversity. Austria was concerned about difficulties posed by coexistence of organic and GM crops. To evaluate the scientific justification for the proposed Austrian ban, the European Commission asked the advice of the EFSA, which consulted the Scientific Panel on Genetically Modified Organisms. The EFSA concluded that the proposed scientific justification did not meet the requirements of Article 95(5). Austria did not provide new scientific evidence related to protection of the environment, prove that its concerns about the coexistence of organic and GM crops were environmental, or show that Upper Austria had unique ecosystems. Subsequently, the European Commission adopted a decision rejecting the ban, on the basis that Austria had provided no new scientific evidence to justify the measure.[167] Both Austria and Upper Austria sued for annulment of the European Commission's decision. The Court of First Instance upheld the decision that no new scientific evidence supported the ban and dismissed the actions.[168] In September 2007, the European Court of Justice affirmed the decision of the Court of First Instance and dismissed the appeals of Austria and Upper Austria.[169]

---

[164] 2928th Council Meeting: Environment, Brussels, 2 March 2009 (Presse 53).
[165] 'Germany Bans Monsanto's GM Maize' (14 April 2009) (available at <http://news.bbc.co.uk/go/pr/fr/-/2/hi/Europe/8008181.stm>, last accessed 14 April 2009). Monsanto sued, but in early May 2009, a German court refused to enjoin the ban: Press Release, *Verwaltungsgericht Braunschweig, Genmais Bleibt Verboten* (5 May 2009) (available from <http://www.verwaltungsgericht-braunschweig.niedersachsen.de>, last accessed 12 May 2009).
[166] EC Treaty, Art 95(5). Art 95(4) includes an analogous provision for existing Member State measures.
[167] Commission Decision (EC) 2003/653 of 2 September 2003 [2003] OJ L230/34.
[168] Joined Cases T-366/03 and T-234/04 *Land Oberösterreich v Commission* [2005] ECR II-4005.
[169] Joined Cases C-439/05P and C-454/05P *Land Oberösterreich v Commission* [2007] ECR I-7141. On this decision, see FM Fleurke, 'What use for Article 95(5) EC? An analysis of *Land Oberösterreich and Republic of Austria v Commission*' (2008) 20 Journal of Environmental Law 267 (concluding that Member State autonomy under Art 95(5) is limited and that the Court sets a high threshold for new scientific evidence of environmental risk to Member States).

4.3.2 GMO-Free Network

In the context of coexistence, a GMO-Free Network has developed. In February 2005, 20 European regions, part of the Network of the GMO-Free European Regions and Local Authorities, signed a document on coexistence.[170] That document, now called the Charter of Florence, reviewed various documents and developments concerning coexistence and set out 10 efforts that would guide the activities of the Network. Those efforts focus, in the main, on promoting rules and activities that will protect conventional and organic crops from contamination by GMOs.[171] Moreover, many of the regions hope to protect their 'uniquely identifiable regional products'.[172]

In November 2005, the Network, by then 40 members strong, issued its Declaration of Rennes,[173] which outlined three motivations: the right to choose GM-free agriculture, the defence of biodiversity, and control and accountability in use of GMOs through full transparency and allocation of costs of segregation and other measures. The Declaration argued for four specific approaches. Because of the geographic nature of agriculture, coexistence rules should be regional or local. The precautionary principle should apply to ensure that GMOs do not harm biodiversity and health. Operators who grow GMOs should be accountable under the polluter pays principle for escape of GMOs and their presence in organic or conventional production chains. Moreover, to ensure a safe supply of GM-free proteins, regional co-operation should occur between areas where GM-free feeds are produced and consumed. Essentially, it seems, the Network 'wants the opportunity to stay GMO-free'.[174]

4.4 Liability

As more authorized GM crops are produced in EU countries, issues of liability will arise, especially for damage caused by commingling GM and other crops. Thus, coexistence and liability are closely linked, and the assignment of liability for damages that might arise from failures of coexistence raises difficult issues. Community measures enacted to govern GMOs and their products do not

---

[170] Charter of the Regions and Local Authorities of Europe on the Subject of Coexistence of Genetically Modified Crops with Traditional and Organic Farming (4 February 2005) ('Charter of Florence'). Many of these regions do not have authority to enact coexistence measures: see Levidow and Boschert (n 11 above) 174, 185–7. The Network grew from a November 2003 initiative by the Regions of Tuscany and Upper Austria: Cenni (n 125 above) 2.

[171] Charter of Florence, 4–6.   [172] Cenni (n 5 above) 4.

[173] GM-Free Regions Network, Declaration of Rennes (30 November 2005).

[174] Cenni (n 125 above) 4. Cenni pleaded for a European Commission strategy of sustainable coexistence at regional or local level that would take into account a number of factors, including 'tolerance threshold in seed equivalent to "*technical zero*"': ibid 6 (technical zero referring to the limit of detection). See also FOEE and the Assembly of European Regions (n 126 above) (setting out 10 principles to guide legislation for compulsory coexistence standards).

resolve these issues. Moreover, neither Council Directive (EEC) 85/374 on the approximation of the laws, regulations and administrative provisions of the Member States concerning liability for defective products[175] nor Directive (EC) 2004/35 of the European Parliament and of the Council on environmental liability with regard to the prevention and remedying of environmental damage ('Environmental Liability Directive')[176] addresses liability from GMOs comprehensively. Because Chapter 8 provides a general description of liability from GMOs, with focus on the Environmental Liability Directive, discussion here is limited to a few observations connected directly with coexistence.

In its general principles, the 2003 Commission Recommendation on coexistence indicated that coexistence measures may affect liability:

> [t]he type of instruments adopted [for coexistence] may have an impact on the application of national liability rules in the event of economic damage resulting from admixture. Member States are advised to examine their civil liability laws to find out whether the existing national laws offer sufficient and equal possibilities in this regard. Farmers, seed suppliers and other operators should be fully informed about the liability criteria that apply in their country in the case of damage caused by admixture.[177]

Member State law will apply to redress traditional damage to persons and property that may result from cultivation of GMOs, because this damage is not covered by the Environmental Liability Directive.[178] National liability regimes are likely to provide remedies through fault-based principles of negligence or nuisance,[179] and some Member State coexistence plans include standards of liability or provisions to indemnify losses.

An ideal scheme to allocate liability and pay compensation for damage from GM crops should distribute risk fairly, in light of producer behaviour, and its requirements should be clear and simple. Access should also be easy, without unnecessary procedural obstacles; likewise, decisions on compensation should be handled efficiently. Administrative costs of the scheme should be reasonable, and the scheme should ensure that money will be available to compensate the producer who suffered harm.[180]

---

[175] [1985] OJ L210/29, as amended by Directive (EC) 99/34 of the European Parliament and of the Council [1999] OJ L141/20. [176] [2004] OJ L143/56.

[177] 2003 Commission Recommendation, Annex, para 2.1.9. The European Commission continued: 'Member States may want to explore the feasibility and usefulness of adapting existing insurance schemes, or setting up new schemes'.

[178] Environmental Liability Directive, Preamble (14) and Art 3.

[179] See MR Grossman, 'Genetically modified crops in the United States: Federal regulation and state tort liability' (2003) 5 Environmental Law Review 86, 97.

[180] BA Koch, 'Economic Damage from GMO Admixture in Non-GM Products: Liability and Compensation Schemes: Freedom of Choice Conference' (Freedom of Choice Conference, n 5 above) 1–2. On allocation of losses to conventional and organic farmers from pollen drift and commingling in the EU, see BA Koch (ed), *Liability and Compensation Schemes for Damage Resulting from the Presence of Genetically Modified Organisms in Non-GM Crops* (Vienna: European Centre of Tort and Insurance Law, 2007).

Tort law will apply, but where failures of coexistence are concerned, several issues arise. One is the nature of loss for which compensation can be provided. That is, should damages be awarded only for actual admixture of GM with other crops, or should mere fear of admixture (and the economic loss suffered when potential customers fear admixture) lead to compensation? Another issue is the proper defendant. A plaintiff must consider whether to sue the nearest farmer, other farmers in the neighbourhood, the seed producer, or even the competent regulatory authority. Proof of causation is also critical. That is, the plaintiff will have to prove that the neighbouring producer actually contaminated the crop and that the admixture did not occur, for example, because of GM content in the seeds. Further, must the GM farmer have engaged in harmful conduct (for example, negligence), or should strict liability be applied? Liability might not be imposed, for example, if the GM farmer followed the applicable rules for segregation of GM crops. The appropriate remedy must also be determined. Monetary compensation is an appropriate remedy, but the possibility of an injunction to prevent sale or cultivation of GM crops must also be considered.[181]

An insurance system—for example, third-party insurance for GM crop farmers or GM seed producers—may help to ensure the availability of funds to cover liability established under tort law. Although such insurance is not generally available, some Member States plan to require insurance. First-party insurance, commonly purchased by farmers for protection against other risks, could cover damage from GM admixture, especially when the farmer–victim cannot recover under tort law. Clear coexistence rules in Member States are likely to facilitate the availability of insurance. Alternatively, a compensation fund, financed either by GM (or even all) farmers, a Member State, or both, can pay for damage. A compensation fund offers 'procedural advantages ... [because] it is designed specifically for a peculiar set of problems, formalities tend to be easier to fulfill for the claimants, and payments are typically faster than under other schemes'.[182]

A critical question, of course, is who should be liable for failures of coexistence. Different sectors, through representative organizations, argue against liability for their constituents. For example, COPA-COGECA, representing professional agricultural organizations and co-operatives, opposed liability for farmers who follow proper practices: '[u]nder no circumstances should a farmer be considered liable for any damage caused by proper use of GMOs and derived products'.[183] Moreover, '[in] the case of adventitious presence, producers must

---

[181] Koch, 'Economic Damage' (n 180 above) 2. Koch refers to 'a tool to ban GM production in advance simply for the fear of admixture that may cause loss in the future'. See also MR Grossman, 'Anticipatory nuisance and the prevention of environmental harm and economic loss from GMOs in the United States' (2008) 18 Journal of Environmental Law and Practice 107.
[182] Koch, 'Economic Damage' (n 180 above) 3 and 4.
[183] COPA-COGECA, *Position of COPA and COGECA on the Use of Gene Technology in Agriculture* (Pr(00)06F1, 21 January 2000) 3.

not be made liable and no financial losses must be imputed to them when they have respected the existing rules'.[184] Instead, '[b]iotechnology companies, responsible for the placing of a product on the market, must have full civil (eg, product liability) and environmental liability for any damages caused by GMOs'.[185] From the seed industry perspective, in contrast, the ESA argued against liability for seed producers. The absence of seed thresholds means that all producers risk liability: because 'it is impossible for the seed industry to guarantee the absence of traces of GM seeds in conventional varieties, the seed industry should not be considered liable for any such presence'.[186] But until Community legislation addresses liability for failure of coexistence or until Member States enact coexistence measures that articulate principles of liability, questions will remain.

5. Member State Measures—the European Commission Reports on Implementation

The Deliberate Release Directive and the 2003 Commission Recommendation direct each Member State to enact coexistence measures suited to its own agricultural conditions. Many Member States have done so, and most others are in the process of preparing legislation. Programmes vary, but many agree on an important principle: 'farmers may use GM crops provided they take precaution and do [what is] necessary to avoid potential negative economic implications that may result from the GMO presence in the neighbours' crops'.[187]

In March 2006, the European Commission issued its 2006 Implementation Report,[188] which compiled information from the legislation of Member States (both enacted and draft measures), responses to a questionnaire completed by the competent authorities of Member States, and information from national experts. At the end of 2005, only four Member States had adopted legislation, although others had notified the European Commission of draft legislation, either at a national or regional level.[189] Some Member States, especially those where maize is not widely cultivated, have little interest in coexistence

---

[184] COPA-COGECA (n 57 above) 4. COPA-COGECA opposed mandatory insurance, a financial guarantee system, or a collective compensation fund.
[185] COPA-COGECA (n 183 above) 3.   [186] ESA (n 109 above) 7.
[187] Commissioner Fischer Boel and Pröll (n 9 above) 2.
[188] 2006 Implementation Report (n. 127 above). The Annex provides detailed information about Member State measures: *Commission Staff Working Document, Annex to the Report on the Implementation of National Measures on the Coexistence of Genetically Modified Crops with Conventional and Organic Farming* SEC(2006)313 ('2006 Annex').
[189] 2006 Implementation Report (n 127 above) 4. Regions have legislative competence for coexistence in Austria, Belgium, Italy, and the United Kingdom. In other Member States, the national level has competence for most aspects of coexistence: 2006 Annex (n 188 above) 9.

strategies.[190] Moreover, some Member States measures have not been notified properly to the European Commission and therefore cannot be enforced against individuals.[191] In April 2009, the European Commission issued the 2009 Implementation Report, which indicates significant progress since 2006.[192] By the end of 2008, 15 Member States had adopted legislation for coexistence, and three others had notified the European Commission of draft legislation.[193]

The Implementation Reports describe measures enacted or proposed in Member States. Their Annexes give more detailed descriptions of individual Member State measures, including a comprehensive list of technical details. The 2006 Implementation Report focused on national measures in light of the principles for coexistence articulated in the 2003 Commission Recommendation, which indicated that coexistence measures must comply with Community legislation. Measures that ban GM crops do not comply and are therefore not legal coexistence measures.[194] Measures that are unnecessarily stringent or impose environmental protection conditions (for example, restrictions on cultivation in protected areas) may also conflict with Community legislation and may not be approved by the European Commission.[195]

Because Chapter 7 discusses selected Member State measures in detail, the following discussion focuses briefly on the Implementation Reports.

5.1 Measures in General

The 2006 Implementation Report noted that national measures have common elements:

[t]hey are designed to protect farmers of non-GM crops from the possible economic consequences of accidental admixture with GMOs. At the same time, GM crop

---

[190] See 2009 Implementation Report (n 131 above) 5–6; and 2006 Annex (n 188 above) 9. Countries where maize is not widely grown are the United Kingdom, Finland, Sweden, Estonia, and Malta.

[191] 2006 Implementation Report (n 127 above) 5. See 2003 Commission Recommendation, Annex, para 2.1.11 (directing Member States to inform the European Commission about national strategies and individual measures). See also Council Directive (EC) 98/34 of the European Parliament and of the Council laying down a procedure for the provision of information in the field of technical standards and regulations [1998] OJ L204/37, Art 9. Draft technical legislation not notified to the European Commission cannot be enforced against individuals in national courts: Case C-194/94 *CIA Sec International SA v Signalson SA* [1996] ECR I-2201.

[192] 2009 Implementation Report (n 131 above). The detailed Annex is *Commission Staff Working Document Accompanying Report from the Commission to the Council and the European Parliament on the Coexistence of Genetically Modified Crops with Conventional and Organic Farming* SEC(2009)408 ('2009 Annex'). [193] 2009 Implementation Report (n 131 above) 5.

[194] 2003 Commission Recommendation, Annex, paras 2.1–2.1.12; and 2006 Implementation Report (n 127 above) 6 and 9.

[195] 2006 Implementation Report (n 127 above) 5. Authorized GM crops have already undergone environmental and health evaluations. The 2006 Implementation Report attracted criticism. FOEE, for example, argued that Member State coexistence measures must protect health and environment, demanded mandatory EU coexistence measures, and wanted more freedom for Member States to ban or restrict GMOs: FOEE (n 10 above) 3–5.

cultivation is not prohibited. While differences in the stringency of the approaches can not be denied, Member States have generally made an effort to allow the different production types—conventional, organic and GM crop cultivation—to coexist within a region. The onus of implementing segregation measures between GM and non-GM crop production has generally been placed on GM crop growers.[196]

Member States differ in their goals for coexistence and the technical measures to achieve those goals. To avoid unnecessary burdens on farmers, seed producers, and others, the 2003 Commission Recommendation pleaded for proportionality; that is, coexistence measures should require practices geared to meet Community thresholds for labelling crops with GM material (generally 0.9 per cent), but not lower thresholds.[197] The Implementation Reports indicate that some Member States have adopted measures geared to the Community threshold; others aim at a lower, more burdensome threshold or do not specify a GM tolerance.[198]

To comply with Community requirements, Member States must establish national registers of GM crops.[199] Thus, all require, or will require, GM crop farmers to notify government authorities of their plans to grow GM crops. Farmers must identify the parcel and report its size, location, and the crop to be grown.[200] Some, but not necessarily all, information from each national register of GM crop production will be available to the public.[201] A few Member States also require farmers to notify neighbours, within certain distances, that they plan to grow GM crops; landowners (in the case of tenancy), purchasers of land, and even bee-keepers may also be entitled to notice.[202]

5.2 On-farm Measures

Some Member States have adopted on-farm measures for cultivation of GM crops. By way of illustration, several require training or other proof of competence for farmers who grow GM crops or even (for example, in Denmark) for all who handle those crops.[203] In addition, 12 Member States have prescribed

---

[196] 2006 Implementation Report (n 127 above) 8–9.
[197] 2003 Commission Recommendation, Annex, para 2.1.4.
[198] 2006 Implementation Report (n 127 above) 9; and 2009 Implementation Report (n 131 above) 6.
[199] Deliberate Release Directive, Art 31 requires Member States to establish registers, both for GM field tests and for GMOs that are placed on the market. On the status of field registers, see FAS (n 162 above) 45–6.   [200] 2006 Annex (n 188 above) 52–3.
[201] In some Member States, GM crop production requires approval at the farm level or even an approval or notification for each GM field: 2009 Annex (n 192 above) 16–19. This approach, according to the European Commission, 'could lead to duplication of authorisation' for use of approved GM crops: 2006 Implementation Report (n 127 above) 6. On confidentiality aspects, see Case C-552/07 *Commune de Sausheim v Azelvandre* (ECJ 17 February 2009); and Chapter 7.
[202] 2006 Annex (n 188 above) 11–12 and 54–5; 2009 Implementation Report (n 131 above) 6; and 2009 Annex (n 192 above) 8–13.
[203] 2009 Annex (n 192 above) 19–21.

technical segregation measures (for example, isolation distances) for GM crops, especially maize; most have not yet prescribed field measures.[204] In some instances, practices have been adapted (or simply duplicated) from those used for production of certified seeds. Isolation distances sometimes vary depending on whether adjacent crops are conventional or organic (though the same labelling thresholds apply). For example, in The Netherlands, separation distances for GM maize are 25 metres from conventional maize and 250 metres from organic or seed maize; in Portugal, distances are 200 metres from conventional and 300 metres from organic maize.[205] Denmark's isolation distance was 200 metres for conventional, organic, or seed crops, but a measure notified in 2007 reduced this to 150 metres.[206] Member States that aim at a threshold lower than required by Community law have imposed particularly burdensome isolation requirements.[207] Reflecting a European Commission recommendation, some Member States permit farmers to agree not to implement segregation measures. This voluntary co-operation means, however, that non-GM crops must be labelled as GM.[208]

Of course, EU experience with cultivation of GMOs is limited (except in Spain). Therefore, even in Member States with coexistence legislation, little GM cultivation has occurred, so the efficacy and feasibility of the measures have not been tested. Importantly, the European Commission insists that coexistence measures must be 'based on sound science taking into account the best available research results and field experience'.[209]

Because the Member States are developing technical measures for coexistence, little is known about their cost. Clearly, however, the use of isolation distances and buffer strips will impose costs on GM farmers, as will extra cleaning of farm equipment. A Commission report suggested that the opportunity costs of isolation distances and buffer strips (that is, not growing GM in those areas) in Spain, for example, are about €84 per hectare, not including additional costs of planting two different crops. In Member States that impose a per-hectare levy to finance a compensation fund, the cost of coexistence will increase further.[210] Because farmers who plant GM crops will bear these additional costs, they may affect farmers' choice of GM or non-GM crops.

---

[204] 2009 Implementation Report (n 131 above) 7. For GM maize, which can be planted legally in the EU, scientific evidence indicates that segregation 'can be achieved with technical measures that are applicable at the level of individual farms or in coordination between neighbouring farms': 2006 Implementation Report (n 127 above) 9.
[205] 2009 Annex (n 192 above) 26–7. Member States have identified separation distances for oilseed rape, maize, beet, wheat, and potato.
[206] Order on the cultivation of genetically modified crops (TRIS 2007/0598/DK). Information about cultivation of maize must be given to those within 225 metres.
[207] 2006 Implementation Report (n 127 above) 6; and 2009 Implementation Report (n 131 above) 7.   [208] 2006 Implementation Report (n 127 above) 7.
[209] Ibid 8 and 10.
[210] Gómez-Barbero and Rodríguez-Cerezo (n 39 above) 33 and 37; and see also Chapter 2.

5.3 Liability

The allocation of liability for economic damage from the adventitious presence of GMOs in conventional or organic crops is central to national coexistence policy. Member State civil liability rules apply, and some Member States will rely on existing civil liability (tort) law, which varies considerably.[211] Other Member States have enacted or proposed liability rules specifically to govern damage from GMOs. This legislation generally focuses on damage to conventional or organic crops that results from cultivation of GM crops on neighbouring land. No reported court cases have yet required application of national liability rules for GM admixture.[212]

Member State measures to redress damage from adventitious presence normally allocate liability to the grower of GM crops. Liability may be fault-based, as in Denmark and The Netherlands, where growers of GM crops are liable only if they have failed to follow practices required for GM cultivation. By contrast, Germany and France have imposed strict liability on the GM grower for damage to non-GM crops.[213] Alternatively, Member States have enacted, or plan to design, compensation schemes funded by contributions from GM growers (Denmark) or even all crop farmers (The Netherlands). Those schemes often authorize compensation only if the GM content in the non-GM crop exceeds the 0.9 per cent Community threshold for labelling. Still other Member States require or recommend third party insurance, but the present unavailability of insurance coverage for adventitious presence in the EU makes mandatory insurance schemes inappropriate.[214]

6. Conclusion

6.1 The Community and Member States

The Community and its Member States are co-operating to reach an effective and economical solution to the problem of coexistence of GM crops with conventional and organic crops. The European Commission continues to act as co-ordinator and advisor to Member States. The guidelines in its 2003

---

[211] 2006 Implementation Report (n 127 above) 7–9. Spain, for example, relies on general civil liability (tort law) rules.

[212] 2009 Implementation Report (n 131 above) 5. Member States have not proposed liability measures for damage from impure seed or from shared planting or harvesting implements: 2006 Annex (n 188 above) 17.

[213] 2006 Annex (n 188 above) 18. On legal liability for cultivation of GM crops in Germany, see SJ Smyth and DL Kershen, 'Agricultural biotechnology: legal liability regimes from comparative and international perspectives' (2006) 6(2) Global Jurist Advances 32.

[214] 2006 Annex (n 188 above) 18–19; and 2009 Annex (n 192 above) 39–45. If Member States mandate insurance coverage where insurance is not available, farmers will be unable to grow GM crops.

Conclusion

Commission Recommendation provided a starting point, and as the Member States develop and implement their coexistence measures, the European Commission plans a number of actions to support development and implementation of effective coexistence measures. These include efforts to share information with Member States, support research to fill gaps in scientific knowledge, and review scientific and economic information about segregation measures. The European Commission will also conclude its economic impact assessment of seed thresholds and propose appropriate legislation. To develop recommendations for specific crops, it will work with Member States to discover best segregation practices, using experience in Spain and other Member States where GM varieties are now cultivated. The European Coexistence Bureau, a special office of the JRC, will help to develop guidelines for effective crop-specific coexistence measures, designed to aid Member States in developing or refining their coexistence measures. The European Commission will report to the Council and European Parliament again in 2012 on its progress, as well as on national implementation of coexistence measures.[215]

For their part, many Member States have made progress in developing coexistence measures in light of their own agricultural conditions. Many, but certainly not all, have notified measures or draft measures to the European Commission. Among those to have enacted legislation or concluded voluntary agreements, approaches vary. It is clear, however, that coexistence in Europe will be governed by both *ex ante* requirements for production of GM crops (for example, isolation distances and other on-farm measures) and some degree of *ex post* liability rules (fault-based or strict liability) for adventitious presence. Some Member States can be expected to resist cultivation of GM crops on their territory, perhaps by enacting coexistence measures that permit regionally imposed or voluntary prohibitions.

For the time being, at least, it seems likely that the responsibility for ensuring coexistence will remain with Member States. In April 2006, Commissioner Fischer Boel stated that '[w]e do not think it would be helpful to propose binding EU-wide rules on co-existence at present, either in terms of segregation methods or in terms of liability'.[216] Current technical expertise for segregation does not yet justify binding rules. Moreover, if segregation techniques are developed and applied effectively, damage from adventitious presence, and the resulting liability issues, can be avoided. Even if damage occurs, Commissioner Fischer Boel noted, Member States have authority to govern civil liability within their territories, so they must develop rules for liability or compensation. To harmonize national law would apply a 'blunt instrument', instead of the

---

[215] 2009 Implementation Report (n 127 above) 10; and Council Conclusions on Genetically Modified Organisms, 2912th Council Meeting: Environment, Brussels, 4 December 2008, para 17.
[216] Commissioner Fischer Boel, Speech/06/223, 'Co-existence of Genetically Modified Crops with Conventional and Organic Farming: Conference on Co-existence', Vienna, 5 April 2006.

'detailed and delicate' work required to achieve coexistence.[217] The European Commission reaffirmed this approach in April 2009.[218]

6.2 Principles of Coexistence

One might ask what characterizes an effective and economical coexistence programme. On the basis of a number of studies of coexistence in North America and Europe, researchers have identified five key principles for best management practices for coexistence: context, consistency, proportionality, equity, and practicality. Context (which influences proportionality) focuses on the relative importance (planted area and value) of crop production systems and the size of non-GM markets. Without demand for non-GM products, the coexistence issue is less significant. If the level of non-GM demand (that is, for certified or organic products) is low, it is less likely that GM and non-GM crops that must be protected will be planted in proximity to each other.[219]

Consistency suggests that adventitious presence of GM and other unwanted material should be treated alike, with the realization that 100 per cent purity for any crop is unrealistic. Therefore, thresholds for adventitious material should be 'proportionate to the risks attached to the presence of the unwanted material'. Low thresholds should apply to limit adventitious presence of materials with health and safety risks, and higher thresholds should apply for risks (for example, of EU-authorized crops) that affect only 'product integrity, purity, quality and functionality'.[220] Proportionality, related to context, insists that coexistence measures should be 'proportionate, non discriminatory and science-based'.[221] That is, coexistence measures should not impose burdensome stewardship conditions where GM varieties pose little risk to organic or traditional crops.

Equity focuses on the issue of liability for adventitious presence of unwanted material in crops. If new legislative provisions are enacted, measures that impose liability for adventitious presence should apply to all farmers and all types of unwanted material (perhaps also including pests, weeds, and plant disease). That is, if conventional or organic growers are eligible for compensation for adventitious GM material, GM growers should be eligible for compensation for adventitious material from conventional crops, if that material causes economic loss. Practicality means that coexistence measures 'should be based on legal, practical and scientific realities' and, in particular, the reality that a zero tolerance level is 'virtually impossible'.[222]

---

[217] Ibid. [218] 2009 Implementation Report (n 131 above) 9–10.
[219] Brookes (n 23 above) 13–14. [220] Ibid 14 (both quotations).
[221] Ibid 16. See also Brookes and Barfoot (n 140 above) 19 (discussing context and proportionality).
[222] Brookes (n 23 above) 16–17. Historically, growers of specialty crops were responsible for protecting the integrity of their crops and were rewarded by price premiums.

Conclusion

Do coexistence measures in the Member States measure up to these suggested principles? It is probably too early to decide. The fact that Member States are developing measures appropriate for their agricultural situation helps to meet both the context and proportionality principles, assuming that the measures impose obligations only when the protection of non-GM crops requires them. Both consistency and equity focus in part on similar treatment for damage from adventitious presence in all types of crops. So far, Member States have concentrated on damage from GM crops to other crops, with little legislative recognition that other types of damage are possible. Finally, it would seem that both the Community and most Member States recognize, albeit reluctantly, that 100 per cent purity is impossible and that some adventitious presence should be tolerated.

Effective coexistence in the EU will require more experience with GM crops and more research, especially with crops other than maize. European farmers cultivated only about 108,000 hectares of GM maize in 2008, in seven of the 27 Member States.[223] As more European farmers grow this crop, and if additional crops are authorized for cultivation, coexistence provisions should be reviewed and adapted. Against this background, it is possible (though perhaps unlikely) that the focus of coexistence may change. That is, the fact that GM farmers are now the 'newcomers' may justify putting the burden of coexistence on those farmers; they are in a position similar to US growers of organic or specialty crops, who bear the burden of identity preservation in exchange for a price premium. But if GM cultivation ever becomes as widely prevalent in the EU as in the United States, where GM varieties in 2009 made up 91 per cent of soya and 85 per cent of maize,[224] that balance may shift and require a different approach to coexistence.

Meanwhile, as Member States develop measures for coexistence, it seems likely that they will attempt to 'protect traditional and organic farmers in an efficient and cost-effective way'. As the Agriculture Commissioner noted:

> [w]ell-designed co-existence measures should guarantee that practical and feasible production arrangements will emerge for all crops and production types, including organic farming. In areas where agricultural structures and farming conditions are such that farm-level co-existence cannot be achieved for a given crop, other sustainable solutions should be explored.[225]

Perhaps surprisingly, the Commissioner seemed open to the idea of partial bans of GM crops, when that is essential to prevent harm, but she rejected complete Member State or regional bans: '[w]e might have to consider excluding an

---

[223] C James, *Global Status of Commercialised Biotech/GM Crops: 2008 (ISAAA Brief No 39)* (Ithaca, NY: International Service for the Acquisition of Agri-biotech Applications, 2009) Executive Summary, 10–11.
[224] National Agricultural Statistics Service (NASS), Acreage (Washington, DC: USDA, 2009) 24–5. [225] Commissioner Fischer Boel and Pröll (n 9 above) 2 (both quotations).

individual GM product from a given area if, for scientific reasons, it genuinely could not co-exist with non-GM crops in that area. *But no, we cannot simply ban all GM crops from an entire region because of hostility to GM products per se*.[226] Indeed, as Commissioner Fischer Boel indicated, '[t]he debate on co-existence must be about ensuring co-existence, not preventing it'.[227]

---

[226] Commissioner Fischer Boel (n 216 above) 4 (emphasis added). But see Commissioner Fischer Boel, Speech/09/474, 'GMOs: Letting the Voice of Science Speak', Brussels, 15 October 2009. Accepting the possibility of a new approach toward cultivation of GM crops, the Commissioner noted, 'On the one hand, we would keep a European-wide authorization system based on science. On the other hand, we could allow Member States to decide whether or not they wish to cultivate GM crops on their territory'.

[227] Ibid. In December 2008, the Environment Council indicated that regions of particular agricultural or ecological sensitivity might require special measures, including prohibition of GMOs, and that producers can create GM-free zones by voluntary agreement: Council Conclusions on Genetically Modified Organisms, 2912th Council Meeting: Environment, Brussels, 4 December 2008, paras 15 and 18.

See JM Barroso, Political guidelines for the next Commission (3 September 2009) 39: 'In an area like GMOs,..., it should be possible to combine a Community authorization system, based on science, with freedom for Member States to decide whether or not they wish to cultivate GM crops on their territory.'

7

Coexistence of Genetically Modified, Conventional, and Organic Crops in the European Union: National Implementation

Luc Bodiguel, Michael Cardwell,
Ana Carretero García, and Domenico Viti

1. Introduction

'Coexistence' implies two or more things existing in the same place at the same time; and, in an ideal world, coexistence should be regarded as essentially 'passive'. It is difficult to argue that this can be achieved when the one thing prejudices the viability of the other; and inevitably the question arises whether the simultaneous growing of genetically modified (GM), conventional, and organic crops can satisfy such a definition. If one is to believe numerous scientific reports and policy documents, there must be severe doubts.[1] In particular, while it is certainly possible to vary the scale and speed of 'cross-contamination', what remains a constant is the fact of cross-contamination itself.[2] As stated in the

---

[1] See, eg Institut National de la Recherche Agronomique (INRA), *Coexistence Entre Cultures OGM et Non OGM en Europe* (Paris: INRA, 2006) (available at <http://www.inra.fr/les_recherches/exemples_de_recherche/coexistence_entre_cultures_ogm_et_non_ogm_en_europe>, last accessed 24 July 2009); and *Assemblée Nationale Française, Rapport de M. Christian Ménard sur les Enjeux des Essais et de l'Utilisation des Organismes Génétiquement Modifiés*: No 2254 (2005) Vol 1, 57 (available at <http://www.assemblee-nationale.fr/12/dossiers/ogm.asp#mission_ogm>, last accessed 25 November 2009). On the other hand, research for the European Commission has taken a more optimistic view of coexistence: see, eg A Messean et al, *New Case Studies on the Coexistence of GM and non-GM Crops in European Agriculture: Tech. Rep. EUR 22012* (Joint Research Centre, European Commission, 2006); but, for strong criticism of this report, see, eg J Milanesi, *Analyse des Coûts Induits sur les Filières Agricoles par les Mises en Culture d'OGM* (2008) 12–27 (available at <http://blog-s.greenpeace.fr/documents/ogm/Rapport-CREG-OGM.pdf>, last accessed 24 July 2009).

[2] Use of the word 'contamination' may be considered somewhat pejorative, in that it likens GMOs to a disease or pollution: see, eg *Assemblée Nationale Française* (n 1 above).

United Kingdom *GM Science Review First Report*, '[t]he complete genetic isolation of crops grown on a commercial scale, either GM or non-GM, is not practical at present': rather, it is a matter of minimizing gene flow.[3] And, as also highlighted in the same document, the extent to which gene flow is to be minimized then becomes effectively a political decision.[4]

Against this background, as has been seen,[5] the Community legislature supports the idea that coexistence can be assured on the basis that 'minor' cross-contamination up to a threshold of 0.9 per cent is acceptable.[6] It may also be reiterated that, as the law stands, Member States do not enjoy the option of simply preventing GM production, as made clear in Commission Recommendation (EC) 2003/556 on guidelines for the development of national strategies and best practices to ensure the coexistence of genetically modified crops with conventional and organic farming ('2003 Commission Recommendation').[7] The first recital unequivocally stipulates that '[n]o form of agriculture, be it conventional, organic or agriculture using genetically modified organisms (GMOs), should be excluded in the European Union'; and Commissioner Fischer Boel affirmed in 2006 that '[t]he debate on co-existence must be about ensuring co-existence, not preventing it'.[8] Nevertheless, during the autumn of 2009, there were indications of a potential shift in policy. In his *Political Guidelines for the Next Commission*, President of the European Commission José Manuel Barroso stated his intention to promote diversity within the Community and, more specifically, that '[i]n an area like GMOs, for example, it should be possible to combine a Community authorisation system, based on science, with a freedom for Members States to

---

[3] GM Science Review Panel, *GM Science Review: First Report—an Open Review of the Science Relevant to GM Crops and Food Based on Interests and Concerns of the Public* (London: 2003) 18.
[4] Ibid. [5] See Chapter 4.
[6] For the 0.9% threshold in Community legislation, see Directive (EC) 2001/18 of the European Parliament and of the Council on the deliberate release into the environment of genetically modified organisms ('Deliberate Release Directive') [2001] OJ L106/1, Art 21, as amended by Regulation (EC) 1830/2003 of the European Parliament and of the Council concerning the traceability and labelling of genetically modified organisms and the traceability of food and feed products produced from genetically modified organisms ('Regulation 1830/2003') [2003] OJ L268/24; Regulation (EC) 1829/2003 of the European Parliament and of the Council on genetically modified food and feed ('Food and Feed Regulation') [2003] OJ L268/1, Arts 12(2) (GM food) and 24(2) (GM feed); and Regulation 1830/2003, Art 4(7) (by reference to the Deliberate Release Directive) and (8) (by reference to the Food and Feed Regulation). See also, generally, eg M Lee, *EU Environmental Law: Challenges, Change and Decision-making* (Oxford: Hart Publishing, 2005) 255–9; MR Grossman, 'The coexistence of GM and other crops in the European Union' (2007) 16 Kansas Journal of Law and Public Policy 324; and L Bodiguel and M Cardwell, 'La Coexistence des Cultures GM et Non GM: Approche Comparative Entre l'Union Européenne, le Royaume-Uni et la France' in G Parent (ed), *Production et Consommation Durables: de la Gouvernance au Consommateur Citoyen* (Québec, Canada: Éditions Yvon Blais, 2008) 325.
[7] [2003] OJ L189/36.
[8] Commissioner Fischer Boel, Speech/06/223, 'Co-existence of Genetically Modified Crops with Conventional and Organic Farming: Conference on Co-existence', Vienna, 5 April 2006 (speeches of the Commissioner for Agriculture and Rural Development are available at <http://europa.eu/rapid/>).

Introduction

decide whether or not they wish to cultivate GM crops on their territory'.[9] While no more than an indication, it may be regarded as significant that Commissioner Fischer Boel swiftly backed this approach.[10]

For the time being, however, there is a strong argument that the existing regime creates a 'fiction juridique', in the sense that it supposedly offers farmers an unfettered choice on the form of agriculture they wish to adopt, free from interference from other forms of agriculture. The 2003 Commission Recommendation recites that '[c]oexistence refers to the ability of farmers to make a practical choice between conventional, organic and GM-crop production'.[11] Moreover, the question of coexistence is confined by the Community legislature to matters of potential economic loss, on the basis that matters of environmental protection and human health have already been addressed at the authorization stage.[12] Yet, importantly, the risk assessment conducted at the authorization stage does not consider whether or not coexistence is feasible. In consequence, it could be argued that all that remains for any coexistence regime is the reduction (rather than total prevention) of cross-contamination. As again stated in the 2003 Commission Recommendation, it is simply a matter of finding the most appropriate way to *minimize* admixture.[13]

On the other hand, even within this Community framework, considerable discretion has been accorded to the Member States when implementing national regimes. Notably, the authority for such regimes in the Deliberate Release Directive would itself appear discretionary, in that 'Member States *may* take appropriate measures to avoid the unintended presence of GMOs in other products',[14] while the 2003 Commission Recommendation lays great emphasis on subsidiarity.[15] There is recognition of great diversity in the conditions under which European farmers work, and coexistence measures 'must be specific to the farm structures, farming systems, cropping patterns and natural conditions in a region'.[16] In addition, considerable weight is accorded to the participation of

---

[9] José Manuel Barroso, *Political Guidelines for the Next Commission* (Brussels: European Commission, 2009) 39 (available at <http://ec.europa.eu/commission_barroso/president/pdf/press_20090903_EN.pdf>, last accessed 20 October 2009).
[10] Commissioner Fischer Boel, Speech/09/474, 'GMOs: Letting the Voice of Science Speak—Policy Dialogue at "European Policy Centre"', Brussels, 15 October 2009 (speeches of the Commissioner for Agriculture and Rural Development are available at <http://europa.eu/rapid/>).
[11] 2003 Commission Recommendation, Preamble (3) (and see also Annex, para 1.1).
[12] Ibid Preamble (4) and (5) (and see also Annex, para 1.2). See, generally, M Lee, 'The governance of coexistence between GMOs and other forms of agriculture: a purely economic issue?' (2008) 20 Journal of Environmental Law 193.
[13] 2003 Commission Recommendation, Preamble (5); and see also, generally, M Lee, *EU Regulation of GMOs: Law and Decision Making for a New Technology* (Cheltenham: Edward Elgar, 2008) 105–27.
[14] [2001] OJ L106/1, Art 26a, as amended by the Food and Feed Regulation (emphasis added).
[15] 2003 Commission Recommendation, Annex, para 1.4. This aspect would be greatly reinforced if President Barroso's proposal to increase diversity comes to fruition.
[16] 2003 Commission Recommendation, Annex, para 1.4.

local stakeholders and account is to be taken of 'national and regional factors' (although such factors remain unspecified). By contrast, the European Commission is to focus on gathering and co-ordinating information based on ongoing studies at Community and national level, offering advice, and issuing guidelines. In the carrying-out of this role, two substantial reports have been issued: *Report on the Implementation of National Measures on the Coexistence of Genetically Modified Crops with Conventional and Organic Farming*;[17] and *Report from the Commission to the Council and the European Parliament on the Coexistence of Genetically Modified Crops with Conventional and Organic Farming* ('2009 Implementation Report').[18] These reveal that the discretion accorded to Member States has been extensively exploited; and also that Member States are progressing at very different speeds (since, by April 2009, only 15 had adopted specific legislation on coexistence).[19] Although there would seem to be some similarities of approach (for example, in all cases coexistence is being regulated by legislative means), different forces at play have produced a variety of models and these will be explored in the context of four Member States: France, Italy, Spain, and the United Kingdom.

2. France

2.1 Basic Principle: Freedom of Cropping

France has been subject to enforcement actions in consequence of its failure to transpose Community GM legislation.[20] Even though on 25 June 2008 Parliament enacted the biotechnology law,[21] this did not prevent the imposition of a €10 million fine for delay.[22] Further, the current position remains somewhat contradictory. The growing of the maize variety MON810 is forbidden,[23] yet the French Government has adopted an overall stance which is close to that

---

[17] COM(2006)104.   [18] COM(2009)153.

[19] The Member States to have adopted specific measures by that date were: Austria, Belgium, the Czech Republic, Denmark, France, Germany, Hungary, Latvia, Lithuania, Luxembourg, The Netherlands, Portugal, Romania, Slovenia, and Slovakia: ibid 6.

[20] Case C-296/01 *Commission v France* [2003] ECR I-13909; and Case C-429/01 *Commission v France* [2003] ECR I-14439.

[21] Law No 2008-595 of 25 June 2008 relating to GMOs, *Journal Officiel de la République Française* (JORF) No 148 of 26 June 2008, NOR: DEVX0771876L (codified at Arts L.531-1ff of the Code de l'Environnement).

[22] Case C-121/07 *Commission v France* (ECJ 9 December 2008).

[23] Ministerial Decree of 7 February 2008 Suspending the Cropping of Genetically Modified Maize Seed (Zea Mays L. Line MON810), JORF No 34 of 9 February 2008, NOR: AGRG0803466A, amended 13 February 2008 NOR: AGRG0803888A. This ban under the safeguard clause has been maintained notwithstanding pressure from the European Commission and notwithstanding reports from the Agence Française de Sécurité Sanitaire des Aliments (AFSSA), which concluded that there were no health safety problems: for AFSSA reports on biotechnology, see <http://www.afssa.fr/Documents/BIOT> last accessed 20 August 2009.

defended by the Community,[24] favourable towards GM crops in spite of the possible risks to health and the environment, and in spite of the impossibility of ensuring fully 'GM-free' coexistence in practice.[25] As indicated, this stance is based upon a 'fiction juridique', the principle of freedom of cropping, which finds expression in the legislation as the right to consume and produce with or without GMOs.[26] In other words, producers should be allowed to grow what they wish and then consumers should be allowed to choose. Emphasis is thus placed on the consumer and the principle of free circulation of goods, but one doubt persists. If fully 'GM-free' coexistence is not possible in practice, for how long will the consumer, distributor, and non-GM producer retain genuine choice?

The principle of freedom of cropping may constitute a pragmatic way of ensuring the development of GM production and of rendering effective the authorization of GM crops. However, conferring equality on GM and non-GM production at law does not recognize the reality that GM production has the capacity to impact detrimentally on non-GM production, but not vice versa. Further, the French legislation expressly provides that the definition of 'without GMOs' is of necessity to be understood by reference to the Community legislation,[27] which only requires labelling where the adventitious or technically unavoidable presence of authorized GMOs exceeds a threshold of 0.9 per cent. Thus, as already noted, a level of cross-contamination is to be accepted.

On the other hand, the French legislation does contain provisions which nuance or limit the principle of freedom of cropping, namely: respect for the precautionary principle and the principles of prevention, information, participation, and liability;[28] and respect for the environment, public health, agricultural structures, local ecosystems, and non-GM production.[29] These factors should, at least to an extent, have been taken into account in the risk assessment procedure at the authorization stage, but there is also the possibility that they could be invoked with regard to coexistence measures. Nonetheless, it is hard to

---

[24] See, eg European Commission, *Communication from the Commission to the European Parliament, the Council, the Economic and Social Committee and the Committee of the Regions: Life Sciences and Biotechnology—a Strategy for Europe* COM(2002)27 [2002] OJ C55/3, para 1.

[25] *Assemblée Nationale Française* (n 1 above). See also, eg S Mahieu, 'Le Contrôle des Risques dans la Réglementation Européenne Relative aux OGM: vers un Système Conciliateur et Participatif' in P Nihoul and S Mahieu (eds), *La Sécurité Alimentaire et la Réglementation des OGM—Perspectives Nationale, Européenne et Internationale* (Brussels: Larcier, 2005) 153; C Lepage, 'OGM: où en est-on?' [2006] *Gazette du Palais*, 28 July, 3; and Milanesi (n 1 above).

[26] Art L.531-2-1, para 5 of the Code de l'Environnement: '[l]a liberté de consommer et de produire avec ou sans organismes génétiquement modifiés'.

[27] Art L.531-2-1, para 1 of the Code de l'Environnement: '[l]a définition du "sans organismes génétiquement modifiés" se comprend nécessairement par référence à la définition communautaire'.

[28] It should be emphasised, however, that use of the word 'principle' denotes lower status in the legal hierarchy and correspondingly weaker effect.

[29] Art L.531-2-1, paras 1 and 5 of the Code de l'Environnement.

conceive that they could form the basis of an action seeking to prevent cropping of a GM variety which had already been authorized, since the authorization procedure itself applies a precautionary and preventive approach and incorporates an analysis of health and environmental risks. Indeed, only the references to local ecosystems and non-GM production could arguably introduce considerations not taken into account at the earlier authorization stage. However, as shall be seen, such a claim would in all likelihood fall foul of French law as constituting an attempt to impose a partial ban on GMOs.

2.2 Failure of Preventive Actions Against GM Cropping

According to the decisions of the French courts, it would seem near impossible to impose even targeted bans on GM production. This line of authority has been adopted whether the cases involve organic producers or local authorities.

2.2.1 Organic producers

Organic producers have called for local bans on GM crops, with a view to marketing their produce free from any trace of GMOs. Impetus for such initiatives has been supplied by research indicating that, in the case of organic production, coexistence with GM crops is generally impossible.[30] As a consequence, if organic producers are themselves to enjoy freedom of cropping, the only realistic way in which this can be achieved is to impose a local ban on GM production: otherwise, organic agriculture may face potential extinction as it becomes harder and harder to secure the desired 'GM-free' status. Indeed, the fact that organic products can no longer be expected to remain 'GM-free' is arguably recognized by their being subject to the general 0.9 per cent threshold for cross-contamination.[31] A matter of some interest is that, during the passage of the GM legislation of 25 June 2008, amendments were proposed to provide exceptions from the general threshold in respect of both organic products and designations of origin and quality, but (perhaps not surprisingly) these were not accepted.[32]

In pursuing the option of legal challenge, the organic farmer is faced by the hurdle that, in order to protect his commercial position, he must invoke 'emergency' ('urgence'), even though GM production has not yet commenced. In 2007, the court had to address this issue when a claim was made by an organic bee-keeper, whose hives were situated near to land which was to be cultivated

---

[30] See, eg INRA (n 1 above).
[31] Council Regulation (EC) 834/2007 on organic production and the labelling of organic products [2007] OJ L189/1, Preamble (9) and (10) and Art 9. For the dangers of this generating consumer dissatisfaction, see, eg L Boy, 'Production et étiquetage des produits "bio" en droit communautaire' [2007] *Droit de l'Environnement* 294.
[32] P Billet, 'Index raisonné de la loi OGM du 25 juin 2008' (2008) 368 *Revue de Droit Rural* 9, 10.

with GM maize.[33] The court accepted that this could be catagorized as an emergency,[34] and that there could be damage;[35] but it still turned down the application of the beekeeper on three grounds. First, in cases of emergency, any measures ordered by the court must not impact upon a material dispute and, in the present case, there was found to be, at a general level, a material dispute over GMOs. Secondly, the court was not impressed with the approach adopted by the bee-keeper, who could easily have moved his hives away from the GM fields. Thirdly, it pointed to the fact that the GM crop concerned was properly authorized and that its production was, accordingly, lawful. The 'nomadic' character of the organic activity and the lack of consensus on GMOs may in part explain this decision. Nonetheless, any reading of the judgment reveals the key importance of prior authorization. As a result, in the absence of a judicial *volte-face* (which is unlikely, but could be based upon the principle of freedom of cropping), any claim for preventive action would seem doomed to failure, since no local consideration would appear capable of swaying the court when the GM production itself is lawful. As shall be seen, the same would seem to apply where a ban is sought by a local authority.

2.2.2 Local authorities

Mayors have on occasion endeavoured to demonstrate their rejection of modern biotechnology by banning GMOs within their territory. The effect would be to create 'GM-free' zones. However, these have systematically been subject to actions for annulment, instigated by the Government or seed companies. As highlighted in the *Land Oberösterreich* litigation, such zones have found no legitimacy in Community law, by reason of the principle of proportionality, the objectives of the internal market, and the obligation to operate on a case-by-case basis through scientific risk assessments.[36] Likewise, French law forbids local

---

[33] CA Agen, Civ 1, 12 July 2007, No 07/00842. The decision was taken on the basis of Art 808 of the *Code de Procedure Civile*, which allows a judge to order in all cases of emergency (dans tous les cas d'urgence) any measures which do not impact upon material disputes (contestations sérieuses), and on the basis of Art 809 of the same Code, which authorizes a judge to intervene in preventive fashion to maintain the status quo and avoid the realization of imminent damage.

[34] The judges did not contest that it would be necessary to make any intervention before the maize flowered.

[35] It was claimed that economic loss would arise where conventional or organic farmers, whose crops were contaminated by GMOs, could no longer sell their production in consequence of the contamination.

[36] See Joined Cases T-366/03 and T-234/04 *Land Oberösterreich v Commission* [2005] ECR II-4005; and Joined Cases C-439/05 P and C-454/05P *Land Oberösterreich v Commission* [2007] ECR I-7141; and see, generally, eg P Trouilly and A Gossement, 'Commentaire de l'affaire *Land Oberösterreich v Commission*' [2005] *Revue Environnement* (November) 30; FM Fleurke, 'What use for Article 95(5) EC? An analysis of *Land Oberösterreich and Republic of Austria v Commission*' (2008) 20 Journal of Environmental Law 267; and S Poli, 'Restrictions on the Cultivation of Genetically Modified Organisms: Issues of EC Law' in H Somsen (ed), *The Regulatory Challenge of Biotechnology: Human Genetics, Food and Patents* (Cheltenham: Edward Elgar, Cheltenham, 2007) 156.

bans on GMOs, except in the case of national parks or regional nature parks. In these instances, it is possible to exclude the cultivation of GMOs over all or part of the land within the parks, but in practice the provisions are of no use, in that such land is ill-suited to commodity crops.[37] As for the procedure adopted by the French Government, it has invoked the special police power of the Minister of the Environment to authorize GMOs, a power which trumps, without exception, the exercise of any general police powers held by mayors.[38]

Over the period 2004–06, in accordance with this hierarchy of police powers, the administrative judges consistently annulled municipal decisions to declare 'GM-free' zones. They refuted the idea that there was imminent danger ('péril imminent'),[39] even when it was argued that there was risk of mutation through cross-contamination from the neighbouring GM crops.[40] In consequence, since damage from their release may be irreversible, Gossement queries whether pollution from GMOs can ever qualify as imminent danger.[41] However, perhaps cross-contamination is not pollution.

These decisions were confirmed in 2007 in cases where exceptional circumstances were pleaded (for example, the presence of strong winds, wild boars (with their habit of destroying field crops), a nature reserve, and a zone of special interest for its fauna and flora).[42] Further, the fact that the 'GM-free' zones were limited in time or physical extent did not influence the judges,[43] nor did argument based on the precautionary principle. This was considered by the courts as simply a 'modalité d'application' in relation to the special police power, as opposed to a principle which of itself could justify intervention by the mayor.[44] Accordingly, by that date it could be concluded that any such intervention by a municipality in matters relating to GMOs would amount to an abuse of power, leading to annulment of the decision to impose a local ban.[45]

---

[37] Art L.335-1 of the Code de l'Environnement; and see generally Billet (n 32 above).
[38] Art L.2212-2 of the Code Général des Collectivités Territoriales.
[39] CAA Bordeaux, 22 September 2004, No 04BX011452, *Préfet de la Haute Garonne*; TA Pau, 6 April 2005, No 0401315, *Préfet du Gers c/ Département du Gers*; CAA Lyon, 26 August 2005, No 03LY000696, *Commune de Ménat*; and TA Rennes, 3 October 2005, No 0502631, *Préfet du Morbihan*.
[40] CAA Bordeaux, 22 September 2004, No 04BX011452, *Préfet de Haute Garonne*; CAA Bordeaux, 12 October 2004, No 04BX01691, *Commune de Saint-André-sur-Sèvre*; and CAA Versailles, 18 May 2006, No 05VE00098, *Commune de Dourdan*.
[41] A Gossement, 'Commentaire sur la jurisprudence OGM' [2004] *Revue Environnement* (December) 20.
[42] CAA Bordeaux, 12 June 2007, No 05BX01360, *Commune de Tonnay-Boutonne*; and CAA Bordeaux, 26 June 2007, No 05BX00570, *Commune de Montgeard*.
[43] CAA Bordeaux, 15 May 2007, No 04BX02001, *Commune de Mouchan*.
[44] Ibid; and CAA Nantes, 26 June 2007, No 06NT01032, *Commune de Carhaix-Plouger*.
[45] CAA de Bordeaux, 15 May 2007, No 06BX01555, *Commune de Londigny*; and CAA Bordeaux, 14 November 2006, No 04BX00265, *Commune d'Ardin*. The same strict approach was also taken in the case of other forms of local authority: see CAA Bordeaux, 15 May 2007, No 05BX02259, *Département du Gers*.

On the other hand, in 2008 there were those who advanced the view that there had been a change in judicial direction, based on a judgment of the Administrative Tribunal of Nîmes on 5 December 2008.[46] In effect, the tribunal did not annul deliberations made by the Municipal Council of the Commune of Le Thor: the elected officials had stated that they were opposed to all GM cropping in the territory of the commune, expressed the wish that the mayor should use his prerogatives to ban such cropping, and authorized the mayor to execute all documentation for that purpose. However, the better view would seem to be that the tribunal did not bring into question the earlier judgments relating to competing police powers. On the contrary, it simply found that a deliberation made by a municipal council is no more than a statement of principle (as opposed to an executive decision).

In consequence, no preventive measure grounded in a decision of locally elected officials would seem capable of surviving judicial scrutiny, the French courts being guided by the spirit of the GM legislation of 25 June 2008 and by Community law. A further consequence is that any coexistence regime would seem reduced to a bundle of technical measures, designed to assure (as far as possible) cohabitation between crops generally considered incompatible both environmentally and economically.

2.3 Detailed Measures

The GM legislation of 25 June 2008 imposes, *inter alia*, obligations to provide information and technical rules.[47]

2.3.1 Obligations to provide information

In accordance with the proposal of the Association Générale des Producteurs de Maïs (AGPM),[48] and for the purposes of ensuring coexistence, the GM legislation of 25 June 2008 requires the authorization holder or the farmer growing GM crops which have been authorized for placing on the market to declare to the competent administrative authority where the crops are being cultivated.[49]

---

[46] TA Nîmes, 5 December 2008, No 0802882, *Préfet de Vaucluse c/ Commune de Le Thor* For confirmation of this judgment by the Conseil d'État in *Préfecture du Gers coutre conseil Général du Gers* (30 December 2009), see <http://www.infogm.org/spip-php?article4282>, last accessed 13 January 2010.

[47] For a survey of the French rules by the European Commission, see European Commission, *Commission Staff Working Document Accompanying Report from the Commission to the Council and the European Parliament on the Coexistence of Genetically Modified Crops with Conventional and Organic Farming* SEC(2009)408.

[48] AGPM, *Livrets de Bonnes Pratiques pour à l'Usage de Maïs Transgénique Bt* (2007) (available at <http://agpm@agpm.com>, last accessed 25 February 2008).

[49] Art L.663-1, para1 of the Code Rural. Most of the rules regarding information in the Code de l'Environnement relate to the authorization procedure (under the Deliberate Release Directive, Arts 24 and 25).

The data contained in the declaration are to be entered into a national register, with the information made available at Departmental level. It may be noted at this preliminary stage that in France the location of releases of GMOs has proved highly controversial, such information being of potential assistance to anti-GM protestors seeking to destroy the crops. Not least, in the context of the authorization procedure (as opposed to coexistence measures), the European Court of Justice has already been required to rule on the issue in *Commune de Sausheim v Azelvandre*.[50] It was argued by the French Government that, when determining the 'location of release' for the purposes of the confidentiality rules in Article 25 (4) of the Deliberate Release Directive, it was sufficient to identify an area larger than the registered parcel itself (such as possibly the commune or canton). This argument did not succeed, the European Court deciding that the determination was to be made instead by reference to all the information submitted by the notifier to the competent administrative authority. Further, the European Court held that no exception could be justified by reason of protection of public order, the exceptions specified in Article 25(4) being exhaustive.

With specific regard to coexistence measures, the French legislator has also provided that the authorization holder or the GM farmer must inform farmers on the surrounding land prior to sowing.[51] This common-sense rule is to be carried into effect by implementing decree.[52] But who is to receive this information? Should some form of modelling be undertaken to find out the maximum possible distance at which cross-contamination may occur, so as to identify the farmers concerned; or can the GM producer simply provide information to those farmers whose land is physically adjacent to the land where he is to grow the GM crops? Moreover, it may be suggested that the provisions are not totally coherent. If the objective is to ensure transparency, then the message which it sends out is somewhat negative, ambiguous, and even contradictory. In effect, when a farmer is informed of the presence of GM cropping nearby, there is the distinct possibility that he will take this as warning of potential cross-contamination. In other words, the information is more likely to generate anxiety than confidence, and to cause conflict. Further, there is the question of the consequences of the information and its publication in a register. In principle, such publication is presumably to allow a farmer to take some form of court action. Yet, as has been seen, in the case of GMOs, the courts have not entertained preventive actions. Accordingly, a person has the right to be informed of the possibility that he may suffer something bad; but he cannot take action against this.

In this context, three additional sources of difficulty may be highlighted. First, the obligation to provide information covers only authorization holders

---

[50] Case C-552/07 (ECJ 17 February 2009).   [51] Art L.663-1, para 2 of the Code Rural.
[52] Ibid para 3. The decree will lay down all the information to be communicated to the competent administrative authority, in particular the land to be cultivated, the dates of sowing, and the variety of GMO grown.

and producers, not those involved in the transportation, processing, or distribution of GMOs. The extent of their obligation depends, therefore, upon the law of contract. Secondly, it may be questioned whether the information has any practical use when it is already settled that GM production is to take place next to conventional or organic crops. In such cases, any issues would fall to be dealt with through a damages action, which would amount to failure of the coexistence regime. Finally, what is the sanction where information is not given or the information given is incorrect? If the authorization has been granted and the other coexistence rules have been observed, there would seem to be no other sanction beyond the fine expressly imposed for non-compliance with the obligation to provide information.[53]

2.3.2 Technical rules

The technical rules in the legislation will override any professional codes of good practice;[54] and the new Article L.663-2 of the Code Rural stipulates that the cropping, harvest, storage, and transport of plants authorized under Article L.533-5 of the Code de l'Environnement (relating to GMOs) or under Community law are subject to technical rules (and, in particular, rules governing isolation distances), with a view to avoiding the adventitious presence of GMOs in other crops. Specific provision is made for crops which benefit from designations of origin and quality, with the relevant organization, or its national institute, to propose to the administrative authority any specific measures required to reinforce protection against GMOs.[55]

Nevertheless, the GM legislation of 25 June 2008 only lays down principles. The efficacy, and the flexibility or rigour, of the measures will not be known until the regulatory texts are issued;[56] and so far the procedure has been somewhat 'stop-go'.[57] The fundamental question today remains the detailed content of the rules and, in particular, whether the Government will simply adopt professional codes of good practice or propose greater restrictions. For example, will the texts distinguish between crops which have been authorized for research purposes (certain of which require a 400-metre isolation distance) and those which, on completion of the research stage, have been authorized for placing on the market? It also goes without saying that the efficacy of any coexistence regime is dependent on the methods employed by inspectors and control mechanisms.[58] That

---

[53] Art L.671-14 of the Code Rural.
[54] For measures within a professional code, see, eg AGPM, *Livrets de Bonnes Pratiques pour à l'Usage de Maïs Transgénique B: Zones Refuges, Distance d'Isolement, Nettoyage Matériel, Semis Échelonné* (available at <http://agpm@agpm.com>, last accessed 25 February 2008).
[55] Art L.642 of the Code Rural.   [56] Ibid Art L.663-2.
[57] A proposal was issued in December 2008, but not pursued by the Government following the reactions that it generated. In any event, a decision of the Conseil des Biotechnologies must be awaited before the Government instigates a new proposal.
[58] Arts L.251-18 and 663-3 of the Code Rural.

said, the overall regime does look relatively robust. Even though the burden falls only on the producer, there is the compensating factor that several links in the production chain are covered, from sowing to transportation. Further, the technical rules are to be backed by administrative controls which can go so far as the destruction of crops,[59] together with specific penal sanctions.[60]

2.4 Conclusion

Although the technical rules are relatively robust, there must remain some doubt as to the efficacy of the French coexistence regime, so long as GMOs are perceived to be a threat to the environment and/or health. Yet, it has to be admitted that the scope for manoeuvre enjoyed by Member States is very limited in light of Community law, which lays down precise rules and under which the policy choice is clearly to accord priority to the free circulation of goods (once the protection of the environment and/or health has been addressed at the authorization stage). As a consequence, it is easy to understand the stance taken by the French legislature, providing that GM and non-GM products should be treated equally in terms of producer and consumer choice, while at the same time treating only GM crops as a potential source of harm. The position adopted by the judiciary (whether in civil or administrative courts) is consistent with the legal texts. Thus, GMOs are the winners in law, even though there remains an internal contradiction: both the Community and the French legislation recognize that, even after authorization, harm may occur through cross-contamination (hence the need for coexistence measures), but the French courts refuse preventive action and, in their judgments, reject any idea of real danger.

3. Italy

3.1 Introduction

Coexistence between GM crops and conventional or organic crops has triggered a major institutional struggle in Italy, involving several levels of government. In particular, there has been a clash between central government on the one hand and several regions on the other, this clash extending to the very principle of coexistence. Moreover, many regions have courted popularity by enacting Bills to declare themselves 'GM-free', without deep or genuine civil debate. For example, when the Charter of Florence was signed in 2005, signatories included the Regione Emilia-Romagna, Regione Lazio, Regione Marche, Regione Sardegna, Regione Toscana, and the Provincia Autonoma di Bolzano.[61] This

---

[59] Art L.663-3 of the Code Rural.   [60] Ibid Art L.671-15 and 16.
[61] Charter of the Regions and Local Authorities of Europe on the Subject of Coexistence of Genetically Modified Crops with Traditional and Organic Farming (4 February 2005).

conflict can in part be explained by the constitutional settlement for Italy, with the republic consisting of communes, provinces, metropolitan cities, regions, and the state.[62] Accordingly, it falls midway between a unitarian and federal state; and the regions have progressively gained powers since the 1970s. No longer are they considered mere administrative outposts of the state, being elevated instead to fully fledged institutional bodies, capable of passing legislation which, in their own territory, has full force of law. There are 20 regions in total, but five of these enjoy privileged status, with greater legislative powers and a higher share of support from central government.[63]

Article 117 of the Constitution, as originally enacted, placed agriculture under regional control (but subject to fundamental principles, which are reserved to state law). By contrast, the environment was not mentioned at all in the original text (although Article 9 cited the protection of landscape as a fundamental constitutional principle). However, major reform was effected in 2001, which had the capacity to impact upon the regulation of GMOs.[64] Under Article 117 (as amended), the protection of the environment is included within the list of matters in respect of which the state has exclusive powers, while agriculture is not listed and thus falls within the exclusive legislative powers of the regions. As GM issues have the capacity to be characterized as both 'environmental' and 'agricultural', the stage was set for a battle of competence between a central government respectful of Community obligations, and regions more than willing to declare themselves 'GM-free'.

3.2 Regional Legislation and the First Decision of the Constitutional Court

This legal battle commenced following issue of the 2003 Commission Recommendation,[65] which prompted certain regions to declare themselves 'GM-free'. In particular, two regions where small farmers' associations enjoyed a strong presence, Puglia and Marche, banned the use of GMOs, with the sanction that operators using them could lose regional financial support for farm businesses (practically all of which was Community-funded).[66] Central government challenged these regional measures; and the Constitutional Court subsequently issued its first decision on GM issues, adopting a very formal approach.[67] The basis

---

[62] Art 114 of the Constitution.
[63] Fruili-Venezia Guilia, Sardinia, Sicily, Trentino-Alto Adige, and Val d'Aosta.
[64] For this Constitutional reform, see, eg A Jannarelli, 'L'agricoltura tra materia e funzione: contributo all'analisi del nuovo art. 117' in A Jannarelli (ed), *Il Diritto dell'Agricoltura nell'Era della Globalizzazione* (Bari: Cacucci, 2003) 179; and A Germanò (ed), *Il Governo dell'Agricoltura nel Nuovo Titolo V della Costituzione: Atti dell'Incontro di Studio (Firenze, 13 Aprile 2002)* (Milan: Guiffrè, 2003).
[65] [2003] OJ L189/36.
[66] Statute of the Regione Puglia of 4 December 2003 (No 26); and Statute of the Regione Marche of 3 March 2004 (No 5).
[67] Judgment 150/2005, issued on 12 April 2005 (*Gazzetta Ufficiale*, 20 April 2005).

of the challenge was that the measures were contrary to Article 117 of the Constitution, it being argued that trade in GMOs was to be governed at national level as part and parcel of the implementation of the Deliberate Release Directive;[68] and that a general ban of GMOs within a region was inconsistent with the national legislation.[69] The court did not accept that a challenge could be mounted under Articles 22 and 23 of the Deliberate Release Directive. In its view, the regional measures were confined to the agricultural sector and did not relate to trade in GMOs, since they covered only farmers who lived within Puglia and Marche.

3.3 The National Legislation

At the same time as this challenge before the Constitutional Court, central government took the initiative to establish a coexistence regime, the impulse being provided by the then Minister of Agriculture (as opposed to the Minister of Environment).[70] Government Legislative-Decree No 279/2004 was designed to prevent proliferation of regional measures seeking to impose 'GM-free' zones in a manner inconsistent with Community law. Securing passage of the Government Legislative-Decree proved tortuous. It was originally proposed as a form of emergency legislation, which is immediately binding, but temporary and subject ultimately to approval by Parliament. In the event, nonetheless, it was passed by Parliament at the eleventh hour with a large majority: most of the opposition abstained, with the Green Party alone remaining defiant to the end, advocating even more restrictive legislation. (It may also be noted that the Government Legislative-Decree has now been superseded by Statute No 5 of 2005, which contains some minor changes, such as the express prohibition of GM cropping until management plans have been approved.)

An important aspect of the national legislation was acceptance of a large role for the regions in determining coexistence rules. In so doing, it accommodated the position of 'moderate' regions, such as Emilia-Romagna, which (while a signatory to the Charter of Florence) had stopped short of declaring itself 'GM-free'. The overall stance adopted was that GM crops should be considered legal throughout the national territory, but should not compromise other forms of agriculture. Article 1 of the Government Legislative-Decree affirmed that

---

[68] [2001] OJ L106/1.

[69] It may be noted that 18 NGOs, representing farmers and environmentalists, asked to be heard by the court, in support of the position adopted by Puglia and Marche, but the court rejected all such requests. For the role of non-governmental organizations (NGOs) and farmers' associations generally, see, eg L Levidow and K Boschert, 'Coexistence or contradiction? GM crops versus alternative agricultures in Europe' (2008) 39 Geoforum 174, 183–4.

[70] It would be easy to underestimate the personal impact of Gianni Alemanno, who, during his time in office, supported an Italian agriculture which was not confined to commodity crops and which generated numerous protected designations of origin and geographical indications.

transgenic farming was farming that used GMOs, while Article 2 laid down the general principle of coexistence, whereby the introduction of GMOs should not damage organic or conventional farming. The same article also provided that farmers, traders, and consumers should enjoy freedom to choose between GM and non-GM production and, accordingly, the GM food chain was to be kept entirely separate. Under Article 3, central government was to issue by decree a legal framework, to be respected by the regions, but, under Article 4, it was affirmed that the regions enjoyed the right to draw up their own coexistence plans, these to contain technical rules for the purposes of keeping GM crops within bounds. On the other hand, the deadlines for approval of the coexistence plans remained vague; and this vagueness would seem to have been deliberate, with the consequence that central government could formally bring to end all moratoria, yet in practice allow the pre-existing situation to continue.[71]

Some regions objected to this balancing act, for reasons both substantive and formal. The substantive reason was outright opposition to GMOs, which could not contemplate coexistence and which found expression in the declaration of 'GM-free' zones. The formal reason was, to an extent, supplied by central government itself. During the passage of the legislation through Parliament, the main source of debate had been agriculture, with public health and the environment attracting less attention. Yet, as indicated, agriculture falls within the competence of the regions, so opening up the opportunity to mount a legal challenge before the Constitutional Court.

3.4 The Second Decision of the Constitutional Court

The Region Marche, already a protagonist in the first legal battle before the Constitutional Court, commenced an action to have the new legislation declared unconstitutional.[72] Thus, in this action it was the claimant, seeking to render national legislation void, whereas in the earlier action it had been the defendant, seeking to uphold regional measures. There were several grounds to the challenge. For example, it was argued that, under Italian constitutional law, the state should have secured the agreement of the regions, which was not possible when enacting a Government Legislative-Decree; and that, in consequence, there was an abuse of power and breach of Article 117 of the Constitution. Further, the Government Legislative-Decree was alleged to contravene the power of the regions to issue rules on agriculture (as defined in Article 32 of the EC Treaty and Article 2135 of the Italian Civil Code).

---

[71] See, eg F Fragale, 'La coesistenza dei sistemi di agricoltura convenzionale, biologica e OGM' in *Agrobiotecnologie nel Contesto Italiano* (Rome: INRAN, 2006) 317.
[72] See generally, eg P Borghi, 'Colture geneticamente modificate, ordinamenti e competenze: problemi di coesistenza (nota a Cort. Cost. 17 Marzo 2006 n. 116)' [2006] *Le Regioni* 961.

The State Attorney rejected these arguments. In his view, the Deliberate Release Directive gave force to the principle of free circulation of GMOs, subject to precautionary measures. Thus the Italian State was obliged to issue the Government Legislative-Decree, with the aim of protecting the environment, while at the same time mitigating the effect of the regional legislation which had declared 'GM-free' zones. The state had exclusive powers with regard to the environment; and, even if the Constitution provided for the state and the regions to have concurrent powers in the case of food and health law,[73] it was evident that the national legislation implemented merely a general legal framework, to be supplemented by further regulations at regional level.

The Constitutional Court issued a novel decision.[74] It recognized that the Deliberate Release Directive formed the legal basis for regulation of environmental and health concerns arising from the deliberate release of GMOs; and that, by contrast, the 2003 Commission Recommendation was directed to their economic consequences (even though it might be argued that there was a link to the environment or health): as stated in its Preamble, '[t]he issue of coexistence addressed in this Recommendation concerns the potential economic loss and impact of the admixture of GM and non-GM crops'.[75] The Constitutional Court also formally recognized the legitimate power of central government to affirm the principle of coexistence. The application of this principle, however, was considered to be a matter for the regions, in that the detailed rules were concerned with the management of agriculture. Accordingly, the judgment affirmed the constitutional validity of Articles 1 and 2 of the Government Legislative-Decree, but it was accepted that the realization of coexistence management plans fell under the exclusive powers of regions. A difficulty with the Government Legislative-Decree was that Article 3 gave to central government the power to enact framework rules by a measure of uncertain legal status, while Article 4 confined the role of the regions to their management plans. Both those provisions, therefore, were found to be contrary to the constitutional rights of the regions to enact their own agricultural legislation, not least because such 'subsidiarity' was directed to accommodating an Italian countryside of great diversity.[76]

The result is a legal mess. Regions are to write their own management rules, but must observe the general principle of coexistence, which precludes 'GM-free' legislation. However, without clear authority some regions persist in claiming 'GM-free' status.[77] Perhaps unsurprisingly, the general public has begun to show a degree of indifference towards what has become a complex battle for

---

[73] Art 117(3).
[74] Judgment 116/2006, issued on 17 March 2006 (Gazetta Officiale, 22 March 2006).
[75] 2003 Commission Recommendation, Preamble (5); and see also Preamble (4) and Annex, paras 1.1 and 1.2.
[76] With Arts 3 and 4 declared invalid, other rules flowing from them have suffered the same fate.
[77] An example is Puglia.

competence between the regions and central government.[78] For the time being, a *de facto* moratorium has been established, with the Italian State unable to impose a coexistence regime on reluctant regions, yet the position may still change. In its judgment No 183, issued 0n 19 January 2010, the Council of State has declared that the Ministry of Agriculture should within 90 days authorize a farmer from Friuli to grow GM maize. Accordingly, central government proposals for legislation are expected in the near future, probably as part of the Agrarian Code, but the approval of Parliament is still to be secured.

4. Spain

4.1 Introduction

Spain is the Member of State which has the highest proportion of its agricultural area given over to GM crops. First introduced in 1998 (when 22,000 hectares were grown), by 2008 they accounted for some 100,000 hectares[79] and the variety most widely cultivated has been maize for animal feed. However, there has been practically no form of control or monitoring; and, in light of this surge in GM production, it is hard to justify the delay in drawing up a specific coexistence regime. For the moment, Spain has only adopted two pieces of legislation to govern agricultural biotechnology. The first is Law 9/2003 of 25 April 2003, which established the regime for the confined use, release into the environment, and placing on the market of GMOs;[80] the second is Royal Decree 178/2004 of 30 January 2004, which approved general rules for the implementation of Law 9/2003.[81] Both these had as their objective the adaptation of the Spanish legal order to comply with Council Directive (EEC) 90/219 on the contained use of genetically modified micro-organisms,[82] and the Deliberate Release Directive.[83]

A matter of some importance is that, in addition to laying down general principles, Law 9/2003 and Royal Decree 178/2004 specify the role both of the General State Administration and of the autonomous communities for the purposes of authorizing confined use, deliberate releases into the environment, and placings on the market of GMOs, including their powers of monitoring and enforcement. Thus, the General State Administration has competence to: authorize commercial use of GMOs and products containing GMOs; grant

---

[78] See, further E Sirsi, 'OGM e coesistenza con le colture tradizionali' [2006] *Agricoltura Istituzioni Mercati* 391.

[79] C James, *Global Status of Commercialized Biotech/GM Crops: 2008 (ISAAA Brief No 39)* (Ithaca, NY: International Service for the Acquisition of Agri-biotech Applications (ISAAA), 2009) Table 1. The growing of GM crops is not evenly distributed across Spain, with most being found in Aragon and Catalonia. [80] *Boletín Oficial del Estado* (BOE) No 100 of 26 April 2003.
[81] BOE No 27 of 31 January 2004.
[82] [1990] OJ L117/1 (as substantially amended by Council Directive (EC) 98/81 [1998] OJ L330/13, to which the Spanish legislation refers). See now Directive (EC) 2009/41 of the European Parliament and of the Council on the contained use of genetically modified micro-organisms [2009] OJ L125/75. [83] [2001] OJ L106/1.

180   *Coexistence of Crops: National Implementation*

permits to import or export GMOs or products derived from GMOs; and implement monitoring and enforcement. The General State Administration enjoys like powers in respect of the confined use and deliberate release of medicines; and it is also in charge of the promotion and general co-ordination of scientific and technical assessments, together with the tests required for registration of commercial varieties. Again in respect of these matters it has competence for monitoring and enforcement. By contrast, the autonomous communities are in charge of granting authorizations for deliberate releases of GMOs, other than placings on the market; and they undertake monitoring and enforcement for those tasks that do not fall within the competence of the state.

In addition, Law 9/2003 and Royal Decree 178/2004 provide for the Inter-Ministerial Council on GMOs ('Consejo Interministerial de Organismos Modificados Genéticamente') and the National Biosafety Commission ('Comisión Nacional de Bioseguridad'). The former body enjoys competence to grant or refuse authorizations for confined use, deliberate releases, and placings on the market. By contrast, the latter is a consultative body, whose main function is to prepare requests for authorizations for consideration by the General State Administration or the autonomous communities (as the case may be). The legislative framework also establishes a central register for GMOs, managed by the Ministry of the Environment and situated in Madrid. The Inter-Ministerial Council on GMOs, the National Biosafety Commission, the relevant central government departments, and the autonomous communities must send data in their possession to this central register, which, for the purposes of monitoring effects on the environment, contains information on the location of GM crops and their distribution within the autonomous communities.

4.2 The 2006 Draft Legislation on Coexistence

4.2.1 *General: purpose and scope*

At present, the coexistence regime consists only of a draft Royal Decree.[84] This is based on the general principles set out in the 2003 Commission Recommendation;[85] and comprises 12 articles divided into three chapters, together with further measures and annexes. For the purpose of the draft Royal Decree, coexistence is concerned with the economic effects and repercussions of mixing GM and non-GM crops, together with implementation of the management measures most appropriate to bring cross-contamination with the Community

---

[84] Proyecto de Real Decreto por el que se Aprueba el Reglamento sobre Coexistencia de los Cultivos Modificados Genéticamente con los Convencionales y Ecológicos. For a survey of the Spanish rules by the European Commission, see European Commission, *Commission Staff Working Document Accompanying Report from the Commission to the Council and the European Parliament on the Coexistence of Genetically Modified Crops with Conventional and Organic Farming* SEC(2009)408.
[85] [2003] OJ L189/36.

threshold. Accordingly, Article 1 defines coexistence as preserving the possibility for farmers to choose between conventional, organic, and GM crops, in compliance with the legal obligations applicable to their chosen system of production.[86] Article 2 then confines the scope of the legislation to plant varieties which have been authorized under Community law and whose production may impact upon other forms of cropping.

4.2.2 Obligations on farmers, coexistence measures, and their application

While Article 3 of the draft Royal Decree respects the labelling threshold in the Food and Feed Regulation,[87] it also provides that, where the threshold falls below 0.1 per cent, the product can be labelled as totally 'GM-free'. Anti-GM campaigners have suggested that this is somewhat economical with the truth, especially when consumers are purchasing organic food.[88] However, it must be reiterated that, although the use of GMOs is prohibited in organic production, the new umbrella regulation, Council Regulation (EC) 834/2007 on organic production and labelling of organic products, permits products to remain 'organic' despite the adventitious presence of authorized GMOs up to the general 0.9 per cent threshold.[89]

Under Article 4.1 of the draft Royal Decree, production requirements to guarantee coexistence *could* include provisions relating to: sowing and cropping;[90] harvesting;[91] transport and storage;[92] and co-operation between neighbouring farmers.[93] However, under Article 4.2, farmers who introduce GM production *must* apply the necessary farm management measures to limit the flow of pollen. Annex II contains these measures in the case of maize (the only crop with GM varieties yet authorized) and other Annexes will contain like measures to be applied for other crops as and when they too are authorized. As indicated, by contrast with Article 4.1, these provisions would seem compulsory, but the specific rules governing maize, as set out in Annex II, are not onerous. They address, *inter alia*: minimum isolation distances between GM and

---

[86] In this regard, see 2003 Commission Recommendation, Preamble (3).
[87] [2003] OJ L268/1.
[88] See, eg *Observaciones de COAG* (Coordinadora de Organizaciones de Agricultores y Ganaderos) on the draft Royal Decree (Madrid, 26 June 2006) (available at <http://www.coag.org/rep_ficheros_web/781a3c6e8e5c944ccec146873f30b1c8.pdf>, last accessed 7 August 2009).
[89] [2007] OJ L189/1. See also European Commission, *European Action Plan for Organic Food and Farming* COM(2004)415.
[90] For example, buffer zones, pollen barriers, crop rotation, control of volunteers, use of sterile varieties or varieties producing little pollen, and seed segregation.
[91] For example, the cleaning of machinery both before and after harvest and the retention of machinery for use specifically in the production of either GM or non-GM crops.
[92] For example, physical separation of GM and non-GM crops once severed from the ground.
[93] For example, information on sowing plans and the conclusion of agreements between farmers (which, in order to limit cross-pollination, might provide for: a voluntary zone dedicated to one form of production; the use of varieties which would flower at different times; and/or a sowing programme which would result in flowering at different times).

non-GM maize (220 metres being the distance envisaged); buffer zones (with at least four rows of non-GM maize to be planted round the GM crop, this non-GM maize to be labelled 'GM'); different flowering periods;[94] the cleaning of equipment during sowing and subsequently; and the physical separation of GM and non-GM crops for the purposes of transport, drying, and storage. Moreover, Article 9 provides two exceptions to the specific rules for maize contained in Annex II: first, when adjacent plots are sown with different crop types, whether GM, conventional, or organic; and, secondly, when all crops within the same 'zone' are GM. Such exceptions may present difficulties. For example, in the former case, no account is to be taken of the long-term effects on the soil (with the potential for generating a GM seed-bank); and, in the latter case, there are no criteria by which to define a 'zone'.

Article 5 sets out a series of general coexistence obligations. First, farmers wishing to grow GM varieties on their land must notify, in writing and one month prior to the commencement of production, any neighbouring farmers and any other farmer whose holding falls within the minimum isolation distances established for each crop. Secondly, the farmer must also notify the competent body of the relevant autonomous communities, again one month before sowing. Annex I specifies the minimum data to be contained in such notification, including: the variety concerned; the form of genetic modification; the location of the crop sown; the area sown; and any coexistence measures the farmer is implementing.[95] Annex I does not indicate whether the coexistence measures are obligatory or merely optional; however, it would appear that they are optional, since an explanatory memorandum suggests that farmers have a choice as to which measures to adopt. Thirdly, a farmer growing GM varieties must use seeds subject to official controls, so as to guarantee the purity and quality of the crops. Fourthly, the farmer must keep for five years the labels on the seed packaging which denote their GM nature. He must also notify in writing the person to whom the severed crop is sold, transferred, or delivered. This notification should address the variety, the form of genetic modification, and its unique identifier. Fifthly, in every case good agricultural practice is to be employed. Sixthly, farmers must notify the competent body of the relevant autonomous community of every problem or difficulty encountered when implementing the coexistence measures. Finally, any farmer growing GM crops must collaborate fully in the carrying-out of any inspections, controls, collections of samples, or gathering of information necessary for monitoring programmes (and he must do so as soon as requested by the competent administrative body).

---

[94] The competent bodies of the autonomous communities are able to authorize, as an alternative to minimum distances between GM and non-GM crops, a planting programme designed to avoid simultaneous flowering.
[95] These are: minimum isolation distances; buffer zones; refuge zones; different flowering periods; segregation at time of sowing; separate crop management; separate treatment for the purposes of transport and storage; and the existence of any agreements with neighbouring farmers.

By contrast, moving beyond obligations imposed on farmers, Article 6 imposes a far lighter burden on companies or individuals producing or selling GM seeds or other GM material. Their only obligation, on any sale to farmers, is to supply in writing the information necessary to comply with the coexistence and traceability regimes imposed by Community law (and, not least, the rules laid down in Regulation 1830/2003).[96] Further, Article 7 governs the role of public authorities, which are required to consult with and promote training for farmers in order to improve information on growing GM crops, coexistence practices, and tracing and labelling requirements.

A significant provision is Article 8, which addresses agreements between farmers, a feature of the Community coexistence regime promoted in Spain. Not least, as highlighted in the 2003 Commission Recommendation, a significant reduction in segregation costs can be achieved where farmers in a neighbourhood co-ordinate their production.[97] Article 8 provides that, where farmers within the same geographic zone enter into a coexistence accord, it will take priority over the specific regime established for maize under Annex II, or to be established for other crops under their respective Annexes.[98] In particular, accords may include voluntary undertakings not to grow GM crops within a defined area. However, the text does not make clear whether they are contractual and, consequently, whether it would be possible to sue for breach of contract. There is also doubt as to how they should fall to be treated under the law governing agricultural tenancies.[99] Importantly, Article 8 of the law governing agricultural tenancies renders void any agreements which seek to restrict the tenant's freedom of cropping,[100] except where the agreements are directed to preventing the impoverishment of the soil, in compliance with Community law or other regulatory provisions. A further, practical, difficulty is that accords between farmers will be very difficult to realize in areas where there is mixed GM, non-GM, and organic production based upon entrenched and divergent interests.[101]

To conclude, the draft Royal Decree introduces in large part an *à la carte* menu from which farmers can pick and choose; and even when the technical rules are compulsory, they may still enjoy latitude. For example, the compulsory provisions in the various Annexes may be circumvented either by accords between farmers (under Article 8) or, in the case of maize, by the two exceptions to the rules in Annex II (under Article 9). This is a relatively light regulatory

---

[96] [2003] OJ L268/24. [97] 2003 Commission Recommendation, Annex, para 3.3.3.
[98] In the case of maize, Annex II indicates that it will not be necessary to have minimum isolation distances between neighbours where they have entered into an accord to that effect and no third party will suffer prejudice.
[99] Law 26/2005 of 30 November 2005, amending Law 49/2003 of 26 November 2003 on agricultural tenancies: BOE No 287 of 1 December 2005.
[100] For similar provision in England and Wales, see the Agricultural Holdings Act 1986, s 15.
[101] See, eg MJ Cazorla González, 'Aspectos jurídicos del proyecto de coexistencia entre cultivos' (2005) 46–7 *Revista de Derecho Agrario y Alimentario* 113.

burden when the consequences of the coexistence regime failing is risk of contamination to third parties. It may thus be questioned why the draft Royal Decree retreats from imposing compulsion across the board (and, indeed, compulsion without scope for derogation).

4.2.3 Monitoring and control

Article 10 provides that the Ministry of Agriculture, Fisheries, and Food should draw up each year a programme of supervision and control, in conjunction with the Ministry of the Environment and in accordance with proposals from the autonomous communities.[102] In compliance with the national programme of supervision and control, the autonomous communities must then inspect a minimum percentage of land declared as sown to GM crops.[103] These inspections must be undertaken at least once at the following points: first, during the growing of the crop, to verify that the coexistence rules are being observed (namely, isolation distances, buffer zones, use of seed and other officially controlled inputs, information to neighbours, and good agricultural practice); and, secondly, at harvest, to verify proper use of machinery and physical separation of crops severed from the ground according to their method of production. In this latter context, account is to be taken of the fact that different forms of production should be stored and treated separately and that accidental contamination should be avoided during transport. The national programme should also provide for appropriate controls and testing and ensure monitoring for adventitious presence of GMOs both on land farmed near to GM crops and on other adjacent areas which remain uncultivated, in order to assess the effectiveness of coexistence measures.

All data submitted by farmers (as specified in Annex I) must be included by the autonomous communities in the register of land under GM crops;[104] and the autonomous communities must forward such data to the Ministry of Agriculture, Fisheries, and Food, together with all information relating to monitoring and controls undertaken and all details supplied by farmers on difficulties arising from the application of the coexistence rules. Finally, in accordance with Article 12, the Ministry of Agriculture, Fisheries, and Food, in collaboration with the Ministry of the Environment, must: first, draw up an annual report on the results of monitoring and evaluation programmes undertaken by the autonomous communities (the report to be sent to the European Commission);[105] and, secondly, make aware the National Biovigilance Commission

---

[102] No clear statement has yet been made as to the content of these plans or as to the precise obligations of the autonomous communities.
[103] No minimum percentage has yet been specified.
[104] Art 11. At present, companies producing GM seed are under no obligation to submit data.
[105] In the first annual report, particular attention is to be paid to the possible economic effects of coexistence measures, these being relevant to the compensatory instruments which might be employed.

('Comisión Nacional de Biovigilancia') of the information received by the autonomous communities.[106] It may be noted that no provision is made for the supply of information in relation to land cultivated with GM crops near the border of another autonomous community or another Member State.[107]

4.2.4 Breach of coexistence rules and competence

The draft Royal Decree also addresses the position where there is breach of its coexistence rules, but with penalties to be imposed under the administrative regime laid down by Law 30/2006 of 26 July 2006, on seeds, nurseries, and phyto-genetic resources.[108] Thus, importantly, it does not contain a specific enforcement mechanism. This may have material consequences, since the scope of Law 30/2006 is less broad in ambit and, in particular, it does not extend to the acts of husbandry which are so central to the coexistence rules. That said, Law 30/2006 does distinguish between breaches which are light, severe, or very severe; and examples of severe breaches include falsifying the data required in GM monitoring programmes, preventing inspections, or importing or exporting GMOs without the appropriate authorizations.[109]

As a general rule, the autonomous communities enjoy competence in relation to agriculture;[110] but the draft Royal Decree is based on Article 149.1.13a, 16a, and 23a of the Constitution, which confers on the state competence to enact framework legislation in relation to economic strategy, health, and environmental protection. The underlying importance of such matters justifies rules imposed at national level, which in turn avoids the risk of different regimes in each of the 17 autonomous communities. As a result, the Ministry of Agriculture, Fisheries, and Food and the Ministry of the Environment, consistent with their respective competences, must adopt the measures necessary to implement the draft Royal Decree, establish the national monitoring programme, and modify or add to the Annexes laying down detailed coexistence rules as and when required.[111]

---

[106] The National Biovigilance Commission is a consultative body on GMO issues for the Ministry of Agriculture, Fisheries, and Food. Its purpose is to provide input on the establishment, development, and application of monitoring programmes for GM varieties, as well as on coexistence between GM, conventional, and organic crops. For its creation, see Royal Decree 1697/2003 of 12 December 2003: BOE No 310 of 27 December 2003.
[107] 2003 Commission Recommendation, Annex, para 2.1.7.
[108] BOE No 178 of 27 July 2006. Such penalties are without prejudice to civil action.
[109] Breaches which qualify as very severe incur a fine of between €30,001 and €300,000. Further, it is provided that the farmer must reimburse any support payments received; and that he may be banned for up to 10 years from acting as a supplier of GMOs.
[110] Art 148.1.7a of the Constitution.
[111] It must be highlighted that the Ministry of Agriculture, Fisheries, and Food, and the Ministry of the Environment have recently been combined into the Ministry of the Environment and Rural and Marine Affairs. In consequence, references in the draft Royal Decree to the Ministry of Agriculture, Fisheries, and Food and the Ministry of the Environment must now be interpreted as reference to the new Ministry: see <http://www.marm.es>, last accessed 3 June 2009.

4.3 Controversial Issues

Several Spanish social, organic, and farmer organizations consider that the text of the draft Royal Decree does not go far enough.[112] According to these organizations,[113] labelling thresholds (whether they be 0.1 per cent or 0.9 per cent) have no place in a coexistence regime; and, by accommodating such thresholds, the draft Royal Decree not only fails to prevent risk of cross-contamination, but, on the contrary, legitimizes the presence of GMOs in conventional and organic agriculture.[114] Accordingly, legal rules designed to avoid cross-contamination provide for a level of contamination: in other words, coexistence is addressed by setting thresholds in the final product, rather than by putting in place a management scheme to ensure that cross-contamination does not occur. However, as has been seen, the approach adopted in the draft Royal Decree is consistent with Community legislation, since the 2003 Commission Recommendation provides that '[n]ational strategies and best practices for coexistence should refer to the legal labelling thresholds and to applicable purity standards for GM food, feed and seed'; and that '[i]solation distances should minimise but not necessarily eliminate gene flow by pollen transfer'.[115] In reality, just as in other Member States, all that is to be achieved in Spain is a level of adventitious presence below a specified tolerance level, with the question of coexistence reduced to a commercial problem.[116]

Since cross-contamination has the capacity to impact along the whole food chain (from cropping, to transport, storage, manufacture, and distribution), this creates an imperative to reduce cross-contamination to the minimum at each link of the chain, beginning with agricultural production; and, in this regard, technical deficiencies in the proposed regime may be highlighted. As indicated, in the draft Royal Decree an isolation distance of 220 metres is stipulated in the case of maize. This is an increase from an earlier proposal of 50 metres, but even the increased distance would seem to be insufficient, on the basis that it has been shown that maize pollen can be carried far further.[117] Annex II has also been

---

[112] See, generally, eg Levidow and Boschert (n 69 above); and R Binimelis, 'Coexistence of plants and coexistence of farmers: is an individual choice possible?' (2008) 21 Journal of Agricultural and Environmental Ethics 437.

[113] See, eg Friends of the Earth in Spain, at <http://www.tierra.org/spip/spip.php?rubrique70>, last accessed 24 March 2009.

[114] In this context, see F Casero Rodríguez, 'En Defensa de la Agricultura Ecológica' in R Herrera Campos and MJ Cazorla González (eds), *Sociedad de Consumo y Agricultura Biotecnológica* (Almería: University of Almería, 2006) 162. For this author, the objective of the coexistence rules is the preservation of an agricultural and food sector free from GMOs, with recognition of the right of farmers to produce GM-free and of consumers to choose GM-free.

[115] 2003 Commission Recommendation, Annex, paras 2.2.3 and 3.2.1.

[116] See also ibid Preamble (5), which limits the scope of the Recommendation to the potential economic loss and impact of the admixture of GM and non-GM crops.

[117] See, eg Agriculture and Environment Biotechnology Commission (AEBC), *GM Crops? Coexistence and Liability* (London: AEBC, 2003) 52.

criticized on the basis that it permits the competent authorities of the autonomous communities to authorize, as an alternative to isolation distances, sowing programmes to avoid adjoining maize crops flowering at the same time. A clear difficulty is that isolation distances are a key tool in the reduction of the risk of cross-contamination, and the autonomous communities should not be able to decide to replace them with a less-effective measure. Similarly, the rules provide that refuge zones are only to be established where there is resistance to corn borers, whereas it would be more effective in terms of protection if they were obligatory for all farmers cultivating GM maize, whether or not there was such resistance. Indeed, perhaps not surprisingly, environmental organizations believe that all the rules in Annex II should be compulsory, as opposed to capable of displacement by accords between farmers or subject to the derogations in Article 9. It may also be noted that any deficiencies in the technical rules have the capacity to be especially acute for Spain, since a defining feature of this Member State is the extent of field trials already undertaken. These now cover not only maize, but cotton, rice, potatoes, sugar beet, and tomatoes;[118] and, with each authorization, it becomes more difficult to preserve a truly organic agriculture, so compromising consumer choice.

Importantly, the draft Royal Decree makes no mention of 'GM-free' zones, nor does it refer to the possibility of banning or restricting the cultivation of GM crops in conservation areas or areas nearby. Spanish environmental organizations see the opportunity to do so as 'measures of a regional dimension' under paragraph 2.1.5 of the Annex to the 2003 Commission Recommendation. Yet there remains the difficulty that the European Commission has repeatedly affirmed that 'GM-free' zones are not legitimate under Article 26a of the Deliberate Release Directive;[119] and the 2003 Commission Recommendation expressly states that '[s]uch measures should apply only to specific crops whose cultivation would be incompatible with ensuring coexistence, and their geographical scale should be as limited as possible'.[120] Nevertheless, the autonomous communities of Asturia, the Basque Country, the Balearics, and the Canaries have all declared themselves 'GM-free' zones, together with numerous communes.[121] If these zones are indeed invalid, then the only alternative would seem to be for farmers to make a voluntary decision to renounce GM crops within a defined area; but doubt remains as to the weight to be attached to such accords (in particular,

---

[118] See, eg R Binimelis, 'Diez años de transgénicos en España: la imposible coexistencia' (available at <http://www.landaction.org/spip/spip.php?article275>, last accessed 24 March 2009).
[119] For a statement to this effect by the European Commission, see, eg *Report on the Implementation of National Measures on the Coexistence of Genetically Modified Crops with Conventional and Organic Farming* COM(2006)104, 5; and see further Chapter 6.
[120] 2003 Commission Recommendation, Annex, para 2.1.5. Besides, the same paragraph requires that such measures 'should only be considered if sufficient levels of purity cannot be achieved by other means' and would 'need to be justified for each crop and product type'.
[121] See, eg Friends of the Earth in Spain, <http://www.tierra.org/spip/spip.php?rubrique70>, last accessed 24 March 2009.

whether they are enforceable contracts), and as to their legitimacy if purporting to operate at regional level.

However, the major criticism must be the complete omission of any specific provisions to cover liability for damage as a result of the presence of GM material in excess of the threshold permitted for conventional or organic agriculture. Nowhere does the draft Royal Decree specify who incurs liability or how farmers suffering cross-contamination should be compensated. Since, as already observed, management measures are apprehended to reduce rather than fully eliminate cross-contamination, damage would seem inevitable, and it may be considered incomprehensible that no regime to compensate for economic loss is to be provided. In addition, there is as yet no indication that liability insurance or a guarantee fund will be required. Environmental organizations and some farmer groups have called for a civil liability regime which would impose on those holding authorizations for deliberate releases into the environment (namely, seed producers) responsibility both for economic loss and for possible injury to health and the environment (including agricultural biodiversity);[122] and they appeal to the Government to include specific liability rules in the draft Royal Decree so as to fill these lacunae. Any such rules would, nonetheless, face the lingering difficulty of identifying the authors of the damage and demonstrating causation. Thus, for the present, seed producers continue to benefit from sales of their products in a market free from liability to provide compensation; and the legislative framework in Spain as yet fails to resolve the problems which arise.

5. United Kingdom

5.1 Introduction

Progress with implementing a coexistence regime in the United Kingdom has not been swift. As a result, when the European Commission issued the 2009 Implementation Report, the United Kingdom could be categorized as one of the Member States where 'the development of a regulatory framework is not envisaged in the near future as the cultivation of GM crops on their territory has been deemed unlikely to take place'.[123] In many ways this tardiness may be considered surprising, in that the Government has been a strong advocate of modern biotechnology as a cornerstone of the 'knowledge economy'. For example, in his 2006 speech to the Royal Society, 'Our Nature's Future—Science', Prime Minister Tony Blair articulated a vision of the United Kingdom

---

[122] In the case of environmental organizations, see, eg Friends of the Earth in Spain, Greenpeace, Ecologistas en Acción, and Sociedad Española de Agricultura Ecológica; and, in the case of farmer groups, see, eg COAG.
[123] COM(2009)153, 6.

as 'a magnet for scientific endeavour', noting that the country was the location for just under half of all public biotechnology companies in Europe.[124] He also asserted that '[t]he 'anti-science brigade threatens our progress and our prosperity', while at the same time conceding the need to be 'honest about the risks'.[125] In addition, within the context of the Community regulatory framework, the United Kingdom has consistently voted 'pro-GM'. For example, in March 2009 it was one of the distinct minority of Member States to support all the European Commission proposals to overturn national safeguard measures banning MON810 and T25.[126]

Against this policy background, the absence of any coexistence regime may perhaps be best explained by the general understanding that such a regime is not needed at present. Only one GM variety has been authorized for commercial use in the United Kingdom: Chardon LL; and it was almost immediately withdrawn by Bayer CropScience (in 2004), so restrictive were the conditions imposed.[127] More recently, in November 2007, the Minister for the Environment affirmed that '[n]o commercial cultivation is expected in England for several years'.[128] Besides, the future for GM crops in Northern Ireland and, in particular, Scotland and Wales, would seem even less auspicious. Agriculture falls within the competence of the devolved administrations, and the 'Celtic fringe' has adopted a significantly more anti-GM stance as compared to that adopted for England by the Department for Environment, Food and Rural Affairs (DEFRA). For example, the 2007 Manifesto of the Scottish National Party contained a commitment to maintain a moratorium on the planting of GM crops in Scotland;[129] and in April 2009 the Scottish Environment Minister reaffirmed that Scotland should remain 'GM-free'.[130] Moreover, in only slightly less robust fashion, the Rural Affairs Minister of the Welsh Assembly Government declared in June 2009 that it continued to be the position of the Welsh Assembly Government 'to adopt the most restrictive policy on GM crops that is compatible with European Union and UK legislation', with clear intention for co-existence to be

---

[124] Prime Minister Tony Blair, 'Our Nation's Future—Science', 3 November 2006 (available at <http://www.number10.gov.uk/Page10342>, last accessed 16 June 2009). For an earlier statement to this effect, see, eg Prime Minister Tony Blair, 'PM Speech: "Science Matters"', 23 May 2002 (available at <http://www.number10.gov.uk/page1715>, last accessed 4 May 2009).

[125] Prime Minister Tony Blair, 'Our Nations Future-Science' (n 124 above).

[126] 2928th Council Meeting: Environment, Brussels, 2 March 2009 (Presse 53); and see, eg *Agra Europe Weekly* No 2351, 6 March 2009, EP/1.

[127] See, eg House of Commons Environment, Food and Rural Affairs Committee, *Eleventh Report of Session 2003–04: GM Planting Regime* (HC 607) paras 4–5.

[128] *Hansard*, HC vol 467, col 17 WS (8 November 2007) (Written Statement by the Minister for the Environment (Phil Woolas)).

[129] Scottish National Party, *Manifesto 2007: SNP—It's Time* (Edinburgh: Scottish National Party, 2007) 72 (the Scottish National Party currently has the largest number of Members of the Scottish Parliament). See also generally G Lean, '"Celts" Revolt Against Westminster Over GM Crops' Independent on Sunday, 28 September 2008.

[130] Scottish Government News Release, *Genetic Modification (GM)*, 24 April 2009 (available at <http://www.scotland.gov.uk/News/Releases/2009/04/24150325>, last accessed 29 June 2009).

190　　　*Coexistence of Crops: National Implementation*

tightly regulated in Wales'.[131] Indeed, the gap between the 'Celtic fringe' and DEFRA may yet widen in the light of more recent pronouncements by the Minister for the Environment that the 2008 global food crisis should stimulate GM plantings.[132]

Whatever the prospects for agricultural biotechnology in the United Kingdom, and however little the sense of urgency, there is nonetheless clear commitment by the UK Government to implement a coexistence regime prior to commercial cultivation. Thus, in what may be regarded as the main policy statement to date (that delivered by the Secretary of State for Environment, Food and Rural Affairs on 9 March 2004), it was anticipated that 'co-existence measures will be in place before any GM crops are grown commercially';[133] and this approach was reaffirmed by the Minister for the Environment in 2007.[134] For the time being, however, as indicated, no coexistence regime has been implemented, and even draft regulations remain to be issued. Instead, the Government has preferred to await further research evidence (in particular on crop separation distances), together with clarification on the thresholds which will trigger the labelling of adventitious GM presence in conventional seeds.[135] Nevertheless, insights into the shape and content of any future UK coexistence regime may be gleaned both from policy statements and from research and consultation already undertaken. These insights may be considered in relation to three key issues: first, the form of implementation; secondly, tolerance thresholds; and, thirdly, 'GM-free' zones.

5.2 Key Issues

5.2.1 *Form of implementation*

A significant feature of the 9 March 2004 policy statement was clear assertion that farmers who wished to grow GM crops should be required to comply with a code of practice and, importantly, that the code should have statutory backing.[136] This followed advice of the AEBC in its report, *GM Crops? Coexistence and Liability*

---

[131] Welsh Assembly Government Press Release, *Announcement of Plans for Tight Regulation of GM Crops in Wales*, 30 June 2009. See also Levidow and Boschert (n 69 above) 181–2.
[132] See, eg 'Genetically Modified Crops "May be Answer to Global Food Crisis"' Telegraph, 19 June 2008 (Phil Woolas). For recent emphasis on food security, see, eg DEFRA, *UK Food Security Assessment: Detailed Analysis* (London: DEFRA, 2009).
[133] *Hansard*, HC vol 418, cols 1380 (9 March 2004) (Secretary of State for Environment, Food and Rural Affairs (Margaret Beckett)).
[134] *Hansard*, HC vol 467, col 17WS (8 November 2007) (Written Statement by the Minister for the Environment (Phil Woolas)).
[135] Ibid. For the proposed implementation of Community measures to regulate seed thresholds, see Chapter 6.
[136] *Hansard*, HC vol 418, col 1381 (9 March 2004) (Secretary of State for Environment, Food and Rural Affairs (Margaret Beckett)).

(a report that substantially informed the 9 March 2004 policy statement).[137] The AEBC considered the options of voluntary protocols and protocols developed and policed by an independent body, and consideration was also given to pre-existing voluntary zoning schemes to protect seed purity (which admittedly did not address GM cross-contamination). All of these were judged to be less effective than measures with statutory backing. Not least, a statutory scheme would cater for any lack of financial incentive on GM growers to observe protocols, and would also be more likely to inspire public and stakeholder confidence.[138] Interestingly, such preference for mandatory measures would seem to have run contrary to, for example, continued use of a voluntary code of practice for leasing business premises.[139]

That said, the 2006 consultation exercise on coexistence measures countenanced a limited retreat from a statutory scheme.[140] DEFRA proposed that crop separation distances for oilseed rape and maize should indeed be legally binding, as should an obligation to notify neighbouring producers. On the other hand, measures relating to, for example, the control of volunteers or the cleaning of farm machinery would be set out in a non-statutory code of practice. These measures were considered to be of more marginal significance in meeting the 0.9 per cent threshold, to be normal farming practice in any event, and to be very difficult to reduce to legislation and enforce.[141] While the proposals may be regarded as a watering-down of the rigour found in the 9 March 2004 policy statement, it must be recognized that the use of differing forms of implementation measure is expressly countenanced in the 2003 Commission Recommendation, which provides that '*[a] priori* there is no particular policy instrument that can be recommended for coexistence. Member States may prefer to explore the use of different policy instruments, e.g. voluntary agreements, soft-law approaches and legislation'.[142] That said, there was far from unanimous support for this change of approach when the responses to the consultation exercise were collated.[143] In particular, reference was made to the poor record of voluntary regimes in agriculture, the need for statutory control to underpin good practice, and the importance of statute in restoring public confidence.[144]

---

[137] AEBC (n 117 above). Prior to being wound up in 2005, the AEBC provided the Government with independent strategic advice on biotechnology issues. [138] Ibid 65–9.
[139] See, eg *Hansard*, HC vol 458, col 87WS (28 March 2007) (Minister for Housing and Planning (Yvette Cooper)) (although it may be noted that legislative options were identified, in case the market failed to deliver). See also, eg DEFRA, *Code of Good Practice for Agri-environment Schemes and Diversification Projects within Agricultural Tenancies* (London: DEFRA, 2004).
[140] DEFRA, *Consultation on Proposals for Managing the Coexistence of GM, Conventional and Organic Crops* (London: DEFRA, 2006). [141] Ibid 21.
[142] 2003 Recommendation, Annex, para 2.1.8.
[143] DEFRA, *Summary of Responses to Defra Consultation Paper on Proposals for Managing the Coexistence of GM, Conventional and Organic Crops* (London: DEFRA, 2007).
[144] Ibid 4. For similar arguments, see AEBC (n 117 above) 50–1; and, for criticism of voluntary compliance in the case of GMOs in the United States, see R Bratspies, 'Myths of voluntary compliance: lessons from the StarLink Corn Fiasco' (2003) 27 William and Mary Environmental Law and Policy Review 593.

5.2.2 Tolerance thresholds

In accordance with the 2003 Commission Recommendation,[145] it is the intention of the UK Government that coexistence measures should be based on the Community 0.9 per cent labelling threshold.[146] This aspect of the proposed regime has proved especially controversial, and it has even been questioned whether Article 26a of the Deliberate Release Directive[147] permits the labelling thresholds to constrain national coexistence regimes.[148] Certainly, they are not mentioned in Article 26a, only in the 2003 Commission Recommendation. Moreover, Article 26a states that 'Member States may take appropriate measures to avoid the unintended presence of GMOs in other products': it does not on its face permit unintended presence up to 0.9 per cent, aiming instead for as low a level of cross-contamination as appropriate measures can attain.[149]

In any event, the argument has become fiercer when the interests of organic farmers are considered. Organic accreditation bodies have advocated a 0.1 per cent threshold, on the basis that, in effect, it is the lowest practical limit for reliable detection of GMOs.[150] This position has been robustly maintained, notwithstanding the decision also to operate a 0.9 per cent threshold under the new Community organic regime, as laid down by Regulation 834/2007.[151] However, there has also been considerable hostility to the imposition of the lower percentage, it being widely regarded as a de facto 'zero' threshold; and the diversity of views is amply reflected in the AEBC report, *GM Crops? Coexistence and Liability*. Some members of the AEBC saw organic producers as responding to consumer demand, believing that in such context '0.1% is a rational, realistic and reasonable threshold'.[152] Other members were of the opinion that organic farmers should be fully entitled to aim for a threshold below 0.9 per cent, but that the extra measures involved should be a matter for the organic farmer. In addition, they were concerned that a 0.1 per cent

---

[145] 2003 Commission Recommendation, Annex, para 2.2.3.
[146] *Hansard*, HC vol 418, col 1381 (9 March 2004) (Secretary of State for Environment, Food and Rural Affairs (Margaret Beckett)).
[147] As amended by the Food and Feed Regulation [2003] OJ L268/1.
[148] See, eg the Advice of Paul Lasok QC and Rebecca Haynes to a consortium of NGOs, delivered in 2005 (available at <http://www.greenpeace.org.uk/media/press-releases/european-gm-crop-co-existence-recommendations-legally-flawed>, last accessed 26 June 2009).
[149] See, eg the discussion of thresholds in House of Commons Environment, Food and Rural Affairs Committee, *Eleventh Report of Session 2003–04: GM Planting Regime* (HC 607) paras 9–21; and see, generally, eg Lee (n 12 above).
[150] See, eg Soil Association, *Proposals for Managing the Co-existence of GM, Conventional and Organic Crops: Soil Association Response to Defra Consultation* (Bristol: Soil Association, 2006) 7. For discussion of the degree of sensitivity that can be achieved when testing for GM presence, see, eg DEFRA (n 140 above) 40 and 43.
[151] [2007] OJ L189/1, Art 9(2). For such hostility, see, eg Soil Association Press Release, *Organic Consumers Shouldn't Pay for GM Contamination—Representatives of 70 Organic Companies Tell Miliband*, 21 June 2007 (Soil Association press releases can be obtained at <http://www.soilassociation.org/>).
[152] AEBC (n 117 above) 58.

threshold would render successful coexistence unachievable, leading to 'a suspicion that the *de facto* "zero" threshold of 0.1% is being used by some—though perhaps not all—interested parties as a way *de facto* to rule out the introduction of the option of growing GM crops'.[153] The 9 March 2004 policy statement noted this diversity of views and undertook to 'explore further with stakeholders whether a lower threshold should be applied on a crop-by-crop basis'.[154]

A significant part of this further exploration was the 2006 consultation exercise. In essence, DEFRA inquired whether or not a 0.1 per cent threshold should be operated when, on its understanding, such a threshold could not be enforced through lack of accurate testing.[155] At the same time, there seemed to be acceptance that any attempt to legislate for very low thresholds was doomed to fail, for three reasons: imported material had already penetrated the food chain;[156] the very long separation distances necessary to eliminate cross-pollination would effectively amount to a ban on GM crops, which breached Community law; and there could be major difficulties in locating seed of the required purity. In addition, the 2006 consultation exercise expressly considered whether a threshold of between 0.1 and 0.9 per cent could be operated. The view of DEFRA was that thresholds should only be adopted if they could be reliably monitored or enforced, with the result that 'any specific GM threshold for organic production could not reasonably go below a level of, broadly, 0.5%'.[157] Responses to these questions revealed a distinct lack of consensus, with some favouring a production system which allowed for pragmatic tolerances, and others emphasizing that organic food derived its competitive advantage from being 'GM-free', as opposed to containing "a bit less" GM'.[158]

5.2.3 'GM-free' zones

The issue of 'GM-free' zones has proved a lively one in the United Kingdom.[159] Although there has been nothing to parallel the litigation flowing from the proposed ban on GMOs in Upper Austria,[160] it has been seen that the Scottish Government is operating a policy of maintaining Scotland 'GM-free'. Further, both the Scottish Region of the Highlands and Islands and the Region of Wales were signatories to the 2005 Charter of Florence (envisaging the protection of

---

[153] Ibid.
[154] *Hansard*, (HC) vol 418, col 1381 (9 March 2004) (Secretary of State for Environment, Food and Rural Affairs (Margaret Beckett)). [155] DEFRA (n 140 above) 41.
[156] It was noted that a UK study conducted in 2004 had found that 10 out of 25 samples of soya-based organic and health food products had detectable GM content: ibid 40.
[157] Ibid 44. [158] DEFRA (n 143 above) 8–11.
[159] See, eg Levidow and Boschert (n 69 above) 185–7.
[160] Joined Cases T-366/03 and T-234/04 *Land Oberösterreich v Commission* [2005] ECR II-4005; and Joined Cases C-439/05P and C-454/05P *Land Oberösterreich v Commission* [2007] ECR I-7141. See further Fleurke, (n 36 above).

conventional and organic crops against GMOs on a regional basis).[161] Moreover, even a year earlier, over 40 administrative areas in the United Kingdom had indicated that they wished to have no part in the GM revolution.[162] Such positions would seem to run contrary to the 2003 Commission Recommendation, with its clear statement that neither GM, conventional, nor organic crops should be excluded from the European Union.[163] Admittedly, the 2003 Commission Recommendation does accept that measures of a regional dimension can be considered. Yet, as has been seen, these may be justified only where the cultivation of specific crops would be incompatible with ensuring coexistence (and then the geographical scale must be as limited as possible).[164] By contrast, the approach of the Scottish Government and other regions in favour of 'GM-free' status would seem to be based more upon principle, and 'blanket' as opposed to targeted.

In this light, the UK Government in its 9 March 2004 policy statement undertook to provide guidance to farmers interested in establishing voluntary 'GM-free' zones, but in a manner 'consistent with EU legislation'.[165] Nonetheless, soon afterwards the weaknesses of voluntary arrangements were well-articulated by Professor Malcolm Grant, Chair of the AEBC. In his evidence to the House of Commons Environment, Food and Rural Affairs Committee, he stated that a 'voluntary GM free zone is only as good as the volunteers who sign up to it and are willing to abide by it', noting also the danger of cross-contamination by vehicles passing through the area concerned.[166] Further, within the context of the 2006 consultation exercise, DEFRA affirmed that it was not advocating 'GM-free' zones, nor did it see them as necessary.[167] Importantly, with regard to the legal position, its view was that mandatory measures would be disproportionate: only voluntary arrangements between farmers were permissible.[168] In addition, the term 'GM-free zone' was itself considered to be inaccurate: such a level of purity was not feasible in practice, with the result that a 'non-GM cultivation zone' was the best that could realistically be achieved. As regards the relationship between participants in such an agreement, this was described as 'similar to establishing a local co-operative' and attention was rightly directed to difficulties which might arise if a farmer sought to withdraw or if a successor (through, for example, retirement or insolvency) did not wish to participate.[169] Perhaps not surprisingly, responses to the 2006

---

[161] Charter of the Regions and Local Authorities of Europe on the Subject of Coexistence of Genetically Modified Crops with Traditional and Organic Farming (4 February 2005).
[162] *Hansard*, HC vol 418, col 1383 (John Whittingdale).
[163] 2003 Commission Recommendation, Annex, para 1.1.
[164] Ibid Annex, para 2.1.5.
[165] *Hansard*, HC vol 418, col 1381 (Secretary of State for Environment, Food and Rural Affairs (Margaret Beckett)).
[166] House of Commons Environment, Food and Rural Affairs Committee, *Eleventh Report of Session 2003–04: GM Planting Regime* (HC 607) para 33. [167] DEFRA (n 140 above) 58.
[168] Ibid. [169] Ibid 60.

Conclusion

consultation exercise (including responses from local authorities) placed little confidence in voluntary arrangements. Instead, with considerable emphasis on the benefit of local decision-making, there was broad support for a change in the Community legislation.[170] Accordingly, once again there would seem to be an uneasy mismatch between the legislative framework and what is happening on the ground. As elsewhere in the European Union, dissatisfaction with the Community coexistence regime is being expressed not only by environmental NGOs, but also by some public bodies. Declarations of 'GM-free' status are being made, to use the words of the AEBC, 'as signs of political intent';[171] and it is clear beyond doubt that the decisions of the European Court of Justice in the *Land Oberösterreich* case have not ended the debate.[172]

6. Conclusion

Under the Deliberate Release Directive, it is hard to deny that, in legal terms, the question is no longer whether to introduce GM crops, but how to carry out their production and how to manage the consequences. To address these issues, 'coexistence' measures are beginning to emerge, but, of the four Member States considered in this chapter (all of them significant agricultural producers), not one has completed implementation. This remains the case notwithstanding that the enabling Community legislation was enacted and the European Commission guidelines issued as far back as 2003. In Italy, the separation of powers between the state and the regions has, as yet, produced a legislative log-jam; and, as for the United Kingdom, implementation is moving forward hesitantly in the absence of any immediate intention to grow GM crops commercially.[173] In France and Spain the procedure is more advanced. In compliance with the 2003 Commission Recommendation, both governments have proposed a series of technical measures which extend down the food chain and seek to guarantee the economic segregation of GM and non-GM crops: for example, buffer zones, isolation distances, staggered sowing times, cleaning obligations, and segregation during transportation and storage. Nevertheless, a material difference may be highlighted between the two countries. In Spain private arrangements between farmers are seen as the best way to achieve an orderly introduction of GMOs, whereas in France a more centralized approach has so far been adopted, with

---

[170] DEFRA (n 143 above) 13–14. It should be stated, however, that approximately 80% of the responses to the 2006 consultation exercise came in the form of stock letters, pre-printed forms, or petitions drawn up by anti-GM groups or individuals.

[171] AEBC (n 117 above) 46.

[172] See Joined Cases C-439/05P and 454/05P *Land Oberösterreich v Commission* [2007] ECR I-7141.

[173] It is of interest that the European Commission has pursued France with considerable zeal to ensure implementation of GM legislation (see nn 20 and 22 above), but this zeal has not been equalled in the case of the United Kingdom.

coexistence measures being imposed 'top down'. That said, no firm conclusions can be reached until the detailed French rules and the Spanish Royal Decree come into force.

Although Community law is unambiguous and national implementation has already commenced, the legislative framework would still seem to have its limits. Indeed, it is very much honoured in the breach by certain national/local actors within the four Member States considered; and the question of coexistence would seem to have become a focus for opposition to GMOs, since coexistence arguably constitutes the last rampart against the full commercialization of the GM crops that have now passed through the authorization stage. As a consequence, there is some disjuncture between, on the one hand, a Community regime that is confident that coexistence between GM and non-GM crops can be achieved and, on the other hand, greater caution at national level. This finds expression, not least, in conflict between central government and its branches (which may enjoy substantial autonomous rights). Thus, in Italy, the regions have challenged the state on the basis of their competence in agricultural matters; and, to seek to resolve this crisis, the state has adopted a decree which lays down the principle of coexistence (and thus validates GM production), while at the same time leaving it to the regions to flesh out detailed coexistence measures, so generating a form of legislative paralysis. Further, in the United Kingdom, the stance adopted in Scotland (and, to a lesser extent, Wales) is avowedly 'anti-GM', as compared with that adopted by DEFRA for England. Similar considerations arise through the actions of many municipalities in France and certain autonomous communities in Spain.

Such conflict is perhaps most marked in the case of declarations of 'GM-free' zones. As has been seen, numerous local authorities in Italy have made such declarations; and even in Spain, with its generally more benign approach to GM crops, the autonomous communities of Asturia, the Basque Country, the Balearics, and the Canaries have all followed the same route. Although these initiatives would seem doomed to fail under Community law, they have had real impact in Italy (although, by contrast, it should be emphasized that they have been systematically struck down by the courts in France).

Many of these difficulties would seem to flow from the fact that precaution and prevention are integrated into the authorization stage of the Community regime for GMOs, but, once an authorization has been granted, precaution and prevention are not at issue, since the legislation works on the basis that the risks have already been evaluated.[174] In other words, the Community coexistence regime would seem confined to regulating the development of authorized GMOs through technical rules; and, strictly speaking, these are not preventive

---

[174] There is the significant exception that the safeguard clause may come into play where new or additional information is made available or where there has been a reassessment of existing information on the basis of new or additional scientific knowledge: Deliberate Release Directive, Art 23.

measures, since they expressly envisage cross-contamination and seek only to limit its economic impact. Such logic may also be seen in the refusal by the French courts to entertain 'GM-free' zones introduced, for preventive reasons, to protect other forms of agriculture. Indeed, it may be argued that the main role for the courts is now to decide on liability where there has been cross-contamination which is not adventitious or technically unavoidable or which exceeds the 0.9 per cent threshold. The time for prevention has passed and, for the moment, the question left is how to deal with liability in circumstances where damages will struggle to provide *restitutio in integrum*. On the other hand, in the case of those Member States where GM crops have not yet been extensively cultivated (and that is the vast majority), the President of the European Commission has intimated that they may yet be able to take control of their own destiny and secure a genuinely 'GM-free' environment.

8
Implementing the Community Environmental Liability Directive: Genetically Modified Organisms and the Problem of Unknown Risk

Christopher Rodgers

1. Introduction

The likely long-term impact on the natural environment of the release of genetically modified organisms (GMOs) is hotly contested and, to a certain extent, largely unknown. The problems are not limited to gene drift and alterations in the genetic make-up of wild species of plant and microorganism, or the possible displacement of natural species. Much will depend upon how the biotechnology is used, and whether (for example) the introduction of broad-spectrum pesticides and herbicides changes the way in which farmers manage growing crops.[1] Against a backdrop of considerable scientific uncertainty, a key question concerns whether the law should develop liability mechanisms to provide for the remediation of environmental damage resulting from GMO releases to the environment.[2] Developing a liability regime for damage to natural

---

[1] This was one of the principal conclusions of the farm scale trials conducted for the UK Department for Environment, Food and Rural Affairs (DEFRA) between 1998 and 2003. The extensive research undertaken on GMHT maize, oilseed rape, and beet produced two surprising results: (i) that GMHT beet and spring rape would produce adverse effects on farmland biodiversity due to the absence of weeds in crops treated with generic glyphosate (eg weed biomass was found to be 85% less in GMHT beet after treatment); and (ii) GMHT maize had greater benefits for biodiversity than conventional maize if managed with contact herbicides like glufosinate-ammonium: see DEFRA, *Managing GM Crops with Herbicides: Effects on Farmland Wildlife* (London: DEFRA, 2003).

[2] The position paper on GMOs released by the statutory conservation advisory body for the United Kingdom, the Joint Nature Conservation Committee, calls for the introduction of legislation to establish liability for environmental harm caused by the deliberate release of GMOs as one of 14 criteria to be satisfied before the commercial release of specific GMOs is authorized: Joint Nature Conservation Committee, *Position Statement on Genetically Modified Organisms in the Environment* (2003) (available at <www.jncc.gov.uk/>, last accessed 2 October 2009).

resources raises a number of extremely difficult issues—for example, what is 'environmental damage', why should the law make provision for it when GMOs are already subject to strict regulatory procedures and risk assessment prior to their release, and should it be governed by public or private law mechanisms (or a combination of the two)? This chapter will address some of these issues, focusing principally on Directive (EC) 2004/35 of the European Parliament and of the Council on environmental liability with regard to the prevention and remedying of environmental damage ('Environmental Liability Directive');[3] and on the Council of Europe's Lugano Convention on Civil Liability ('Lugano Convention').[4]

2. What Role for Environmental Liability?

Before an appraisal of the liability mechanisms can be attempted, we must first establish the conceptual framework for their role and function in relation to environmental damage and GMO releases. A clear distinction must be made between, on the one hand, regulatory measures targeted to minimizing risk by applying structured environmental risk assessment (ERA), in order to evaluate and minimize risk prior to the release of a GMO to the environment (*ex ante* regulatory measures); and, on the other hand, liability measures applicable after a GMO release has occurred, in order to provide for the restoration of damage resulting from the release or for the payment of compensation (*ex post* measures). The two are closely linked, both conceptually and in practice, in that the existence of a thorough and effective *ex ante* regulatory framework will reduce or eliminate the environmental risks posed by GMO releases, thus minimizing the likelihood of damage flowing from the release or eliminating that risk altogether. Where an effective *ex ante* regulatory framework exists, therefore, there is arguably either no need for *ex post* liability mechanisms at all; or, if there is, their importance will be commensurately reduced.

Since 1990, Community law has had a complex *ex ante* regulatory framework for both GMO field trials and the commercial exploitation of GM crops, and for the labelling of products made with or containing GM materials. In that year were enacted both Council Directive (EEC) 90/219 on the contained use of genetically modified organisms and Council Directive (EEC) 90/220 on the deliberate release into the environment of genetically modified organisms;[5] and both required technocratic authorization processes to be established by each Member State, integral to the operation of which is a scientific risk assessment. In the case of deliberate releases into the environment, separate authorization procedures

---

[3] [2004] OJ L143/56, as amended by Directive (EC) 2006/21 of the European Parliament and of the Council [2006] OJ L102/15.
[4] Lugano Convention on Civil Liability for Damage Resulting from Activities Dangerous to the Environment (Council of Europe, 1993). The Lugano Convention has not yet entered into force.
[5] Respectively, [1990] OJ L117/1 and [1990] OJ L117/15.

were established for field trials ('Part B authorizations') and for placings on the market ('Part C authorizations'). Central to the regulatory framework was an ERA that enabled the competent authorities to satisfy themselves that the release was 'safe'. Directive 90/220 was replaced by Directive (EC) 2001/18 of the European Parliament and of the Council on the deliberate release into the environment of genetically modified organisms ('Deliberate Release Directive'), adopted in 2001;[6] and under the revised legislation, the ERA required before a GMO can be authorized for market release must now take account of its 'direct, indirect, immediate and delayed' potentially adverse effects.[7] Further, each Part C authorization will have a limited period of 10 years, after which it must be reviewed.[8] This will enable new scientific knowledge and data on the impact of genetic material to be brought into play.

Compliance with the regulatory regime for GMO releases should reduce the risk of environmental impacts flowing from the release to, if not zero, then, at most, acceptable levels, with the result that any serious long-term environmental impacts that later emerge should have been unforeseeable at the time the authorization was granted. In the event that an authorized GMO release is later shown to produce long-term impacts in terms of damage to biodiversity or damage to property, a key question for the purposes of fixing liability under either civil law or a public liability mechanism (such as the Environmental Liability Directive) will be the relevance of the regulatory authorization, and the ERA on which it was based, in establishing foresight of damage. Where the terms of the regulatory authorization have been met, the allocation of unforeseen risk will either lie with the authorities or remain unallocated, as most civil law systems preclude relief where the damage in question was not reasonably foreseeable.[9] If, on the other hand, the regulatory requirements have not been met, then civil law may have a role to play: this might be the case, for example, where the authorization was itself obtained without providing full information about the GMO in question, or where a farmer ignores the terms of land management protocols when managing the GM crop.

As far as general liability in tort is concerned, the law of negligence would cover traditional damage (damage to persons or property), but not loss of biodiversity. Public liability mechanisms could therefore have a supplementary role to play in this situation in making biodiversity damage recoverable. Many civil liability regimes also provide for strict liability in the case of hazardous activities, but limit recovery to foreseeable harm.[10] Again, in the case of a non-compliant

---

[6] [2001] OJ L106/1. The Deliberate Release Directive came into force on 17 October 2002.
[7] For interpretation of these words, see ibid Annex II.   [8] Ibid Art 15(4).
[9] See, eg the tests established in English law for remoteness of damage in negligence: *Overseas Tankship (UK) Ltd v Miller Steamship Co Pty Ltd (The Wagon Mound (No 2))* [1967] 1 AC 617 (and subsequent decisions); and for recovery of damages in nuisance: *Cambridge Water Co v Eastern Counties Leather Ltd* [1994] 2 AC 264 (and subsequent decisions).
[10] This is the case, for instance, in English Law under the rule in *Rylands v Fletcher* (1868) LR 3 HL 330: see *Cambridge Water Co v Eastern Counties Leather Ltd* [1994] 2 AC 264.

release or the irresponsible management of the GMO, once a significant risk is recognized this may also qualify for the application of strict liability under a public liability regime such as the Environmental Liability Directive. Where this is the case, there may be overlapping remedies in public and civil law, but the focus of each is different and so will be the remedies available to the harmed party. Civil liability is focused on the economic damage to the wronged party, and does not necessarily link compensation to the restoration of habitats. The application of an appropriately structured public liability mechanism can address this deficiency in the available civil law remedies by introducing remedies requiring the restoration of damaged natural habitats and natural resource services.

Advocates of environment liability stress several economic functions that it can perform, including the allocation and/or spreading of risk, and the internalization of environmental externalities.[11] European environmental policy heavily stresses the cost internalization and deterrence effects of environmental liability; a stance that is explicitly based on the polluter pays principle of European environmental law.[12] As stated in the European Commission's *White Paper on Environmental Liability*, '[e]nvironmental liability makes the causer of environmental damage (the polluter) pay for remedying the damage that he has caused'.[13] The European Commission's argument in that document, one that underpinned the subsequent Environmental Liability Directive, is that without liability a failure to comply with the existing norms and procedures for GMO releases may result only in penal or administrative sanctions. Whereas, if liability is added to *ex ante* regulation, potential polluters will also face the prospect of having to pay for restoration of the natural environment or compensation for the damage they have caused. This will, it is claimed, have the effect of internalizing environmental costs: if polluters have to pay for the damage they cause, they will cut back on their polluting activities up to the point where the marginal cost of abatement exceeds the compensation thereby avoided.[14] Inasmuch as the cost of the polluting activity will increase, this will be reflected in output costs, thereby increasing the price of—and reducing demand for—the products of the polluting process, a factor that will lead in turn to a reduction in pollution.

In the case of damage allegedly flowing from the release of GMOs into the environment, however, it is difficult to see how the cost internalization or deterrent effects of a liability mechanism could operate. The release will be

---

[11] See, eg L Bergkamp, 'The Commission's White Paper on Environmental Liability: a weak case for an EC strict liability regime' (2000) 9(4) European Environmental Law Review 105. Bergkamp is critical of the public liability model posited in the Environmental Liability Directive as a mechanism for delivering any of these objectives.
[12] EC Treaty, Art 174(2). See also, in particular, the Environmental Liability Directive, Preamble (2) and (18).
[13] European Commission, *White Paper on Environmental Liability* COM(2000)66, 3.
[14] Ibid 11.

permitted by an ERA and licence based on the supposition that there are no foreseeable environmental risks attached to the release. The only deterrent effect would therefore be in the case of unauthorized or irresponsible releases, where liability would simply provide an added sanction to the penal or administrative sanctions attaching to the regulatory breach. There would be no additional deterrent effect beyond that already provided by the regulatory offences governing the application for an authorization and breaches of the terms of the licence granted. It is also difficult to apply economic theories of cost internalization in the case of GMO-related environmental damage, especially in cases where the liability mechanism is primarily concerned with damage to natural resources and biodiversity.[15] The damage in such cases will be specific to individual, high nature-value sites and highly individualized. The additional costs of compensatory measures imposed through the liability regime may well impact on the overall costs of the polluter's enterprise, but it will not have the economic impact on price and markets that environmental economists posit in cases of (for example) water and air pollution from factory processes.

The potential role of liability regimes in this situation is perhaps best understood if we consider their function in resource allocation. Writing almost 50 years ago, RH Coase suggested a way to take decisions about environmental harm by harnessing the market in property rights. The Coasean theorem posits that the nature of the problem of allocating costs arising from environmental damage is always reciprocal. If A's factory pollutes B's land, measures taken to avoid the harm to B will also inflict harm on A. If we are to arrive at an optimum allocation of resources, it is desirable that both parties should take into account the harmful effect (the nuisance) when deciding their course of action. In this regard, a smoothly operating pricing system will, furthermore, ensure that the fall in the value of production due to the harmful effect would be a cost for both parties.[16] The problem we face when dealing with harmful effects, according to Coase, is not one of simply restraining those responsible for them. What has to be decided is whether the gain from preventing the harm is greater than the loss that would be suffered elsewhere as a result of stopping the action that produces the harm. Because there are costs associated with the rearrangement of property rights established by the application of a liability mechanism to 'rectify' damage, the courts are (in nuisance cases) making a decision on the economic problem of how to allocate and use resources.[17]

In this connection, land is usually thought of as a physical entity, but a Coasean analysis requires us to view property rights instead as a factor of production. This resonates with more modern interpretations of property theory, which stress the

---

[15] This is the case with the Environmental Liability Directive (see below).
[16] RH Coase, 'The problem of social cost' (1960) 3 Journal of Law and Economics 1, 13. See also RH Coase, *The Firm, the Market and the Law* (Chicago: University of Chicago Press, 1986). *Cf* welfare economics approaches (criticized by Coase): eg AC Pigou, *The Economics of Welfare* (London: 4th edn, Macmillan, 1932). [17] Coase (n 16 above) 27.

resource allocation role of property institutions, and the role of property rights in giving access to a resource or stream of benefits.[18] When a landowner owns land, what he in fact owns is access to an economic resource and the right to carry out a circumscribed set of actions upon it. If factors of production are conceived of as property rights, it is easier to understand that the right to do something which has a potentially harmful effect (for example, to release GMOs into the environment) is also a factor of production. One can use a piece of land in such as way as to prevent someone else having a view or having unpolluted air. The cost of exercising a right is therefore always the loss that is suffered elsewhere in consequence of the exercise of that right—the inability to enjoy a fine view, or to breathe clean air, for example. In choosing which social arrangements to permit, Coase would argue that we should (on this analysis) consider the total effect of the reallocation of property rights required to remedy the damage at issue.[19]

Applying a Coasean analysis, the potential role of liability mechanisms in internalizing different types of environmental cost arising from GMO releases is best explained by two examples. In *scenario A*, Farmer Giles cultivates GM herbicide-resistant maize on fields adjacent to fields owned by his neighbour, Farmer Jones.[20] Jones grows organic sweetcorn on his farm, and claims that his product will be cross-fertilized, or 'contaminated', by GM material from Giles' GM fields. He will no longer be able to grow organic sweetcorn and his Soil Association accreditation as an organic farmer will be in danger. In *scenario B*, Farmer Giles instead cultivates GM herbicide-resistant oilseed rape on fields adjacent to Farmer Jones' land. The area of Jones' farm immediately adjacent to Giles' land is a rare habitat for slipper orchids, a protected plant under UK and Community law. Giles' GM rapeseed cross-pollinates with wild brassica weed species on Jones' land, and the herbicide-resistant weed outcrosses then establish themselves and aggressively colonize the surrounding area. This leads in turn to a considerable reduction in the number and quality of protected slipper orchids on Jones' land and deterioration of the habitat.

Scenario A is a classic example of a relationship with bilateral effects in the Coasean sense.[21] When considering whether to give Farmer Jones damages for the impact on his sweetcorn crop, the court will have to balance the increased costs of GM production if Giles has to pay damages (and the loss of additional

---

[18] See, eg KJ Gray, 'Property in Thin Air' (1991) 50 Cambridge Law Journal 252; K Gray and SF Gray, 'The Idea of Property in Land' in S Bright and K Dewar (eds), *Land Law: Themes and Perspectives* (Oxford: Oxford University Press, 1998) 15; J Waldron, 'What is Private Property?' (1985) 5 Oxford Journal of Legal Studies 313; and CP Rodgers, 'Nature's Place? Property Rules, Property Rights and Environmental Stewardship' (2009) 68 Cambridge Law Journal 538.
[19] Coase (n 16 above) 44.
[20] This example is closely analogous to the facts of the only case in the English courts in which some of the issues have been raised: *R v Secretary of State for the Environment, ex p Watson* [1999] Env LR 310. *Scenario B* is purely hypothetical.
[21] For a discussion of the implications of *R v Secretary of State for the Environment, ex p Watson* [1999] Env LR 310 within a Coasean framework of analysis, see D Campbell, 'Of Coase and corn: a (sort of) defence of private nuisance' (2000) 63 Modern Law Review 197.

crop yield if GM cropping is restricted as a result), against the loss to Jones and other organic farmers if the law fails to protect their right to grow organic crops without GM cross-contamination. Liability therefore has, in this scenario, a very clear resource allocation function.

Scenario B is, by contrast, altogether more difficult to interpret within a Coasean framework of analysis. Indeed, it is questionable whether environmental liability has a meaningful resource allocation function in this type of case. Coase would argue that all environmental damage is inherently reciprocal in its economic effects. Whether the harmful effects of Giles' GM cropping are bilateral will, however, depend upon the facts. The damage in *scenario B* is biodiversity damage, and as such is normally uncompensated by civil law mechanisms such as nuisance and negligence, which limit the range of available remedies to damage to the 'owned' environment.[22] If a public liability mechanism restricts the right of landowners to grow GM crops in a manner that damages biodiversity, this will have the effect of reallocating Giles' property rights (it will restrict his right to grow GM crops in a manner that will have this effect and increase his production costs). But this is arguably a unilateral effect, because it is not balanced by a reciprocal alteration in Jones' property rights.

If, on the other hand, Jones is running an eco-tourism business on his land, the impact of the liability mechanism may be bilateral. Viewed as a resource-allocation mechanism, the imposition of environmental liability here will protect his access to a resource—the protected habitat—that will generate an economic stream of benefits. In so doing, it will prioritize this over Giles' right to produce GM crops in a manner which is damaging to neighbouring farmers. But if Jones' land use is not contingent upon the habitat in this sense (he simply has a site of special scientific interest (SSSI) on his land and does not run an eco-tourism business), the imposition of environmental liability will arguably have a unilateral effect on property rights. In other words, it will restrict Giles' rights without necessarily amplifying those of Jones.[23]

Or will it? Arguments concerning whether remedies for environmental harm have bilateral or unilateral effects to some extent miss the point. Even if we accept that a utilitarian analysis is necessary,[24] what is required is a consideration

---

[22] See, eg *Marquis of Granby v Bakewell UDC* (1923) 87 JP 105; *Pride of Derby and Derbyshire Angling Association Ltd v British Celanese* [1953] Ch 149; and *Hunter v Canary Wharf* [1997] 2 WLR 684.

[23] See, eg L Bergkamp, 'Allocating unknown risk: liability for environmental damages caused by deliberately released genetically modified organisms' (2000) 23–4 (available on the Social Science Research Network).

[24] Many would argue that this is not the case. The law may seek to protect living natural resources for deontological reasons, completely unconnected with considerations of resource allocation or the internalization of environmental externalities. The utilitarian Coasean approach is fundamentally at odds with approaches based on deontological ethics: see, eg A Leopold's 'land ethic' as expounded in *A Sand County Almanac* (Oxford: Oxford University Press, 1949). These approaches emphasize that we protect wildlife and natural resources because it is the right thing to do, irrespective of the economic consequences of so doing or of the impact of liability on questions

of the overall reallocation of property rights and resources brought about by the application of the environmental liability rule.[25] The argument that the imposition of environmental liability for damage to living natural resources has unilateral effect ignores the fact that, if Jones has an SSSI[26] or other legally protected wildlife habitat[27] on his land, his property rights will already have been partially reassigned to the state. When the site was notified by the conservation bodies, he will have been served with a list of operations likely to damage the conservation interest of the site,[28] and it is a criminal offence to carry any of these out without the permission of the conservation body.[29] Property rights in Jones' land having been reassigned to the state by the site notification, it will have taken on a 'quasi public' character as a result.[30] If damage to the protected habitat on Jones' land results in a further reassignment of property rights resulting from the application of environmental liability, this will take place within a tripartite framework of reference, not a bilateral one. Depending on the facts, the environmental harm resulting from Giles' actions may impact upon either Jones' property rights or those reassigned to the state by the SSSI notification, and the imposition of environmental liability will have bilateral effect in each case. Or it may have tripartite effects—for example, if damage to the conservation interest of the SSSI is also coupled with damage to Jones' eco-tourism business.

Imposing environmental liability on Giles to protect the conservation interest of the SSSI on Jones' land will therefore have potentially multiple effects—not only on Giles' property rights, which will be commensurately reduced, but also on Jones' property rights and on those reassigned to the state by the site notification. Any impact on Jones' property rights will, however, be incidental: the primary objective of the liability mechanism is the protection of the property

---

of resource allocation. See also, passim, C Palmer, *Environmental Ethics and Process Thinking* (Oxford: Clarendon Press, 1998).

[25] This is essentially what Coase is arguing for: see Coase (n 16 above) 44.

[26] SSSIs are notified for protection under s 28 of the Wildlife and Countryside Act 1981, as amended by Sch 9 to the Countryside and Rights of Way Act 2000. Notification is undertaken by the public conservation bodies, ie Natural England, the Countryside Council for Wales, or Scottish Natural Heritage: Environmental Protection Act 1990, Part VII; Natural Environment and Rural Communities Act 2006, s 2; and Natural Heritage (Scotland) Act 2003.

[27] For example, a special area of conservation (SAC) or special protection area (SPA) designated for protection in accordance with Council Directive (EEC) 79/409 on the conservation of wild birds ('Wild Birds Directive') [1979] OJ L103/1 or Council Directive (EEC) 92/43 on the conservation of natural habitats and of wild fauna and flora ('Habitats Directive') [1992] OJ L206/7. These are designated as European sites under the Conservation (Natural Habitats & c.) Regulations 1994, SI 1994/2716, as amended, and are subject to similar (but in some respects more stringent) land use controls to those for SSSIs.

[28] Under s 28 of the Wildlife and Countryside Act 1981 (SSSIs) or reg 19(2) of the Conservation (Natural Habitats etc) Regulations 1994 (European wildlife sites).

[29] See s 28E of the Wildlife and Countryside Act 1981.

[30] For a discussion of the wider characterization of 'quasi public' property concepts, see K Gray and S Gray, 'Private Property and Public Propriety' in J McLean (ed), *Property and the Constitution* (Oxford: Hart Publishing, 1999) 18; and Waldron (n 18 above) 327–33.

rights reassigned to the state by the site notification,[31] and any economic benefit accruing to Jones (for example, the protection of his eco-tourism venture) will be coincidental. Viewed within this multi-faceted framework of analysis, and having regard to the overall effect of the social arrangements applied and to the allocation of costs implicit within them, environmental liability would appear to perform a not dissimilar function to civil liability mechanisms in situations where environmental harm has a purely bilateral effect (for example, where Jones' land has not been notified as an SSSI and the only question concerns damage to Jones' eco-tourism business). The principal difference will lie in the different legal remedies available to the harmed party in each case, and by virtue of which the consequent reassignment of property rights is effected.[32]

In the second part of this chapter we will consider how effectively the existing models of environmental liability perform these functions. Attention will be focused primarily on the Environmental Liability Directive and the Lugano Convention.

3. Environmental Liability: The Models

Both of the extant models of environmental liability—the Environmental Liability Directive and the Lugano Convention—impose strict liability for environmental damage caused by 'hazardous' activities. Both also limit their scope and application to 'environmental damage', and give no remedy for economic loss, other commercial losses, or property damage. In the context of Community law on GMO releases, these are separately provided for in the so-called 'co-existence' measures. The legislation on co-existence is considered elsewhere in this work.[33]

The civil law regimes of the EU's Member States mostly fail to provide adequate provision for civil liability for 'environmental' damage that may result from the cross-fertilization of GM crops with wild flora, and from the wider impacts of GMOs on biodiversity.[34] The Environmental Liability Directive attempts to address this issue, and had a lengthy gestation prior to its adoption in 2004. The European Commission published the *Green Paper on Environmental Liability* in 1993,[35] in which it noted the disparity of approach between

---

[31] Or, to be more precise, the protection of the habitat that the modification of Jones' property rights by the SSSI notification seeks to protect.
[32] Damages and injunctions are not available under public liability mechanisms such as the Environmental Liability Directive. Thus, if Jones wishes to gain financial recompense for damage to his eco-tourism business venture in *scenario B*, he will have to sue in nuisance or negligence and seek the imposition of civil liability.   [33] Chapters 6 and 7.
[34] For a review of some of these, see the 2005 *Report of the International Court of Environmental Arbitration and Conciliation*: 'Consultative opinion on liability of public and private actors for genetic contamination of non-GM crops' (2005) 7 Environmental Law Review 253.
[35] European Commission, *Green Paper on Environmental Liability* COM(1993)473.

the Member States' civil liability regimes in their potential application to liability for environmental damage, and stressed the need for harmonization of the liability rules. The *Green Paper* suggested the introduction of a strict liability regime for damage resulting from pollution—one in which the establishment of liability would not require proof of the polluter's fault, as was the case in the civil liability regimes of many Member States.

These proposals were weakened in the subsequent *White Paper on Environmental Liability*,[36] in which the European Commission moved away from its original harmonization proposal based on the introduction of strict liability within the private law of the Member States. The *White Paper* instead proposed the imposition of a public liability mechanism, under which the Member States would be required to recover the cost of remediating environment damage by introducing administrative direction and regulatory cost-recovery mechanisms to identify and allocate responsibility for environmental damage.

It was this model that, in amended form, subsequently formed the basis for the Environmental Liability Directive. The European Commission published its final proposals in 2002,[37] and these were more limited than those in the earlier *White Paper*. The Environmental Liability Directive covers damage arising from hazardous[38] occupational activities, which includes for these purposes the contained use of GMOs and/or deliberate release of GMOs into the environment.[39] This implements the polluter pays principle in Community law, and reflects a policy approach focused on 'internalizing' the environmental costs generated by the polluter's production methods. Strict liability must be applied by the Member States, as explained further below, and the Environmental Liability Directive requires states to establish administrative cost-recovery mechanisms to ensure that the cost of rectifying environmental damage is recovered from the polluter. Liability is limited, however, by identifying the occupational activities covered by the regime by reference to pre-existing Community environmental legislation. Further, the Environmental Liability Directive is not retrospective.[40] It should have been implemented by the Member States by 30 April 2007, and does not apply to any emission event or incident that took place before that date.[41]

A second model is offered by the Lugano Convention. This was adopted by the Council of Europe in 1993, but has yet to be ratified by sufficient signatories

---

[36] European Commission, *White Paper on Environmental Liability* COM(2000)66.

[37] European Commission, *Proposal for a Directive of the European Parliament and of the Council on Environmental Liability with Regard to the Prevention and Remedying of Environmental Damage* COM(2002)17.

[38] The proposal published by the European Commission in 2002 referred to 'dangerous substances or preparations, organisms and micro-organisms, and plant protection and biocidal products and their manufacture, use and release into the environment': n 37 above, 18.

[39] Environmental Liability Directive, Annex III, paras 10 and 11.   [40] Ibid Art 19.

[41] Ibid Art 17.

to enter into force.[42] It offers a wider model of liability than that applied by the Environmental Liability Directive. Like the Environmental Liability Directive, it is based on operator liability,[43] and imposes liability for incidents arising from dangerous activities—including the release of GMOs in circumstances where, as a result of the properties of the organism released, the genetic modification, or the conditions under which the operation is conducted, the release poses a 'significant risk' for man, the environment, or to property.[44] The scope of potentially applicable liability is wider, nonetheless, in that the Lugano Convention applies to property damage, personal injury, and the cost of remediation, as well as to damage to the environment (widely defined to include not only damage to natural resources, but also to the cultural heritage and landscapes).[45]

Under the Lugano Convention, the operator is liable for damage caused by a dangerous activity for which the operator has responsibility, provided that the damage is a result of 'incidents' occurring at a time when he had control of that activity.[46] It also differs from the Environmental Liability Directive in providing that no liability will accrue for pollution 'at tolerable levels'.[47] What will be considered 'tolerable' remains undefined, but it clearly refers to the *effect* of the pollution, not the act causing it. Whether or not a GMO release will result in minor ('tolerable') damage to natural resources, or major damage engaging liability, would be very difficult to predict and outside the control of the operator. This defence arguably weakens the deterrent impact of the liability scheme and its role in applying the polluter pays principle. Not only does the release of GMOs itself pose an unknown risk, but under the Lugano Convention the consequences of that risk, should it materialize, are also unknown, depending upon whether or not the damage that results is tolerable. As a result there is little legal certainty in the potential application of the liability rules.

In the case of GMOs, where impacts on natural resources are likely to be diffuse and have multiple causes, it is also difficult to see how the causative issues could be resolved so as to link individual operators to 'incidents' resulting in pollution, as posited by the liability rules in the Lugano Convention. This is

---

[42] By virtue of Art 32(3), the Lugano Convention requires ratification by three signatories to enter into force. As at 24 November 2009, it had been signed by nine states, but ratified by none, the nine signatories being: Cyprus, Finland, Greece, Iceland, Italy, Lichstenstein, Luxembourg, The Netherlands, and Portugal. The Community has neither signed nor ratified the Lugano Convention.

[43] Liability under the Lugano Convention for GMO-related damage would attach to the individual operating the process from which the GMO was released—typically the farmer—and not the biotechnology corporation developing the product used.

[44] Dangerous activities are defined to include: 'the production, culturing, handling, storage, use, destruction, disposal, release or any other operation dealing with one or more [of the following]: ... genetically modified organisms which as a result of the properties of the organism, the genetic modification and the conditions under which the operation is exercised, pose a significant risk for man, the environment or property': Lugano Convention, Art 2(1)(b).

[45] Ibid Art 2(7) and 2(10).   [46] Ibid Art 6(1).   [47] Ibid Art 8.

also, of course, a major problem under the Environmental Liability Directive itself.[48] Under the Lugano Convention, the multiple causation issue is complicated by the fact that although the rules apply joint and several liability, they also require that, where damage originates in a plurality of sites, liability must be apportioned by reference to the part of the damage caused by the dangerous activity on each site.[49] When a number of farms are growing the GM maize or rapeseed that is alleged to have had a detrimental effect on wildlife, for example, the application of this rule would involve the identification of the proportion of damage originating from releases from each farm—an impossible requirement.

Finally, although the Lugano Convention does require the application of any financial award towards restoration of the environment, restoration is not itself the primary focus of the liability scheme; and it does not contain criteria for the valuation of natural resource damage or its restoration—unlike the Environmental Liability Directive.[50] Indeed, the inherent vagueness of its application to cases of environmental damage was one of the principal reasons cited by the European Commission for not acceding to the Lugano Convention, and for implementing a separate Community liability regime.[51] Many of the difficult issues inherent in both liability regimes will become apparent if we consider the provisions of the Environmental Liability Directive in greater detail.

4. Application of the Environmental Liability Directive

The liability regime under the Environmental Liability Directive is differentially applied to two categories of damage:

- 'environmental damage' occurring as a result of occupational activities listed in Annex III, to which a strict liability regime is applied; and
- 'damage to protected species or natural habitats', to which strict liability applies if caused by occupational activates within Annex III, or to which liability based on negligence or fault applies if caused by occupational activates *other than* those listed in Annex III.[52]

---

[48] *Cf* the relaxation of the causation rules in the law of torts by the English courts in *Fairchild v Glenhaven Funeral Services Ltd* [2002] UKHL 22, [2002] 3 WLR 89 and *Barker v Corus (UK) Ltd* [2006] UKHL 20. This permits a victim to sue when exposed to a material risk of injury by the defendant, even if it is not possible to show that the exposure to risk by the particular defendant was the operative cause that actually led to the damage for which redress is sought; the precise ambit of this extension of liability remains unclear.   [49] Lugano Convention, Art 11.
[50] Although Art 2(7)(c) of the Lugano Convention does provide that 'compensation for impairment of the environment, other than for loss of profit from such impairment, shall be limited to the costs of measures of reinstatement actually undertaken or to be undertaken'.
[51] European Commission, *White Paper on Environmental Liability* COM(2000)66, 25.
[52] Environmental Liability Directive, Art 3(1).

4.1 Defining 'Environmental' Damage

The Environmental Liability Directive adopts complex scientific concepts to define the types of damage for which it provides a remedy. 'Environmental' damage has three possible components: first, damage to protected species and natural habitats; secondly, water damage that significantly adversely affects the ecological, chemical, and/or quantitative status or ecological potential of the waters concerned; and, thirdly, land damage, meaning any land contamination which creates a significant risk of human health being adversely affected as a result of the direct or indirect introduction in, on, or under the land of substances, preparations, or micro-organisms.[53]

4.2 Defining Environmental 'Damage'

The Environmental Liability Directive's scope is also comparatively narrow as to the *types of damage* to which it may potentially apply. It does not apply to damage to the person or goods. Neither does it apply to property damage, unless it falls within the narrow definition of 'land damage' comprised within the wider definition of environmental damage used in the Directive (that is, it creates a significant risk to human health).[54] It also has a very limited scope in its possible application to cases of biodiversity damage. As regards protected species and habitats, liability only mandatorily extends[55] to damage to the conservation status of those protected under specified provisions of the Wild Birds Directive[56] and Habitats Directive.[57] The Environmental Liability Directive gives the Member States the option, however, to extend the scope of liability to apply in cases of damage to conservation sites protected in domestic (but not Community) law.[58] The UK's initial stance was to resist an extension of the basis of liability beyond that mandatorily required, on the basis that this represented 'gold plating'—and therefore as a matter of policy to be avoided in implementing the Directive.[59] In the event, the regulations implementing the Environmental Liability Directive in England have extended the scope

---

[53] Ibid Art 2(1).   [54] Ibid Art 2(1)(c).   [55] Ibid Art 2(3).
[56] Wild Birds Directive, Art 4(2) and Annex 1
[57] Habitats Directive, Annexes I, II, and IV. Note that it is sufficient for the potential application of the Environmental Liability Directive that there has been damage to a listed habitat or species: it is not necessary for the damage to occur in, or relate to, a designated SAC established under the terms of the Habitats Directive. This could give rise to difficulties if damage were to be caused to a site hosting a listed habitat type, but which had not been designated by the Member State under the terms of the Habitats Directive.
[58] Environmental Liability Directive, Art 2(1)(a) and (3) (definitions of respectively 'environmental damage' and 'protected species and natural habitats').
[59] See DEFRA, *Consultation for Options for Implementing the Environmental Liability Directive* (London: DEFRA, 2006). The UK Government's minimalist approach to implementation was criticized by the House of Commons Environment Food and Rural Affairs Committee in 2007: *Sixth Report of Session 2006–2007: Implementation of the Environmental Liability Directive* (HC 694).

of liability so that it will also apply to environmental damage to national conservation sites (in other words, SSSIs notified for protection under the Wildlife and Countryside Act 1981).[60]

'Damage' is defined in general terms to mean any 'measurable adverse change in a natural resource or measurable impairment of a natural resource service'.[61] Although the Environmental Liability Directive is innovative in applying liability to natural resources damage, it applies a restrictive definition to biodiversity 'damage' by closely aligning liability with breaches of existing Community environmental law. Damage to protected species and habitats is defined as 'any damage that has significant adverse effects on reaching or maintaining the favorable conservation status' of the habitats or species in question, a concept that is closely tied to each site designation established under the Wild Birds Directive and Habitats Directive (which generate the *Natura 2000* network).[62] The significance or otherwise of any adverse effects on a site's conservation status is measured by reference to the baseline condition of the habitat or species in question, that is the condition of the natural resources and services that would have existed had the environmental damage not occurred, estimated on the basis of the best information available[63] and taking account of the criteria set out in Annex I.[64]

Recovery is excluded for damage that results from actions that were expressly authorized by the relevant authorities in accordance with provisions implementing the regime for the management of designated sites under the Habitats Directive, or in accordance with provisions of national law having an equivalent effect in relation to habitats or species.[65] In English law, operations likely to damage the conservation interest of either a European site[66] or an SSSI notified under domestic legislation[67] require the consent[68] of Natural England[69] before they can lawfully be carried out. If Natural England has given consent to the planting of GM crops either within, or near, a protected wildlife site, any

---

[60] Environmental Damage (Prevention and Remediation) Regulations 2009, SI 2009/153 (in particular, reg 4(1)(a) (meaning of 'environmental damage') and reg 2(1) (definition of 'natural resource')). [61] Environmental Liability Directive, Art 2(2).
[62] Ibid Art 2(1)(a). [63] Ibid Art 2(14).
[64] Annex 1 sets out various criteria including the capacity of the habitat or species for natural regeneration and its capacity for propagation and for recovery without further legal protection measures; and see also Annex I to Common Position (EC) 58/2003 [2003] OJ C277/E/10.
[65] Environmental Liability Directive, Art 2(1)(a).
[66] Ie a site notified under the Conservation (Natural Habitats & c.) Regulations 1994, SI 1994/2716.
[67] As indicated, SSSIs are notified under s 28 of the Wildlife and Countryside Act 1981, as amended by Sch 9 to the Countryside and Rights of Way Act 2000.
[68] In the case of an SSSI, operational consent is required from Natural England under s 28E of the Wildlife and Countryside Act 1981 for operations likely to damage the conservation interest of a site. In the case of a European site (eg an SAC) consent is required under the Conservation (Natural Habitats & c.) Regulations 1994, SI 1994/2716, reg 20.
[69] Natural England is the public body with regulatory responsibility for nature conservation law in England: see Part 1 of the Natural Environment and Rural Communities Act 2006.

biodiversity damage that later emerges from the cross-pollination of GM and wild plant species will not constitute 'damage' for these purposes and will be unremediated under the Environmental Liability Directive.[70] It is not clear whether adverse effects arising from licensed activities that are not *directly* connected with the management of a protected site itself would be exempt from liability, for example the licensing of the planting of GM crops in fields adjacent to a Natura 2000 site under a management agreement with Natural England.[71]

It should perhaps be noted that by linking recovery of natural resources damage exclusively to legally protected habitats, the Environmental Liability Directive takes a narrower approach than the Convention on Biological Diversity, which defines biodiversity in much wider terms.[72] The European Commission rejected a wide approach, taking the view that the adoption of the concept of variability in living organisms as a qualification for defining biodiversity damage would raise difficult questions as to how such damage would be quantified and what would be the threshold of damage entailing liability.[73] Similar issues arise under the Cartagena Protocol to the Convention on Biological Diversity, which seeks to regulate trade in living modified organisms, and includes provision for the development of a liability mechanism.[74] The working group established under the Cartagena Protocol to develop such a potential liability mechanism has identified the quantification of damage as a key problem area, but has yet to make substantive progress in this direction.[75] It has, nevertheless, considered the Environmental Liability Directive's framework for assessing damages as a possible model for adoption. By focusing on developing plans to restore damaged natural resources and resource services, rather than assessing the monetary value of the damage to the resource, the Cartagena Protocol working group considered that the Environmental Liability Directive ensured not only that the 'polluter' pays for the cost of implementing the compensatory restoration project (rather than for the monetary value of the

---

[70] Environmental Damage (Prevention and Remediation) Regulations 2009, SI 2009/153, reg 4 and Sch 1, para 5.

[71] Natural England has wide powers to offer management agreements to landowners irrespective of whether their land is within a protected European site or SSSI: Natural Environment and Rural Communities Act 2006, s 7.

[72] Convention on Biological Diversity, Art 2 (available at <http://www.cbd.int/convention/convention.shtml>).

[73] European Commission, *Proposal for a Directive of the European Parliament and of the Council on Environmental Liability with Regard to the Prevention and Remedying of Environmental Damage* COM(2002)17, 17.

[74] Cartagena Protocol, Art 27 (available at <http://www.cbd.int/biosafety/protocol/shtml>).

[75] See, most recently, UNEP 2006 Report of the Working Group on Liability and Redress: *Channelling of Liability* (March 2006): UNEP/CBD/BS/COP-MOP/3/10; and UNEP 2007 Report of the Working Group on Liability and Redress in the Context of the Cartagena Protocol on Biosafety on its Third Meeting (March 2007): UNEP/CBD/BS/WG-L&R/3/3.

interim losses), but also had the advantage of making economic valuation techniques less controversial and thereby more acceptable to all parties.[76]

A final point concerns the nature of any environmental damage likely to result from the interaction of GMOs with the natural environment. The *White Paper on Environmental Liability* noted that, in order for a liability regime to be effective, three factors were required: there must be one (or more) identifiable polluter(s); the damage alleged must be concrete and identifiable; and a causal link needs to be established between the damage and identified polluter(s).[77] Biodiversity damage allegedly arising from the introduction of GMOs is unlikely to meet any of these requirements. In the first place it will, if it occurs, take a very long time to become apparent. It is also likely to be diffuse in character. Were GM cropping to become widespread, it may therefore be difficult to identify a single source for cross-pollination and other environmental impacts. This may also have a restrictive effect on the liability mechanism in the Environmental Liability Directive, as the latter only applies to environmental damage (or to an imminent threat of environmental damage) where it is possible to establish a causal link between the damage and the activities of individual operators.[78] It may be impossible to establish a causal link with individual GM producers in cases of widespread and diffuse impacts on biodiversity resulting from alleged interactions with GMOs, and this may well render the Directive's impact minimal.[79]

5. The Basis of Liability

5.1 Strict Liability for 'Hazardous' Activities

The Environmental Liability Directive's policy to internalize external environmental costs is central to the imposition of strict liability for occupational activities causing environmental damage. This is reflected in provisions in Annex III, which require the Member States to ensure that operators whose activities fall within the listed categories bear the cost of taking action to prevent or to clean up such environmental damage as they threaten or cause, irrespective of fault. Member States are required to establish strict liability regimes as regards administrative direction to prevent harm,[80] and to require operators to reimburse costs incurred by public bodies in undertaking remedial action.[81] As far as

---

[76] Including potentially liable parties: see, generally, E Brans, 'Liability for damage to public natural resources' (2005) 7 Environmental Law Review 90.
[77] European Commission, *White Paper on Environmental Liability* COM(2000)66, 11.
[78] Environmental Liability Directive, Art 4(5).
[79] As has been seen, the same problem arises under Art 6(1) of the Lugano Convention, when identifying 'incidents' which cause GM contamination in neighbouring crops or the natural environment.
[80] Environmental Liability Directive, Art 5.
[81] Ibid Art 6.

GMOs are concerned, the Environmental Liability Directive encompasses environmental damage resulting from both the contained use of GMOs (including transport) and from the deliberate release into the environment, transport, or placing on the market of GMOs in accordance with the authorization procedures established under the Deliberate Release Directive. These are both occupational activities listed in Annex III, and will engage strict liability under Article 3 of the Environmental Liability Directive even though the release will be preceded by an ERA as part of the authorization process.[82]

One important point that has so far escaped comment in connection with the place of GMOs within the overall liability scheme is this: if the release of a GMO has been authorized following an ERA that certifies it to be free from significant risk, how can it be said to be a 'hazardous' operation? The European Commission considered that the deliberate release into the environment of GMOs has the potential in certain circumstances to cause unforeseen results, such as health damage or significant environmental damage, and that this justified their inclusion in the Environmental Liability Directive.[83] Whether this is the case must be open to question.

The Environmental Liability Directive's approach to risk allocation is premised on the view that strict liability is a more efficient mechanism for internalizing cost than fault-based liability. Although widely accepted, this view has not gone without challenge.[84] Where the activity potentially attracting liability is preceded by a rigorous risk assessment, as is the case with GMO releases, it is clearly arguable that the imposition of strict liability will do little to change producer behaviour. It may even prove harmful if it proves a deterrent to either research or development, and/or to operators engaging in GM agricultural innovation. This may well be the case if producers perceive that the pre-release public ERA fails to provide a guarantee against the subsequent imposition of strict liability for unforeseen environmental damage. Indeed, it might fairly be argued that in this event the state should underwrite liability because the authorization procedures and ERA undertaken prior to the GMO release must have been flawed. In any event, as foresight of harm is a constituent element of proving causation, the fact that such damage is unforeseeable is in itself likely to mean that it will be impossible to establish liability (whether strict or otherwise) against an individual operator.

5.2 Fault-based Liability for Biodiversity Damage

Enhanced protection is given to biodiversity, in that liability will apply whether damage was caused by a hazardous activity or a non-hazardous activity.[85] The

---

[82] Ibid Annex III, paras 11 and 12; and, in England, the Environmental Damage (Prevention and Remediation) Regulations 2009, SI 2009/153, reg 5 and Sch 2, para 9.
[83] European Commission, *White Paper on Environmental Liability* COM(2000)66, 16.
[84] See, eg Bergkamp (n 11 above) 108–9.
[85] European Commission, *White Paper on Environmental Liability* COM(2000)66, 16.

Environmental Liability Directive therefore extends the potential range of liability for biodiversity damage by providing for liability based on fault or negligence where an operator causes damage to a protected site as a result of any operational activity, including activities *other than* those in Annex III to which strict liability applies.[86] The widened ambit of potential liability would, for example, encompass damage to a wildlife site caused by negligence in carrying out the site-management requirements of the Habitats Directive or domestic legislation (such as ignoring management protocols for buffer strips between GM and non-GM crops and other safeguards).

Fault-based liability under the Environmental Liability Directive has important limitations. It only applies to damage caused by 'operational activities'. This is defined to mean 'any activity carried out in the course of an economic activity, a business or an undertaking, irrespectively of its private or public, profit or non-profit character'.[87] Further, it must be carried out by an 'operator', that is someone who operates or has control of an occupational activity or to whom 'decisive economic power over the technical functioning' of that activity has been delegated.[88] This would clearly cover activities by a farmer or contractor working the land, but would exclude damage caused by a private trespasser or a fly-tipper, unless the latter was engaged in some kind of occupational activity at the relevant time.

The scope of the 'fault' or negligence required for liability is left undefined. This is for each Member State to define and will reflect the liability standards for delict or tort applied in its domestic civil liability regimes. This could be very restrictive. The regulations implementing the liability regime in England, for example, limit fault-based liability to cases where an operator intended to cause 'environmental damage', or was negligent as to whether environmental damage would be caused to a protected habitat or species, or to an SSSI.[89] This will require proof of an intention to cause damage to a site that the operator knows is an SSSI, or negligence as to this effect.[90] The problems of establishing liability in similar circumstances under pre-existing criminal offences protecting SSSIs in English law led to the creation of a new strict liability offence by the Natural Environment and Rural Communities Act 2006 (in other words, one that does not require the perpetrator causing damage to its conservation features to have prior knowledge that a site is an SSSI).[91] This offence will continue to apply to protect SSSIs, and may be of greater utility than the environmental liability regime.

---

[86] Environmental Liability Directive, Art 3(1)(b).   [87] Ibid Art 2(7).
[88] Ibid Art 2(6).
[89] Environmental Damage (Prevention and Remediation) Regulations 2009, SI 2009/153, reg 5(1).
[90] This would follow from the definition of 'environmental damage' in ibid reg 4, which is integral to the new offence in reg 5.
[91] Natural Environment and Rural Communities Act 2006, s 55(3), inserting a new s 28P(6A) into the Wildlife and Countryside Act 1981.

In the context of English law, civil liability for GMO contamination could also raise issues in the law of negligence or nuisance.[92] The scope of potential nuisance liability in the context of GMO releases was considered, *obiter dicta*, by the Court of Appeal in *R v Secretary of State for the Environment, ex p Watson*,[93] and the court's attitude in that case illustrates just how narrow the potential scope of liability is likely to be. Although it had been pleaded as a judicial review application, Buxton LJ doubted whether the courts would entertain a nuisance case for GMO 'contamination' on similar facts to those before it in that case. He described organic farming as a 'hypersensitive' land use, and therefore unprotected in the private law of nuisance. This may also be indicative of the likely attitude of the courts to questions of 'fault' under the Environmental Liability Directive. Foresight of damage is a key element of nuisance liability in English Law.[94] The courts are unlikely to consider that environmental damage is a reasonably foreseeable consequence of a GMO release if the ERA carried out prior to the release has certified the product as 'safe'. As noted above, there would also be difficult problems of causation—for example, in establishing that cross-fertilization of wild flora with a nearby GM crop has occurred, and if so when and by what means. The Environmental Liability Directive's potential to offer redress for biodiversity damage flowing from GM releases is therefore likely to be limited.

6. Administrative Liability and Remediation of Natural Resources

The Environmental Liability Directive is based squarely on an administrative liability model, requiring the Member States to establish administrative mechanisms for preventing and remediating environmental damage where it occurs. They must also establish cost-recovery mechanisms to ensure that the financial cost of rectifying environmental damage is recovered from those responsible for causing it, and that remediation takes place at the polluter's expense. The Directive's framework for assessing damages focuses on developing plans to restore damaged natural resources and resource services, rather than assessing the monetary value of the damage to the resource. In other words, the responsible party pays for the cost of implementing the compensatory restoration project, not the monetary value of the interim losses.[95]

---

[92] See, eg CP Rodgers, 'Liability for the release of GMOs into the environment: exploring the boundaries of nuisance' (2003) 62 Cambridge Law Journal 371.   [93] [1999] Env LR 310.
[94] See, eg *Cambridge Water Co v Eastern Counties Leather Ltd* [1994] 2 AC 264; and *Transco v Stockport MBC* [2003] UKHL 61 [2003] 3 WLR 1467. Damage of the kind suffered must be reasonably foreseeable at the time of the release in question into the environment.
[95] Under Art 6, the primary duty is on the operator to control, contain, remove, or otherwise manage contaminants so as to limit or prevent further damage and to take the necessary remedial measures. Under Art 7, the competent national authorities have jurisdiction to determine the necessary remedial measures, in accordance with criteria set out in Annex II.

Annex II of the Environmental Liability Directive provides for 'primary remediation', being any remedial measure taken to restore damaged natural resources (and natural resource services) to, or towards, their baseline condition. Compensatory remediation includes off-site measures, such as creating a replacement habitat elsewhere, where primary restoration is not possible. The person responsible for the damage can be held liable for the cost of restoring the injured natural habitat to baseline condition, compensation for interim loss of resource services during the restoration period, and (in addition) the costs of assessing damages and legal and enforcement costs.[96] In the context of damage to wildlife habitats caused by GMOs, it is difficult to see how restoration could work in practice, as genetic interactions between GM crops and wild flora will be almost impossible to reverse. It follows that the emphasis in many cases will presumably be on the establishment of an alternative habitat elsewhere[97] for the protected species whose habitat has been degraded or destroyed.[98]

Accordingly, although the Environmental Liability Directive adopts an innovative approach to habitat valuation and restoration, it nevertheless has limitations that will impair its utility as a mechanism for providing for redress and for the restoration of the environment.

6.1 Limitation of Actions

The Environmental Liability Directive applies an absolute limitation period for the application of liability and cost recovery.[99] It does not apply to any damage if more than 30 years has elapsed since the 'emission, event or incident' which resulted in the damage that has occurred.[100] The emergence of environmental damage arising from the introduction of GMOs will potentially have a very long time frame. The replacement of wild non-GM weed species with volunteers or outcrosses from GM plant varieties may, for example, take considerably longer than 30 years to emerge, and the environmental impacts may not be discernible for many years.

---

[96] Environmental Liability Directive, Arts 2 and 8.
[97] This is 'complementary remediation' within Annex II. Where possible the alternative site should be geographically linked to the primary site that has been damaged, taking into account the interests of the affected population of affected species: ibid Annex II, para 1.1.2.
[98] It is perhaps worth noting that the recitals in the Preamble to the Deliberate Release Directive refer to the fact that 'the effects of such releases [ie of GMOs] on the environment may be irreversible': Preamble (4).
[99] As opposed to a qualified limitation period (ie one where the period fixed for barring legal claims starts to run when the damage concerned was discovered, or ought reasonably to have been discovered).
[100] Environmental Liability Directive, Art 17. This rule is applied in England by reg 33 of the Environmental Damage (Prevention and Remediation) Regulations 2009, SI 2009/153, which provides that no enforcement action may be taken '30 years or more after the emission, event or incident concerned'.

The limitation model applied by the Environmental Liability Directive is similar to that used in several other international treaties.[101] The Lugano Convention also applies an absolute limitation period, which provides that no action can be brought more than 30 years from the date of the incident that caused the damage.[102] Its provisions are, nonetheless, arguably somewhat more sophisticated than those in the Environmental Liability Directive. In the first place, the Lugano Convention tempers its absolute limitation rule by providing that, where the incident consists of a continuous occurrence, the 30-year period shall run from the end of that occurrence. Similarly, where the incident consists of a series of occurrences having the same origin, the 30-year limitation period will run from the date of the last of such occurrences. In the context of GMO releases this would be of obvious relevance, as cross-pollination from GM crops will take place over a period of time, and not in a single event. It may, for example, occur through wind drift, transfer by insects or other wildlife, or by human means. The Lugano Convention also applies a qualified limitation rule barring claims made more than three years from the date on which the claimant knew or ought reasonably to have known of the damage and of the identity of the operator.[103] The absolute limitation rule therefore provides a maximum period within which action can be commenced, within which the shorter qualified period will bar 'stale' claims brought more than three years since the damage emerged and/or since the polluter was identified.

The application of an absolute (rather than a qualified) limitation period by the Environmental Liability Directive could give rise to a number of problems that a more sophisticated limitation rule might have avoided. Proving the necessary causal link between a particular GMO release and its genetic effects on biodiversity within a period of 30 years may in practice be almost impossible. A related problem for the purposes of establishing causality will be the necessity of identifying the 'emission, event or incident'[104] by which the GMO was released. This will be necessary for the purposes of calculating when the 30-year limitation period starts to run. Because the limitation period is an absolute one, it will run from the date of the GMO release—not from the date when the damage became apparent, as would be the case if a qualified limitation bar were to apply. In the case of GMO releases, the cross-pollination of wild flora with commercial GM plant varieties will take place continuously over a period of time as a consequence of seasonal cross-pollination. It may be impossible to establish a single 'emission, event or incident' as the causative factor leading to environmental damage.

---

[101] See, eg S Smyth et al., *Regulating the Liabilities of Agricultural Biotechnology* (Wallingford: CABI Publishing, 2004) 94–5.   [102] Lugano Convention, Art 17(2).
[103] Ibid Art 17(1).
[104] As indicated, these are the terms used in Art 17 of the Environmental Liability Directive and reg 33 of the Environmental Damage (Prevention and Remediation) Regulations 2009, SI 2009/153.

6.2 The Problem of Standing

The Environmental Liability Directive imposes restrictions on standing to bring claims before the courts. It gives no right to compensation to individuals or non-governmental organizations (NGOs) for environmental damage.[105] It does, however, give NGOs and individuals who have been affected (or are likely to be affected) by environmental damage a right to convey information to the competent authorities and request that they take action under the Directive; and the same right is conferred where they have sufficient interest in environmental decision-making relating to the damage or where they allege the impairment of a right (if administrative procedural law of the Member State requires this).[106] The Member States must provide for the review of the authorities' decision-making in this regard by a court or other 'independent and impartial public body competent to review the procedural and substantive legality of the decisions, acts or failure to act of the competent authority' under the terms of the Directive.[107] It is for the Member States themselves to determine what sufficient interest and impairment of rights will be required for the operation of this provision. NGOs promoting environmental protection are deemed to have a sufficient interest for these purposes, and to have rights capable of being impaired by activities causing environmental damage. Notwithstanding this concession, it must be open to doubt whether this jurisdiction will be widely used. The reluctance of English courts to interfere with administrative decisions based on risk assessment in the regulatory approval process for GMOs, as clearly demonstrated in *R v Secretary of State for the Environment, ex p Watson*,[108] would suggest otherwise.

6.3 Excluding the Polluter Pays Principle

The Environmental Liability Directive provides that in two situations Member States are permitted to allow an operator *not* to bear the cost of remedial action to rectify environmental damage. First, Member States do not have to impose on an operator the cost of rectifying damage caused by emissions or activities that were not considered harmful according to the state of scientific knowledge at the time when the GMO was released or the activity took place.[109] Secondly, they can provide for the exclusion of remediation by the operator in cases where

---

[105] Environmental Liability Directive, Art 3(3). [106] Ibid Art 12.
[107] Ibid Art 13(1). Implementing these requirements in England, reg 29 of the Environmental Damage (Prevention and Remediation) Regulations 2009, SI 2009/153, gives any person affected or likely to be affected by environmental damage or who otherwise has a sufficient interest or right to notify the appropriate enforcing authority of environmental damage that has been caused, or which is imminent. The Regulations do not provide a specific review mechanism and the subsequent decision of the regulatory body will therefore be subject to judicial review in the English courts, in the same manner as any other administrative decision or act.
[108] [1999] Env LR 310. [109] Environmental Liability Directive, Art 8(4)(b).

damage is caused by an action that has been licensed by permit under national laws implementing specified Community environmental measures.[110] In both cases the operator cannot be exonerated if he has been negligent or was at fault. Establishing fault may be difficult in practice, however, given the likely relevance of technical evidence on scientific issues to questions of fault and negligence in cases involving GMO releases. Moreover, maintaining only a fault basis for liability, where liability has otherwise been excluded by the Member State, is difficult to justify in terms of the precautionary principle of Community environmental law.[111]

The first option provides, in effect, a 'state of the art' defence and has the potential to exclude a wide range of liability in cases involving GMOs—not least because the release will have been sanctioned following an ERA based upon the extant scientific evidence available at the time.[112] Where the state of the art defence is applied, the operator will only be liable if he has conducted his occupational activities in a manner that is negligent in the sense of being outside the terms of the risk assessment on which the GMO authorization was based (for example, by ignoring mandatory crop-separation distances or the recommended frequency of application for herbicides or pesticides). He may also be liable, of course, if he negligently failed to supply full information for the purposes of the ERA itself.

The second option, as indicated, permits the Member States not to impose remediation costs on an operator where environmental damage is caused by 'an emission or event expressly authorised by ... an authorisation conferred by or given under applicable national laws and regulations' implementing specified Community environmental legislation.[113] This would preclude liability from arising where GMO releases have been authorized under national legislation introduced to implement the Deliberate Release Directive. Its potential scope is much wider, however, and it could legitimate damage flowing from activities permitted under a range of regulatory authorizations issued by the national regulatory bodies. Damage resulting from activities covered by an integrated pollution prevention and control permit, waste management licence, water discharge consent, water abstraction licence, or many other statutory environmental authorizations would potentially be within this exception.

The regulations transposing the Environmental Liability Directive in England have taken full advantage of the potential to exclude liability for damage

---

[110] Ibid Art 8(4)(a).

[111] See, eg M Stallworthy, 'Environmental liability and the impact of statutory authority' (2003) 15 Journal of Environmental Law 3, 18.

[112] Consider, eg M Lee, 'Regulatory solutions for GMOs in Europe: the problem of liability' (2003) 12 Journal of Environmental Law and Policy 311.

[113] Environmental Liability Directive, Art 8(4)(a). The specified Community environmental legislation is defined by reference to the list in Annex III (ie occupational activities engaging environmental liability under the Environmental Liability Directive).

resulting from the release of GMOs on 'state of the art' grounds, and also where the release has been permitted under the national measures governing authorization. Liability to remediate has therefore been disapplied where the operator was not at fault or negligent and the activity 'was not considered likely to cause environmental damage according to the state of scientific and technical knowledge at the time when the emission was released or the activity took place'.[114] Likewise, the operator's potential liability to remediate damage is also excluded in cases where he was not at fault or negligent and the damage was caused by an emission permitted under the terms of a consent for the deliberate release of a GMO issued by the Secretary of State,[115] or by a consent given in any other Member State under Regulation (EC) 1829/2003 of the European Parliament and of the Council on genetically modified food and feed.[116] The regulations implementing the Environmental Liability Directive in Wales, on the other hand, exclude the state of the art and permit defences in cases where damage is caused by the deliberate release of a GMO into the environment.[117] Yet both defences are more generally provided for in Wales in the case of environmental damage that does not result from GMO releases.

The potential of these exclusions to limit the scope of the Environmental Liability Directive in England, and to undermine it effectiveness in relation to GMOs, is considerable. Indeed, their impact on the coherence and effectiveness of the new liability regime across the Community will ultimately depend upon how many Member States opt to include them in their domestic legislation. Initial indications are that the majority of the Member States have taken full advantage of the facility to exclude liability by including both a state of the art and permit defence in their transposition measures.[118] The use of these discretionary elements is therefore likely to minimize the impact of the Environmental Liability Directive and has, not surprisingly, been criticized by commentators.[119]

---

[114] Environmental Damage (Prevention and Remediation) Regulations 2009, SI 2009/153, reg 19(3)(e).
[115] Ibid reg 19(3)(d) and Sch 3, para 1(g) (consents under s 111 of the Environmental Protection Act 1990 and the Genetically Modified Organisms (Deliberate Release) Regulations 2002, SI 2002/2443).
[116] [2003] OJ L268/1. For the national provisions implementing this exception, see the Environmental Damage (Prevention and Remediation) Regulations 2009, SI 2009/153, reg 19(3)(d) and Sch 3, para 1(h).
[117] Environmental Damage (Prevention and Remediation) (Wales) Regulations 2009, SI 2009/995, reg 19(e) and Sch 3 (which makes no reference to authorizations of GMOs).
[118] See the review of Member State implementation by K Smedt, 'The implementation of the Environmental Liability Directive' (2009) 18 European Energy and Environmental law Review 2.
[119] See, eg G Winter et al., 'Weighing up the EC Environmental Liability Directive' (2008) 20 Journal of Environmental Law 163. The authors doubt whether such as strategy can be maintained by the Member States in the longer term: ibid 191.

7. Risk Allocation and Insurance

The Environmental Liability Directive requires the Member States to take measures to encourage the development of 'financial security instruments and markets' by the appropriate economic and financial operators, including financial mechanisms in case of insolvency.[120] The aim is to enable operators to use liability insurance to cover their responsibilities under the Directive, by encouraging the development of an insurance market for environmental risks covered by its provisions. The primary aim envisaged for the insurance mechanism would appear to be the securing of sufficient funds to cover environmental restoration, rather than the spreading of risk. Whether a liability insurance model is the most appropriate way of achieving this is, however, debatable:

- In the first place, there are substantial problems in establishing the risk to be insured. It might be thought that the adoption of a restoration model for rectifying environmental damage, as outlined above, might clarify the quantum and nature of insurable risk for the polluter—and thus aid the establishment of a liability insurance mechanism. Where a natural habitat has been damaged as a result of a GMO release, the insurable risk would be that of having to fund restoration work to the satisfaction of the public authorities. Although this model makes it easier to achieve consensus on the valuation of 'damage', it is still unsuitable as a basis for the development of an insurance mechanism. The cost of executing possible remediation work will remain unquantifiable until damage actually occurs. And in the case of GMO releases, quantifying the insurable risk of damage will involve complex scientific issues. It is therefore difficult to see how in this context the establishment of an insurance mechanism can be other than a matter of some considerable difficulty.

- A liability insurance model will suffer from adverse selection problems, especially where strict liability is applicable. Inasmuch as biodiversity damage is damage to a public resource, it is arguable that the risk is already as widely spread as possible, as we all lose when biodiversity is damaged. The necessary conclusion must be that the only realistic aim of an insurance mechanism for environmental damage of this kind is the securing of funds for environmental restoration. If this is the case, there is a strong argument for the adoption of a first party insurance model to secure the necessary guarantees of funding for individuals whose operations generate biodiversity damage.[121]

---

[120] Environmental Liability Directive, Art 14(1).
[121] This point is strongly argued by Bergkamp (n 11 above) 112.

8. Conclusion

The Environmental Liability Directive has serious limitations as a mechanism for preventing and remedying environmental damage arising from GMO releases. It will undoubtedly impose more stringent safeguards in wildlife sites designated for protection under the Wild Birds Directive and the Habitats Directive, as Member States will be required to apply strict liability for biodiversity damage that results from the release of GMOs into the environment. It could be argued that the Environmental Liability Directive offers, at best, a form of regulation ancillary to pre-existing European environmental legislation (such as the Habitats Directive), with an emphasis on restoration and cost recovery; and, at worst, a legislative scheme for releasing private funds for environmental restoration.[122] Its principal contribution will arguably be the addition of an administrative recovery mechanism to existing Community environmental legislation.[123]

The conclusion reached in the first part of this chapter was that the primary objective of the environmental liability mechanism is the protection of the property rights reassigned to the state by the notification of wildlife sites for legal protection. Having regard to the overall effect of the social arrangements applied, and to the allocation of costs implicit within them, environmental liability would appear to perform a similar function to that of civil liability mechanisms in situations where environmental harm has a purely bilateral effect. The principal difference will lie in the different legal remedies available to the harmed party in each case, and by virtue of which the consequent reassignment of property rights is effected: the state will have the benefit of the cost-recovery mechanism established by the Environmental Liability Directive to protect the property rights reassigned to the public by the notification of a site for protection under the Habitats Directive, or as an SSSI under English law,[124] whereas adjoining landowners will only have the benefit of civil liability mechanisms to protect property rights impugned by GM cropping. This reinforces the point made above, namely that one of the principal objectives of the Directive will be to impose the cost of habitat remediation on private landowners in a situation where, through the process of site notification under, for example, the Habitats Directive or the Wildlife and Countryside Act 1981, the state has reassigned land use rights to itself.

---

[122] *Cf*, eg Stallworthy (n 111 above) 18–19.
[123] See the criticism expressed, eg in S Bell and D McGillivray, *Environmental Law* (6th edn, Oxford: Oxford University Press, 2006) 395: '[i]n essence the Directive is neither environmental, not does it provide for civil liability. It is little more than an administrative adjunct to EC environmental legislation'.
[124] This follows from the extension of the scope of the Environmental Liability Directive's application in England by reg 4(1) of the Environmental Damage (Prevention and Remediation) Regulations 2009, SI 2009/153.

Perhaps the most important conclusion on the application of the Environmental Liability Directive to GMOs is, however, that its impact will be minimal. The territorial limits placed upon the scope of the proposed liability will mean that its practical impact on operators cultivating GM crops will be very small. When the Natura 2000 network of protected European wildlife sites envisaged under the Habitats Directive is fully established, it is anticipated that it will extend to at most 20 per cent of the territorial extent of the EU.[125] In practice, most protected habitats will be in wilderness areas (for example, the uplands or wetland areas), far from sites where GM cropping is likely to be envisaged. This raises another, perhaps more fundamental, criticism of the Directive. By focusing on the Natura 2000 network of protected wildlife sites, and giving the Member States limited scope to extend liability to nationally protected sites such as SSSIs, it largely ignores wider farmland biodiversity, which will remain unprotected against the long-term consequences of the introduction of GM cropping. Whatever the merits of its approach to cost recovery as a basis for liability, therefore, the Environmental Liability Directive remains a flawed model for the resolution of the complex and difficult liability issues, and impacts upon property rights, to which the widespread adoption of agricultural biotechnology may give rise.

---

[125] See the *Natura 2000 Barometer* (available at <http://europa.eu/environment/nature/natura2000/barometer/index>, last accessed 17 January 2010). This estimates that at 17 December 2007 approximately 13.2% of the territorial area of the EU had been notified as sites of Community importance under the Habitats Directive, with a further 9.9% notified as SPAs under the Wild Birds Directive.

PART III

REGULATION BEYOND THE EUROPEAN UNION

9
Genetically Modified Organisms in Africa: Regulating a Threat or an Opportunity?

Fikremarkos Merso Birhanu

1. Introduction

Africa has entered the new millennium with diverse problems, including poverty, malnutrition, disease, and conflict, all of which make it the poorest continent in the world. Virtually all countries in Africa remain dependent on agriculture or other primary commodities (such as minerals and fuels) for their export revenues, and dreams of industrialization and technological advancement remain elusive. Indeed, in a number of African countries agriculture accounts for over 50 per cent of the value of exports and up to 80 per cent of employment;[1] and this makes agriculture a critically important economic sector.

Yet agriculture in Africa is in a state of paradox. Although it engages over 70 per cent of the population, the continent remains unable to feed itself. Per capita food production and availability is the lowest in the world, with it being estimated that 33 per cent of the population of Sub-Saharan Africa is undernourished (as opposed to 16 per cent in Asia and the Pacific and 10 per cent in Latin America and the Caribbean).[2] Moreover, Sub-Saharan Africa is the only region where average food production per person has been declining over the past 30 years, putting a large segment of the population at risk of food insecurity;[3] and the same region has the highest incidence of poverty: by 2004, 44 per cent of the population still lived on less than a dollar a day.[4] The number of rural poor has also continued to rise, making poverty a serious rural

---

[1] Food and Agricultural Organization (FAO), *The State of Food and Agriculture 2005—Agricultural Trade and Poverty: Can Trade Work for the Poor?* (Rome: FAO, 2005) Fig 51 and Table A4.
[2] FAO, *The State of Food and Agriculture 2003–2004—Agricultural Biotechnology: Meeting the Needs of the Poor?* (Rome: FAO, 2004) Part II, 1.   [3] Ibid Part II, 3.
[4] Economic Commission for Africa (ECA), *Economic Report on Africa 2007* (Addis Ababa: ECA, 2007) Table 2.8.

phenomenon.[5] Factors behind this phenomenon include rapid population growth, declining farm sizes, falling soil fertility, low use of agricultural technology, and other structural and infrastructural problems.[6]

To the extent that a great majority of Africans depend on agriculture for their livelihood and income, agricultural development is naturally at the heart of economic growth and increased welfare. In consequence, a number of African governments see food self-sufficiency and the increase of agricultural exports as policy imperatives. Likewise, eradication of extreme poverty and hunger is a commitment undertaken by the international community in the context of the Millennium Development Goals (MDGs).[7] Thus, Goal One of the MDGs calls for the halving between 1990 and 2015 of the proportion of people whose income is less than a dollar a day. Given that a large majority of the poor in Africa live in villages relying heavily on agriculture, both agricultural and rural development play key roles in the effort to achieve this goal. Nevertheless, there remains the fear that, on current trends, it might take well over a century for Africa as a whole to attain this MDG target.[8]

Accordingly, the debate about agricultural biotechnology in general and genetically modified organisms (GMOs) in particular must be seen in the context of the continent's drive to attain food security and to bring about economic development, which must start from agriculture. It is interesting to note, however, that despite these well-documented challenges of deep-rooted rural poverty and chronic food insecurity, both of which could potentially be served by a more positive approach to biotechnology, only a few African countries view such scientific developments with the benefit of the doubt which it deserves. This chapter aims to provide some insight into the different approaches which are being pursued throughout Africa in the regulation of GMOs. To that end, it commences with a general discussion of the potential role of biotechnology in Africa, followed by a general overview of the extent to which GM technology has been adopted. It then describes the multi-level approaches to regulation, from that at continental level under the African Model Law on Safety in Biotechnology, down to national level. At national level it outlines different regulatory approaches (to the extent that information is available), with focus on three broad categories of country. The chapter then concludes with a summary of its findings and suggests some future trends.

---

[5] World Bank, *World Development Report 2008: Agriculture for Development* (Washington, DC: World Bank, 2008) 3–4.   [6] Ibid *passim*.
[7] See United Nations Millennium Declaration: Resolution Adopted by the General Assembly (55/2) (8 September 2000).
[8] See United Nations Development Program (UNDP), *UNDP Annual Report 2004* (New York: United Nations, 2004) 12 (stating that Africa might not attain Goal One until 2147) (also available at <http//:www.undp.org/annualreports/2004>, last accessed 7 April 2008).

2. The Potential Role of Modern Biotechnology in Africa: Setting the Agenda

There has been an ongoing debate on the potential role of modern biotechnology in improving African agriculture and the extent to which it may contribute to economic development. Given low productivity and the challenges of severe food insecurity, malnutrition and drought, this is no surprise. Proponents of the technology argue that modern science and, notably, biotechnology are key elements of the solution to these problems. They point to evidence that scientific advances have brought about significant improvements in crop production and, not least, lifted millions of people out of poverty in China and India.[9] On this view, Africa cannot afford to ignore GM crops: they may increase productivity, facilitate job creation in other sectors such as the agri-processing sector, and assist in diversification of the economy away from its current dependence upon agriculture. Africa was essentially bypassed by the 'Green Revolution', which brought increased food production into much of Asia through use of improved seeds, application of chemical fertilizers, and expanded irrigation;[10] and this should not be repeated in the case of the 'Gene Revolution'.

On the other hand, opponents argue that modern biotechnology is not the solution to farming problems in Africa; some go further and claim that it may even bring more ills to the continent than benefits. They argue that GM crops grown in different parts of the world have so far been used mainly for the production of feed and fibre and not for the production of food, which is the immediate priority in Africa. Moreover, they argue that the GM industry is now controlled by a small number of private seed companies, which control the patents for GM crops, so making the seed expensive and thus unsuitable for poor and marginalized farmers.[11] Opposition to GM technology is particularly strong among international non-governmental organizations (NGOs), which have been campaigning against its introduction into Africa. By contrast, the role and engagement of local protest groups have been more limited, with the

---

[9] See, generally, eg R Paarlberg, *Starved for Science: How Biotechnology is Being Kept Out of Africa* (Cambridge, MA: Harvard University Press, 2008) 6–9. For the same line of argument, see African Union (AU) and New Economic Partnership for Africa's Development (NEPAD) (2007), *The Freedom to Innovate: Biotechnology in Africa's Development, Report of the High-Level African Panel on Modern Biotechnology* (available at <http//:www.nepadst.org>, last accessed 19 February 2008); and the views of Monsanto (<http://www.monsanto.com>, last accessed 10 December 2008).

[10] RE Evenson and D Gollin, 'Assessing the impact of the Green Revolution, 1960 to 2000' (2003) 300 Science 758.

[11] See, eg Friends of the Earth International, *Questions and Answers: Who Benefits from GM Crops?* (2008) (available at <http//:www.foe.co.uk/resources/briefings>, last accessed 9 June 2008).

exception of a number of South African-based NGOs, such as Biowatch South Africa[12] and the African Centre for Biosafety, which work with the larger Europe-based NGOs, such as Greenpeace and Friends of the Earth. For example, this alliance worked together at the time of the World Summit on Sustainable Development in Johannesburg, with over 140 local NGOs and other civil society organizations signing an open letter to the World Food Program (WFP) and the US Government protesting against the delivery of food aid containing GMOs.[13]

Several factors account for the resistance to modern biotechnology. First, while the potential benefits of GMOs have been recognized, concerns remain as to their potential risks to human and animal health and to the environment. However, as is well-known, there is still no consensus on the part of the scientific community on the nature and extent of such risks.[14] Besides, if the potential benefits of GMOs are large in Africa, the potential concerns would also seem significant. Most African countries lack the scientific capacity to assess the safety of GMOs, the economic expertise to evaluate their worth, and the regulatory capacity to implement guidelines for safe deployment. Given this reality, if GMO issues are controversial elsewhere, the lack of expertise and resulting fear of the unknown appear to have made them even more controversial in large parts of Africa.

Secondly, apart from the perceived risks to health and the environment, commercial considerations would also seem to contribute to resistance to GMOs.[15] The ability to export is certainly to be factored into the equation when determining their benefit in Africa, in large part reflecting the fundamental opposition to GM technology in areas of the world, such as Europe, to where Africa has a history of selling its conventional agricultural products. The obvious concern is that if agriculture in Africa is 'contaminated', countries may lose markets for their agricultural products. Indeed, European scepticism of GMOs has been blamed as the single most important reason for their rejection in Africa (it is even argued that African resistance is more Western than African). On this view, Europe, having attained food security through the use of agricultural science, is now working hard to block Africa from technology that has great potential to address the chronic problems the continent continues to face.[16] However, some contest the assertion that export markets in Europe would be lost if GM crops were adopted. Paarlberg, for example, having analysed the export data of African countries, concludes that most of the agricultural products that are likely to be 'contaminated' by GMOs are destined for

---

[12] Biowatch South Africa has been a strong critic of GMOs in South Africa, undertaking its own research and publishing a monthly bulletin: see, further, <http://www.biowatch.org.za>, last accessed 3 July 2009.   [13] Paarlberg (n 9 above) 143–4.
[14] See, eg FAO (n 2 above) Part I, 5.
[15] S Zarrilli, *International Trade in GMOs: Legal Frameworks and Developing Country Concerns* UNCTAD/DITC/TNCD/2004/1, United Nations Conference on Trade and Development (UNCTAD) (Geneva: UNCTAD, 2005).   [16] Paarlberg (n 9 above) 16–17.

consumption in Africa and that the possibility of loss of exports to GM-sensitive markets is therefore small.[17] The way in which Paarlberg analysed the data, and his consequent conclusion, could, however, be questioned. The analysis, for example, took into consideration only the products likely to be 'contaminated' or affected by GMOs; yet, once GM material enters the food chain, it is difficult, if not impossible, to detect and, consequently, the whole food chain may be compromised. Even if this is not accepted, African countries may legitimately be concerned not only with current GM-sensitive markets, but also future potential markets. That said, even if only current markets are taken into account, the risk is not negligible. FAO figures show that in 2002, 66 per cent of exports from Sub-Saharan Africa were destined to developed countries, of which 50 per cent went to the European Union, six per cent to the United States and Canada, and eight per cent to Asia and the Pacific.[18]

Thirdly, modern biotechnology, at least in its current state, is economically driven and, for the most part, targets crops grown by commercial farmers in temperate regions, with little emphasis on varieties grown by poor farmers in Africa (such as sorghum, rice, wheat, cowpeas, and cassava, which are critically important to realize food security in the continent). This would seem to replicate the 'Green Revolution' of the 1960s and 1970s, which failed to reach such farmers.[19] The technology as it stands now is simply unaffordable to the poor and marginalized farmers of Africa.[20]

Fourthly, different considerations arguably underpin the drive for modern biotechnology in Africa and in developed countries. In Africa, the use of such technology to enhance food production is for the purposes of addressing widespread poverty, hunger, and starvation, while in developed countries it is largely for economic and commercial purposes. It is not clear therefore how far GM technology addresses the peculiar agricultural problems in Africa; and, indeed, little attention has been directed by larger interests to locally important food crops. One example is the collaboration between the Swiss Federal Institute of Technology and local research institutions in Kenya and Nigeria, relating to African Cassava Mosaic Virus (a serious problem for cassava farmers across Africa).[21] It has also been reported that GM rice, eggplant, cassava, banana, sweet potato, and lentils have been approved for field-testing.[22] But, while initiatives such as these are obviously important in the effort to make GM

---

[17] R Paarlberg, 'Are genetically modified (GM) crops a commercial risk for Africa?' (2006) 2 International Journal of Technology and Globalization 81.   [18] FAO (n 1) above Table 1.
[19] World Bank (n 5 above) 160.
[20] The most often-quoted use of GM crops in Africa by small farmers is cultivation of GM cotton in the Makhatini Flats areas of South Africa. While industry bodies claim that farmers have increased their income, this is contested and, in any event, one case does not allow the making of a general conclusion on the benefits of GM crops for farmers across the continent: see, eg, Friends of the Earth International (n 11 above).
[21] W Sawahel, 'GM Cassava uses viral gene to fight diseases' (2005) (available at <http//:www.scidev.net>, last accessed 10 February 2008).   [22] World Bank (n 5 above) 177.

technology relevant to Africa, much more is required. It may also be noted that such initiatives are generally directed from outside Africa, with the result that the continent does not yet seem to 'own' the technology, let alone the policy.

Nonetheless, as will be explored further, GMOs do have a role in addressing the ills of agriculture in Africa. What would seem evident is that African countries should strive to own and make use of the technology in a way appropriate to their own needs, in a safe and sustainable way, by investing in research and development. This is where informed leadership at governmental level is required, so that fear of sometimes non-existent risks do not keep Africa in the grip of poverty and starvation.

3. The State of GMOs in Africa

The attitude towards GMOs varies across the continent, ranging from outright rejection to cautious reception and even promotion. The issue attracted perhaps greatest attention when, during the 2002 drought in southern Africa, Zambia, Zimbabwe, Malawi, and Mozambique all declined food aid (in the form of maize) coming from the United States through the WFP, on the ground that it might be contaminated with GMOs. Zambia refused to accept even milled grains[23] and its President went so far as to declare that GM food was a poison.[24] Zambia's position was at the most extreme end; other countries accepted the maize on condition that it was first milled, so that the crop would not be replanted by farmers.[25]

Furthermore, in addition to Zambia, several African countries have directly banned or restricted the use of GMOs at different times:

- Algeria banned the import, distribution, commercialization, and use of GM material from December 2000.
- Angola banned import and use of GM foods, except milled, food-aid grains, from 2004.
- Benin declared a five-year moratorium in 2002 on the import, commercialization, and use of GMOs and their products.
- Malawi, Mozambique, Nigeria, and Zimbabwe decided to accept GM food aid only if it was milled prior to distribution.
- Namibia rejected GM food aid in general.
- Sudan required GM-free food aid from 2003, with a waiver until June 2005.[26]

---

[23] For an interesting account of the Zambian position on GM food aid, see R Walters, 'Crime, bio-agriculture and the exploitation of hunger' (2006) 46 British Journal of Criminology 26.
[24] BBC World News, 'Famine and the GM Debate', 14 November 2002.   [25] Ibid.
[26] See African Centre for Biosafety, *GMOs in African Agriculture—Overview* (available at <http://www.biosafetyafrica.net>, last accessed 12 February 2008).

Although approximately 20 African countries have reportedly engaged in research and development into GMOs, only nine countries have so far conducted field trials (Burkina Faso, Egypt, Kenya, Morocco, South Africa, Senegal, Tanzania, Zambia, and Zimbabwe).[27] Further, most research and development is confined to three countries: South Africa, Egypt, and Kenya. Indeed, until recently, South Africa has been the only country in the continent to have approved commercial plantings.[28] The key regulatory challenge, as elsewhere in the globe, is where to strike the balance between the two contending interests of maximizing the potential benefits from modern biotechnology and avoiding, or at least reducing, the potential risks to human health and the environment. Importantly, the regulatory approaches to GMOs in Africa appear to reflect the research and development capacity of the respective countries. A distinct minority recognize the potential opportunities of modern biotechnology and attempt to promote it by adopting a more liberal regulatory regime. Others approach the technology with caution, while still others capitalize on the potential hazards and similar concerns to move towards a stricter regulatory regime. It is curious to observe that this approach is preferred by the large majority of African countries which are chronically food insecure. Some have already enacted a stringent biosafety regulation, while others are in the process of doing so. And amidst this diversity of approaches the African Union and sub-regional, economic-integration organizations are trying to encourage a harmonized approach.

4. Multi-level Approaches to GM Regulation in Africa: Continental, Sub-regional, and National

4.1 General

A number of initiatives have been taken in Africa, both at continental and regional levels, to forge a common understanding and position on GMOs. A major feature of GM regulation in Africa is its focus on ensuring biosafety; and, as noted, several African countries have already put in place a biosafety regulation. In recognition of the potential risks, and with a view to encouraging a regional approach, the Organization of African Unity (OAU, now African Union (AU)) in 2001 adopted the African Model Law on Safety in

---

[27] African Biotechnology Stakeholder Forum, *Biotech Status in Africa* (2008) (available at <http://www.absfafrica.org/pages/biotech_status.html>, last accessed 14 March 2008).

[28] It has now been reported that in addition to South Africa, Burkina Faso and Egypt have also permitted commercialization of GM crops: see C James, *Global Status of Commercialized Biotech/GM Crops: 2008 (ISAAA Brief No 39)* (Ithaca, NY: International Service for the Acquisition of Agri-biotech Applications (ISAAA), 2009) Table 1.

Biotechnology ('African Model Law').[29] As indicated, there have also been attempts to harmonize biosafety approaches to GMOs at sub-regional level, through regional economic-integration arrangements. However, these initiatives have remained largely ineffective, with individual countries in Africa as a rule still pursuing unilateral approaches in their regulation of GMOs.

4.2 The African Model Law

As the name of the instrument indicates, this is not a binding treaty but a 'model law' or template, on the basis of which African countries can develop their national laws in a manner that, on the one hand, reflects their own national specific circumstances, but, on the other hand, also achieves a degree of coherence and harmony at continental level. The OAU from the start urged African Countries to use the African Model Law in the process of formulating their own biosafety framework;[30] and the Third Ordinary Session of the AU, which took place from 4–8 July 2003, reaffirmed this approach, expressly advocating that Member States should use the African Model Law for drafting national legal instruments on biosafety, while taking into account differing national circumstances, so as to create a more harmonized, Africa-wide system for regulating the movement of GMOs.[31] Although there is evidence of its influence on several national biosafety regimes, the harmonization of biosafety regulation, as already noted, is far from a reality, with the biosafety laws of many African countries still maintaining a wide divergence.

The African Model Law is based on the same philosophy as the Cartagena Protocol to the Convention on Biological Diversity,[32] that is, the protection of biodiversity from the potential risks associated with modern biotechnology. It is informed by the provisions of the Cartagena Protocol, from which in some cases it draws its provisions, while at the same time containing significant departures. The salient features may be described as follows.

First, in terms of scope, the African Model Law is more comprehensive than the Cartagena Protocol: under Article 2 it applies to any GMO, 'whether intended for release into the environment, for use as a pharmaceutical,[33] for

---

[29] OAU, The African Model Law on Safety in Biotechnology (2001) (available at <http//:www.nepadst.org>, last accessed 3 July 2009).

[30] OAU Council of Ministers Decision No CM/Dec.623 of July 2001 (available at <http://www.nepadst.org/>, last accessed 3 July 2009).

[31] Decision of the Executive Council of the AU on Biosafety, EX/CL/Dec.26 (III) (available at <http//:www.nepadst.org>, last accessed 23 February 2008).

[32] The text of both the Convention on Biological Diversity and the Cartagena Protocol are available at <http://www.cbd.int/>, last accessed 3 July 2009; and see Chapter 14.

[33] By contrast, Art 5 of the Cartagena Protocol states that it 'shall not apply to the transboundary movement of living modified organisms which are pharmaceuticals for humans that are addressed by other relevant international agreements or organisations'.

food, feed or processing'. It thus subjects the entire range of GMOs and their products to the same authorization requirements; and, unlike the Cartagena Protocol, it also includes non-living products of GMOs.[34] Since GM food has been calculated to represent an average trade value of 42 billion dollars per year,[35] this 'Cartagena-plus' requirement has the capacity to impact heavily on international trade. Nevertheless, as shall be seen, the African Model Law does not seem to have taken into account the difficulties associated with the detection and management of GM products and the cost implications of such a system. Nor does it make any distinction among different activities involving GMOs, covering the import, transit, contained use, release, and placing on the market of any GMOs or their products.[36] Accordingly, while it adopts the advance informed agreement (AIA) procedure—a central building block of the Cartagena Protocol[37]—it applies that procedure to a broader range of GM material and a broader range of activities.

Secondly, while the import of GMOs is subject to prior authorization under both the Cartagena Protocol and the African Model Law, the information requirements under the latter are far more detailed. Article 8(2) stipulates, in particular, that all decisions on applications to import, make contained use of, release, or place on the market a GMO or a product of a GMO should be conditional on a risk assessment, with firm emphasis on the precautionary principle. Importantly, in comparison to the Cartagena Protocol, a stronger version of this principle is adopted: under Article 6(7) it is affirmed that '[n]o approval shall be given unless there is a firm and sufficient evidence that the genetically modified organism or the product of a genetically modified organism poses no risks/significant risks to the environment, biological diversity or human health'.[38] Thus, the implicit assumption is that all GMOs are unsafe and should in principle be prohibited (unless the risks can be avoided).[39] However, there is no guidance on what in practice constitutes a 'significant risk'. Interestingly, proof of safety is not enough for approval; additional socio-economic criteria are also to be taken into account. These include determination of the particular benefits of a GMO to the country concerned in terms of contributing to its sustainable development, not having adverse socio-economic impacts, being

---

[34] For the scope of the Cartagena Protocol, see Art 4: '[t]his Protocol shall apply to the transboundary movement, transit, handling and use of *all living modified organisms* that may have adverse effects on the conservation and sustainable use of biological diversity, taking also into account risks to human health' (emphasis added).

[35] GP Gruere, *Analysis of Trade Related International Regulations of Genetically Modified Food and their Effects on Developing Countries* (Washington, DC: International Food Policy Research Institute, EPT Discussion Paper 147, 2006). [36] African Model Law, Art 4(1).

[37] Cartagena Protocol, Arts 7–10 and 12.

[38] For discussion of the precautionary principle in the context of Cartagena Protocol, see, eg C Redgwell, 'Biotechnology, Biodiversity, and International Law' in J Holder and C O'Cinneide (eds), *Current Legal Problems 2005: Volume 58* (Oxford: Oxford University Press, 2006) 543.

[39] African Model Law, Art 8(5).

in accord with the ethical values and concerns of communities, and not undermining community knowledge and technologies.[40] While these considerations are designed to address genuine concerns, they can also easily lend themselves to abuse and serve as an excuse to deny authorization almost at will.

Thirdly, the African Model Law expressly addresses public participation in decision-making through a notice-and-comment procedure, as well as formal public consultations. The relevant competent authorities are required, in making or reviewing decisions, to take into account the views and concerns of the public.[41] While this is important in making the process transparent and building public confidence in GMOs, it is not clear what kind of concerns should be taken into account and how far or in what way those concerns would affect the decision. Further, it is questionable whether public consultation on individual applications would be practicable in countries where the institutional capacity and other resources are severely constrained; and, even if it were practicable, such consultations would have a negative impact on the cost and speed of decision-making.[42]

Fourthly, provision is made for both access to information and protection of confidential information. On the one hand, the competent authority must make available to the public information on any GMO or a product of a GMO, which has been granted or denied approval for import, contained use, release, or placing on the market and, in particular, any risk assessment report.[43] On the other hand, the African Model Law also requires that access to information be limited when such information is considered confidential. The competent authority is required to determine which information is to be kept confidential following a claim to that effect by the applicant, but may deny the request on account of the public interest, even where the information is commercially confidential.[44] Further, a negative list is provided of the types of information that may not in any case be considered confidential.[45]

Fifthly, Article 11 lays down detailed rules on labelling and traceability. Any GMO or product of a GMO should be clearly identified and labelled as such. The identification should specify the relevant traits and characteristics given in sufficient detail for purposes of traceability. GMOs or products containing a GMO are also required to be clearly labelled and packaged using specific words as set out in Annex II.

Sixthly, liability and redress are covered by Article 14. Liability is placed on the person importing, making contained use of, releasing, or placing on the market GMOs or their products. Importantly, the African Model Law here

---

[40] Ibid Art 6(9).   [41] Ibid Art 5(2)–(4). *Cf* Cartagena Protocol, Art 23.
[42] For proposed revisions to the African Model Law, including the promotion of capacity building see, eg <http://nepadst.org/doclibrary/pdfs/amcost3_17_nov2007.pdf>, last accessed 7 May 2009).   [43] African Model Law, Art 5(5)(i) and (ii).
[44] Ibid Art 12(1) and (3).   [45] Ibid Art 12(2).

adopts a regime of strict liability that foresees full compensation for harm caused, including compensation for:

> harm or damage caused directly or indirectly by the genetically modified organisms or product of the genetically modified organisms to the economy or social or cultural conditions, the livelihood or indigenous knowledge systems or technologies of a community or communities, or incidents of public disorder triggered by the genetically modified organism or the product of a genetically modified organism.[46]

Specific examples of such harm include: 'disruption or damage to production systems, agricultural systems, reduction in yields, soil contamination, damage to the biological diversity, damage to the economy of an area or community, or any other consequential disorder'.[47] Moreover, anyone is entitled to bring an action for compensation in the public interest (in other words, even when the individual or group claimant is not directly affected by the GMO).[48] This is a very expansive interpretation of *locus standi*, and exemption from payment of the costs of failed litigation is awarded where the action was instituted reasonably out of concern for, *inter alia*, the protection of human health and biological diversity. As a consequence, there is potential for unsubstantiated claims by individuals or groups who are opposed to the technology for any reason, and the dampening effect on the development of GM technology in the continent must be unmistakable. A further brake would be the stipulation that the competent authority should, as a condition for approval of a GMO or a product of a GMO, require the applicant to furnish evidence of insurance cover or some other arrangements sufficient to meet his/her obligations under the law.[49] No detail is furnished as to the type of insurance, the risks to be covered, or the amount or length of any cover.

The provisions contained in the African Model Law are complex and it is questionable whether African countries have the capacity and means at present to put the regulatory framework envisaged into effect. Besides, no account is taken of differences in biotechnology capacity of different African countries. For example, as shall be seen, regulation on such a scale might be appropriate in South Africa, but less so in Tanzania. In addition, the African Model Law does not address its relationship with other multilateral treaties (an omission which could give rise to issues such as those raised in the *EC—Biotech* case).[50] Nonetheless, this being only a 'model law', African countries are at liberty to pick and choose from among its rules and principles and introduce their own laws adapted to their individual needs, priorities, and institutional capacities.

---

[46] Ibid Art 14(5).   [47] Ibid.   [48] Ibid Art 14(7).   [49] Ibid Art 6(10).
[50] *EC—Approval and Marketing of Biotech Products* WT/DS291/R, WT/DS292/R, and WT/DS293/R, 29 September 2006; and see Chapters 13 and 14.

4.3 Sub-regional Approaches to Biosafety and Biotechnology

The African Model Law, as described above, has been supplemented by a number of sub-regional efforts launched over the years to develop a harmonized response to the challenge of biotechnology. The main vehicles for this sub-regional effort have been regional, economic-integration agreements, such as the Common Market for Eastern and Southern Africa (COMESA), the Southern African Development Community (SADC), and the Economic Community of Western African States (ECOWAS). Notable in this respect is the SADC, which, partly in response to the 2002 regional food crisis, approved guidelines on GMOs in May 2004, developed by its Advisory Committee on Biotechnology and Biosafety.[51] These guidelines focus on such issues as handling of food aid, policy and regulation, capacity building, and public awareness and participation. They urge each country to develop national biotechnology policies and strategies and to expedite the process of establishing national biosafety regulatory systems, as well as signing and ratifying the Cartagena Protocol. Several references are made to the African Model Law. For example, countries which manage or handle food aid containing GM material are encouraged to make use of the provisions laid down under the African Model Law, in the absence of national legislation. Similarly, the guidelines promote development at a sub-regional level of harmonized policy and harmonized regulation based on the African Model Law and the Cartagena Protocol. They also provide that GM food aid should be clearly identified and labelled, whether under national law or, in its absence, the African Model Law.

Likewise, an attempt is being made by the COMESA to develop a regional policy towards GMOs, including GM food aid. In 2002 COMESA agriculture ministers agreed to develop a regional policy;[52] and in 2006 a COMESA regional workshop on biotechnology and biodiversity was convened, and a draft regional position on biotechnology and biosafety adopted (although the draft common position is yet to be endorsed).[53] Finally, a 2005 ECOWAS Ministerial Conference on biotechnology, held in Bamako, Mali, recommended, *inter alia*, the development and use of biotechnology in the region, advancing a sub-regional approach to biosafety.[54]

---

[51] Science and Development Network (SciDev.Net), 'Southern African Nations Adopt Common GM strategy' (available at <http://www.scidev.net/en/news/southern-african-nations-adopt-common-gm-strategy.html>, last accessed 7 May 2009).

[52] SciDev.Net, 'South and East Africa Draw up GM Policy' (available at <http://www.scidev.net/News/index.cfm?fuseaction=readNews&itemid=708&language=1>, last accessed 17 February 2008).

[53] See African Union, *An African Position on Genetically Modified Organisms in Agriculture* (available at <http://www.africa-union.org/root/au/AUC/Departments/HRST/biosafety/DOC/AfricanPositionOnGMOs.pdf>, last accessed 12 May 2009).

[54] See International Centre for Trade and Sustainable Development (ICTSD), *Biosafety Regulations Necessary, ECOWAS Hears* (2005) 5(13) Bridge Trade BioRes.

That said, these sub-regional harmonization efforts remain at a general level and their practical implementation is as yet untested. Furthermore, the initiatives are limited to the provision of guidelines, thereby leaving the actual formulation of the regulatory regimes to individual states and restricting any harmonizing effect.

4.4 National Regulation of GMOs in Africa

4.4.1 A broad survey

As the previous discussion of the undeveloped state of GMO policies at the continental and sub-regional levels shows, national regulatory frameworks on biosafety in Africa are still evolving. Even in countries where regulations are already in place, they are in most cases little more than a framework, with detailed rules yet to be developed for their full implementation. Almost all biosafety regulations in Africa mention risk to human health an as important consideration, but only in a few cases do they provide detailed safety assessment procedures and mechanisms. In addition, they often address risk to the environment more broadly. This might suggest that the main driver of the regulatory regimes in Africa is to implement the Cartagena Protocol. Indeed, as of October 2009, 45 African countries have ratified the Cartagena Protocol,[55] indicating a willingness to engage in the use of biotechnology with a precautionary approach.

On the basis of their approach towards GMO regulation, African countries could be classified into three broad categories. First, there are a few countries which have already built a relatively modern biotechnological capacity and infrastructure and which have even commercialized GMOs (or are at the final stages of doing so). South Africa and Egypt would most naturally belong to this group. In the second category fall those countries which have acquired limited biotechnological capacity, but aspire to build a more advanced biotechnology infrastructure in the near future. Ghana, Kenya, and Uganda would belong to this group. The third and largest category contains those countries which at present have little or no biotechnological capacity.

As might be expected, the countries in the first category generally encourage their biotechnology industry, with the regulatory regime being more liberal; those in the second category tend to adopt a favourable, but cautious, approach; and those in the third category place emphasis on safety, with their stance being both defensive and restrictive of the new technologies. The next part of this

---

[55] Algeria, Benin, Botswana, Burkina Faso, Burundi, Cameroon, Cape Verde, Central African Republic, Chad, Comoros, Congo, Democratic Republic of the Congo, Djibouti, Egypt, Eritrea, Ethiopia, Gabon, Gambia, Ghana, Guinea, Kenya, Lesotho, Liberia, Libyan Arab Jamahiriya, Madagascar, Malawi, Mali, Mauritania, Mauritius, Mozambique, Namibia, Niger, Nigeria, Rwanda, Senegal, Seychelles, South Africa, Sudan, Swaziland, Togo, Tunisia, Uganda, United Republic of Tanzania, Zambia, and Zimbabwe. More information is available from <http://www.biodiv.org/biosafety/signinglist.aspx?sts=rtf&ord=dt>, last accessed 6 October 2009).

chapter will address the state of GMO regulation in a selected number of countries falling within each of these three categories.

4.4.2 The first category: national regulations promoting GMOs

4.4.2.1 South Africa

South Africa is the most biotechnologically advanced country in the continent and has already acquired an advanced scientific infrastructure for GM research and development. Moreover, it is one of only three African countries to be included in the list of countries which grow GM crops commercially.[56] With a total area of 1.8 million hectares being cultivated with GM maize, cotton, and soya in 2008, it ranked eighth of the 25 countries which have proceeded to commercialization (far ahead of any other country in Africa).[57] Field trials began in 1992, and commercialization commenced in 1998, when the Department of Agriculture (DoA) granted permission for commercial planting of GM maize and GM cotton (both insect resistant).[58] In the same year, permissions were granted for the import of GM soya for animal consumption; and over the period January to October 2000 some 105 out of 111 applications were approved.[59]

The regulatory framework is provided by the Genetically Modified Organisms Act (Act No 15 of 1997) ('GMO Act'), which came into force in 1999 and was subsequently amended by the Genetically Modified Organisms Amendment Act (Act No 23 of 2006).[60] Interestingly, as indicated, the Committee for Genetic Experimentation had, before that date, already begun to receive and evaluate applications and to issue permits.[61] In terms of scope, the GMO Act is limited to viable, living GMOs, and excludes products of GMOs.[62] Thus, it follows the Cartagena Protocol rather than the African Model Law.

According to the measure issued under the GMO Act (Regulation No 1420 of 26 November 1999 ('GMO Regulation')[63]) any import to or export of GMOs from South Africa requires a permit. A permit is also required to develop, produce, use, release, or distribute any GMO.[64] However, no permit is required for GMOs to be used under conditions of contained use in academic and research facilities, nor where they are listed in Table 3 of the Annexure (in other words, those already approved in South Africa).[65] A risk assessment to

---

[56] James (n 28 above); and see, generally, eg GP Gruere and D Sengupta, *Biosafety at the Crossroads: an Analysis of South Africa's Marketing and Trade Policies for Genetically Modified Products* (Washington, DC: International Food Policy Research Institute, Discussion Paper 796, 2008). [57] James (n 28 above) Table 1.
[58] M Mayet, *Analysis of South Africa's GMO Act of 1997* (African Centre for Biosafety, 2000) (available at <http//:www.biosafetyafrica.net>, last accessed 27 February 2008). [59] Ibid.
[60] The amendments effected by Act No 23 of 2006 addressed, *inter alia*, the obligations of South Africa under the Cartagena Protocol.
[61] For a list of early approvals, see, eg DoA, *Genetically Modified Organisms Act, 1997: Annual Report 2004/2005* (Pretoria: DoA, 2005) Table 1. [62] GMO Act, s 2.
[63] As foreseen by s 20 of the GMO Act. [64] GMO Regulation, reg 2(1).
[65] Ibid reg 2(2).

address risks to both human health and the environment is required for all activities involving GMOs.⁶⁶ The GMO Regulation provides that lack of scientific knowledge or consensus on the safe use of GMOs may not be interpreted as indicating a particular level of risk, an acceptable risk, or an absence of risk.⁶⁷ Otherwise, little guidance is given as to the conduct of the risk assessment. There is no specific reference to the precautionary principle and there may be some preference for the 'substantial equivalence' test. Indeed, in relation to human health, the South African Department of Health (DoH) states that:

> [a]ll assessments are done case-by-case and step-by-step. As with all new experiences comparisons with known foods are constantly made. This approach, which is the starting point of risk assessment of genetically modified food, is often called substantial equivalency.⁶⁸

Interestingly, the GMO Act places the liability for damages caused by activities relating to GMOs on their users.⁶⁹ Liability is thus not placed on producers, distributors, and suppliers of GMOs, as in the case of the African Model Law. While the logic of this uniquely South African approach to liability remains unclear, it would seem to reflect a drive to promote the biotechnology industry. Nonetheless, presumably with a view to making the GMO Act compatible with the Cartagena Protocol, its 2006 amendment by Act No 23 of 2006 came up with a definition of 'user' as 'a person who conducts an activity with a genetically modified organism'.⁷⁰

The GMO Act provides strong protection for confidential business information.⁷¹ The Executive Council, the decision-making body under its provisions, makes a decision after consultation with the applicant on which information should be kept confidential. That said, a list is also provided of the types of information that may not be kept confidential, such as the general description of the GMO, name and address of the applicant, purpose and location of the release of the GMO, and methods and plans for monitoring GMOs and for emergency measures in case of accident. The Executive Council could decide that even this information should be kept confidential, however, with a view to protecting the intellectual property rights of the applicant; and the information remains confidential even if the application is later withdrawn for whatever reason. These provisions have proved controversial in the *Biowatch* litigation which reached the Constitutional Court.⁷² In 2005 Biowatch was largely successful before the Pretoria High Court in its request for information on the status of GM crops in South Africa, but was ordered to pay the costs of

---

⁶⁶ Ibid reg 3(1).   ⁶⁷ Ibid reg 3(2).
⁶⁸ DoH, *Guideline Document for Food Safety and Risk Assessment of Genetically Modified Organisms and Products Thereof* (available at <http://www.doh.gov.za/department/foodcontrol/gmo/risk.html>, last accessed 17 March 2008).
⁶⁹ GMO Act, s 17(2).   ⁷⁰ Ibid s 1(m), as amended.   ⁷¹ Ibid s 18.
⁷² *Trustees for the time being of the Biowatch Trust v Registrar Genetic Resources* [2009] ZACC 14 (3 June 2009).

Monsanto. An appeal against this costs order reached the Constitutional Court in 2009, which held that Biowatch was not required to pay the costs of Monsanto and that the governmental authority was required to pay the costs of Biowatch.

Public participation is foreseen by the GMO Act through a notice-and-comment procedure in permit applications for environmental release, but no clear and obligatory procedure exists to ensure public participation in decision-making.[73] According to the procedure issued by the DoA, an application for a GMO permit would be brought to the attention of the Advisory Committee, which consists entirely of scientists appointed by the Minister for Agriculture. The Committee would evaluate the risk assessment undertaken by the applicant and gives its recommendations. The final decision would be made by the Executive Council, which consists of officials from government departments whose number has been raised from six to eight under Act No 23 of 2006.[74] Decisions are to be made by consensus and, in the absence of decision by consensus, the application is considered to have been refused.[75] Anyone aggrieved by the decision could take an appeal to an appeal board consisting of an unspecified number of members to be appointed by the Minister and with the power to confirm, set aside, substitute, or amend the decision or refer the matter back to the Executive Council for reconsideration.[76] However, the Minister may also take such further actions as he considers necessary.[77]

Under foodstuff labelling regulations issued by the DoH, GM food has to be labelled if it differs significantly from the corresponding existing foodstuff in terms of composition, nutritional value, or in mode of storage, preparation, or cooking.[78] Labelling is also required if a plant-derived food contains genetic material derived from a human or an animal, or if animal-derived food contains genetic material derived from a human or from a different taxonomic animal family. Importantly, the DoH claims that GM foods approved in South Africa (maize, cotton, and soybeans) are as safe as their conventional counterparts;[79] and, in consequence, the extent of mandatory labelling in South Africa has been limited. However, the new Consumer Protection Act (Act No 68 of 2008) stipulates that:

[a]ny person who produces, supplies, imports or packages any prescribed goods must display on, or in association with the packaging of those goods, a notice in the prescribed manner and form that discloses the presence of any genetically modified ingredients or components of those goods in accordance with applicable regulations.[80]

---

[73] GMO Regulation, reg s 6.
[74] GMO Act, s 3, as amended, stipulates that the Executive Council consists of representatives from eight government departments, as listed in the Act itself. [75] Ibid s 7, as amended.
[76] Ibid s 19, as amended. [77] Ibid s 19(6), as amended.
[78] Regulations relating to the labelling of foodstuffs obtained through certain techniques of genetic modifications (Notice R 25 of 16 January 2004), made pursuant to the Foodstuffs, Cosmetics and Disinfectants Act 1972 (Act No 54 of 1972).
[79] DoH, *Understanding GMOs* (available at <http://www.doh.gov.za/department/foodcontrol/gmo/brochure/html>, last accessed 8 May 2009).
[80] Consumer Protection Act 2008, s 24(6).

That said, the Consumer Protection Act only operates as from 29 October 2010; and the necessary regulations are yet to be implemented.

4.4.2.2 Egypt

Egypt is another African country with a fairly well-developed biotechnology capacity and infrastructure, research and development having been undertaken for over a decade. GM crops which have been subject to field trials include cotton, potato, squash, tomato, maize, and wheat.[81] In 2008 the first commercialization was reported, this extending to some 700 hectares of GM maize.[82] It may also be noted that Egypt, as a supporter of GM technology, was the only African country to participate in the *EC—Biotech* case (although later withdrawing).

The national biosafety system in Egypt was established in the mid-1990s by two ministerial decrees of the Ministry of Agriculture and Land Reclamation (MALR).[83] Ministerial Decree No 85 of 1995 established a National Biosafety Committee (NBC) and Ministerial Decree No 136 of 1995 adopted both biosafety regulations and guidelines. The NBC is responsible for the formulation of biosafety standards, risk assessment, and licensing. By contrast, the Supreme Committee for Food Safety (SCFS), under the Ministry of Health, is empowered to regulate the safety of food and controls food import permits. The import of GM material requires an advance permit from the SCFS. The biosafety guidelines were largely intended to regulate research and field trials of GM crops and did not deal with commercialization and labelling, because no GMO was being grown commercially at the time. Commercialization of GM crops is governed instead by Ministerial Decree No 1648 of 1998.[84] Accordingly, for GM varieties developed in Egypt, the applicant is required to provide information about the GMO, including the process employed, and evidence to the effect that there would be little environmental risk. If approved in other countries, evidence to that effect is also considered important. The NBC considers the application and sends it on to the Seed Registration Committee for preliminary approval to conduct field trials. If these are successful, an application is made with NBC authorization to the Seed Registration Committee for final approval. If the application relates to GMOs developed outside the country, the applicant would have to obtain a permit to import the GMO from the SCFS, but otherwise the procedure would be the same as that for a locally developed GM variety. By Ministerial Decree No 242 of 1997, the Ministry of

---

[81] JA Thomson, 'The status of plant biotechnology in Africa' (2004) 7(1 and 2) AgBioForum 9.
[82] James (n 28 above) Executive Summary.
[83] See, generally, eg MA Madkour, AS El Nawawy, and PL Traynor, *Analysis of a National Biosafety System: Regulatory Policies and Procedures in Egypt* (The Hague: International Service for National Agricultural Research, Country Report 62, 2000) 6.
[84] Ibid 8–11; and Z El Haleem and A Hamza, *Country Report Proposed by Egypt to FAO/WHO Global Forum of Food Safety Regulators, Marrakech, Morocco, 28–30 January 2002*) (available at <http//:www.fao.org/DOCREP/MEETING/004/AB424E.htm>, last accessed 14 December 2007).

Health prohibits the import of any foodstuff produced from or through GMOs unless their safety is confirmed.[85] GM plants or seed can be imported if previously approved for use in the country of origin and considered safe. It may therefore be suggested that the Egyptian biosafety regulatory framework remains general and fails to include the basic principles of the African Model Law.

4.4.2.3 Malawi

Malawi provides an exception to the general rule that national attitudes towards GMOs in Africa broadly reflect levels of technological development. Compared to South Africa and Egypt, Malawi does not have a developed biotechnology capacity or infrastructure, but has still regulated biosafety. Further, in July 2008 it adopted a National Biotechnology Policy which sees GM crops as the road to food security, wealth creation, and socio-economic development.[86]

The legislative framework under the the Biosafety Act (Act No 13 of 2002) ('Malawi Biosafety Act') provides that the development of GMOs, together with their field trial, commercial planting, manufacture, sale, or import, require a permit from the Ministry of the Environment. Simplified procedures apply in the case of, for example, research and experiment, and food aid.[87] The Ministry of the Environment is responsible for granting, renewal, amendment, suspension, or withdrawal of licences with regard to the import, development, production, testing, release, use, and application of GMOs. All the decision-making power is vested in the Minster of the Environment, with no obligation to consult experts or specialists in the field. The Science and Technology Act (Act No 17 of 2003) also provides that no one should be involved in activities involving GMOs without securing the consent of the Ministry of Science and Technology; but it is not clear whether authority is transferred to that Ministry or whether both the Ministry of the Environment and the Ministry of Science and Technology have concurrent authority.

Even if under the Malawi Biosafety Act the Minster of the Environment is to take into account the safety of GMOs or their products before issuing a licence,[88] there is no specific provision which deals with risk assessment, with the result that there can be no certainty if his decision requires a risk assessment at all (and, even if a risk assessment is required, there is no certainty as to the bases upon which risks are to be assessed). Indeed, the Malawi Biosafety Act does not even specifically oblige the applicant to supply detailed information for the purposes of risk assessment: all that is necessary is to provide basic information, such as a description of the product, which may not be sufficient for proper

---

[85] For the relevant legislation, see <http//:www.fao.org/DOCREP/MEETING/006/Y4839E/y4839e07.htm>, last accessed 8 May 2009).
[86] See, eg James (n 28 above) Executive Summary.
[87] Malawi Biosafety Act, ss 17 and 18. For criticism of the Malawi Biosafety Act, see, eg M Mayet, *Analysis of Malawi's Biosafety Legislation* (2004) (available at <http://www.biosafetyafrica.net>, last accessed 3 July 2009). [88] Malawi Biosafety Act, s 22(a).

evaluation of any risk. This stands in stark contrast to the African Model Law, which, as has been seen, requires all applicants to provide sufficiently detailed information for the making of a decision.

There are other areas where the Malawi Biosafety Act does not conform with the African Model Law. For example, the Minster has the power to waive any of the requirements for licensing (it again being unclear as to the grounds upon which this can be done). Further, no provision is made for public participation in the licensing process and/or decision-making. Instead, there is protection for any information related to the financial or business affairs of any person or undertaking involved in GMO activities, with disclosure of such information being a criminal offence.[89] Similarly, there is no specific mention of liability and redress and, in consequence, only the general civil law rules of liability will be applicable to GMOs and their products. Again, unlike the African Model Law, there is no specific mention of the need to take into account socio-economic impacts of GMOs in decision-making.

By contrast, s 26, which deals with packages, containers, and identifications of GMOs or their products, stipulates that GMOs should be clearly labelled pursuant to regulations prepared by the regulatory bodies involved; and, accordingly, this is one area in which some constraint is imposed. Further, the Minister is authorized to make regulations in relation to the advertisement of GMOs and their products, ostensibly to protect the public from misleading information.[90]

Nonetheless, it may be concluded that the Malawi Biosafety Act significantly departs from the African Model Law, despite the fact that it was issued after the adoption of the African Model Law. In addition, the treatment of risk assessment, risk management, and decision-making do not sit well with the Cartagena Protocol, which Malawi has recently ratified. This probably makes Malawi the country with the most pro-GMO regulatory regime in Africa, in that it grants very broad discretion to the Government, with very little room for participation by other stakeholders.

4.4.3 The second category: national regulations promoting GMOs with caution

4.4.3.1 Kenya

Kenya is one of the few African Countries to have embraced GM technology, which it considers a crucial tool for socio-economic development. From early times such a stance was adopted by political leaders, President Moi reportedly having written a letter in 2000 to the US President, Bill Clinton, to request assistance in developing modern biotechnology.[91] Kenya is also one of the few

---

[89] Ibid s 30.   [90] Ibid s 28.
[91] Letter from Kenyan President Moi to US President Bill Clinton, 21 August 2000 (available at <http://www.biotech-info.net/moi.html>, last accessed 13 March 2008).

African countries to allow food donations with few preconditions, and this has given rise to suspicion that food aid from the United States Agency for International Development (USAID) and WFP, especially maize, might be GM.[92] As at 2008, no commercial release had yet been reported;[93] but Bt cotton was stated to be in the final stages of field trials by May 2009.[94]

As first implemented, the biosafety regulatory system in Kenya was based on the Regulations and Guidelines on Biosafety in Biotechnology, issued in 1998 by the National Council for Science and Technology (NCST).[95] The authority for these measures was the Science and Technology Act 1980. A National Biosafety Committee (NBC) was set up by the NCST, which, *inter alia*, reviews GMO applications. The procedure under the Regulations was that applications to import, release, or conduct confined field trials of a GMO were submitted to the relevant institutional biosafety committee (IBC), where they were scrutinized to ascertain whether they complied with the Guidelines. They were then sent to the NBC for review and a recommendation was made to the NCST Secretary, who had the final decision. However, the Guidelines were not binding (as they were not issued as law), nor were they sufficiently detailed for practical implementation. The result was that several issues pertaining both to labelling and to the handling of GMO movements from laboratory to farm were left unregulated.[96] Further, the Guidelines did not cover food-safety issues.

In an effort to overcome this deficiency, Kenya has developed a comprehensive Biosafety Bill, which was finally approved on 13 February 2009 after fierce resistance from environmental NGOs and other interest groups.[97] This applies to all GMOs, with the exception of those which are pharmaceuticals for human use.[98] Before any decision is made, a new National Biosafety Authority (NBA) is required to take into account information submitted by the applicant, the risk assessment report, relevant representations from the public, and the socio-economic consequences arising from the impact of the GMO on the

---

[92] African Conservation Foundation, *Kenya: Government Accused of 'Secrecy' on GMO Technology Bill*, 17 July 2007 (available at <http://www.africanconservation.org/dcforum/DCForumID35/183.html>, last accessed 12 May 2009). [93] James (n 28 above) Executive Summary.
[94] *Kenyan Cotton Farmers to Adopt Bt Seeds*, 8 May 2009 (available at <http://greenbio.checkbiotech.org>, last accessed 3 July 2009).
[95] See, generally, PL Traynor and HK Macharia, *Analysis of the Biosafety System for Biotechnology in Kenya: Application of a Conceptual Framework* (The Hague: International Service for National Agricultural Research, Country Report 6, 2003); M Harsh, 'Formal and informal governance of agricultural biotechnology in Kenya: participation and accountability in controversy surrounding the draft Biosafety Bill' (2005) 17 Journal of International Development 661; and G Jaffe, *Comparative Analysis of the National Biosafety Regulatory Systems in East Africa* (Washington, DC: International Food Policy Research Institute, EPT Discussion Paper 146, 2006) 17.
[96] H Odame, P Kameri-Mbote, and D Wafula, *Governing Modern Agricultural Biotechnology in Kenya: Implications for Food Security* (University of Sussex: Institute of Development Studies Working Paper 199, 2003).
[97] *Kenya Approves GM After Years of Delays*, 18 February 2009 (available at <http://scidev.net/en/news/kenya-approves-gm-after-years-of-delays-html>, last accessed 8 May 2009).
[98] Biosafety Act 2009, s 3.

environment.⁹⁹ However, no detail is provided as to how such impacts are to be analysed and how they should affect the decision. Separate approval procedures are laid down for applications in respect of contained use, introduction into the environment, import, placing on the market, export, and transit.¹⁰⁰ The Act also provides that a risk assessment may be waived if there exists sufficient information or experience to conclude that a GMO or an activity involving a GMO does not pose a significant risk to the environment.¹⁰¹ Decisions are to be communicated to the applicant within 150 days of receipt of the application;¹⁰² and there is also provision for review and appeal.¹⁰³

The NBA is required to provide notice of each application for the release of GMOs into the environment by publishing information in at least two national newspapers, with the public granted 30 days to react.¹⁰⁴ As indicated, in making a decision, account must be taken of any relevant representations submitted by the public. That said, as the word 'relevant' suggests, not all representations must be taken into account, and it is difficult to ascertain upon which criteria this preliminary question will be resolved.

The Act also creates offences and penalties, but lays down no specific rules on liability and redress.¹⁰⁵ Accordingly, as in Malawi, it would seem that the general civil law rules of liability will also apply to GMOs. The Act requires labelling of GMOs, while leaving the details to future regulation.¹⁰⁶ Thus, as again in Malawi, there are significant departures from the African Model Law, reflecting the drive to promote modern biotechnology.

4.4.3.2 Cameroon

Cameroon is one of the minority of African countries which have put in place a biosafety regulatory system. Law No 2003/006 of 21 April 2003 to Lay Down Safety Regulations Governing Biotechnology in Cameroon ('Cameroon Biosafety Law') was issued in 2003.¹⁰⁷ Any activity in research, development, production, manipulation, and marketing of GMOs or their products requires approval from the competent national authority. The specific procedures are to be laid down by regulations.¹⁰⁸ The intentional release into the environment, contained use, import, export, sale, and placing on the market of a GMO or its products must be preceded by a risk assessment.¹⁰⁹ Any decision involving GMOs 'should take into account' (as opposed to being 'based on') the precautionary principle;¹¹⁰ but there is no clearly stated obligation to take into account

---

[99] Ibid s 29. [100] Ibid ss18–23. [101] Ibid s 28. [102] Ibid s 30.
[103] Ibid Part IV. [104] Ibid s 19. [105] Ibid ss 40–2. [106] Ibid ss 50–1.
[107] For criticism of the Cameroon Biosafety Law, see, eg M Mayet, *Explanation and Comments on the Cameroon Biosafety Law* (2004) (available at <http://www.biosafetyafrica.net>, last accessed 3 July 2009).
[108] Cameroon Biosafety Law, s 25. It may be noted that GMOs for human and animal pharmaceuticals are specifically included, with their own rules: ss 51–3. [109] Ibid s 20.
[110] Ibid s 18.

the precautionary principle in assessing the risks in relation to products of GMOs. Moreover, while risks are classified into four categories (Level 1: no risk; Level 2: minor risk; Level 3: slight risk; and Level 4: high risk), the details of this categorization remain opaque, since the criteria are to be determined by implementation decree.[111] Importantly, classification in this way seems to suggest that there would be different parameters for different levels of risk and that certain levels of risk may be accepted. The lack of clarity in the provisions relating to risk assessment is replicated in the requirement that a thorough study of the ethical and socio-economic impacts on the local population should be carried out before any deliberate release into the environment, in that the impact of such a study in the decision-making process is again not clearly articulated.

Under ss 11 and 37, liability for damages consequent upon the release of a GMO is borne by the 'user', defined as 'any person, institution or body (including companies) responsible for the development or preparation, production, experimentation, marketing and distribution of organisms presenting new traits'.[112] The liability rules apply only to GMOs, not to their products (where it would appear that the general civil law rules of liability will be applicable).

Any GMOs or their products intended for intentional release or marketing on the national territory must be packaged and labelled with the wording 'product based on genetically modified organisms' or 'contains genetically modified organisms'.[113] The packaging and labelling obligations should be carried into effect in accordance with specific supplementary rules defined by the competent authority. There is no requirement for public participation in decision-making, but the authorities should promote public awareness of GMOs;[114] and, although s 12 deals with confidential information obtained by the inspector or controller and the circumstances where such information should and should not be divulged to the public, there is no stipulation as to whether information supplied by the applicant should be considered confidential or made accessible to the public. Finally, ss 60–4 create offences and criminal sanctions for violation of different provisions of the law, and the penalties may extend to imprisonment for up to 10 years.

In conclusion, the Cameroon Biosafety Law draws some of its principles from the Cartagena Protocol (adopting an AIA procedure) and others from the African Model Law. Nonetheless, it remains in some senses a framework document and, although in principle applicable to all GMOs and their GMO products, it provides less-stringent requirements in relation to their products.

---

[111] Ibid s 6(3). [112] Ordinarily, this does not seem to exclude farmers from being liable.
[113] Cameroon Biosafety Law, s 49(1) and (2). [114] Ibid s 35.

4.4.4 The third category: restrictive national regulations

4.4.4.1 Tanzania

Tanzania put in place its National Biosafety Framework (NBF) in March 2005. The Ministry of the Environment is the National Biosafety Focal Point (NBFP) in charge of the approval of applications and general implementation of the NBF.[115] An NBC has been established, comprising representatives from government institutions and NGOs to provide technical advice for the NBFP. The Tanzanian Environmental Management Act 2004 (EMA) also foresaw specific regulation in this area, with the Ministry of the Environment empowered to regulate GMOs; but overall the regulatory regime is far from complete.

The scope of the NBF is generous.[116] Thus, it provides that the AIA procedure applies to a GMO, a product of a GMO, or any product derived from processing of a GMO. No distinction is made between different activities involving GMOs such as for contained use, contained field trials, or commercialization: all need a risk assessment in accordance with the AIA procedure. On the other hand, although both the NBF and EMA recognize GM risks to human health, they do not provide detailed risk assessment procedures or parameters for the assessment of such risks.

Perhaps most significantly, the NBF draws heavily on the African Model Law. For example, no approval is granted unless it is established that the GMO will contribute to sustainable development, does not have adverse socio-economic impacts, and does not undermine community knowledge and technologies. The fact that the particular GMO is safe in terms of the environment and human health is not sufficient for approval. Yet, it may also be mentioned that, for the purposes of decision-making, account should be taken not only of the potential risks, but also of the benefits of the GMO—a difficult balancing act. Further similarities with the African Model Law may be found in the requirement for disclosure of confidential business information in the public interest, while public participation is required at all stages of the biosafety decision-making process, whether this relates to contained use, contained field trials, or commercial release. It is also provided that public opinion as expressed through public consultations should be taken into account, although there may be some doubt as to the kind of public opinion that would impact upon or affect decisions.

Again in line with the African Model Law, the NBF lays down strict rules on labelling. It requires the use of specific wording, such as 'This product contains GMOs', when there is a GMO in the product, or 'This product may contain GMOs', when the existence of GMO in the product cannot be ruled out. The known reactions or side-effects of the GMO, if any, should also be

---

[115] See, generally, eg Jaffe (n 95 above) 18–19.
[116] See, generally, eg Government of Tanzania, *National Biosafety Framework for Tanzania* (Dar Es Salaam: Division on Environment, 2005).

indicated. Indeed, the NBF even adopts the strict liability regime of the African Model Law.

4.4.4.2 Ethiopia

Ethiopia is in the process of establishing a regulatory framework for GMOs, and the draft law has now reached an advanced stage.[117] As it currently stands, the most important biosafety principles in relation to scope, risk assessment and management, socio-economic impacts, public participation, confidential business information, labelling and traceability, liability, and redress are all taken directly from the African Model Law. Thus, the draft law applies to GMOs for direct release into the environment, for food, feed or processing, and for use as pharmaceuticals; it also applies to GM products.[118] Moreover, it covers the import, export, transit, contained use, release, transport, or placing on the market of any GMO or their products.[119] All the above activities require AIA or authorization as the case may be.[120] The overarching principle in risk assessment and decision-making is the precautionary principle, as stated not just in the Preamble but a separate article (Article 5). If the draft is approved in this form, Ethiopia will join the list of African countries that have adopted the provisions of the African Model Law almost in their entirety.

5. Conclusion

In an attempt to address the Africa-wide GMO controversy following the 2002 drought in southern Africa (when, as has been seen, several countries rejected maize food aid from the United States through fear that it contained GMOs), the AU Executive Council called on the Chair of the AU Commission to convene a meeting of experts and civil society organizations to develop proposals for an African Common Position on Biosafety for adoption by the AU.[121] Following this, in 2005 the AU Commission Chair appointed a High-level African Panel on Modern Biotechnology (APB) to look into the issue; and in 2007 the ABP produced its draft report entitled *The Freedom to Innovate: Biotechnology in Africa's Development*.[122] Emphasis was laid upon the fact that modern biotechnology had failed to take root in Africa, as with the Green Revolution, and suggested that, if there was a will to invest in human resources and infrastructure, African countries could reap the benefits offered by GM crops. The

---

[117] The author has served as one of the experts tasked with the responsibility of reviewing the 2005 draft law under the auspices of the Ethiopian Environmental Protection Authority; the draft is on file with the author. [118] Draft law, Art 2.
[119] Ibid Art 3. [120] Ibid Art 8(1).
[121] Decision of the AU Executive Council on Biosaftey EX.CL/Dec 26(III) (2003).
[122] C Juma and I Serageldin (lead authors) (Addis Ababa and Pretoria: AU and New Partnership for Africa's Development, 2007).

extensive use of modern biotechnology was advocated, with science-based risk assessment underpinning regulation. The draft report also called upon African countries to work together at regional level to enhance the scale of development and use of modern biotechnology. The draft report has not yet been approved by the AU and it remains unclear whether it will prove the basis of a pan-African position on biotechnology in general and biosafety in particular. That said, if approved, it will definitely represent a liberalization of policy as compared with the more defensive stance to be found in the African Model Law.

Whatever may transpire in the future, the discussion in this chapter has shown that, to date, Africa has remained largely ambivalent, if not completely hostile, towards GMOs. Indeed, a number of African countries have openly rejected GMOs, even at times of severe starvation; and, although a few African countries have been attempting GM-related research and development, only South Africa, Egypt, and Burkina Faso have proceeded to commercialization (and in the case of Egypt and Burkina Faso the proportion of GM crops is very limited).[123] Further, with the exception of South Africa, such limited research and development that has taken place has been initiated from outside Africa, with the result that the continent is far from owning the technology and using it for its own needs.

The regulatory approaches to GMOs in Africa are generally a reflection of this ambivalence, and of the research and development capabilities of the various countries. Importantly, the examination of emerging biosafety regulations in Africa has shown that legislative frameworks can broadly be grouped into three categories.

First, some African countries (such as South Africa, Egypt, and Malawi) have chosen to design their biosafety regimes in a liberal fashion to promote biotechnology. Their regulatory regimes emphasize science-based risk assessment and provide strong protection for confidential business information, while giving less attention to principles incorporated in the African Model Law (for example, public participation in decision-making, labelling and traceability requirements, and strict liability rules). Secondly, other African countries (such as Kenya and Cameroon) treat GMOs with caution and have based their biosafety regimes in large part upon the Cartagena Protocol. Indeed, compliance with international obligations would seem to be their main objective (even if some elements of the African Model Law have also been taken on board). In consequence, they tend to rely on the precautionary principle as enshrined in the Cartagena Protocol. Thirdly, inspired by the African Model Law, several African countries (such as Tanzania and Ethiopia) have adopted or are in the process of adopting a biosafety regime which subjects to stringent requirements all GMOs and their products and all activities involving GMOs or their products. A matter

---

[123] James (n 28 above) Executive Summary (in 2008 Egypt grew 700 hectares of Bt maize and Burkina Faso 8,500 hectares of Bt cotton).

of some interest is that several of the most chronically food-insecure countries belong to this group. Regulations would typically include: a requirement for risk assessment based on strict application of the precautionary principle; provision for public participation in decision-making; the inclusion of socio-economic impacts as a factor in decision-making; stringent labelling and traceability obligations; and strict liability rules. Not only are these measures restrictive, but they also create a regulatory burden such that it would be difficult for the counties concerned to have the capacity to enforce them.

Several concerns would seem to explain this widespread ambivalence of African countries towards modern biotechnology; and three may be emphasized. First, there is obvious concern as to the perceived risks associated with GMOs. However, while caution is certainly in order in this new field, African countries cannot afford to continue to be hostile to modern biotechnology. A fear of the unknown is useful only in so far as it forces these countries to take every possible precaution; it should not prevent their citizens from benefiting from technologies which that have gone through rigorous tests and which are widely used in countries that have some of the highest standards of environmental safety. If anything, Africa needs more proactive adoption of these advances in science in areas where the safety of products has already been proven. It cannot afford to miss the potential benefits of the safe use of modern biotechnology, so long as this fits into its agro-ecological and socio-economic needs. It is worrying that Europe, known for its high standards and the reluctance of its institutions to embrace GM crops, is approving certain GM varieties that are not allowed in countries where the increased yields could make the difference between life and death for large numbers of their citizens.

Secondly, as indicated, there is also concern as to the relevance of the technology to Africa, and this has two aspects: that little attempt has so far been made to develop GM varieties that address the real needs of Africa; and that the technology is owned by the private seed industry, often protected by intellectual property rights, so making the varieties expensive and inaccessible to poverty-stricken farmers in Africa. While such apprehension may be legitimate, the solution may not necessarily be rejection of the technology altogether. Rather, African countries should strive to own and make use of the technology in relation to their own needs in a safe and sustainable way by investing in public research and development for agriculture.

Thirdly, as again indicated, there is concern that adopting GMOs would result in the loss of exports to markets where there is even greater hostility to modern biotechnology. On the other hand, commercialization of GMOs has continued to increase in the last few years, and there now 25 countries growing GM crops (14 of these growing 50,000 hectares or more).[124] In view of

---

[124] Ibid.

this trend, the long-term economic wisdom of being GMO-free appears questionable.

There is much to be said, therefore, for the shift in policy at AU level, with the APB draft report taking a decidedly pro-GM approach. It may be reiterated that it is high time for Africa, almost the only continent still grappling with starvation and food insecurity, to gain ownership of safe and sustainable modern biotechnology, with emphasis on investing in public research and development. This is the important objective that GM regulatory regimes in Africa should strive to promote. It is also high time that development aid to Africa should focus on building the research and development capacities of African agricultural institutions, as opposed to food aid. A focus on food aid has not been able to bring about a lasting solution to the agricultural problems of Africa—it is capacity-building which will bring about sustainable agriculture in the long term.

10

The Coexistence of Genetically Modified and Non-genetically Modified Agriculture in Canada: A Courtroom Drama

Jane Matthews Glenn

1. Whereof What's Past is Prologue ...[1]

The Canadian prairies, and notably its central province of Saskatchewan, are a major breadbasket of the world. Images of endless fields of golden crops rippling in the wind and elevators bursting with grain awaiting shipment to world markets are part of our landscape and our heritage. For many years, wheat ruled supreme, but canola (or oilseed rape) joined its ranks as a major crop in the 1960s. Canola was developed for Canadian farmers by Canadian scientists in publicly funded research facilities. It became the Cinderella crop of the Canadian prairies and elsewhere, and its name (an acronym for 'Canada oil low acid') reflects its origins. However, this publicly provided resource has now been virtually privatized as transnational biotechnology companies made relatively simple genetic changes to it and patented these changes. Almost 80 per cent of Canada's six million hectares of canola are now planted with genetically modified (GM) seeds, and Monsanto dominates this market.[2] Development of GM wheat has been high on the biotechnology companies' agenda, but opposition to commercialization from the Canadian Wheat Board (CWB) (which fears the impact this might have on the international market

---

[1] William Shakespeare, *The Tempest*, Act 2, Scene 1.
[2] M Foster and S French, *Market Acceptance of GM Canola* (Australian Bureau of Agricultural and Resource Economics Research Report 07.5, prepared for the Australian Department of Agriculture, Fisheries, and Forestry, 2007) 17 (available at <http://www.abare.gov.au/publications_html/crops/crops_07/GM_Canola.pdf>, last accessed 10 July 2008); and Agriculture and Agri-Foods Canada, 'Canada: Grains and Oilseeds Outlook: 2008–09' (27 May 2008) (available at <http://www.agr.gc.ca/pol/mad-dam/pubs/go-co/pdf/go-co_2008-05-27_e.pdf>, last accessed 10 July 2008). See, generally, E Moore, 'The new direction of federal agricultural research in Canada: from public good to private gain?' (2002) 37(3) Journal of Canadian Studies 112.

for Canadian wheat generally) caused Monsanto to defer further development of GM wheat in 2004.[3]

The possibility, read likelihood, of commingling of GM and non-GM crops is now incontestable, and concerns about coexistence are front and centre in the debate about GM agriculture in Canada. Unlike Europe, however, where these concerns are being addressed by the development of appropriate regulatory regimes prior to the commercial release of GM seeds, in Canada they are being addressed outside the regulatory framework and after the release of GM seeds for cultivation. Why this difference? One reason might simply be timing, as Canada gave regulatory approval for unconfined release of canola in 1995 and commercialization commenced in 1996, the first country in which this occurred.[4] European regulators thus have the benefit of hindsight from this decade-long Canadian experience and its attendant problems of coexistence. A second reason might be the general enthusiasm with which the Canadian government greeted biotechnology and its concern not to fall behind other countries in adopting and promoting it.[5] This has meant 'concerted government effort to promote the biotechnology industry through policies that placed primary emphasis on technological innovation, economic competitiveness and a narrow framing of potential hazards as strictly scientific issues'.[6] A third reason might be the virtual exclusion of key players other than the biotechnology companies (such as

---

[3] D Berwald, CA Carter, and GP Gruère, 'Rejecting new technology: the case of genetically modified wheat' (2006) 88 American Journal of Agricultural Economics 432, 434. The CWB is particularly concerned about post-harvest commingling during shipping and handling: eg I Huygen, M Veeman, and M Lerohl, 'Cost implications of alternative GM tolerance levels: non-genetically modified wheat in Western Canada' (2003) 6 AgBioForum 169; WW Wilson and B Dahl, 'Costs and risks of segregating GM wheat in Canada' (2006) 54 Canadian Journal of Agricultural Economics 341; and T Demeke, DJ Perry, and WR Scowcroft, 'Adventitious presence of GMOs: scientific overview for Canadian grains' (2006) 86 Canadian Journal of Plant Science 1 (suggesting, at 18, that biotechnology companies could facilitate identification through incorporation of a specific, biologically neutral, DNA sequence in all GM plants—'a single identification tag system'—but that this has not been done). Canadian opposition to GM wheat has meant that it has not been released in North America, as Monsanto had agreed to US farmer demands that it be released simultaneously in Canada and the United States, or not released at all: Berwald, Carter, and Gruère (above) 433.   [4] Demeke, Perry, and Scowcroft (n 3 above) 10.

[5] Canadian grain farmers also willingly adopted the new technology, as the rapid increase in acreage planted attests. Müller attributes this willing adoption to a self-image of progressiveness: B Müller, 'Infringing and trespassing plants: patented seeds at dispute in Canada's courts' (2006) 48 Focaal-European Journal of Anthropology 83: 'Also, among the conventional farmers I talked to, the idea of progress was prevalent. It was part of their self-image that they had to outdo themselves every year, embrace new technologies, experiment with new crops that their neighbours didn't have'.

[6] E Abergel and K Barrett, 'Putting the cart before the horse: a review of biotechnology policy in Canada' (2002) 37(3) Journal of Canadian Studies 135, 136. See also Moore (n 2 above) (describing the public philosophy of the 1980s and 1990s as one of technological neoliberalism functioning in a context of fiscal restraint); and P Andrée, 'The biopolitics of genetically modified organisms in Canada' (2002) 37(3) Journal of Canadian Studies 162 (analysing language—eg manageable risk, familiarity, substantial equivalence—used to narrow scope of assessment to one that is science-based).

representatives of conventional and organic farmers, and shippers and handlers of Canadian grain) and of the public in general from the development and implementation of the regulatory framework, again in striking contrast to the United Kingdom.[7] A final reason might be that people were lulled about the possible impact of commingling by the fact that Monsanto had put in place an 'identity preservation program' to ensure the segregation of GM and conventional canola for export when GM canola was first commercialized, which it subsequently abandoned when Japan accepted the importation of GM canola.[8]

In the result, the Canadian regulatory framework takes what might be described as a hard-edged, market-driven approach to third party economic loss occasioned by GM. It has no specific provisions about coexistence, nor does it address the problems of economic and other fallout when cross-pollination or commingling occurs. It accepts the recommendation of the Canadian Biotechnology Advisory Committee (CBAC), apparently unsupported by any in-depth background studies, that general tort law suffices:

[i]n our view, Canadian law adequately addresses issues of liability and compensation for damages through the common law of negligence and the civil law of obligations, which are based on principles of accountability and responsibility. Specific provisions for the damages caused by the products of biotechnology, patented or not, are not required.[9]

The Committee expressed this opinion in spite of, and without addressing, misgivings stated earlier in the same report that the 'practical difficulties' of suing in negligence 'may make this remedy illusory'.[10] The practical difficulties identified in the report—proving a duty of care and a breach of that duty—are fundamental. Moreover, reliance on litigation as a dispute-resolution mechanism exacerbates the regulatory disadvantage already experienced by conventional and organic producers and other excluded key players. Empirical evidence demonstrates that 'repeat players' (governments and big businesses) are more successful in litigation than 'one-shotters' (individuals and small businesses): not only can they afford more experienced counsel but they can also pick and choose the cases they pursue, defend, settle, and appeal, which enables them to develop

---

[7] See, particularly, S Hartley and G Skogstad, 'Regulating genetically modified crops and foods in Canada and the United Kingdom: democratizing risk regulation' (2005) 48 Canadian Public Administration 305 (distinguishing between 'representative', 'functional', and 'participatory' democracy, and criticizing Canada on all three fronts). Even in contrast to the United States, Canadian risk-regulatory style has been described as 'closed, informal and cooperative' rather than 'open, adversarial, formal and legalistic': RS Turner, 'Of milk and mandarins: rBST, mandated science and the Canadian regulatory style' (2001) 36(3) Journal of Canadian Studies 107, 107–8.

[8] *Hoffman and Beaudoin v Monsanto Canada and Bayer Cropscience Inc* [2005] 7 Western Weekly Rep 665 (Sask QB) paras 21 and 82–8; and Saskatchewan Organic Directorate, Media advisory, 9 September 2004 (available at <http://www.saskorganic.com/oapf/pdf/press-rel-9sept04.pdf>, last accessed 10 July 2008).

[9] CBAC, *Patenting of Higher Life Forms and Related Issues* (June 2002) 17 (available at <http://cbac-cccb.ca/epic/site/cbac-cccb.nsf/en/h_ah00094e.html>, last accessed 6 March 2008).

[10] Ibid 14.

a stable of useful precedents and to influence policy decisions. As one author put it, '[i]n general terms, we must think of judicial decisions—particularly appellate decisions, with their stronger and more wide-ranging precedent-setting implications—as constituting a valuable resource for which a variety of parties compete'.[11]

In the absence of a regulatory structure adequately addressing questions of coexistence, therefore, these issues are being dealt with by the courts. Some of the court decisions are past, with *Monsanto v Schmeiser* and *Hoffman v Monsanto* (discussed below) being well-known outside Canada. The question now is, 'What is the next act to which these past cases might be prologue?'

2. What's Past ...

Both *Schmeiser* and *Hoffman* illustrate the propensity of GM genes to wander, which is at the heart of the problem of coexistence. The two cases have together strengthened the position of biotechnology companies, the first in recognizing their right to enforce patents over adventitious canola, and the second in denying their responsibility for damage caused.

2.1 Recognition of Biotechnology Company Rights: *Schmeiser*

Monsanto Canada Ltd v Schmeiser[12] is a very early case about the cultivation of GM crops in Canada, as the facts took place in 1997 and 1998, when GM canola was first introduced. Schmeiser, who had some 50 years' experience as a conventional canola farmer, noticed that some of the canola he had planted in the spring of 1997 from seeds saved from 1996 resisted spraying with Roundup. He then saved the seeds, inadvertently or otherwise, from the resistant plants and seeded some 1,000 acres with them the following year, 1998. Monsanto obtained evidence, one way or another, of the resistant quality of his crops. Unlike other farmers, Schmeiser did not accept an out-of-court settlement for the unwanted presence of GM canola and sign the usual letter agreeing to remain silent about it.[13] Monsanto therefore sued him for breach of patent and

---

[11] P McCormick, 'Party capability theory and appellate success in the Supreme Court of Canada, 1949–1992' (1993) 26 Canadian Journal of Political Science 523, 523; and C Rothmayr and A L'Espérance, 'Courts and the biotechnology revolution: policy-making in Canada, the USA and Switzerland' (paper presented at 2006 Canadian Political Science Association Annual Conference, Toronto, 2006). Both articles illustrate the thesis of M Galanter, 'Why the haves come out ahead: speculations on the limits of social change' (1974–75) 9 Law and Society Review 95.

[12] [2004] 1 SCR 902 ('*Schmeiser* SCC'); affirming (2002) 218 DLR (4th) 31 (Federal CA); affirming (2001) 12 Canadian Patent Rep (4th) 204 (Federal Ct, Trial Div).

[13] A Clark, 'The implications of the Percy Schmeiser decision', (2001) 26 Synthesis/Regeneration (available at <http://www.greens.org/s-p/26/26-08.html>, last accessed 18 January 2010). In a separate incident, GM canola again appeared in Schmeiser's fields in 2005; he again refused

was successful at all three levels—at trial (2001), on appeal (2003), and before the Supreme Court of Canada (2005).

The principal issue before the Supreme Court was the validity of Monsanto's patent in light of the Supreme Court's earlier decision (2002) in *Harvard College v Canada (Commissioner of Patents)* that higher life forms (in this case, cancer-prone mice) are not patentable.[14] This decision was reached by the narrowest of margins (5–4), and the seemingly contrary decision supporting the patent in *Schmeiser* was reached by an equally narrow margin (5–4). Much learned ink has been spilled over this decision[15] and, at the risk of oversimplification, both the majority and minority agreed that Monsanto's patent extended only to the genes and modified cells in the plants and not the plants themselves,[16] but they disagreed over whether Schmeiser had 'used' the patented object (since he had not sprayed with Roundup herbicide). The majority held that he had nevertheless used it since the genes and cells had 'stand-by' utility,[17] whereas the minority felt that this approach was tantamount to extending patent protection to the plant as a whole, which was contrary to the decision in *Harvard Mouse*.[18]

Monsanto's offer to pay the costs associated with removing it because he did not want to sign the accompanying non-disclosure agreement; instead, he brought a small claims action against Monsanto for the clean-up costs, and was awarded 660 Canadian dollars in March 2008, without any non-disclosure restraints: M Hartley, 'Grain Farmer Claims Moral Victory in Seed Battle', The [National] Globe and Mail, 20 March 2008, A3. The non-disclosure clause makes the number of such gene-wandering incidents difficult to estimate.

[14] [2004] 4 SCR 45 ('*Harvard Mouse*'). More precisely, the Supreme Court held that the process by which the oncogenic mice were produced was patentable, but that the end product of the process (the 'founder' mice and their oncogenic offspring) was not.

[15] See, eg N Siebrasse, 'The innocent bystander problem in the patenting of higher life forms' (2004) 49 McGill Law Journal 349; R Burrell and S Hubicki, 'Patent liability and genetic drift' (2005) 7 Environmental Law Review 278; P Cullet, '*Monsanto v Schmeiser*: a landmark decision concerning farmer liability and transgenetic contamination' (2005) 17 Journal of Environmental Law 83; AD Morrow and CB Ingram, 'Of transgenic mice and Roundup Ready canola: the decisions of the Supreme Court of Canada in *Harvard College v Canada* and *Monsanto v Schmeiser*' (2005) 38 University of British Columbia Law Review 188; Müller (n 5 above); M Phillipson, 'Giving away the farm: the rights and obligations of biotechnology multinationals: Canadian developments' (2005) 16 King's College Law Journal 362; B Ziff, 'Travels with my plant: *Monsanto v Schmeiser* revisited' (2005) 2 University of Ottawa Law and Technology Journal 493; J de Beer, 'The rights and responsibilities of ag-biotech patent owners' (2007) 40 University of British Columbia Law Review 343; and I Mgbeoji, 'Adventitious presence of patented genetically modified organisms on private premises: is intent necessary for actions in infringement against the property owner?' (2007) 37 Bulletin of Science, Technology and Society 314.

[16] This part of the decision is consistent with *Harvard Mouse*, and Monsanto had not in fact claimed a more extensive reach for its patent.

[17] 'Whether or not a farmer sprays with Roundup herbicide, cultivating canola containing the patented genes and cells provides stand-by utility. The farmer benefits from that advantage from the outset; if there is reason to spray in the future, the farmer may proceed to do so': *Schmeiser* SCC para 84.

[18] 'The plants containing the patented gene can have no stand-by value or utility as my colleagues allege. To conclude otherwise would, in effect, confer patent protection on the plant:' ibid para 160.

Two factors perhaps explain the difference between the two Supreme Court decisions. One is the differing burden of proof. *Harvard Mouse* was an appeal from a refusal to award a patent and the burden of proving patentability rested with the proponent of the technology, whereas *Schmeiser* was a challenge to the validity of an existing patent and the burden thus fell on Schmeiser to disprove patentability. The second is the change in composition of the nine-member Supreme Court. Two judges who formed part of the five-member 'unpatentability' majority in *Harvard Mouse* retired, and were replaced by two members who were in the 'patentability' majority in *Schmeiser*. This makes the decision of Madam Justice Arbour, writing for the minority in *Schmeiser*, particularly interesting, as she had been in the minority favouring patentability in *Harvard Mouse*. In other words, her conclusion of unpatentability in *Schmeiser* did not reflect an opposition per se to the patentability of higher life forms, but was rather, to her, a logically consistent application of the majority decision of unpatentability in *Harvard Mouse*.

Schmeiser was thus about the rights of biotechnology companies. What of their responsibilities?

2.2 Denial of Biotechnology Company Responsibilities: *Hoffman*

Hoffman and Beaudoin v Monsanto Canada and Bayer Cropscience Inc[19] was a class action commenced in 2002 by two Saskatchewan organic grain farmers against two biotechnology companies for compensation for the economic losses flowing from their inability to cultivate canola as a result of cross-fertilization from surrounding GM crops. The action was brought on behalf of certified organic producers in Saskatchewan, and was framed in tort (trespass, nuisance, strict liability, and negligence) and breach of statutory duty. After some procedural skirmishes, certification of the class action was denied by both the Saskatchewan Court of Queen's Bench (2005) and the Saskatchewan Court of Appeal (2007); leave to appeal to the Supreme Court of Canada was refused on 12 December 2007.[20]

The statutory requirements for certification of a class action in Saskatchewan are generally similar to those of other jurisdictions across Canada, and require a

---

[19] (2007) 283 DLR (4th) 190 (Sask CA) ('*Hoffman* CA'); affirming [2005] 7 Western Weekly Rep 665 (Sask QB) ('*Hoffman* trial').

[20] See, eg J Matthews Glenn, 'Footloose: civil responsibility for GM gene wandering in Canada' (2004) 43 Washburn Law Journal 547; MZP Olszynski, '*Hoffman v Monsanto Inc.*: looking for a generous approach to the elephant in the garden' (2005) 16 Journal of Environmental Law and Practice 53; K Garforth, 'When worlds collide: biotechnology meets organic farming in *Hoffman v Monsanto*' (2006) 18 Journal of Environmental Law 459; N Craik, K Kulver, and N Siebrasse, 'Genetically modified crops and nuisance: exploring the role of precaution in private law' (2007) 27 Bulletin of Science, Technology and Society 202; and H McLeod-Kilmurray, '*Hoffman v Monsanto*: courts, class actions, and perceptions of the problem of GM drift' (2007) 27 Bulletin of Science, Technology and Society 188. See also Phillipson (n 15 above) and de Beer (n 15 above).

plaintiff to satisfy the court that: a) the pleadings disclose a cause of action; b) there is an identifiable class; c) the claims of the class members raise common issues; d) a class action is the preferable procedure; and e) the plaintiff would fairly and adequately represent the interests of the class.[21]

Failure to satisfy the court of any one of the certification requirements means that the class action will not go forward. The trial judge in *Hoffman* held that none of the requirements had been met except for a possible cause of action under two environmental statutes,[22] and the three Court of Appeal judges unanimously agreed with this assessment (although they doubted even the existence of a statutory cause of action).

Perhaps the strongest ground for refusing to certify the class action was the determination that the plaintiffs would not 'fairly and adequately represent' the interest of the class because they had signed an agreement giving control of the action to the Saskatchewan Organic Directorate's Organic Agricultural Protection Fund. Smith J, at trial, saw this as 'a major issue that is sufficient, in my view, to dispose of the question';[23] Cameron JA, on appeal, labelled her analysis 'unimpeachable' and endorsed her conclusion:

> The representative plaintiff under The Class Actions Act has the responsibility to prosecute the lawsuit, once certified, in the interests of the members of the class. Their duty is akin to that of a fiduciary. They must have adequate knowledge and ability to instruct counsel and they must act in the interests of the members of the class. They are answerable to the Court for the adequate performance of these obligations. These are duties that cannot, in my view, be delegated to another party who is not answerable to the Court.[24]

As for the requirement relating to class composition, both levels of court agreed that, although the members of the class were objectively ascertainable,[25] the class was over-inclusively drawn both substantively (by including all organic grain farmers, not just those who had grown or intended to grow organic canola) and temporally (by including organic canola farmers operating before the adoption of standards prohibiting genetically modified organisms (GMOs) in organic canola). As the Court of Appeal put it, '[i]t is not as though the difficulty lies in an individual assessment of *damages*; it lies in an individual determination of *liability*'.[26] A determination that the class was not properly identifiable led in

---

[21] Class Actions Act, Statutes of Saskatchewan (SS) 2001, c C-12.01, s 6.
[22] Environmental Assessment Act, SS 1979–80, c E-10.1; and Environmental Management and Protection Act, SS 2002, c E-10.21.   [23] *Hoffman* trial para 330.
[24] Ibid para 337.
[25] The class was described as composed of 'all organic grain farmers in Saskatchewan who were certified organic farmers at any time between January 1, 1996, and the date of certification as a class action by any one of six named private certification organizations': ibid para 2; certification standards restricting GMOs 'were only introduced gradually, over a number of years following 1995 ... [and thus] vary widely from year to year and from certifier to certifier': ibid para 218.
[26] *Hoffman* CA para 85.

turn to conclusions that the claims of the class members did not raise common issues[27] and that a class action was therefore not the preferable procedure.

The bulk of the two judgments addressed the thorny substantive issue of whether the pleadings disclosed a cause of action in tort. The usual test to determine this in a certification application is similar to the test applied in a motion to strike pleadings, which is that the defendant applicant demonstrate that it is 'plain and obvious' that no cause of action exists.[28] However, the Court of Appeal in *Hoffman* did not think this test appropriate in a certification application where the burden is on the plaintiff, and favoured a 'plausible basis' test:

there is in our judgment no more effective and balanced and functionally appropriate way of setting the tenor and tone of the matter than to expect the representative plaintiff to satisfy the judge that there exists a plausible basis in principle and presumed fact for supposing the defendant to be held liable.[29]

The trial judge, for her part, thought that the novelty of the claims required the plaintiffs to show 'a reasonable prospect for success'.[30] Whether or not these formulations are more demanding than the 'plain and obvious' test usually required is open to debate.

The claims in trespass, nuisance, or strict liability (under *Rylands v Fletcher*[31]) were all relatively readily ruled out for the same general reason that the immediate source of the damage was the activities of neighbouring farmers in planting GM crops on their lands, not the sale of GM seeds by the biotechnology companies.[32] The claim in negligence was examined exhaustively, applying the principles set out by the House of Lords in 1978 in *Anns v Merton London Borough Council*[33] and refined by the Supreme Court of Canada in 2001 in *Cooper v Hobart*[34] and *Edwards v Law Society of Upper Canada*.[35] Cameron JA on appeal agreed with the trial judge that the pleadings did not disclose a relationship of sufficient proximity to give rise to a duty of care and that, in any event, the fact that the defendants had received Canadian government regulatory approval for the unconfined release of their GM canola varieties prior to release

---

[27] 'Bearing in mind the need for a rational connection between the class, the cause or causes of action, and the common issues, it is clear that some of the members of the class would not share the alleged cause or causes of action and that accordingly there would be no common issues to be resolved in relation to them': ibid para 87.
[28] *Hollick v Toronto (City)* [2001] 3 SCR 158 para. 25.
[29] *Hoffman* CA para 53. The 'plain and obvious' test in a certification application requires the plaintiff to prove what is in effect a double negative, that it is *not* plain and obvious that the facts do *not* disclose a cause of action. [30] *Hoffman* trial para 36.
[31] (1868) LR 3 HL 330. [32] *Hoffman* trial paras 89–133; *Hoffman* CA paras 63–4.
[33] [1978] AC 728 ('*Anns*'). In other words, Canadian courts have not followed the House of Lords' subsequent rejection of its own approach in *Murphy v Brentwood DC* [1991] AC 398.
[34] [2001] 3 SCR 537 ('*Cooper*').
[35] [2001] 3 SCR 562 ('*Edwards*'). See also J Neyers, 'Distilling duty: the Supreme Court of Canada amends *Anns*' (2002) 118 Law Quarterly Review 221.

was 'a powerful policy reason for not fastening them with the duty of care as pleaded'.[36]

Because the Supreme Court refused leave to appeal, the case of *Hoffman* is now closed. What might be the next act?

3. ... Is Prologue

Several possible recourses are open to affected non-GM farmers in light of the decision in *Hoffman*. One possibility might be to sue neighbouring GM farmers for the unwanted presence of GM genes in non-GM fields. However, this hardly seems to be an acceptable option, not just because of the difficulties of proof (which neighbouring GM farmer is responsible?[37]) and the difficulty of collecting damages, but mainly because of the devastating effect such suits would have on the social fabric of a community (already sorely tried by Monsanto's enforcement tactics).[38]

As for possible further suits against the biotechnology companies, one option might be to look outside tort law entirely and to explore the suggestion of Elizabeth Judge to look instead to patent law as a source of liability—a liability '*as* intellectual property owners and grounded *within* intellectual property law'.[39] Another option might be to commence a second class action in tort against the biotechnology companies, after tweaking the procedural aspects relating to class composition and role of the plaintiffs to respond to the criticisms in *Hoffman* and after reflecting on the courts' analysis of the causes of action. Their analysis has not been approved by the Supreme Court of Canada,[40] and leave could have been refused solely for procedural reasons (the circumscribed role of the representative plaintiffs, for example). In particular, it is not clear how the tests applied by the two lower levels of court to determine whether the pleadings disclosed a cause of action—that is, 'reasonable prospect

---

[36] *Hoffman* CA para 60. For a discussion of the impact of prior regulatory approval in the European context, see Chapter 8.
[37] This raises the difficult issue of multiple tortfeasors, although courts sometimes relax the traditional 'but for' test of causation if necessary to do justice: see, eg *Shell v Farrell* [1990] 2 SCR 311; *Fairchild v Glenhaven Funeral Services Ltd* [2002] UKHL 22, [2002] 3 WLR 89; and *Barker v Corus (UK) Ltd* [2006] UKHL 20, [2006] 2 AC 572. [38] Müller (n 5 above)
[39] E Judge, 'Intellectual property law as an internal limit on intellectual property rights and autonomous source of liability for intellectual property owners' (2007) 27 Bulletin of Science, Technology and Society 301, 301 (emphasis in original) (advocating a principled application of the 'doctrine of exhaustion' based on a sharp distinction between the different rights in intellectual property and physical property).
[40] The Supreme Court does not give reasons in leave applications; and it recently clarified (in *Western Bank v Alberta* [2007] 2 SCR 3 para 88) that 'refusal of leave should not be taken to indicate agreement with the judgment sought to be appealed from any more than the grant of leave can be taken to indicate disagreement. In the leave process, the Court does not hear or adjudicate a case on the merits'.

for success' (trial) and 'plausible basis in principle and presumed fact' (appeal)—measure up to the usual 'plain and obvious' test. Nonetheless, the existence of prior regulatory approval remains a difficult, but not impossible, hurdle in such an action.

The hurdle of prior regulatory approval suggests another possible course of action, that of a suit against the government for negligence in regulating and controlling GM technology, particularly in failing to provide adequately for the coexistence of GM and non-GM agriculture. This will be explored in more detail in the rest of this chapter, looking first to the general principles that would apply and then illustrating these principles with several class actions currently before Canadian courts relating to bovine spongiform encephalopathy (BSE, or 'mad cow disease').[41]

3.1 Principles of Liability: *Anns*, Canadian-style

Canada, like other common law jurisdictions, has moved away from the historical position of Crown immunity and adopted legislation subjecting governments and government agencies to the same rules, generally speaking, of tortious liability as private persons.[42] This means that in an action against the Crown for negligent regulation of GM technology, the courts would apply, as in *Hoffman*, the principles set out in *Anns* as refined in *Cooper* and *Edwards*. *Anns* applies a two-stage test to determine the existence of a duty of care in relation to the damages suffered.

3.1.1 *Prima facie duty of care: the first stage in* Anns

The first stage of the *Anns* test looks at factors arising from the relationship between the parties to determine whether it gives rise to a prima facie duty of care. This requires that the plaintiff demonstrate not just foreseeability of harm but also proximity. Proximity describes a relationship that is 'close and direct ... such that it is just to impose a duty of care in the circumstances',[43] one which is 'of such a nature that the defendant may be said to be under an obligation to be mindful of the plaintiff's legitimate interests in conducting his or her own affairs'.[44] Proximity is presumed in a number of established or analogous categories, which include causing foreseeable physical harm to the plaintiff or

---

[41] The analysis draws upon J Matthews Glenn, '"Government wrongs": civil liability for GMO regulation in Canada' (2008) 18 Journal of Environmental Law and Policy 169.
[42] See, eg Crown Liability and Proceedings Act, Revised Statutes of Canada (RSC) 1985, c C-50; and PW Hogg and PJ Monahan, *Liability of the Crown* (3rd edn, Toronto: Carswell, 2000).
[43] *Edwards* para 9.
[44] *Odhavji Estate v Woodhouse* [2003] 3 SCR 263 para 48 (quoting *Hercules Management v Ernst & Young* [1997] 2 SCR 165 para 24).

plaintiff's property and misfeasance in public office.[45] The plaintiff must prove proximity in novel cases falling outside the recognized categories. The evidentiary burden is on the plaintiff at the first stage of *Anns*.

A particular difficulty in establishing a duty of care in negligence actions against public authorities is the requirement that proximity be grounded in the legislative intent of the governing statute. What does this mean? On the one hand, it is not necessary that the statute contain a specific legislative intention to impose a private law duty of care in favour of someone in the position of the plaintiffs (which would rarely be found) but, on the other, it is not enough that the statute simply create duties owed to the public. What is required is something between these two extremes—that the statute show, even implicitly, an intent 'to benefit a particular class of individuals through the provision of protective services',[46] even though it might also include duties owed to the public. Existence of a private law duty of care under the relevant legislation has been denied by the courts in a number of recent cases[47] and accepted in a more limited number of others.[48] One possible way of reconciling the cases is to suggest that those in which a private law duty is denied concern what might be called 'risk-reduction' measures and those in which a duty is recognized involve 'risk creation' measures.[49] Insistence on legislative intent as the source of proximity is one way of limiting the scope of government liability, and one would expect that the need for an intention to benefit a particular class of people would be stressed more in regard to 'risk-reduction' measures than 'risk-creation' ones. It is very similar to the differing treatment the law accords to acts of nonfeasance and acts of misfeasance. Approving GM technology for commercial release would fall at the 'risk-creation' end of the spectrum.

---

[45] Other examples include nervous shock, negligent misstatement, breach of duty to warn, negligent inspection of housing developments, and negligent execution of road maintenance policy: *Cooper* para 36, citing authority. 'The reference to categories simply captures the basic notion of precedent: where a case is like another case where a duty has been recognized, one may usually infer that sufficient proximity is present and that if the risk of injury is foreseeable, a *prima facie* duty of care will arise': *Childs v Desormeaux* [2006] 1 SCR 643 para 15.

[46] K Horsman and G Morley, *Government Liability Law and Practice* (Aurora, Ontario: Canada Law Book, 2007) 5:30:60; as quoted in *McMillan v Canada Mortgage and Housing Corporation* [2008] 3 Western Weekly Rep 505 (BC Sup Ct) para 52.

[47] See, eg *Cooper* and *Edwards* (class actions on behalf of investors suffering financial losses, brought against statutory financial regulators for negligent monitoring of trust accounts); *Williams v Canada (AG)* (2005) 257 DLR (4th) 704 (Ont Sup Ct) ('*Williams*') (class actions on behalf of Toronto SARS victims, brought against all levels of government for negligent control of spread of disease); and *Eliopoulos v Ontario* (2006) 82 Ontario Rep (3d) 321 (Ont CA), leave to appeal refused (SCC 2007) (one of some 40 individual actions by West Nile virus victims, brought against Ontario government for failure to implement plan to control spread of disease).

[48] See, eg *Ring v R* (2007) 268 Nfld & PEI Rep 204 (Nfld & Labrador Sup Ct, Trial Div) (one of several class actions on behalf of army base residents brought against federal government for negligent spraying of herbicides). [49] Matthews Glenn (n 41 above) 180–2.

3.1.2 Residual policy considerations: the second stage in Anns

The second stage of the *Anns* test looks at factors outside the relationship between the parties, and asks whether there are any residual policy considerations that negate the imposition of a duty of care. *Edwards* identifies these considerations as including 'the effect of recognizing a duty of care on other legal obligations, its impact on the legal system and, in a less precise but important consideration, the effect of imposing liability on society in general'.[50] *Cooper* asks, 'Does the law already provide an effective remedy? Would recognition of the duty create the spectre of unlimited liability to an unlimited class? Are there other reasons of broad policy that suggest that the duty of care should not be recognized?'[51] The evidentiary burden shifts to the defendant at the second stage of *Anns*.

The main residual policy consideration invoked to deny liability in actions against the Crown is the general principle of maintaining separation between the activities of the legislature and the judiciary, coupled with the inherent difficulties of having judges decide questions involving scarce resources. This concern is captured in the oft-applied distinction between 'policy' decisions, which are said to be immune to judicial scrutiny, and 'operational' activities, for which government authorities can be held responsible.[52] The policy/operational distinction also serves to reinforce the proximity requirement:

[s]tatutory duties of Ministers to make policy decisions are quintessentially duties owed to the public and not to private individuals but, as the cases where liability is found in respect of operational decisions illustrate, an implementation of decisions made in the exercise of such powers or duties can create a relationship of proximity with persons affected.[53]

A main difficulty with the policy/operational distinction is to determine where to draw the line in a given case. The Supreme Court rightly warns in *Just v British Columbia* that 'complete Crown immunity should not be restored by having every government decision designated as one of "policy"'.[54] Where might the line be drawn in an action against the government for negligent regulation of GM technology?[55] Canada follows an American 'product-based' approach to the

---

[50] *Edwards* para 10.   [51] *Cooper* para 37.

[52] Ibid para 38 ('The basis of this immunity is that policy is the prerogative of the elected legislature'); and see also J-D Archambault, 'La distinction politique-opérationnel fournit-elle une assise judiciaire solide à la responsabilité civile de l'État consécutive à une décision de politique administrative?' (1999) *Revue du Barreau* 579.   [53] *Williams* para 72.

[54] [1989] 2 SCR 1228, 1239.

[55] For a description of the Canadian regulatory framework, see M Phillipson, 'Are genetically modified crops in Canada under-regulated?' (2008) 18 Journal of Environmental Law and Practice 195; and JA Chandler, 'The Regulation of Genetically Modified Plants: Authorization of Unconfined Environmental Release' in Institute of the Environment, University of Ottawa, *Practicing Precaution and Adaptive Management: Legal, Institutional and Procedural Dimensions of Scientific Uncertainty* (Ottawa: SSHRC & Law Commission of Canada, 2005) (available at <http://www.ie.uottawa.ca/English/Reports/JBPP_Final_Report.pdf>, last accessed 22 February 2008). See also Royal Society of Canada, *Elements of Precaution: Recommendations for the Regulation of Biotechnology in Canada* (Report of the Expert Panel on the Future of Food Biotechnology, 2001)

approval of GMOs, rather than the European 'process-based' one. This approach assimilates GM plants as much as possible to their conventional equivalents, describing them simply as 'plants with novel traits' (PNTs). Within this general framework, responsibility for the safety of plants falls mainly on the Canadian Food Inspection Agency (CFIA). The CFIA applies a 'substantial equivalence' analytical framework, first determining the extent to which the novel plant is substantially equivalent, except for defined differences, to its conventional counterpart, and then concentrating the analysis on the defined differences (for example, the novel gene). If this is demonstrated to be harmless, the novel plant is predicted to have no greater impact on health or the environment than its traditional counterpart.[56] The CFIA limits the scope of assessment to scientific, especially agronomic,[57] matters and ignores economic, social, or ethical questions. One might argue that the decision to approve GM technology for release is a policy decision, but the choice of regulatory approach (product- or policy-based, scientific or holistic) is an operational decision.

Even if the decisions to use a product-based, substantial equivalence, agronomic-focused approach are held to be policy decisions and thus immune to judicial scrutiny, other aspects of the procedure are clearly operational. One is a perceived conflict between the CFIA's regulatory role and Agriculture and Agri-Food Canada's enthusiastic promotion of biotechnology. The extent of CFIA autonomy within the government department is moot,[58] and the recent dismantling of its advisory body, the CBAC, and its replacement by a more generally focussed Science, Technology and Innovation Council, housed in Industry Canada, gives pause for thought.[59] A second, clearly operational aspect of the procedure is possible undue industry influence and lack of transparency, with the attendant danger of 'regulatory capture'.[60] The CFIA relies heavily on data and information provided by the biotechnology companies themselves in

(available at <http://www.rsc.ca/foodbiotechnology/GMreportEN.pdf>, last accessed 24 March 2008); CBAC, *Improving the Regulation of Genetically Modified Foods and Other Novel Foods in Canada* (Report to the Government of Canada Biotechnological Ministerial Coordinating Committee, 2002) (available at <http://www.cbac-cccb.ca>, last accessed 19 October 2007); A Singh, 'Proceed with precaution: the statutory, legal and consumer influence on genetically modified foods in Canada' (2005) 4 Canadian Journal of Law and Technology 181; and Andrée (n 6 above).

[56] Royal Society of Canada (n 55 above) 182.

[57] Ie reproductive biology of species, possibility of cross-pollination, propensity to 'weediness', etc.

[58] Recent government intervention into the activities of other autonomous agencies—notably the Canadian Nuclear Safety Commission, the Canadian Wheat Board, and the Chief Electoral Officer—demonstrates the fragility of the CFIA's autonomy.

[59] The 18-member Council consists of a mix of scientists, federal deputy ministers, business leaders, and university presidents, all very busy people in their own right; five have legal training (two university presidents, two deputy ministers, and one business leader) (see <http://www.stic-csti.ca>, last accessed 6 March 2008). This is part of a 2007 'Science and Technology Strategy' strongly supportive of scientific and technological development. See Industry Canada, *Mobilizing Science and Technology to Canada's Advantage* (2007) (available at <http://www.ic.gc.ca/cmb/welcomeic.nsf/vRTF/PublicationST/$file/S&Tstrategy.pdf>, last accessed 24 March 2008).

[60] See, eg D Mortimart, 'The life cycle of regulatory agencies: dynamic capture and transaction costs' (1999) 66 Review of Economic Studies 920.

making its scientific assessment; this data is evaluated by CFIA scientists but not made available for peer-reviewing. The CFIA also relies on the biotechnological companies for post-release monitoring. The public is almost totally excluded from the pre- and post-release processes, and the information made available to it is sketchy.[61]

All of these aspects of the approval process are strongly criticized, and could found an action in tort against the federal government.

3.1.3 Liability for economic loss

A claim by non-GM producers against the government for losses suffered as a result of the authorization of unconfined release of GM seeds and plants would, as in *Hoffman*, be a claim for 'pure' economic loss, represented by their inability to receive the market premiums attached to organic and conventional products because the level of commingling exceeds the GM threshold levels adopted in response to market pressures. Canadian courts are more open than American and English courts to claims for pure economic loss,[62] but they are more circumspect about compensating for economic loss than for physical damage.

Pure economic loss claims are decided using the same two-stage *Anns* analysis as are general actions in negligence, with the court looking first to whether the damage is sufficiently foreseeable and proximate to come within the scope of the duty of care—that is, a duty of care to avoid causing foreseeable pure economic loss—and then to whether there are any residual policy considerations that might negate or limit the scope of the duty. The major hurdle at the first stage, here as well as in negligence actions generally, is to establish proximity.[63] As with general questions of duty of care, this is facilitated in economic loss claims by the judicial recognition of a number of established or analogous categories in which proximity is presumed. These categories include liability of public authorities.[64] The strongest second-stage, or policy, arguments against liability for pure economic loss have been a preference for contract over tort as a vehicle for attributing responsibility[65] and a concern about indeterminacy of liability. Contractual allocation of risk is not possible between parties who are not in a

---

[61] Chandler (n 55 above); and Hartley and Skogstad (n 7 above).
[62] See, eg B Feldthusen, *Economic Negligence: the Recovery of Pure Economic Loss* (4th edn, Toronto: Thomson Carswell, 2008); and J Stapleton, 'Comparative economic loss: lessons from case-law-focused "middle theory"' (2002) 50 University of California at Los Angeles Law Review 531.
[63] This would be a particularly difficult hurdle if courts accepted the approach of the trial judge in *Hoffman*, who suggested that economic losses are '*[b]y definition* . . . not the direct result of the defendant's action': *Hoffman* trial para 73 (emphasis added).
[64] Other categories are negligent supply of dangerous or defective goods or structures, negligent misrepresentation, negligent performance of a service, and relational economic loss (ie economic loss suffered by the plaintiff as a result of injury to another person's person or property): *Canadian National Rly v Norsk Pacific Steamships Co* [1992] 1 SCR 1021.
[65] See, generally, J Blom, 'Tort, contract and the allocation of risk' (2002) 17 Supreme Court Law Review 289.

contractual relationship, as is the case with the relationship between organic and conventional farmers, on the one hand, and federal regulatory authorities, on the other. Indeterminacy of liability—or 'liability in an indeterminate amount for an indeterminate time to an indeterminate class'[66]—similarly ought not to be a problem, although this issue is compounded in class actions, as *Hoffman* shows, where the plaintiff must demonstrate an identifiable class whose claims raise common issues.

Convincing the courts to accept a claim for pure economic loss is therefore an additional obstacle facing a plaintiff like Hoffman in a suit against the government for negligent regulation and control of GM technology. The plaintiff's position is strengthened by the recognition of 'liability of public authorities' as one of the established or analogous cases in which proximity is presumed. But the issue of indeterminacy is an important residual policy consideration which must be overcome. The amounts claimed are often very large, and courts are generally loath to render decisions affecting government spending priorities in a significant way.

3.2 Example of Possible Government Liability: Mad Cow Disease and *Sauer*

Sauer v Canada (Attorney General)[67] is one of four class actions born out of the recent BSE crisis in Canada.[68] It is a case that warrants close watching, as the facts and issues echo those of *Hoffman*.

On 18 May 2003 a single cow in Alberta was diagnosed with BSE, leading to the immediate closing of international borders to Canadian cattle and beef products, with 'catastrophic economic consequences for the commercial cattle industry in Canada'.[69] The source of the BSE was eventually attributed to cattle feed manufactured by Ridley (a Canadian subsidiary of an international feed conglomerate based in Australia) incorporating ruminant meat and bone meal (RMBM), as was permitted by Canadian regulation at the relevant time. The approximately 135,000 Canadian cattle producers responded in April 2005 by filing four class actions against both the feed manufacturer and the federal government—in Quebec, Ontario, Saskatchewan, and Alberta. The Ontario action also encompassed cattle producers in the remaining provinces, and, in February 2008, it was enlarged to include Saskatchewan and Alberta producers. There are thus now two active BSE class actions, *Sauer* in Ontario and *Bernèche v Canada (Procureur Général)* in Quebec. The defendants applied to have the Quebec action stayed until final judgment in the Ontario action, but this

---

[66] An oft-cited statement of Cardoza J in *Ultramares Corp v Touche* (1931) 255 NY 170, 179.
[67] (2007) 225 Ontario App Cases 143 (Ont CA) ('*Sauer* CA'); affirming (2006) 79 Ontario Rep (3d) 19 (Ont Sup Ct) ('*Sauer* trial'), leave to appeal refused (SCC 2008).
[68] See generally <http://www.bseclassaction.ca/>, last accessed 9 January 2009.
[69] *Sauer* CA para 1.

... *Is Prologue* 269

application was denied by the Quebec courts both at trial and on appeal.[70] This is an interesting development, as Quebec's class action procedure is more favourable to plaintiffs than that of other provinces.[71] The Quebec action has in fact now been certified, with the causes of action being limited to gross negligence, *faute lourde* and bad faith.[72] A motion to strike pleadings outside these areas was recently granted.[73]

The actions allege negligence by the feed manufacturer for including RMBM in the feed and for failure to warn of its presence, and by the federal government for its negligence in regulation of the cattle industry. The regulatory negligence is said to consist of gross negligence in enacting a regulation under the Feeds Act[74] in 1990 (in other words, after the serious outbreak of BSE in the United Kingdom in the 1980s) that expressly permitted the continued incorporation of RMBM (including brains and spinal cords) in cattle feed,[75] and ordinary negligence in failing to enact a regulation prohibiting RMBM until October 1997.[76] The actions are for damages for pure economic loss, and the amount claimed exceeds 20 billion Canadian dollars.[77]

Both Ridley and the Crown brought pre-certification motions in *Sauer* to strike the pleadings for failure to disclose a cause of action. This required the defendants to demonstrate that it is 'plain and obvious' that there is no cause of action, and the defendants met with only limited success in discharging this burden.[78] The analysis followed the same two-stage *Anns* test as was applied at the certification stage in *Hoffman*.

The trial judge concentrated on the case against Ridley, and one of the holdings was to reject Ridley's argument that the 1990 regulation authorizing the incorporation of RMBM constituted 'a "radical defect" in the plaintiff's claim that serves as a bar to the action at the pleadings stage'.[79] The Court of

---

[70] *Bernèche v Canada (Procureur Général)* 2006 QCCS 3046 (CanLII), leave to appeal refused: *Ridley Inc v Bernèche* 2006 QCCA 984 (CanLII).
[71] Mainly because only plaintiffs may appeal certification decisions (Code of Civil Procedure, Revised Statutes of Quebec, c C-25, art 1010), and this appeal asymmetry is usually interpreted to encompass interim motions.
[72] *Bernèche v Canada (Procureur Général)* 2007 QCCS 2945 (CanLII).
[73] 2008 QCCS 2815 (CanLII) ('une lecture même minimaliste du jugement d'autorisation amène à conclure que le recours n'a été autorisé que dans la mesure où le requérant puisse établir *la grossière négligence, la faute lourde ou la mauvaise foi* des défendeurs et qu'en l'absence de telles allégations, le recours n'aurait jamais été autorisé': para 11 (emphasis added), leave to appeal refused: 2008 QCCA 1581 (CanLII). [74] RSC 1985, c F-9.
[75] Canada did not require cattle imported from the United Kingdom to be certified as free from BSE until 1987, and did not impose a ban on the importation of cattle from the United Kingdom and Ireland or monitor the already imported cattle until 1990; 191 head of cattle were imported from the United Kingdom between 1982 and 1990, and at least 80 of them could have entered the feed system: *Sauer* CA para 16. [76] Enacted under the Health of Animals Act, SC 1990, c 21.
[77] *Sauer* trial para 27.
[78] Ridley succeeded in having the pleadings concerning a failure to warn struck.
[79] *Sauer* trial para 53. The judge quoted the Supreme Court in *Ryan v Victoria (City)* [1999] 1 SCR 201 para 29: '[l]egislative standards are relevant to the common law standard of care, but the two are not necessarily co-extensive. The fact that a statute prescribes or prohibits certain activities

Appeal agreed, and expressly declined to follow the decision in *Hoffman* on this point: '[s]o far as *Hoffman* ... applies a similar government policy to negate a *prima facie* duty of care, I would respectfully decline to follow it'.[80]

As for the Crown's motion to strike, the trial judge simply held, without elaborating, that it was not 'plain and obvious' that the plaintiff's claims were based on policy rather than operational decisions.[81] The Court of Appeal's treatment was more extensive. At the first stage of *Anns*, it accepted the plaintiff's argument that proximity was demonstrated by 'the many public representations by Canada that it regulates the content of cattle feed to protect commercial cattle farmers among others', and held that it was not plain and obvious that the plaintiff's claim of a prima facie duty of care would not succeed.[82] The bulk of the appeal decision focused on the second stage of *Anns* and the possible existence of residual policy reasons negating the existence of a duty of care. The Crown argued that the two regulatory decisions—one to regulate in a certain way and the other not to regulate until a certain date—were legislative decisions which plainly and obviously could not attract tort liability. The plaintiff, as would be expected, responded that they were operational rather than policy decisions, and that, even if they were policy decisions, they were enacted in bad faith and therefore did not have the immunity normally attaching to policy decisions. Bad faith was described as 'encompassing legislative decisions so markedly inconsistent with the relevant legislative context that it cannot be concluded that they were performed in good faith'.[83] The Court of Appeal, like the trial court, felt that it was not plain and obvious that the plaintiff's pleadings on these points would fail.

In June 2007 the defendants sought leave to appeal the denial of the motion to strike to the Supreme Court. The Supreme Court stayed the application for leave until it had released its decision on the merits of a motion to strike in a somewhat similar case, *Holland v Saskatchewan*.[84] *Holland* is a class action brought by a number of game farmers against the provincial government for economic loss in relation to a monitoring programme designed to prevent the spread of chronic wasting disease (CWD) in elk herds. The plaintiffs successfully brought a public law action in 2004 against the government for acting outside its jurisdiction in requiring herd owners to sign a very broad release and indemnity clause in favour of the government, and downgrading the herd status (and thus the value) of all herd owners who refused to sign.[85] When the

---

may constitute evidence of reasonable conduct in a given situation, but it does not extinguish the underlying obligation of reasonableness'.   [80] *Sauer* CA para 48.

[81] *Sauer* trial para 91.   [82] Ibid para 62.

[83] Ibid para 64. See also the certification decision in *Bernèche v Canada (Procureur général)* 2007 QCCS 2945 (CanLII) para 100.

[84] 2008 SCC 42 ('*Holland* SCC'); reversing in part (2007) 281 DLR (4th) 349 (Sask CA); reversing (2006) 277 Saskatchewan Rep 131 (Sask QB).

[85] *Holland et al v Saskatchewan* (2004) 258 Saskatchewan Rep 243 (Sask QB).

government failed to revise the herds' status following this decision, the affected farmers sued the government in tort, notably negligence (as well as misfeasance in public office and intimidation). The government moved to strike the pleadings, and succeeded in having the claim of intimidation struck at trial and the claim in negligence struck on appeal. The Supreme Court, for its part, was of the view that the claim of negligence, as framed, was more properly a claim for breach of statutory duty (as the plaintiff's argument was that *ultra vires* actions are *ipso facto* negligent), which is not a recognized tort; but it held that the action could go forward for negligent failure to implement a judicial decision, as this was clearly an operational activity.[86] The claim of misfeasance in public office will also go forward, as the trial judge had granted permission to amend the pleadings and this was not appealed.

The Supreme Court released its decision in *Holland* on 11 July 2008 and refused leave to appeal in *Sauer* almost immediately thereafter (on 17 July). This meant that the Court of Appeal's decision in *Sauer* refusing the motion to strike the proceedings against the two defendants, Ridley and the federal government, stood. The trial court's decision in the ensuing application for certification was released on 3 September 2008, with the court holding, notably, that the defendants were precluded from arguing that the proceedings do not disclose a cause of action, as in *Hoffman*, because this had been 'conclusively and finally determined' in the motion to strike, and that the issue of Crown negligence was one that was common to the class.[87]

In the meantime, Ridley, although still denying liability, entered into a partial settlement with all plaintiffs under which its potential liability is capped at six million Canadian dollars, in return for which it agrees to co-operate with the plaintiffs (while still remaining as a defendant in the case), and this agreement has been approved by both the Quebec and Ontario courts.[88]

4. Conclusions

Conflict prevails over coexistence in Canada between GM and non-GM agriculture, and this conflict is being played out in the courts. Both *Schmeiser* and *Hoffman* favour GM agriculture, and *Hoffman*, in particular, demonstrates the difficulties facing non-GM farmers in trying to obtain redress through the courts for the presence of unwanted GM plants or seeds. An action by them, whoever the defendant, would almost certainly be a class action,[89] which could involve

---

[86] *Holland* SCC paras 9–10 and 14.
[87] 2008 CanLII 43774 (Ont Sup Ct) paras 7 and 43–53.
[88] Ibid; and *Bernèche v Canada (Procureur Général)* 2008 QCCS 2248 (Can LII). Under this agreement, the monies are to be paid into a trust fund, administered by a court appointee, to support the ongoing actions against the Crown. The text of the agreement is available at <http://www.bseclassaction.ca/pdfs/ridley_docs/Notice%20of%20Settlement%20Hearing.pdf>, last accessed 20 July 2008).   [89] Except actions against neighbouring GM farmers.

three separate procedural steps: a motion to strike the pleadings, certification of the class action, and a hearing on the merits. The need to prove a cause of action arises at each step: it must be properly pleaded at the first two steps[90] and properly proved at the third. If the cause of action is negligence, as would most likely be the case, its existence is assessed at all three procedural steps using the two-stage test in *Anns*, and this is the case both in regard to the existence of a duty of care and to responsibility for damages for economic loss. Proving proximity is a key issue in both cases.

Residual policy issues are raised at the second stage of *Anns*. Both the trial judge and the Saskatchewan Court of Appeal in *Hoffman* stressed the government's prior regulatory approval of the unconfined release of GM canola as a 'powerful policy reason'[91] for denying the existence of a duty of care on the part of the corporate defendants. But this is not necessarily fatal to a second attempt at obtaining redress from them, as the Ontario Court of Appeal in *Sauer* expressly declined to follow *Hoffman* on this point in the action against the feed manufacturer.[92]

Actions against the federal government for negligent regulation and control of GM technology raise particular difficulties within the *Anns* framework. A first difficulty would be to establish the necessary proximity to support a duty of care, which would require the plaintiff to demonstrate the existence of a private law duty of care under the applicable legislation. We have suggested that this is easier to do when the activity complained of is a 'risk creation' measure, as with the unconfined release of GM technology, than when it is a 'risk reduction' measure.[93] A second difficulty would be to establish what government activities are operational in nature and whether the government has acted negligently in regard to these, and to do so in the context of a regulatory process that is strongly criticized for its lack of transparency. Regulatory decisions are normally regarded as policy rather than operational in nature, and thus immune to a court's jurisdiction. However, the mad cow (*Sauer* and *Bernèche*) and wasting elk (*Holland*) cases open the door to arguing that, although the decision to regulate (for example, to permit the unconfined release of GM technology) is policy, the manner in which this decision is reflected in the regulation is operational. Moreover, both sets of cases stress that government immunity for policy decisions is relative, not absolute. It therefore does not encompass legislative decisions made in bad faith, and bad faith can extend to regulatory decisions 'so markedly inconsistent with the relevant legislative context that it cannot be concluded they were performed in good faith'.[94] A third difficulty would be to

---

[90] Although a decision that a cause of action is properly raised at the first stage cannot then be put into question at the second stage: see text to n 87 above.   [91] See text to n 36 above.
[92] See text to n 80 above.   [93] See text following n 49 above.
[94] See text to n 83 above.

establish liability for economic loss, with judicial concern about the impact of the decision on the public purse lurking in the background.

These difficulties suggest that looking to tort law to provide a remedy is misplaced, and that basic fairness to affected non-GM farmers calls for the provision of some other remedial regime. Insurance does not seem an option as the risks of commingling are undoubtedly too high to attract insurers, at least at a premium non-GM farmers could afford to pay. And why should they be required to pay, when others profit?[95] A fairer possibility would seem to be an industry-funded compensation scheme, as was suggested in 2000 by Canada's National Farmers Union (NFU), many members of which are organic farmers: '[t]he federal government must compel companies which own patents on GM seeds or livestock to set up contingency funds to compensate for product liability and legislate efficient and accessible mechanisms to enable liability claims to be effectively pursued'.[96] Financing such a scheme would undoubtedly be hotly contested.

Or perhaps one should turn one's attention away from glyphosate-resistant GM crops themselves, such as canola, to look more closely at the safety of the herbicides to which they are resistant. Recent studies cast doubt on the claim that they are benign to the environment, some even linking them to DDT in effect.[97] If almost five million hectares of the Canadian prairies are now being planted with GM canola, the same area—and more—is necessarily being sprayed with glyphosate herbicides, year after year. What can be the impact of this?

---

[95] See, eg United Kingdom Agricultural and Environment Biotechnology Commission (AEBC), *GM Crops? Coexistence and Liability* (London: Department of Trade and Industry, 2003) (available at <http://www.aebc.gov.uk/aebc/reports/coexistence_liability.shtml>, last accessed 25 June 2008): 'Aside from the unavailability of insurance, there is also an issue over whether it would be appropriate that farmers growing organic or exclusively non-GM crops should have to take out extra cover to protect themselves against the impact of other farmers' GM crops': ibid 89.

[96] NFU, 'Policy on Genetically Modified (GM) Foods' point 9 (available at <http://www.nfu.ca/policy/GM_FOOD_POLICY.misc.pdf>, last accessed 24 March 2008). The preamble describes the policy as an attempt 'to introduce precaution and prudence into a process of GM food proliferation driven by profit'; and point 8 of the policy stresses that it is 'unreasonable to allow genetic modification companies to privately reap profits and not require that they also assume all costs'.

[97] See, eg MTK Tsui and LM Chu, 'Aquatic toxicity of glyphosate-based formulations: comparison between different organisms and the effects of environmental factors' (2003) 52 Chemosphere 1189; RA Relyea, 'The lethal impact of Roundup on aquatic and terrestrial amphibians' (2005) 15 Ecological Applications 1118; J Marc, M Le Breton et al, 'A glyphosate-based pesticide impinges on transcription' (2005) 203 Toxicology and Applied Pharmacology 1; A Martínez, I Reyes, and N Reyes, 'Citotoxicidad del glifosato en células mononucleares de sangre periférica humana' (2007) 27 *Biomédica* 594; and L Mamy, B Gabrielle, and E Barriuso, 'Measurement and modelling of glyphosate fate compared with that of herbicides replaced as a result of the introduction of glyphosate-resistant oilseed rape' (2008) 64 Pest Management Science 262.

11
Genetically Modified Organisms in MERCOSUR

Rosario Silva Gilli

1. Introduction

1.1 The Process of Integration

MERCOSUR was formed through a regional process of integration, commenced in 1991 under the Treaty of Asunción. The original members were Argentina, Brazil, Paraguay, and Uruguay; and, since then, two associate members have joined, namely Bolivia in March 1997 and Chile in October 2006.[1] The Treaty of Asunción did not, as such, create a common market, but provided for staged implementation over quadrennial cycles. This process of integration, essentially economic and commercial in nature, has led to a customs union, as yet incomplete, which in the future will be based on a common frontier tariff. Fresh life was breathed into the project in 2000, when a declaration by the presidents reaffirmed the political will to take matters forward, but indicated that the realization of a common market remained a distant prospect.[2]

MERCOSUR, accordingly, comprises two major Latin American economies (those of Argentina and Brazil), together with two in the second rank (those of Paraguay and Uruguay). Argentina and Brazil also share the characteristic of being federal states, while Paraguay and Uruguay are unitary states. All four

---

[1] Venezuela has applied for full membership, but the application had not been ratified by Paraguay as at January 2010. Columbia, Ecuador, and Peru have entered into commercial agreements with the four full members of MERCOSUR, but are not subject to MERCOSUR rules. For the Treaty of Asunción see 30 ILM 1041 (1991); and both the Treaty and its Protocols are available at <http://www.mercosur/int/> last accessed 3 July 2009. See also, eg *Guide to MERCOSUR Legal Research: Sources and Documents*, available at <http://www.nyulawglobal.org/globalex/mercosur.htm>, last accessed 15 February 2009. See, generally, eg JAE Vervaele, 'MERCOSUR and regional integration in South America' (2005) 54 International and Comparative Law Quarterly 387.
[2] See, generally, eg J Carrera, *El Dilema del Mercosur: Avanzar o Retroceder?* (Argentina: Capital Intelectual, 2005).

countries are significant agricultural producers and exporters, but a swift look at the map of South America shows the clear disparity in their size.[3] Thus, Argentina extends to 2,780,400 kilometres^2, Brazil to 8,514,880 kilometres^2, Paraguay to 406,750 kilometres^2, and Uruguay to 176,220 kilometres^2.[4] This difference in size is replicated in their economies. In 2005 the GDP of Argentina was 183.3 billion dollars, that of Brazil 794.1 billion dollars, that of Paraguay 8.2 billion dollars, and that of Uruguay 16.8 billion dollars.[5] In addition, compared to Argentina and Brazil, Paraguay and Uruguay would seem less developed economically. Both producers groups and the number of consumers are smaller; and, with a smaller internal market, they are more dependent on world markets and their prices. Further, they export largely primary products.[6] Nevertheless, it must be recognized that entry by Paraguay and Uruguay into a broader international grouping has had advantages. Not least, they have acquired the ability to block initiatives, since consensus is in effect required for all decisions.[7]

Common characteristics of the four countries may also be highlighted. They all make intensive use of their natural resources, with large areas dedicated to livestock and crops; and the emphasis is on production rather than processing. Further, all four countries are in the course of development, seeking to increase exports of products best suited to such natural resources and physical advantages, with little reliance on skilled labour.[8] There are also particularly close similarities in terms of topography and production. For example, in the south of Brazil (the states of Rio Grande do Sul and Santa Catarina), the centre of Argentina (the provinces of Buenos Aires, Córdoba, Entre Rios, Santa Fe, La Pampa, and San Luis), and in Uruguay (in respect of the pampas), there are vast plains, few hills, and a temperate climate.

Finally, it should be observed that there are extensive common frontiers between members of MERCOSUR. Thus, Argentina and Brazil both border Paraguay, while Uruguay borders Brazil and is separated from Argentina by a river which in places may easily be crossed. This has inevitable repercussions in the case of the introduction and subsequent widespread cropping of genetically modified organisms (GMOs). In particular, their initial cultivation in Argentina has had the capacity to impact upon neighbouring members.

---

[3] For useful data on MERCOSUR countries, based on World Bank World Development Indicators, see House of Commons Trade and Industry Committee, *Seventh Report of Session 2006–07, Trade with Brazil and Mercosur* (HC 208) Vol I, Fig 2.   [4] Ibid.
[5] Ibid.
[6] P Krugman, 'La nueva teoría del comercio internacional y los países menos desarrollados' [1988] *El Trimestre Económico* 41.
[7] See, generally, eg L Bizzozero, *El Comportamiento de Socios Pequeños en el Inicio del Proceso de Integración* (Montevideo: Ediciones Trilce, 1992).
[8] G Schötz et al, *Innovación y Propiedad Intelectual en Mejoramiento Vegetal y Biotecnología Agrícola* (1st edn, Buenos Aires: Heliasta, 2006) 119–20.

1.2 The Legal Order of MERCOSUR

The Protocol of Ouro Preto of 17 December 1994 grants legal personality to MERCOSUR and defines its institutions. It must be emphasized, however, that the Protocol does not impose any legal sanction.[9] There are three bodies which may make inter-governmental decisions: the Council of the Common Market ('Consejo de Mercado Común'), which issues decisions; the Common Market Group ('Grupo Mercado Común'), which regulates by resolution; and the Commercial Commission ('Comisión de Comercio'), which operates through directives.[10] It is the combination of these decisions, resolutions, and directives which comprises the detailed legal framework. MERCOSUR is not yet constituted as a supranational organization, but Article 42 of the Protocol of Ouro Preto makes provision for its rules to become binding, since, when necessary, the members must incorporate the rules into their national legal order. The question then arises as to when this incorporation will take place; and any answer to that question must rest upon the nature of the process of integration and the rules that are emerging.

1.2.1 Provisions not requiring transcription into national law

One aspect would seem tolerably clear. As MERCOSUR is primarily economic and commercial in nature, the process of integration focuses on customs and trade between its members. In this context, use has been made of partial-scope agreements, under the umbrella of the Latin-American Integration Association ('Asociación Latinoamericana de Integración') (ALADI).[11] The core elements of these agreements apply directly in national legal orders without the need for transcription. In addition, there are other rules which do not require national implementation: first, those governing the organizational structure and internal workings of MERCOSUR (which enter into force as soon as they are approved); and, secondly, pre-existing provisions of national law which are in identical terms to MERCOSUR law.[12] Against this background, any rules governing GMOs at MERCOSUR level would seem to require national

---

[9] R Silva Gilli, 'Diez años de MERCOSUR: una visión Uruguaya' [2001] *Anuario 2001 Globalización Económica y Derechos en el MERCOSUR* 188.

[10] For the organizational structure of MERCOSUR, see the Treaty of Asunción, Arts 9–18.

[11] The statutory authority for such agreements is provided by the Treaty of Montevideo 1980, Art 7. All the Latin-American countries participate in the ALADI process of integration (a process which commenced in 1960 with the Latin-American Free Trade Association ('Asociación Latinoamericana de Libre Comercio') (ALALC). Despite the fact that ALADI has not reached its objective of securing integration, it has provided a framework for reinforcing commercial ties between members. From this were born the partial-scope agreements which bind their signatories (for these purposes, the countries of MERCOSUR). [12] CMC: Dec 20/02.

implementation (except where national law already covers the rules to be introduced).

1.2.2 National implementation

Even if MERCOSUR provisions are binding as from their approval, their entry into force is suspended until their incorporation within the law of each of the four members according to an established procedure. Entry into force is simultaneous, but not immediate. The mechanisms for incorporation are complex and there is no certainty as to when the rules become effective. In particular, there is no uniform criterion, common to all four members, as to the legal procedure. Nor are there criteria for deciding which rules will be incorporated, so that the decision rests with each member, with it being impossible to find out the reasons why incorporation did or did not take place.[13] As a rule nothing is done, through lack of sanction.

In case of conflict, resolution is to be achieved through ad hoc arbitration tribunals for each case, with appeal to a permanent conflicts tribunal, the Tribunal Permanente de Revisión. The competence of these tribunals extends to conflicts between members as to the interpretation, application, or compliance with the Treaty of Asunción and its Protocols or the rules of MERCOSUR. That said, their powers are limited, in that they do not have the ability to impose sanctions in respect of their decisions.[14] For example, the tribunal decided in favour of Uruguay and against Argentina in respect of the blocking of the international bridge over the River Uruguay, a key trade route for MERCOSUR.[15] Notwithstanding this decision, Argentina continues to block the bridge in protest against the installation of a cellulose factory by a Finnish company based in Uruguay and the tribunal does not have the ability to raise this blockade.

Consequently, the lacuna in MERCOSUR as an organization is not so much the absence of institutions to drive the process of integration (such as a court of justice), but the weakness of the institutions and mechanisms already created, from which flows a lack of legal certainty.

---

[13] R Silva Gilli, *Hacia la Protección del Ambiente en el MERCOSUR, Rio + 10* (Brasilia: ESPMU, 2002) 56.

[14] Nevertheless, temporary compensatory measures have been imposed in favour of a member which succeeds in arbitration and against a member who fails to comply with the decision: see, eg the Protocol of Olivos (available at <http://www.mercosur.int/>, last accessed 3 October 2008).

[15] Laudo del Tribunales Arbitrales Ad Hoc, 6 September 2006, 'Omisión del Estado Argentino en Adoptar Medidas Apropiadas para Prevenir y/o Hacer Cesar los Impedimentos a la Libre Circulación Derivados de los Cortes en Territorio Argentino de Vías de Acceso a los Puentes Internacionales Gral. San Martín y Gral. Artigas que unen la República Argentina con la República Oriental del Uruguay' (available at <http://www.mercosur.int>, last accessed 3 October 2008).

2. GM Crops in the Members of MERCOSUR

2.1 Statistics

The four countries which comprise MERCOSUR are key players in the GM revolution.[16] While the United States remains the largest producer of GM crops in the world, the countries of MERCOSUR occupy the second, third, seventh, and ninth places in terms of area cultivated (respectively Argentina, Brazil, Paraguay, and Uruguay). The various statistics available show, on the one hand, an overall increase in the amount of GM crops grown in MERCOSUR and, on the other hand, a 'shift to soya' in all four countries, with a gradual encroachment on land previously used for livestock.[17]

Table 1 Area Cultivated in 2008 by Country[18]

Country	Area cultivated	Type of crop
1st United States	62.5 million hectares	Soya, maize, cotton, oilseeds, squash, papaya, alfalfa, and sugarbeet
2nd Argentina	21 million hectares	Soya, maize, and cotton
3rd Brazil	15.8 million hectares	Soya, maize, and cotton
7th Paraguay	2.7 million hectares	Soya
9th Uruguay	0.7 million hectares	Soya and maize

2.2 Differences in GM Production

The question of GM crops is not susceptible of a 'yes-no' answer: to authorize or not to authorize. Rather, it turns on the question of regulation. In effect, all four members of MERCOSUR have opted in favour of GM crops and, accordingly, the issue today is the regulatory process for their authorizations.

---

[16] See, eg P Aerni, 'Stakeholder attitudes to agricultural biotechnology in developing countries: political polarization as a result of the private management of the public trust', Proceedings of the Seventh ICABR International Conference on Public Goods and Public Policy for Agricultural Biotechnology (Ravello, Italy, 29 June to 3 July 2003).

[17] See, eg J Dellacha et al, *La Biotecnología el MERCOSUR* (Argentina: Universidad Nacional del Litoral, 2003); and J Carullo, *Regulación de la Bioseguridad Agroalimentaria: Nuevos Escenaríos en el 2006* (Argentina: Universidad de las Naciones Unidas: UNU/BIOLAC, 2006).

[18] C James, *Global Status of Commercialized Biotech/GM Crops: 2008 (ISAAA Brief No 39)* (Ithaca, NY: International Service for the Acquisition of Agri-biotech Applications (ISAAA), 2009) Table 1.

2.2.1 Argentina, Uruguay, and Paraguay

Both regionally and globally, Argentina was at the forefront of the introduction of GM crops and has subsequently promoted their expansion. Indeed, it is the country which has argued most forcefully for their use and general commercialization.[19] It is also the second-largest producer and exporter in the world.[20] Since 1996, when glyphosate-tolerant soya was first made available on global markets, Argentina has authorized its commercialization; and today 99 per cent of the soya crop is GM.[21]

A different road has been followed in Uruguay, where the GM debate has largely been confined to government ministries. To begin with, the country adopted the same pro-GM posture as Argentina: cultivation developed without authorization, since there were no authorization procedures. However, through pressure from environmental groups, and through the indecision of the ministries concerned (those for health, the economy, agriculture, and the environment), a moratorium was subsequently established, as from July 2008 (although this is now finished). Notably, the debate between the social actors involved (farmers in favour and environmentalists against) had failed to produce a consensus.[22] This caused the Government to enact a decree in July 2008 to provide for prior authorization, an institutional framework, voluntary labelling, and the making available of information to the public should new circumstances arise.[23]

Paraguay has been even more cautious. Although the introduction of GM crops has not given rise to social discord, the academic world has been alert to their negative repercussions, looking to their impact on the environment and biodiversity, and to the economic dependence which they may create through restricted access to seeds and the imposition of payments for use of intellectual property rights.[24]

2.2.2 Issues specific to Brazil

Brazil has also been a strong proponent of GM crops and has now become the third largest producer in the world, after the United States and Argentina.[25]

---

[19] R Schnepf et al, 'Agriculture in Brazil and Argentina: Developments and Prospects for Major Field Crops' (Washington DC: United States Department of Agriculture, Economic Research Service, 2001).   [20] See James (n 18 above).
[21] See <http://www.argenbio.org/>, last accessed 3 July 2009.
[22] At the same time questions have been raised as to whether Uruguay should regulate agricultural biotechnology strictly and position itself as a country producing quality agricultural products. This would be a difficult objective to achieve since, as has been seen, Uruguay is a small country, with the result that international collaboration may be required.
[23] Decree 353/08, Official Journal of 28 July 2008.
[24] See, eg *En Paraguay los transgénicos son ilegales* (available at <http://www.grain.org/biodiversidad/?id=264>, last accessed 17 February 2009).
[25] J Wilkinson, 'GMOs: Brazil's export competitiveness and new forms of coordination' (2005) 1 *Estudos Sociedade e Agricultura* (English edition).

In 2007, out of 15 million hectares of soya, some 14.5 million hectares were 'Roundup Ready'. That said, it is in Brazil that the controversy over the introduction and regulation of GM crops has been the most intense. There has been no rush to authorization, this reflecting divergent positions held at federal and state level and the varying speeds at which the regulators are proceeding. Indeed, it is difficult to speak of a single regulatory approach towards agricultural biotechnology: different approaches are adopted by different authorities to take account of the particular circumstances which pertain in each state of Brazil.

The use of GM crops has also provoked conflict, as may be illustrated by cases brought before tribunals.[26] Thus, in 1996 Greenpeace instigated successfully an action to prevent the unloading of 'Roundup Ready' soya in a port of the state of Rio Grande do Sul; and in 1998, when the competent authority, the National Technical Biosafety Committee ('Comissão Técnica Nacional de Biossegurança') (CTNBio), issued a favourable report for commercial authorization of 'Roundup Ready' soya, this was successfully challenged by the Brazilian Institute for Defence of the Consumer ('Instituto Brasileiro de Defesa do Consumidor') (IDEC), on the basis that that there had not been a prior environmental impact assessment.[27] Further, in 1998 an action was brought to block the unloading of animal feed from Argentina containing GM maize, again in a port of Rio Grande do Sul. Having regard to the provisions of the Federal Constitution and the legislation of the state concerned, the federal tribunal, noting the presence of transgenic material, prohibited the importation of the disputed products and again accepted the necessity of a prior environmental impact assessment.[28]

3. The Regulation of GM Crops in the Countries of MERCOSUR

3.1 Common Characteristics

Regulation in the four countries of MERCOSUR shows different levels of development and importance in the respective national legal orders. A common characteristic is that all the regulatory frameworks are based upon pre-existing national rules governing seeds and crops and their regimes for registration.[29] With the exception of Brazil, the regulation of GM crops is carried into effect by decrees or resolutions enacted by the executive. The principal legislative focus in all four countries is the institutional structure for the creation, adaptation, or

---

[26] See, generally, eg J Bell, 'Brazil's transgenic-free zone', Seedling, September 1999 (available at <http://www.grain.org/seedling/?id=55>, last accessed 24 February 2009).
[27] Case 1998-34-00.027682-8; and see E Astorga et al, *Evaluación de Impacto Ambiental y Diversidad Biológica* (Gland, Switzerland: IUCN Environmental Policy and Law Paper No 64, 2007) 64–89.
[28] See ibid; and see also Case 1998–00.36170-4 (relating to technical advice from the CTNBio on GM cattle feed). [29] See, generally, eg Schötz et al (n 8 above).

co-ordination of multidisciplinary bodies open to both public and private participation. These bodies have been granted special competence in matters of technical evaluation of transgenic material and are located within ministries of agriculture, environment, and health, but retain a degree of autonomy. The ability of countries to assure independent studies and evaluations (predicated upon independent finance) would merit in-depth analysis. It is the main challenge for the countries of MERCOSUR, but it is not the main question here, where the focus is primarily legal. Nevertheless, it is necessary to highlight that the success or failure of the system of regulation depends more upon sufficient capacity, both technical and economic, to guarantee independence than upon laying down an ambitious body of law.

It is also necessary to highlight that a number of features which would seem fundamental for cereal-exporting countries await incorporation within a legal framework (for example, rules on coexistence, labelling, and traceability). Further, the rules which have been enacted remain weak, such as those relating to the regulation of authorizations prior to release into the environment, medium- and long-term evaluation, strategic planning, environmental concerns held by the public, and the protection of the rights of consumers and farmers.

3.2 Argentina

3.2.1 The general framework

Argentina was not only the first country to introduce GM crops, but also the first to impose a system of regulation. 1991 saw the creation of the National Advisory Commission on Agricultural Biotechnology ('Comisión Nacional Asesora en Biotecnología Agropecuria') (CONABIA) within the Ministry of Agriculture, Livestock, Fisheries, and Food ('Secretaría de Agricultura, Ganadería, Pesca y Alimentos') (SAGPyA').[30] Its role extends to matters of evaluation and consultation, both multidisciplinary and institutional in nature; and it includes representatives from both the public and private sectors. That said, no provision is made for representation by consumer organizations.[31] The regulation of GMOs is conducted under the umbrella of the regime governing agricultural production and, more particularly, the measures which govern seeds and the health and safety of agricultural production and veterinary products.[32] On the basis of these measures, the executive at national level legislates by way of resolutions which apply in all provinces.

---

[30] See <http://www.sagpya.mecon.gov.ar/new/0-0/programas/biotecnologia/index_en.php>, last accessed 3 July 2009. See also P Newell, 'Bio-hegemony: the political economy of agricultural biotechnology in Argentina' (2009) 41 Journal of Latin American Studies 27.
[31] Ministerial Resolution No 124.
[32] Decree-Law 6740/63; Law 20247/63; Decree 2183/91; and Law 13636/49.

The broad definition of seeds under Law 20247/63 is sufficiently wide to cover transgenic seeds, even though they are not expressly mentioned, since Article 29 brings within its compass all vegetable matter destined for sowing or propagation.[33] Such inclusion of GMOs is reinforced by the general purpose of the legislation, which, under Article 1, is to promote the efficient production and commercialization of seeds, to guarantee for farmers the identity and quality of the seed itself, and to protect ownership of phytogenetic innovations. Accordingly, the general regime governing seeds envisages their commercial exploitation, even when transgenic. Further, classification, registration, and the institution of competent authorities are all regulated by the same law, together with its subsequent detailed rules.

Since 1991, the SAGPyA has given its opinion on whether to authorize or reject requests for the import or export of seeds, having assessed whether there is compliance with requirements relating to: registration, hygiene, certification of origin for the variety concerned, and end use (whether that be for general release, breeding, or a trial). The registration procedure is twofold: first, there is a national register of 'cultivars'; and, secondly, since 2004 it has been compulsory for persons who deal with GMOs to enter their names on a separate national register.[34]

3.2.2 *The specific regime*

In addition to the general framework, GM seeds are subject to a complementary, yet specific, regime. Following the example of the United States, in Argentina the focus is directed to the characteristics and any identifiable risks in the biotechnological product, as opposed to the production process. When assessing the proposed use, and whether or not to grant a permit, the only considerations taken into account are risks to health, environment, and agricultural production.[35] In the plant sector,[36] Resolutions 39 of 2003 and 46 of 2004 constitute the key provisions for the evaluation of risk and the imposition of conditions upon any releases of GMOs, it being axiomatic that releases or trials without authorization are not permitted. These resolutions provide that, when a request is made for authorization to commercialize GM crops, the evaluation procedure requires the involvement of three different bodies at three different stages: the CONABIA; the National Service of Agrifood Health and Quality ('Servicio Nacional de Sanidad y Calidad Agroalimentaria') (SENASA); and the National Direction of Agricultural Food Markets ('Dirección Nacional de Mercados Agroalimentarios') (DNMA).

---

[33] See, generally, eg JP Kesan and AA Gallo, 'Property rights and incentives to invest in seed varieties: Governmental regulations in Argentina' (2005) 8 (2 and 3) AgBioForum 118.
[34] Resolution 46/2004.
[35] See <http://www.sagpya.mecon.gov.ar/new/0-0/programas/biotecnologia/index_en.php>, last accessed 3 July 2009.
[36] The animal sector is subject to a different regime (Resolution 57/2003).

The CONABIA evaluates on a case-by-case basis requests for the purposes of research, whether in a laboratory or in the field, this being the first step on the road to release into the environment. The evaluation is made by consensus and the provisions in question require observance of the precautionary principle, respect for confidentiality of information, and transparency[37] (although it is not easy to envisage great transparency where information is confidential). The first phase of the evaluation seeks to ascertain the probability of any negative impact on the farming eco-system. It is then necessary to discover whether any such impact is clearly different from that produced by non-GM crops. The second phase seeks to analyse all the information in the dossier on the GMO.

Although it does not amount to final authorization, a favourable decision by the CONABIA on the safety of release of a GMO into the environment acts as a precursor to other evaluations and the subsequent work of the other bodies. Accordingly, recommendation does not remove the need for authorization by the SAGPyA, but it does fix any conditions to be imposed and allows to be set in motion a procedure conferring considerable latitude, in that the one dossier is required for both the initial and subsequent releases of the GMO concerned. Moreover, the details to be included in the dossier are limited: these are the date of sowing and harvest, the area sown, and its location, so that the CONABIA can carry out its inspections. By contrast, if the decision is unfavourable, this brings to an end the evaluation process and prevents the procedure from continuing.

In addition to, and subsequent to, a favourable evaluation by the CONABIA, the SENASA (an agency of the SAGPyA) tests for toxicity in GM products destined for human food or animal feed. The criterion employed is that of substantial equivalence, as laid down by the Organisation for Economic Co-operation and Development in 1993 and taken up by the Food and Agriculture Organization (FAO).[38] This criterion allows GM crops or the food derived from them to be treated in the same way as those produced by conventional breeding, provided that they are shown to carry no risk. To determine whether or not this is the case, an evaluation is conducted of both the direct and indirect effects which may flow from use of the GM material. However, in line with the position of the FAO, the evolution of agricultural biotechnology makes probable that in the future there will be novel products without conventional counterparts, so rendering it impossible to maintain the criterion of substantial equivalence.[39]

---

[37] P Godoy, 'Marco Regulatorio de OGM en Argentina, SAGPyA', Conference on 'Building Capacity for Effective Participation in the Biosafety Clearing House (BCH) of the Cartagena Protocol on Biosafety' UNEP-GEF (Montevideo: May 2006).

[38] See, eg <http://www.fao.org/ag/agns/biotechnology_safety_equivalence_en.asp>, last accessed 1 March 2009. See also, in particular, FAO, *Genetically Modified Organisms, Consumers, Food Safety and Environment* (Rome: FAO, 2001); and FAO, *Report of the FAO Expert Consultation on Environmental Effects of Genetically Modified Crops* (Rome: FAO, 2003).

[39] See <http://www.fao.org/ag/agns/biotechnology_safety_equivalence_en.asp>, last accessed 1 March 2009.

The DNMA is responsible for the evaluation of economic risks connected with the introduction of GMOs. These risks could flow from direct loss—damages or reparation costs—or loss of competitiveness. It is apprehended that productivity is generally higher for GM crops, but, in the context of competitiveness, the DNMA does specifically give its opinion on the effect of introduction of the GMO on the Argentinian export market.[40] Finally, the SAGPyA, the executive authority in this context, authorizes any release of GMOs, subject to conditions imposed on a case-by-case basis; and the National Seed Institute ('Instituto Nacional de Semillas') (INASE) monitors the various requirements stipulated for commercialization of each seed, by means of samples after harvest, with a view to minimizing the risks of cross-contamination. Taking together this involvement of different bodies and the difficulties of implementation, it is hardly surprising that the whole process of evaluation can last for several years.

3.3 Brazil

3.3.1 Federal and state competence

In the case of environmental protection, pollution control in all its forms, the conservation of flora and fauna, the development of agricultural production, and public health, competence falls to be shared between the Union, the states, the Federal District of Brasilia, and the communes. The Union and the states have joint competence to legislate on liability for damage to the environment, claims by consumers, rights connected with the countryside, and the protection of health. In cases of such joint competence, the Union is restricted to laying down general rules, which it is then for each state to implement in detail. Where federal law conflicts with state law, federal law prevails. If there is no relevant federal law, the states enjoy full legislative competence.

3.3.2 State regulation in Brazil

Before the Biosafety Law of 2005,[41] each state could regulate the authorization, use, commercialization, and export of GMOs or products containing GMOs.[42] In consequence, Rio Grande do Sul, the most southern state of Brazil, made provision for notification with the public authorities and an environmental assessment in advance of any activity involving GMOs.[43] In 2003, Decree 42618/03 amended the earlier measures and, notably, removed the need for a

---

[40] MA Rapela, *Plantas Transgénicas Bioseguridad y Principio Precautorio* (Argentina: Universidad de la Plata, 2005) 125–6. [41] For which, see below.
[42] E Gudynas and G Evia, 'Brasil: la "política provisional" vuelve a liberar los transgénicos' (2004) (available at <http://choike.org/nuevo/informes/2227.html>, last accessed 1 March 2009).
[43] Law 9453/99 and Decree 39314/99.

prior environmental impact assessment. In 2002, the state of Santa Catarina, also situated in the south of Brazil, imposed a five-year moratorium on GM seeds and crops, and their derivatives.[44] This included an obligation to package and label in such a way that it was clear to consumers that transgenic techniques had been used in the production process. Further, in 2000, the state of Para had already banned for two years GM seed and cropping for the purposes of commercial use in human food and animal feed.[45]

Contrary to the position in Rio Grande do Sul, the state of Paraná has developed and maintained a 'GM-free' stance. Taking action against states producing GM soya, and applying its policy of zero tolerance towards GMOs, in 2004 it prohibited use of Paranagua (the key port of departure for exports from the centre-west of Brazil) for export of GM crops.[46] Paraná also banned the sowing and commercial use of GM soya until the end of 2006, when presidential decrees provisionally authorized GM crops. Adopting a slightly different approach, the states of Acre and Amapá restricted access to genetic resources in their territories.[47] At a broad level, safeguards were provided for the traditional knowledge of local communities and indigenous peoples, with recognition of their right to share in the economic and social benefits of access to such material. Further, in both states the legislative framework developed practices for the conservation and management of genetic resources and all activity in this sphere became subject to prior approval by the competent authorities.

3.3.3 The Biosafety Law

Brazil is the only Member of MERCOSUR which has specific rules for GMOs set out in a major statute. In 1995 it enacted the law which established the CTNBio;[48] and this remained in force until the enactment in 2005 of Law 111/05 ('Biosafety Law'), together with its implementing provisions in Decree 5591/05. These together reformed the regulation, registration, and security provisions applicable to GMOs and their derivatives. The Biosafety Law is based upon the Environmental Chapter of the 1998 Federal Constitution, which recognizes the right to an environment in ecological balance. On its approval, the legislator provided an, at least partial, answer to the problem of regulating GMOs, through an ambitious framework governing all GMOs and their derivatives. However, not all difficult issues are covered. Thus, while a system was implemented to regulate the introduction of GM crops, no provision was made for GM crops which were already being cultivated.

---

[44] Law 12128/02 and Decree 6092/02.   [45] Law 6328/00.
[46] Personal interview with Roberto Requiâo, Governor of the State of Paraná, Buenos Aires, 12 March 2004.   [47] See, respectively, Law 1238/97 and Law 388/97.
[48] Law 8974/95.

The Biosafety Law and its implementing Decree establish the regime applicable to genetic modification of living organisms (whether as receptors or donors). This regime covers control mechanisms for their manufacture, cultivation, production, handling, transport, transfer, import, export, storage, research, commercialization, consumption, and release into the environment, together with disposal of their residue. Significantly, any dealings with GMOs cannot be undertaken by individuals: rather, such activities are restricted to institutions or companies.

At an institutional level, the system is organized around different bodies, both public and private, and a combination of the two: the National Biosafety Council ('Conselho Nacional de Biosegurança') (CNBS); the CTNBio; and the Internal Biosafety Commission ('Comissão Interna de Biossegurança') (CIBio). The CNBS, a creature of the Presidential Council, is charged with the formulation and execution of national biosafety policy; and it lays down guidelines for administrative action by the Federal bodies with competence in this arena. In addition, it conducts analysis, at the request of the CTNBio, of the appropriateness, socio-economic importance, and national interest of applications for commercial use of GMOs and their derivatives. To it falls the final decision whether or not permission should be given for commercial use, on the basis of recommendations by the CTNBio and the other bodies involved in the approval process. In like vein to the CONABIA in Argentina, the CTNBio (as reformed by the Biosafety Law) is a multi-disciplinary body which provides expert advice to the Federal Government. It adopts a position on the biosafety of GMOs on a case-by-case basis and, where its advice is against approval, the application fails. More precisely, it enjoys responsibility for research, classification of the level of risk, safety measures, and any restrictions. It also lays down the criteria for assessment and monitoring of any risks associated with a GMO and its derivatives; and it issues biosafety certificates, prerequisites for any activities involving GMOs and their derivatives undertaken by public or private institutions in a laboratory. The CNBS is a political body, whereas the CTNBio is understood to perform an expert function.[49] The ramifications of reports by the CTNBio have generated considerable debate.[50]

Civil society has sought the right to intervene in the regulatory process and, in particular, in the drawing-up of clear rules to govern health and environmental risks, matters falling with the province of the CTNBio. Such demands have been channelled through a number of non-governmental organizations (NGOs), concerned at the lack of control mechanisms to stop GM crops cross-contaminating conventional and organic crops. In response, there have been

---

[49] See Gudynas and Evía (n 42 above).

[50] See, eg AH Benjamín, *Engenharia Genetica: Implicaçoes Ambientais e na Proteçao do Consumidor—Vol 3* (Curitiba: Livro de Tesses do 13º Congresso Nacional do Ministério Público, 2000) 182 (cited by Astorga et al (n 27 above) 85).

instances of public consultation, notably consultation with the National Health Surveillance Agency ('Agência Nacional de Vigilância Sanitária') (ANVISA), the objective being to define and bring transparency to the rules on risk assessment employed by the CTNBio. By contrast, the CIBio comprises institutions which carry out genetic engineering or which conduct research into GMOs. Its responsibility extends to individual projects, drawing up control programmes in accordance with models and norms established by the CTNBio.

The Biosafety Law has been challenged by the Procurator-General before the Federal Supreme Court, on the ground that it is unconstitutional, but the court had not yet reached a decision by 2009.[51] The central thrust of the claim is directed to Article 25 of the Federal Constitution, upon which the Biosafety Law is based. It is argued to be unconstitutional that the CTNBio has exclusive competence for risk evaluation,[52] this being contrary to the powers the Federal Constitution confers on federal bodies in matters of environmental protection (for example, the power to require the undertaking of a prior environmental impact assessment).

3.3.4 Labelling obligations

The Biosafety Law imposes labelling obligations in respect of food or food ingredients destined for human or animal consumption once they contain GMOs or qualify as products derived from GMOs.[53] On the other hand, this labelling requirement does not appear to be taken seriously in practice and there is no evidence of it being enforced. In addition, at an international level, Brazil has ratified the Cartagena Protocol to the Convention on Biological Diversity, also proceeding to its incorporation within the national legal order; and this offers the possibility of using the labelling term 'may contain' living modified organisms.[54]

3.4 Paraguay

In 1997, Paraguay commenced a process of reform in the field of biosafety, setting up the Biosafety Commission ('Comisión de Bioseguridad') (COMBIO).[55] Its mandate was to carry out analysis and to respond to questions in relation to the introduction of GMOs, with particular reference to field trials, research, releases into the environment, the establishment of a register, and the laying-down of a procedure for evaluation.[56] The COMBIO encompassed both the public sector

---

[51] ADI 3526/005 STF. [52] Biosafety Law 2005, Art 14; and Decree 5591/05, Art 5.
[53] Identical obligations had earlier been enacted by Federal Decree 4680/03 (applicable since 2003).
[54] Cartagena Protocol on Biosafety, Art 18 (available at <http://www.cbd.int/biosafety/protocol/shtml>).
[55] See now the Biosafety Commission for Agriculture and Forestry ('Comisión de Bioseguridad Agropecuaria y Forestal') (as from 13 August 2008: Decree 12706/08).
[56] Decree 18481/97.

and civil society, the latter being represented by NGOs which have as their objective the protection of the environment and which are specifically concerned with biotechnology.[57] Use of the conjunctive 'and' in this context raised a question: did it mean that the organization must have as its objective the protection of the environment and, in addition, that the organization must have some link with biotechnology? A cumulative reading in this way would seem to exclude a broad interpretation of civil society which accommodates organizations without technical expertise, often composed of citizens and consumers, concerned by lifestyle and food issues, matters of health, effects on the environment, and their impact (whether positive or negative) on future generations.

In 1998, 'Roundup Ready' soya began to be introduced from Argentina. Subsequently, between 1999 and 2001, the executive enacted a series or resolutions and decrees banning commercial use of GM material, but these did not contain enforcement or control provisions.[58] Subsequently, between 2004 and 2005, the executive adopted provisional measures which permitted commercialization of GM soya seeds, although not yet formally authorized, and which even allowed the introduction of and trade in unregistered varieties. During the same period, the first varieties of 'Roundup Ready' soya were registered, commercial use having been authorized by the Minister for Agriculture and Livestock Farming. Further, in 2004, with the creation of the National Service for Quality and Plant Health and for Seeds ('Servicio Nacional de Calidad y Sanidad Vegetal y de Semillas') (SENAVE), Paraguay acquired a body with general competence, derived from other bodies, in the field of support for and development of agricultural production, together with specific competence in matters of biotechnology.[59]

3.4.1 Risk evaluation

The COMBIO enjoyed exclusive responsibility for risk evaluation and was required to assess the following factors: the possibility of cross-contamination; escapes of GM material subject to trial; insect movements; and other variables. More precisely, it was required to take account of the ecosystem where the field trial took place, the biological characteristics of the GMO, whether there was the capacity for gene transfer or distribution, levels of toxicity, whether there were allergens which might affect humans or other organisms, as well as probable damage to other organisms within the environment. The text of Decree 18481/97 addressed damage to organisms within the environment, as distinct from the environment itself, and prioritized the former (to the detriment of the latter 'public good'). Thus, when carrying out risk assessments, the Decree required evaluation of only the 'probable' adverse effects of the GMO; and this does not

---

[57] Ibid Art 2.
[58] Decree MAG 10661/2000; Resolutions MAG 62/99, 82/9, 07/99, 554/99, and 201/00; and Resolution 10/22/04.   [59] Law 2459/04.

appear to conform to the Cartagena Protocol, which in the same context uses the word 'possible'.[60] This introduced potential conflict, since something can be possible, but not probable, and an evaluation is more rigorous if conducted on the basis of possibility, as opposed to probability, of risk. And yet, according to legal hierarchy, Law 2309/03, which ratifies the Cartagena Protocol, should trump the specific regime under Decree 18481/97. Nonetheless, doubts still remain as to how the two regimes can be reconciled, as will now be examined in greater detail.

3.4.2 The Cartagena Protocol and Decree 18481/97

The Cartagena Protocol has as its objective to:

contribute to ensuring an adequate level of protection in the field of the safe transfer, handling and use of living modified organisms resulting from modern biotechnology that may have adverse effects on the conservation and sustainable use of biological diversity, taking also into account risks to human health, and specifically focusing on transboundary movements.[61]

Law 2309/03, as has been seen, ratifies the Cartagena Protocol in Paraguay and applies to transboundary movements of GM plants and animals which could have adverse effects on biological diversity or entail risks to human health (pharmaceutical products, however, being excluded). Decree 18481/97, on the other hand, is more limited in its compass, since it deals only with evaluation of GMOs by the COMBIO. Accordingly, as indicated, there was a conflict, in that both the provisions implementing the Cartagena Protocol and the national provisions governing risk assessment concerned GMOs, yet differently; and the question was, which provisions should apply on the import or export of a GMO. The answer, which would seem equally valid in the case of other countries, arguably depends upon whether or not the trade is with a country which is itself a party to the Cartagena Protocol. Where it is a party, the rules of the Cartagena Protocol would seem to apply; but, where it is not, any bilateral or regional agreement should arguably comply with the Cartagena Protocol. (Also, if neither country is a party, the import requirements to be met would seem to be those imposed by the buyer-importer). As the level of risk differs in the case of evaluation under Law 2309/03 and Decree 18481/97 (the former dealing with possibility and the latter dealing with probability), one could suppose that the criteria applicable in the case of the former apply equally to trials, research, and commercialization of living organisms that are animal in nature, even outside the sphere of transboundary movements.

---

[60] Cartagena Protocol, Art 15.   [61] Ibid Art 1.

3.5 Uruguay

3.5.1 Before 2000

In 1998, 'Roundup Ready' soya was introduced into Uruguay without information, evaluation, scientific debate, or participation by civil society. Indeed, when first cultivated, Uruguay could not look to any specific regime governing agricultural biotechnology. The first traces of regulation of GMOs were to be found instead in the legislation governing plant health (enacted in 1911), public health (enacted in 1934), the establishment of the Ministry of the Environment (enacted in 1990), and in the Seeds Law of 1997. Nonetheless, in 2000 a specific regime was laid down by Decree 249/00 of the executive. This regulated research into, and the introduction and use of, plant GMOs. It also gave rise to the Commission for the Evaluation of Risks from Genetically Modified Plants ('Comisión de Evaluación de Riesgo de Vegetales Genéticamente Modificados') (CERV), an institutional body which comprises both public authorities and academics, as well as civil society. The CERV enjoys substantial capacity for conducting analysis of GMOs and, in addition, has the benefit of support from the executive.

Prior authorization is compulsory in the case of the introduction, use, and handling of GM plants and their derivatives, such authorization involving an evaluation of risks to the environment, biological diversity, and any possible consequences to human, animal, and plant health. Originally, the function of the CERV was to lay down rules for the conduct of risk evaluations, but it also became responsible for evaluating completed analyses, for advising the authorities on authorizations, and for drawing up risk management and communication measures on a case-by-case basis. The regime covered a variety of different situations: confined experiments; field trials carried out in accordance with specific conditions to address biosafety; seed propagation; and first import or export of a GM product destined for direct consumption or transformation.[62] Public participation is assured by the making available and circulation of documentation, both that required for the initial application and that flowing from the evaluation process. Participation is also achieved through publication in the *Official Journal* and in two national, mass-circulation newspapers, together with provision for the holding in public of information and consultation fora.

The competent authorities responsible for the authorization of GMOs are several. The Ministry of Livestock, Agriculture and Fisheries and the National Seed Institute undertake: the determination of biosafety conditions where use has been agreed; the authorization of experiments and field trials (or other forms of cultivation where a lesser degree of protection is required); the conduct of national evaluation; and the authorization of seed propagation. The Ministry of

---

[62] D Bayce, 'Los cultivos transgénicos, su regulación y situación actual en Uruguay' [2006] *Tribuna País Agropecuario* 86.

Livestock, Agriculture and Fisheries and the Ministry of Economics are the competent authorities for the production or first import of GMOs when destined for direct consumption or transformation. And the Ministry of Health and the Ministry of the Environment retain their competence in, respectively, the sphere of sanitary policy and the protection of the environment.

In this context, the position of maize presents a paradox. Commercial use of fodder varieties of cereals and oilseeds destined for animal feed requires an agronomic evaluation of the crops concerned and prior entry on the national crop register. Yet maize is categorized as a fodder crop *and* a crop destined for human consumption. In consequence, it escapes the requirement of prior registration, despite the fact that certain varieties (such as sweetcorn) are destined for the plate. As a result, animals would seem to receive greater protection than humans.[63]

3.5.2 After 2000

After Decree 249/00 came a general law in 2000 on the protection of the environment, Law 17283/00. This confers on the Ministry of the Environment the responsibility for measures to prevent and control environmental risks incurred by the creation, handling, use, or release of GMOs when such operations have the capacity to affect conservation and sustainability in respect of both biodiversity and the environment. The ensuing rules governing GM seeds were enacted by resolution. This allowed decisions to be made within ministries and avoided the debates which would have been necessary had superior legislation been employed. The resolutions for each seed variety impose an obligation to create refuges which amount to 10 per cent of the area sown on each farm growing crops from conventional seed, with a defined buffer zone.

In 2003, by Resolution 267/03, conditions were approved and fixed for the introduction, use, production, and commercialization of maize RM236; and the following year a second variety was also authorized. The year 2005, however, was distinguished by enhanced evaluation criteria, which applied not only to new applications, but also to authorizations already agreed. This shift in policy was the result of more active intervention by the National Direction of the Environment ('Dirección Nacional de Medioambiente') (DINAMA), a ministerial body. Further, the shift in policy was carried into effect by decrees, higher up the legislative hierarchy than resolutions, which require detailed studies on the environmental, economic, and social impact of GMOs. In 2006 the use and commercialization of sweetcorn was banned, with a moratorium beginning in January 2007 and continuing until June 2008. Accordingly, the analysis of new GMO applications by the CERV was suspended,[64] pending a national biosafety regime, harmonized at the level of MERCOSUR.[65]

---

[63] I Carcamo, 'Transgénicos en Uruguay: animales mas protegidos que los seres humanos' [2006] *Red de Acción Plaguicidas y sus Alternativas América Latina* 1. [64] Decree 37/07.
[65] 2004–09 PNUMA-GEF.

In July 2008 a new regime was implemented by Decree 353/08, which requires prior authorization for the introduction, use, and handling of GM plants, whether in respect of confined use or release to the environment, and whether for research, trade, import, or export. Provision is also made for a new institutional structure, comprising a National Biosafety Cabinet ('Gabinete Nacional de Bioseguridad') (GNBio), a Commission for Risk Management ('Comisión para la Gestión del Riesgo') (CGR), and a Coordination Committee ('Comité de Articulación Institucional') (CAI). This new structure should ensure the transparency, reliability, and independence necessary for the evaluation of risks and to guarantee coexistence of GMOs with organic and conventional production. Proposals have also been made for a machinery both to provide information and to carry out public consultation. On the other hand, the new body to carry out these functions is not yet in place, and the relevant legal texts have not been published (in particular, those relating to inspection). Further, labelling remains voluntary and control still comes under different ministries. Accordingly, the new regime looks very much like a paper exercise.[66]

4. Towards Harmonized Regulation of GMOs in MERCOSUR?

4.1 Difficulties in Reconciling Interests

Against this legal and institutional background, a preliminary question is whether it is legitimate to hope that a 'law of MERCOSUR' may emerge. Is this a necessity for members? Can one envisage a harmonized regime and, if so, on what basis? Difficulties in reconciling interest may be illustrated by the fact that Argentina and Uruguay formed (with Australia, Canada, Chile, and the United States) the Miami Group, which operated as a block when negotiating the Cartagena Protocol. In particular, it adopted a contrary position to that of the European Union (EU), arguing against the adoption of measures relating to environmental security or human and animal health where these had the capacity to affect trade in, and the transfer and management of, GMOs. The Miami Group therefore opposed seed segregation and compulsory labelling. The only things to be accepted were the primacy of scientific proof when conducting risk analysis and submission to the rules of the WTO. By contrast, Brazil and Paraguay were members of the Ideas Afines Group, together with China and 77 developing countries. The Ideas Afines Group disagreed with the views of the Miami Group and was closer in position to that of the EU. Accordingly, the stances adopted by the different members of MERCOSUR have differed, with members pursuing their own economic agenda. Argentina and Brazil, two of the

---

[66] DINAMA-PNUMA-FMAM UU 04-009, *Propuesta de marco nacional de bioseguridad* (available at <http://www.unep.org/Biosafety/files/UYNBFrepSP.pdf>, last accessed 12 March 2009).

great cereal producers of the world, protect their own trade, while Paraguay and Uruguay have taken a more conservative approach, having appreciably increased their GM production, but without any realistic ability of competing in terms of volume on world markets.

In such circumstances, there are clear difficulties in accommodating the conflicting interests of the various members of MERCOSUR. Not least, it is also a valid question whether it would be wise for those countries which are relatively small agricultural producers to compete in the same market as the large producers, in that they may gain little economic benefit. A better course might be to distance themselves from the large producers by directing their focus to quality production and protected designations of origin. By opting for niche markets, farmers could participate effectively in world trade, where there is demand for assurance in matters of health and the environment (as found principally in Japan and the EU). For example, Uruguay has enjoyed success in exporting certified meat products (such as Natural Meat, Hereford Beef, Angus Beef, and *Orgánica*), these exports being largely directed to the United States and the EU.[67] In such context, there is obvious advantage in the fact that most livestock farming takes place on grass. However, grazing may be supplemented by the feeding of grain, which may include maize; and, in consequence, there is a risk that cattle may eat GM maize, which has the capacity to jeopardize at least a proportion of the export trade, notably that to the EU with its strict labelling requirements.[68] Similar considerations apply in Paraguay, although on a smaller scale.

4.2 A Single Market or Harmonization?

4.2.1 The Internal Market and the External Market

The trade rules laid down in MERCOSUR are a product of the relations between its members. It is a fact that trade in GM products and their derivatives is not a cause for concern to any of the four members, owing to their respective trading partners. For example, Paraguay exports 54.56 per cent of its soya to Argentina, 13.59 per cent to Uruguay, and 3.3 per cent to Brazil.[69] Similarly, 58.42 per cent of maize exports are destined for Brazil, 19.90 per cent for Uruguay, and 5.53 per cent for Argentina. Uruguay, on the other hand, exports 1.83 per cent of its soya meal to Brazil, but neither soya nor maize to Argentina or Paraguay. In this context, there is no pressing urgency to lay down a specific regime to cover GMOs in MERCOSUR. Besides, in light of the legal and institutional background, the strict application of any such regime could not be guaranteed.

---

[67] See statistics at <http://www.ine.gub.uy/>, last accessed 3 July 2009. [68] See Chapter 4.
[69] Economic Department United States Embassy in Uruguay—2008.

If one rules out the possibility of a legislative framework at MERCOSUR level in the short term, the issue then arises whether harmonization of existing rules can be achieved. Taking into account the fact that all four countries are agricultural exporters competing on world markets (while not forgetting the structural asymmetries already mentioned), a good starting point would be to inquire into the attitudes of consumers outside MERCOSUR. In turn, this inquiry may begin with the consumers of the EU—the second export market for MERCOSUR cereals— whose attitudes continue to show a level of mistrust, concern at lack of information on GM food, and a desire to avoid damage to health and environment.[70] Consequently, to secure entry into the EU market, it is simply not possible to put aside indefinitely the implementation of detailed rules for the segregation, labelling, certification, and traceability of GM crops, notwithstanding that these, as yet, may not be recognized as absolute requirements for world trade.[71] Certainly, the EU continues its struggle to incorporate such requirements within the legal order for world trade.[72] Nevertheless, if MERCOSUR is to achieve a balance, account must also be taken of the more lenient approaches adopted by other major players in the world cereals market (for example, the United States, China, and India).

Accordingly, MERCOSUR will be (and already has been) confronted by pressures generated by GMOs which affect its agricultural economies. If each member retains its own legislative framework, distortions in competition are likely to arise, which will only accentuate pre-existing inequalities in terms of scale of production. It would thus appear necessary to contemplate a regime at MERCOSUR level which must satisfy two main conditions: first, the growing of GM crops must be carried out in a way that is compatible with the requirements of importing countries; and, secondly, the members of MERCOSUR must be able to meet the demands of world trade in GMOs on terms that ensure fair competition for exports.

4.2.2 *The normative position*

As has been seen, MERCOSUR does not as such have a regime governing GMOs. Yet one could be derived from the Treaty of Asunción, its Preamble reciting that the optimum use of available resources and the preservation of the environment are two objectives of integration. Moreover, resort could also be had to the Framework Agreement on the Environment, approved in 2001 and in force since

---

[70] See, eg G Gaskell et al, *Europeans and Biotechnology in 2005: Patterns and Trends—Final Report on Eurobarometer 64.3* (Brussels: European Commission, 2006); and Chapters 1, 4, and 7.

[71] In particular, it may be necessary to impose labelling requirements that are stricter than the 'may contain' GMOs obligation in the Cartagena Protocol, Art 18.

[72] See, eg the WTO disputes which followed the decision of the Appellate Body in *EC— Measures Concerning Meat and Meat Products (Hormones)* WT/DS26/AB/R and WT/DS48/AB/R, 16 January 1998 (eg *US—Continued Suspension of Obligations in the EC—Hormones Dispute* WT/DS320/AB/R, 16 October 2008; and *Canada—Continued Suspension of Obligations in the EC— Hormones Dispute* WT/DS321/AB/R, 16 October 2008).

July 2004, which reaffirms the undertakings contained in the Rio Declaration on Environment and Development.[73] In essence, this Agreement is generic in its wording and lays down guidelines, underpinned by principles of proportionality, flexibility, and fairness. Its purpose is to promote environmental protection and the management of resources by means of co-ordinated, sectoral policies, with the aim that commercial policy and environmental policy should work together without distortion in competition. Its coverage extends to the management of natural resources, biodiversity, and biosafety.

Under the Framework Agreement it is provided that, when the parties consider it necessary, they may adopt common policies for, *inter alia*, the protection of the environment, conservation of natural resources, and the promotion of sustainable development; and such regional co-operation in environmental matters should comply with international agreements signed by the members of MERCOSUR.[74] That said, no compulsion is imposed on the parties to the Framework Agreement, nor is there a formal strategy. Rather, everything is dependent upon the exercise of political will. It may be emphasized, nonetheless, that specific actions to be undertaken include: the encouragement of harmonization of legal and institutional guidelines for the purpose of preventing, controlling, or mitigating environmental impact; and the consideration of cultural aspects, where appropriate, in environmental decision-making.[75] Reference to cultural aspects arguably opens the door to future rules which could, for example, accommodate the dominant cultural position in the EU (namely, mistrust of GMOs). That said, there can, as yet, be no certainty on this point.

4.2.3 The Convention on Biological Diversity and the Cartagena Protocol

Argentina entered into and ratified the Convention on Biological Diversity[76] by Law 24375/94. Brazil did the same by Decree 2519/98, and likewise Paraguay by Law 253/93 and Uruguay by Law 16408/93. Brazil incorporated the terms of the Cartagena Protocol into the national legal order in 2006 by Decree 5705/06. Paraguay proceeded to ratification by Law 2309/03. However, neither Argentina nor Uruguay has yet done so.[77] The national legal orders of all four members of MERCOSUR recognize the precautionary principle, as found in both the Convention on Biological Diversity and the Cartagena Protocol,[78] which opens up the possibility of full risk assessments before introducing GMOs. Nonetheless, difficulties in bringing to bear modern technology to improve the detection and evaluation of risks puts members of MERCOSUR at

---

[73] CMC: Dec. 2/01. [74] Framework Agreement on the Environment, Art 5.
[75] Ibid Art 6(j) and (m). [76] Available at <http://www.cbd.int/biosafety/protocol/shtml>.
[77] For details of the countries which have signed and ratified the Cartagena Protocol, see <http://www.cbd.int/>, last accessed 3 July 2009.
[78] See further Z Drnas de Clément, 'El "principio de precaución" en materia ambiental: nuevas tendencias' (2000) (available at <http://www.acaderc.org.ar/doctrina/derecho-internacional-publico/>, last accessed 19 March 2009).

a disadvantage as compared to more scientifically developed countries, such as the United States and the EU. Further, scientific and technological advances grant developed countries an edge when invoking the precautionary principle to limit imports.

4.2.4 The Technical Regulation on the Labelling of Packaged Foods

At the level of MERCOSUR, consideration must also be given to the Technical Regulation on the Labelling of Packaged Foods.[79] Under this Regulation, food is any substance which can be eaten in its natural state or which can be partially or fully processed for human consumption (excluding cosmetics, tobacco, and medicines); and packaged food is any food wrapped or packaged for consumer use. These definitions cover, for example, GM crops marketed in packaged form and products processed from GM crops, such as oil, mayonnaise, and soy sauce. Labelling on packaged foods must not contain any terms, indications, denominations, symbols, or other elements which could provide false or incorrect information or confuse the consumer as to the true nature, composition, origin, variety, quality, quantity, shelf-life, or characteristics of the product. Under these rules, the absence of a 'GM' label could be regarded as failure to provide sufficient information, so generating confusion for the consumer. But it must be noted that, whether through the indifference of government or consumer organizations, such an interpretation has not yet gained currency.

4.3 Recent Developments

4.3.1 The ad hoc Biotechnology Group

In 2004, MERCOSUR established an ad hoc Biotechnology Group, with the following four objectives: first, to harmonize the regulatory regimes on biodiversity in the four countries; secondly, to analyse whether it would be possible to co-ordinate commercial authorizations of GMOs; thirdly, to consider the implications of introducing labelling at both regional and international levels; and, fourthly, to carry out consultations with a view to co-ordinating a MERCOSUR position in international negotiations (whether within the WTO or in relation to the Codex Alimentarius or Cartagena Protocol). At present this Group is working on the drawing up and harmonization of biosafety rules among the full members of MERCOSUR.

---

[79] GMC/Res 26/03.

4.3.2 *The Declaration on Biodiversity Strategy for MERCOSUR*

In March 2006, an extraordinary meeting of environment ministers resulted in the Declaration on Biodiversity Strategy for MERCOSUR, to be implemented in 2010. The joint basis for the Declaration was the Framework Agreement on the Environment, national laws governing international agreements to which a member is signatory, and national strategies for biosafety.[80] The concept of biodiversity for the purposes of the Declaration is wide, extending to sociocultural aspects as well as intrinsic value. Other matters addressed include: the approval of conditions to guarantee the continuation of ecological processes; state sovereignty over biological resources within national territories; fair and equitable sharing in the benefits derived from the use, knowledge, and practices of indigenous and local communities; the imposition on states of a duty to ensure that biodiversity is so managed nationally that damage to the environment of other states is prevented; and co-operation in biodiversity protection.

4.3.3 *The drawing-up of legal regimes for biosafety and the provision of management tools for biosafety*

Following the entry into force of the Cartagena Protocol on 11 September 2003, and with the objective of risk management, MERCOSUR has developed a regulatory framework for transboundary movements of GMOs. Since 2007, with the assistance of the FAO, all the members of MERCOSUR (including Bolivia and Chile) have participated in a project to develop a common biosafety management policy.[81] Its purpose is to put in place agreed technical requirements to manage biosafety at national and regional level; and it is apprehended that these should be recognized externally as a body of criteria which, through their strength and integrity, can inform decisions.[82] Another component vital for MERCOSUR is the development of the skills necessary to carry out risk assessments, monitoring, and the detection of GMOs in complex products.

5. Conclusion

From the legal standpoint, the four full members of MERCOSUR have not opted to follow the same road. Argentina would appear to be an advocate of the

---

[80] These national strategies for biosafety are not only to be found in legal rules. It is necessary also to consider a wider range of sources, such as statements by presidents or others in high office, research agendas of both public and private institutions, and even media communications. Analysis of their precise wording provides precious insights into national approaches towards biodiversity.
[81] Project TCP/RLA/3109.
[82] UNEP-GEF Project on Development of National Biosafety Frameworks—Report 2007 (available at <http://www.unep.org/biosafety/>, last accessed 3 July 2009).

development of GMOs. Admittedly, there are an increasing number of regulations governing agricultural biotechnology, but these regulations come low in the legal hierarchy, leave considerable discretion to the executive, and do not provide for participation by civil society. By contrast, Brazil has legislated by law of Congress, but whether this law is constitutional remains in doubt. Further, the ambitious scope of the Biosafety Law allows uncertainty to persist as to its implementation, at least in the short term, owing to the lack of effective control mechanisms at different levels. The Paraguayan Parliament is in the process of scrutinizing a biosafety law, while Uruguay is drawing up its own text. MERCOSUR has plans to establish a harmonized regime, whose objective is to ensure the successful coexistence of conventional and organic crops with GM crops. The question remains whether this is a possibility. Any response will require patience and a close watch on future developments within MERCOSUR, but, for the time being, no concrete or effective step has been taken.

12

Genetically Modified Crops and Food in the United States: The Federal Regulatory Framework, State Measures, and Liability in Tort

Margaret Rosso Grossman*

1. Introduction

In 2008, the United States planted 62.5 million hectares of genetically modified (GM) crops, 50 per cent of the world's total of 125 million hectares.[1] In that year, GM varieties in the United States were estimated to make up 92 per cent of soya, 80 per cent of maize, and 86 per cent of upland cotton.[2] Developers have commercialized these GM varieties under procedures required by federal statutes and regulations and after consultations recommended by federal agency guidelines.

This chapter describes the regulation of GM crops and their food products in the United States. It focuses primarily on federal policy, laws, and regulations that govern crops and food produced through biotechnology. In addition, it gives a brief account of some state laws that apply to biotechnology. Finally, it discusses potential liability under state tort law for harm caused by GM crops.

---

* This chapter is based on work supported by the National Institute of Food and Agriculture, United States Department of Agriculture (USDA), under Hatch Project No ILLU-470-309, and a grant from the Danish Social Science Research Council. The chapter is derived in part from a draft article by MR Grossman and HT Anker and from articles cited in n 155 below. The author thanks the Institute of Food and Resource Economics, Copenhagen University and the Law and Governance Group, Wageningen University.

[1] C James, *Global Status of Commercialized Biotech/GM Crops: 2008 (ISAAA Brief No 39)* (Ithaca, NY: International Service for the Acquisition of Agri-biotech Applications (ISAAA), 2009) Executive Summary.

[2] National Agricultural Statistics Service (NASS), *Acreage* (Washington, DC: USDA, 2008). Estimates for 2009 are: soya 91 per cent; maize 85 per cent; and upland cotton 88 per cent: NASS, *Acreage* (Washington, DC: USDA, 2009).

2. Federal Policy

2.1 Coordinated Framework—1986

The US government focus on genetically modified organisms (GMOs) dates from the early 1980s, and in 1986 the Coordinated Framework for Regulation of Biotechnology ('Coordinated Framework'), drafted by the White House Office of Science and Technology Policy (OSTP), with the co-operation of several administrative agencies, established US policy for GMOs. The Coordinated Framework recognized the exciting possibilities and commercial opportunities that biotechnology offered, but acknowledged questions about the adequacy of existing laws, regulations, and review processes for products of biotechnology.[3]

Ultimately, the Coordinated Framework concluded that biotechnology products are not fundamentally different from conventional products; that the product, rather than the process, should be regulated; and that regulatory jurisdiction over products of biotechnology (like traditional products) should be based on their use. The document articulated two basic principles: regulatory agencies should, insofar as possible, adopt consistent definitions of the organisms subject to review, and agencies should carry out 'scientific review of comparable rigor'.[4] As the OSTP later explained:

[t]he Coordinated Framework provides a regulatory approach that is intended to ensure the safety of biotechnology research and products, using existing statutory authority and building upon agency experience with agricultural, pharmaceutical, industrial, and other products developed through traditional genetic modification techniques.[5]

The Coordinated Framework relied on existing federal laws to govern GMOs, and three federal agencies carry out the most important responsibilities. In general, the USDA ensures that GMOs are safe to grow; the Environmental Protection Agency (EPA) ensures that they are safe for the environment; and the Food and Drug Administration (FDA, along with the EPA) ensures that they are safe to eat. Policies articulated in the Coordinated Framework continue to influence agency action, though laws and regulations enacted since 1986 also apply.

2.2 2002 Policy

The Coordinated Framework had acknowledged that, as biotechnology advanced, new regulatory measures might be required. The development and field testing of many new GM varieties increased the risk of cross-pollination and commingling

---

[3] OSTP, Coordinated Framework for Regulation of Biotechnology Products, 51 Fed Reg 23,302 (26 June 1986). [4] Ibid 23,303–4.
[5] OSTP, Proposed Federal Actions To Update Field Test Requirements for Biotechnology Derived Plants and To Establish Early Food Safety Assessments for New Proteins Produced by Such Plants, 67 Fed Reg 50,578, 50,578 (2 August 2002) ('2002 Policy').

and triggered the need for new regulation. In 2002, therefore, the OSTP, with the agreement of the USDA, the EPA, and the FDA, proposed federal measures to update field testing requirements for plants derived from biotechnology and to establish early food safety assessments for food or feed proteins from these new plants. These proposals were designed to reduce the unintended presence of low levels of GM material in seeds, commodities, and food until safety standards had been met, and thus to protect public health and the environment and to increase public confidence in the effectiveness of regulatory oversight of GM foods.

Three principles form the basis of the 2002 regulatory proposals. First, the level of confinement for a field test should be consistent with the level of risk to the environment and health. Secondly, if a GM trait or protein presents an unacceptable or unknown risk, confinement measures should be strict, and GM materials from the tests should be prohibited in seeds, commodities, and other products. Thirdly, even without unacceptable risks, out-crossing and commingling should be minimized, but low levels of biotechnology-derived gene presence from field tests might be found acceptable.[6]

The 2002 Policy outlined the plans of the lead agencies for enhanced regulatory measures. The USDA had already made field-testing requirements for permits more stringent for GMOs intended for pharmaceutical or industrial products (rather than commodity crops) and also planned to amend and update its GM regulations. The FDA planned to publish guidelines to encourage the early evaluation of crops with new non-pesticidal proteins so that new crops would not raise food safety issues—for example, toxins, or allergens that might escape to seeds, commodities, or food. The EPA planned to publish guidance on safety review of low-level residues and containment controls during field trials for those conducting field testing on plant-incorporated pesticides and to review its requirements for experimental use permits and containment controls to minimize gene flow from field trials.[7]

3. USDA Regulation under the Plant Protection Act

The USDA regulates GM plants—including interstate movement, import, field testing, and eventual release—through the Animal and Plant Health Inspection Service (APHIS) and its Biotechnology Regulatory Services (BRS). APHIS uses a 'science-based regulatory framework that allows for the safe development and use of genetically engineered (GE) plants'.[8]

---

[6] Ibid 50,578–9.

[7] Ibid 50,579–80. This chapter does not focus extensively on crops that produce pharmaceutical or industrial compounds, which are subject to stricter regulations.

[8] APHIS, USDA, APHIS Policy on Responding to the Low-Level Presence of Regulated Genetically Engineered Plant Materials, 72 Fed Reg 14,649 (29 March 2007).

The 2000 Plant Protection Act (PPA) gives the USDA authority to control plant pests, and the USDA uses this authority to regulate GM plants.[9] The PPA authorizes the USDA to restrict the import, export, and movement in commerce of plant pests, and APHIS determines whether the organism is a plant pest. Under the broad definition of plant pest in APHIS regulations,[10] most GMOs are 'regulated articles' (materials that could harbour a plant pest)[11] and therefore subject to USDA jurisdiction. APHIS rules govern movement of regulated articles in interstate commerce or introduction into the environment through field tests. GMOs that are regulated articles must be evaluated and determined to be 'unregulated' before they can be sold in commerce.

3.1 Field Trials

APHIS governs field trials of GMOs, which take place while the new crop is still a regulated article under the PPA, through two processes—notification or permit.

The majority of field trials follow the notification process,[12] introduced in 1993 for plants that do not present novel plant risks. APHIS regulations prescribe the procedure for notification and the information required, including the location and size of the field test and technical data about the regulated article.[13] APHIS informs state regulatory officials and acknowledges within 30 days that environmental release is appropriate under the notification procedure; the acknowledgement applies to field testing for one year.[14] Performance standards for field trials under the notification procedure help to ensure that the field trial will not cause environmental or economic harm. Standards require careful shipment and storage of plant material, planting to avoid inadvertent mixture with non-regulated plant material, and identity preservation. The regulated article or its offspring may not persist in the environment, and post-trial volunteers must be prevented or managed. Anyone conducting a field trial must allow APHIS and state regulators to inspect facilities and records and must report results of field tests and any unusual occurrence to APHIS.[15]

---

[9] 7 USC ss 7701–72; s 7711 authorizes regulation of plant pests. The PPA replaced the Plant Pest and Plant Quarantine Acts, on which the USDA had relied for authority.

[10] See 7 CFR 340.1; and 7 USC s 7702(14).   [11] 7 CFR 340.

[12] See BRS, APHIS, USDA, *User's Guide: Notification* (5 February 2008) (available at <http://www.aphis.usda.gov/brs/pdf/Notification_Guidance.pdf>). APHIS authorized almost 19,000 notifications and 4,300 permits between 1987 and 2007: Government Accountability Office (GAO), *Genetically Engineered Crops: Agencies Are Proposing Changes to Improve Oversight, but Could Take Additional Steps to Enhance Coordination and Monitoring* (GAO-09-60, November 2008) 11.

[13] 7 CFR 340.3(b)(1)–(6); and 340.3(d)(2). Multiple environmental releases can be combined in the same notification: BRS, APHIS (n 12 above) 16.

[14] 7 CFR 340.3(e)(4). All releases for more than one year will require a permit: APHIS, *Policy Statement Regarding Releases of Perennials Under Notification* (no date).

[15] 7 CFR 340.3(c), (d)(5) and (6).

The permit procedure applies to experimental releases of GM plants that may carry higher risks (for example, plants modified with certain human or animal genetic material and those that produce industrial or pharmaceutical compounds).[16] The permit requires detailed technical information and specific information about experimental design, geographic locations, plans to prevent escape and dissemination, and final disposal.[17] APHIS sends the completed permit application (omitting trade secrets or confidential business information), along with its initial review, to the relevant state department of agriculture. To comply with the National Environmental Policy Act (NEPA),[18] APHIS must prepare an environmental assessment and, if required, an environmental impact statement. After review of the application and accompanying data, APHIS will grant or deny (with reasons and subject to appeal) the permit. The permit will include conditions, some specified by regulation and others required by APHIS, for introduction of the plant. Regulatory conditions include maintenance and disposal of the regulated material (and its packing materials) to prevent dissemination and establishment of plant pests, separation from other organisms, restriction to areas specified in the permit, identification by label, remedial measures, and inspection. The permit holder must report the results of field tests and notify APHIS promptly of accidental or unauthorized releases, unexpected characteristics of the organism, and other unusual occurrences.[19]

3.2 Petitions for Deregulation

During field testing, a GM crop is still a regulated article, subject to APHIS oversight.[20] When field tests indicate that the variety is not a plant pest and poses no threats to agriculture or the environment, the variety is ready to be commercialized. The next step is a petition for determination of 'nonregulated status'.[21] If the petition is granted, the organism can move freely in commerce,

---

[16] Ibid 340.3(b)(4)(iii). A person whose notification results in a denial of permission may apply for a permit: ibid 340.3(e)(5). [17] 7 CFR 340.4(b)(1)–(14).

[18] 42 USC ss 4321–70f. Federal agencies must prepare a detailed environmental impact statement for 'major Federal actions significantly affecting the quality of the human environment': 42 USC s 4332(2)(C).

[19] 7 CFR 340.4(b) and (f). More stringent requirements apply for GM crops with pharmaceutical and industrial uses: see BRS, USDA, *Guidance for APHIS Permits for Field Testing or Movement of Organisms Intended for Pharmaceutical or Industrial Use* (July 2008).

[20] A BRS programme helps applicants approved to introduce GM organisms 'to develop sound management practices and to enhance compliance' with APHIS regulations and permit requirements: BRS, APHIS, *Biotechnology Quality Management System* (Factsheet, September 2007). Under a pilot project, five volunteer participants will develop quality management systems to manage field releases: APHIS, News Release, *USDA Launches Biotechnology Quality Management System Pilot Project* (16 January 2009). [21] 7 CFR 340.6.

provided that it also meets the regulatory requirements of the FDA and EPA. As of February 2009, APHIS had granted 75 petitions, and 14 were pending.[22]

The petition, submitted in a structure prescribed by APHIS, must explain why the organism should no longer be regulated. The applicant must provide detailed information about the organism, including information that would be 'unfavorable to a petition'.[23] Data requirements include experimental data, unpublished studies, and scientific literature; results of field tests are an important part of the petition. The applicant must describe the conventional plant, the regulated article (the GM plant), and differences between the two. Data must substantiate that the GM plant is 'unlikely to pose a greater plant pest risk than the unmodified organism from which it was derived'.[24]

APHIS, acting through the BRS, will review the risk of the regulated variety. Part of that review includes a preliminary environmental assessment, required under NEPA and made available for public comment. After review of comments, APHIS prepares the final environmental assessment. If that assessment results in a finding of no significant impact, no environmental impact statement is required, but if the variety may have a significant impact, APHIS must prepare an environmental impact statement.[25] On the basis of the assessment or environmental impact statement, which helps to determine whether the GM crop can be released safely into the environment, and other documentation, APHIS makes the determination of non-regulated status or, less frequently, denies the petition.[26] Because non-regulated status means that a variety poses no environmental or agricultural risk, APHIS no longer has authority over that variety. Should it later become a plant pest, however, it will again be subject to regulation.[27]

---

[22] Varieties, with links to relevant documents, are listed at <http://www.aphis.usda.gov/brs/not_reg.html>, last accessed 24 March 2009. The earliest determination, in 1992, was Calgene's Flavr Savr tomato. [23] 7 CFR 340.6(b).
[24] Ibid 340.6(c) (listing data to meet the standard).
[25] A recent court decision held that the failure of APHIS to prepare an environmental impact statement for glyphosate-tolerant alfalfa did not comply with NEPA. The court vacated APHIS' determination of 'nonregulated status', enjoined planting of the GM alfalfa, and ordered APHIS to prepare an environmental impact statement: *Geertson Seed Farms Inc v Johanns* 2007 WL 518624, 2007 WL 1302981 (ND Cal); injunction affirmed in *Geertson Seed Farms Inc v Johanns* 541 F3d 938 (9th Cir 2008). See MR Grossman, 'Genetically modified food and feed and the US National Environmental Policy Act' (2007) 3 European Feed and Food Law Review 373. See also *Delaware Audubon Soc Inc v Secretary US Dep't Interior* 612 F Supp 2d 442 (D Del 2009) (enjoining cultivation of GM crops at Prime Hook National Wildlife Refuge until an environmental assessment or environmental impact statement is completed). [26] 7 CFR 340.6(d) and (f).
[27] The GAO recommended that the USDA, along with the EPA and FDA, develop a strategy to monitor GM crops that have been commercialized for effects on the environment, food safety, and traditional (non-GM) agriculture. The agencies, agreeing only in part with the recommendation, plan to discuss a co-ordinated strategy for monitoring deregulated crops: GAO (n 12 above) 24–34 and 48.

3.3 APHIS Policy for Low-level Gene Mixing

In March 2007, APHIS noted that plant breeding, both in conventional and GM plants, sometimes results in 'low-level mixing of genes and gene products from unintended plant sources'.[28] Natural processes (pollen movement) or human actions (for example, field testing and seed production) may introduce unauthorized GMOs into other products. In reviewing its biotechnology regulations, APHIS will consider new criteria to define the acceptability of low levels of regulated materials in seeds and grain. In the meantime, the agency articulated a policy developed on the basis of practical experience.

Because regulatory requirements minimize the likelihood of low-level presence of regulated GMOs, APHIS, in close co-operation with the EPA and the FDA, responds to occurrences with 'remedial action that is appropriate to the level of risk and warranted by the facts in each case'.[29] When an incident would result in introduction or dissemination of a plant pest or would threaten plant health or the environment, APHIS will require remediation measures, using its authority under the PPA.[30] APHIS will generally not require remedial measures if the regulated material comes from a plant that meets requirements for notification of field tests (and thus poses little risk) or if the plant is sufficiently similar to another plant that APHIS has already deregulated and therefore poses no significant safety risk to plant health or the environment.[31]

3.4 Proposed USDA Regulation

In July 2007, APHIS requested comments from the public on possible amendments to its regulations for environmental release of GMOs.[32] APHIS outlined and discussed possible regulatory alternatives in a lengthy Draft Programmatic Environmental Impact Statement.[33] In connection with its regulatory review, it compiled a list of lessons learned from its experience in regulating biotechnology.[34] In the 2008 Farm Bill, enacted in May 2008,

---

[28] APHIS (n 8 above) 14,649.
[29] Ibid 14,650. The few reported unauthorized releases of GM crops have led to costly food recalls and lost trade: GAO (n 12 above) 3 and 14–24.   [30] 7 USC s 7714.
[31] APHIS (n 8 above) 14,651. For application of the policy, see *USDA, EPA and FDA Statement on Genetically Engineered Corn 'Event 32'* (22 February 2008) (finding no safety risk from low levels of a GM hybrid maize seed). See also *FDA, EPA and USDA Conclude That Accidental Release of Genetically Engineered Cotton Poses No Safety Risk to Humans or Animals* (3 December 2008) (finding no harm from small quantities of an unauthorized GM cotton variety harvested with a similar, deregulated GM variety).
[32] APHIS, Introduction of Organisms and Products Altered or Produced Through Genetic Engineering, 71 Fed Reg 39,021, 39,022 (17 July 2007).
[33] APHIS, *Introduction of Genetically Engineered Organisms, Draft Programmatic Environmental Impact Statement—July 2007*, Docket APHIS-2006–0112.
[34] USDA, *Lessons Learned and Revisions under Consideration for APHIS' Biotechnology Framework* (October 2007) (listing nine lessons and considerations for enhancing regulation).

Congress then directed the USDA to take action on issues identified in the document within 18 months. These issues include the quality and completeness of records, the availability of representative samples, maintaining identity and control of regulated material, and other actions. In taking action, the USDA must consider establishing a system of risk-based categories for regulated articles, means of identification (including retention of seed samples), and standards for isolation and containment. The USDA must also consider requiring permit holders to maintain a 'positive chain of custody' and to keep records.[35]

In response to this congressional mandate, APHIS published proposed regulations in October 2008, the first comprehensive regulatory review of these regulations in two decades. Its proposals, already under consideration when Congress required prompt regulatory action, are designed:

to respond to emerging trends in biotechnology, to address the current and future needs of the agency, to continue to ensure a high level of environmental protection, to improve regulatory processes so that they are more transparent to stakeholders and the public, to more efficiently use agency resources and to eliminate unnecessary regulatory burdens.[36]

APHIS believes that its regulatory oversight, under authority of the PPA, should be expanded beyond plant pests to include noxious weeds and biological control organisms,[37] and its proposed regulations would extend its regulatory authority accordingly.[38] In light of the increased variety of GM organisms, APHIS proposed to revise its permit system, establishing four permit categories (based on risk of persistence and potential harm) for environmental releases of GM plants. Plants in the same category will have similar oversight and permit conditions. The proposed regulations would eliminate the notification procedure, and the permit category for GM plants with lowest risk would impose regulatory oversight similar to current notifications. New regulations would revise procedures for permit applications and establish permit conditions and obligations.[39]

Proposed amendments would adapt the system of petitions for non-regulated status to the broadened scope of regulation and focus on whether the GMO is 'unlikely to be a plant pest or noxious weed'.[40] APHIS also proposed a procedure to revoke an approval of non-regulated status and new measures to

---

[35] Food, Conservation and Energy Act of 2008, Pub L 110–246, s 10204.
[36] APHIS, USDA, Importation, Interstate Movement, and Release into the Environment of Certain Genetically Engineered Organisms, 73 Fed Reg 60,008, 60,009 (6 October 2008), corrected at 73 Fed Reg 66,563 (10 November 2008), to be codified at 7 CFR part 340. The comment period ended on 29 June 2009: 74 Fed Reg 16,797 (13 April 2009).
[37] PPA, 7 USC s 7702(2) and (10), defines biological control organism and noxious weed.
[38] APHIS (n 36 above) 60,011–15.   [39] Ibid 60,016–23.
[40] Ibid 60,023–25 and 60,047. Organisms already deregulated will be approved automatically under the proposed new regulations.

strengthen compliance (including stricter requirements for record-keeping) and enforcement. In addition, it proposed a regulation on low-level presence of regulated materials, designed to incorporate its 2007 policy statement, which sets out criteria for situations in which low-level presence will not be subject to remediation.[41]

4. EPA Regulation of GMOs

In 1992, the FDA and the EPA agreed that GM plants that express pesticidal substances are pesticides, subject to EPA regulatory authority under the Federal Insecticide, Fungicide, and Rodenticide Act (FIFRA).[42] In addition, the EPA regulates pesticide residues in foods under the Food, Drug, and Cosmetic Act (FDCA).[43] The agencies agreed that the EPA would address food safety issues associated with pesticides, and that the FDA would address other food safety issues (for example, compositional changes in food).[44] Some instances, however, would require expertise of both the EPA and the FDA.

The EPA issued its 1994 Policy Statement and Proposed Rules[45] and then started to register plant pesticides under FIFRA and to regulate their residues on foods under the FDCA. In response to public comments, the EPA adopted the term 'plant-incorporated protectant'—PIPs. The agency established its regulatory framework for PIPs only in 2001.[46] These rules, a step toward full implementation of the Coordinated Framework, established criteria, procedures, and requirements for PIPs under both FIFRA and the FDCA. In 2007, a new guidance document[47] focused on small-scale field studies and low-level presence of PIPs in food and elaborates on the policies described in the OSTP's 2002 Policy.

---

[41] Ibid 60,025–26. See text at nn 28–31 above. APHIS also proposed to exempt privileged or confidential information from release under the Freedom of Information Act: ibid 60,026.
[42] 7 USC ss 136–136y.
[43] 21 USC s 346a. At the time of the 1992 Policy Statement (see n 44 below), the FDCA regulated pesticide residues as food additives.
[44] FDA, Statement of Policy: Foods Derived From New Plant Varieties, 57 Fed Reg 22,984, 23,005 (29 May 1992) ('1992 Policy Statement'). The FDA has authority for substances intended to enhance plant resistance to chemical herbicides, like glyphosate.
[45] EPA, Plant Pesticides Subject to the Federal Insecticide, Fungicide and Rodenticide Act and the Federal Food, Drug, and Cosmetic Act; Proposed Rule, 59 Fed Reg 60,496 (23 November 1994).
[46] EPA, Regulations Under the Federal Insecticide, Fungicide, and Rodenticide Act for Plant-Incorporated Protectants (Formerly Plant-Pesticides), 66 Fed Reg 37,772 (19 July 2001), codified at 40 CFR parts 152 and 174).
[47] EPA, *Pesticide Registration (PR) Notice 2007–2: Guidance on Small-Scale Field Testing and Low-Level Presence in Food of Plant-Incorporated Protectants* (PIPs) (EPA Doc EPA-HQ-OPP-2007-0654-0001, 2007). The guidance document does not bind the EPA or others: ibid 7.

4.1 FIFRA and PIPs

4.1.1 Pesticide registration

The EPA regulates pesticides moved into plants from other organisms (including plants not sexually compatible) under FIFRA. A common example is the pesticide frequently used in GM crops, Bt (*bacillus thuringiensis*).[48] PIPs fit within the FIFRA definition of pesticides because they are introduced into plants as a means of 'preventing, destroying, repelling, or mitigating any pest'.[49]

FIFRA provides that no pesticide can be marketed and used in the United States until it has been registered.[50] Registration is a complex process, and applicants must provide data on safety and efficacy to the EPA. A pesticide can be registered if its composition warrants the registrant's claims and its labelling complies with FIFRA. Moreover, when used in accordance with normal practice, it must not 'cause unreasonable adverse effects on the environment'.[51]

The EPA uses the FIFRA registration system to collect data on the efficacy and environmental effects of PIPs. Field tests of unregistered PIPs are important, and experimental use permits allow the developer to gather information for its own use and to support registration.[52] Further, if a food residue is reasonably expected to result from the field test, the EPA may establish a temporary food tolerance (a legal limit on the maximum amount of the substance in or on food) under the FDCA before issuing the permit.[53]

Although a permit is generally required for testing any unregistered pesticide (including a PIP), some tests 'are presumed not to involve unreasonable adverse effects'[54] and therefore do not require a permit. Laboratory or greenhouse tests are presumed to be exempt. A small-scale test on a cumulative total of no more than 10 acres of land per pest does not require an experimental use permit, but any food or feed crops involved in the test must be destroyed or eaten by experimental animals unless the EPA has established a food tolerance (or exemption) for residues.[55] The EPA encourages those conducting field trials to consult with the agency to discuss confinement of the tested PIP; physical or biological controls that comply with APHIS requirements will generally satisfy the EPA. As a result of the consultation, the EPA may recommend that the

---

[48] EPA, *Introduction to Biotechnology Regulation for Pesticides: Guidance Document* (EPA Doc EPA-HQ-OPP-2007-0827-0001, 2007) 4.

[49] 7 USC s 136(u). See also 40 CFR 174.3 (defining PIP); and EPA (n 47 above) 1.

[50] 7 USC s 136a. Establishments where pesticides are produced must also be registered under FIFRA s 136e. The EPA can exempt from regulation a pesticide that is regulated by another federal agency or 'of a character which is unnecessary to be subject to' FIFRA: ibid s 136w(b).

[51] 7 USC s 136a(c)(5). See 7 USC s 136(bb) (defining 'unreasonable adverse effects on the environment').

[52] 7 USC s 136c; and 40 CFR part 172. The EPA must determine that the experimental use will not result in 'unreasonable adverse effects' on human health or the environment: 7 USC s 136c(d); and 40 CFR 172.10(a).   [53] 7 USC s 136c(b).

[54] 40 CFR 172.3(a).   [55] Ibid 172.3(b) and (c).

developer seek a temporary food tolerance, or the agency may require an experimental use permit.[56]

4.1.2 Future regulation?

In April 2007, the EPA issued an advance notice of proposed rule-making to seek public comments on possible new regulations for PIPs, which are governed in part by general regulations that apply to all pesticides.[57]

PIPs are produced and used in a living plant, so they differ from pesticides intended for external physical application.[58] EPA regulations that govern the registration of pesticide establishments and pesticide production are not completely appropriate for PIPs, which present unique regulatory issues. For example, regulations that require reporting of volume or weight of pesticides are inappropriate for PIPs in living plants.[59] Therefore, the EPA plans to codify new data requirements for pesticide registration of PIPs. These will reflect current scientific advances and 'improve the Agency's ability to make regulatory decisions about human health and environmental effects of PIP pesticides to better protect wildlife, the environment and people'.[60]

4.2 Pesticide Residues under the Food, Drug, and Cosmetic Act

The FDCA governed pesticide residues like other food additives until enactment of the 1996 Food Quality Protection Act,[61] which removed pesticide residues from the definition of food additives and from the controversial food additive Delaney Clause.[62] A special section of the FDCA now governs pesticide residues in foods. Under that section, raw or processed food or feed that contains pesticide chemical residues is considered adulterated and cannot be moved in interstate commerce, unless the residue complies with an established tolerance or has been exempted from the tolerance requirement.[63] Residues of PIPs in foods (including food produced during field tests) are subject to this requirement.

By regulation, the EPA sets pesticide tolerances for foods (or establishes an exemption), and the FDA enforces those pesticide tolerances. The EPA may establish a tolerance only if it is 'safe', which means 'a reasonable certainty that

---

[56] EPA (n 47 above) 2.
[57] EPA, Plant-Incorporated Protectants; Potential Revisions to Current Production Regulations, 72 Fed Reg 16,312 (4 April 2007).
[58] Ibid 16,312–13. The EPA believes that PIPs usually present lower risk than chemical pesticides: ibid 16,313. [59] 40 CFR part 167.
[60] EPA, Statement of Priorities, 72 Fed Reg 69,922, 69,939 (10 December 2007). The planned regulations, with no legal deadline but a May 2009 completion goal, will amend 40 CRF parts 158 and 174. [61] Pub L 104–170, 110 Stat 1489, 1513 (3 August 1996).
[62] Delaney Clauses prohibit FDA approval of substances that contain carcinogens: 21 USC ss 348(c)(3)(A) (food additives), 360b(d)(1)(H) (colour additives), and 379e(b)(5)(B) (animal drugs). See *Les v Reilly* 968 F2d 985 (9th Cir 1992).
[63] 21 USC s 346a(a)(1). See EPA (n 47 above) 3.

no harm will result from aggregate exposure to the pesticide chemical residue, including all anticipated dietary exposures and all other exposures for which there is reliable information'.[64] The EPA may grant exemptions, either temporary or permanent, from the tolerance requirement if tests indicate that the pesticidal protein is neither toxic nor allergenic and there is 'a reasonable certainty that no harm will result' from aggregate exposure.[65] Scientific data must support requests for tolerances or exemptions.[66] On a case-by-case basis, the EPA has granted both temporary and permanent exemptions to PIPs in GM crops (for example, various proteins associated with *Bt* crops).[67]

5. FDA Regulation under the FDCA

Under the FDCA,[68] the FDA regulates the safety of food and feed, including non-pesticidal GM foods from new plant varieties. The FDA relies primarily on provisions that prohibit adulteration of food and govern food additives.[69] It has authority to require a pre-market safety review of new foods,[70] but insists that those who introduce a new food have a legal responsibility to evaluate its safety and to ensure that it meets the requirements of the FDCA.[71] Moreover, the FDA believes that GM varieties and their food products 'are as safe and nutritious as their traditional counterparts'.[72]

5.1 FDA Policy

5.1.1 1992 Policy Statement

The FDA's 1992 Policy Statement on bioengineered foods specified how the FDCA would apply to GM foods and provided guidance to industry for food derived from new plant varieties. Because new plant varieties had been developed safely, the FDA did not routinely conduct pre-market safety reviews of new foods derived from plants, and the agency adopted a similar approach to GM foods.[73] It indicated that it would focus on the food product, rather than the process by which the food was produced:

---

[64] 21 USC s 346a(b)(2)(A)(i) and (ii). Section 346(a) of the FDCA does not consider occupational or environmental risks: EPA (n 46 above) 37,774.
[65] 21 USC s 346a(c)(2)(A)(i) and (ii). The risk to infants and children must be considered, but benefits from the use of the pesticide cannot be considered. [66] EPA (n 47 above) 6–7.
[67] See 40 CRF 174.501–28 for exemptions.
[68] 21 USC ss 301–99. Meat and poultry products are regulated, in the main, by the USDA.
[69] 21 USC ss 342 and 348.
[70] FDA, 1992 Policy Statement (n 44 above) 22,988, citing 21 USC s 342(a)(1) as the source of authority for regulating safety of whole foods.
[71] 21 USC s 342(a)(1); see FDA, 1992 Policy Statement (n 44 above) 22,990.
[72] KT Atherton, 'Safety assessment of genetically modified crops' (2002) 181–182 Toxicology 421, 421. [73] FDA, 1992 Policy Statement (n 44 above) 22,988.

[t]he regulatory status of a food, irrespective of the method by which it is developed, is dependent upon objective characteristics of the food and the intended use of the food (or its components). ... [t]he key factors in reviewing safety concerns should be the characteristics of the food product, rather than the fact that the new methods are used.[74]

The FDA relied on its existing regulatory regime to ensure the safety of foods and food ingredients from new plant varieties. Although it expected most transferred genetic material to be safe, genetic modification could cause changes that require evaluation to ensure food safety.[75] If necessary to protect public health (for example, when new foods differ in structure, function, or composition), the FDA has authority to require pre-market review and approval of new foods.[76]

In its 1992 Policy Statement, the FDA indicated that the scientific concepts described 'are consistent with the concepts of substantial equivalence of new foods' articulated by the Organisation for Economic Co-operation and Development (OECD) and principles for assessment of food safety established by the United Nations Food and Agriculture Organization (FAO) and the World Health Organization (WHO).[77] Substantial equivalence is 'regulatory shorthand for defining those new foods that do not raise safety issues that require special, intensive, case by case scrutiny'.[78] It is 'an internationally recognized standard that measures whether a biotech food or crop shares similar health and nutritional characteristics with its conventional counterpart'.[79]

Substantial equivalence is not a safety assessment. Instead, it is a *'comparative* approach and embodies the idea that existing traditionally produced foods can serve as a reference to evaluate the safety of genetically modified foods'.[80] When

---

[74] Ibid 22,984–5. A recent National Academy of Sciences (NAS) report agrees with the FDA: a 'policy to assess products based exclusively on their method of breeding is scientifically unjustified': NAS, *Safety of Genetically Engineered Foods: Approaches to Assessing Unintended Health Effects* (Washington, DC: NAS, 2004) 9.

[75] FDA, 1992 Policy Statement (n 44 above) 22,986–8. Animal feeds raise particular issues, because a single food type may make up a majority of animals' diets and because animals consume parts of plants that humans do not eat: ibid 22,988.

[76] Ibid 22,989–90. The Flavr Savr tomato was the first genetically engineered food reviewed by the FDA. At Calgene's request, the FDA reviewed the Flavr Savr tomato stringently and determined that it was as safe as other tomatoes: FDA, Calgene, Inc.; Availability of Letter Concluding Consultation, 59 Fed Reg 26,647 (23 May 1994). See also ibid 26,700 (approving the tomato's marker gene as a food additive).

[77] FDA, 1992 Policy Statement (n 44 above) 22,992 (referring to OECD and FAO/WHO documents).

[78] HI Miller, 'Substantial equivalence: its uses and abuses' (1999) 17 Nature Biotechnology 1042, 1042. Both concept and name were borrowed from a FDA definition connected with medical devices.

[79] Council for Biotechnology Information, *Substantial Equivalence in Food Safety Assessment* (2001) 1. Substantial equivalence is 'the most practical way to address the issue of food safety at this time': OECD, *Safety Evaluation of Foods Derived by Modern Biotechnology: Concepts and Principles* (Paris: OECD, 1993) 6.

[80] HA Kuiper et al, 'Substantial equivalence—an appropriate paradigm for the safety assessment of genetically modified foods?' (2002) 181–2 Toxicology 427, 427.

a GM food product and its conventional counterpart do not differ in nutritional (or anti-nutritional) components, the GM product is considered to be substantially equivalent.[81] Accordingly, the FDA subjects to pre-market review only foods with characteristics that carry higher risk (for example, higher toxin levels or a new substance) and therefore lack substantial equivalence.[82]

5.1.2 2001 Policy Proposal

In 2001, the FDA proposed to require a pre-market biotechnology notice at least 120 days before commercial distribution of most bioengineered foods, including those derived from new GM plants with pesticidal substances. This measure would allow the FDA to ensure that plant-derived bioengineered foods comply with the FDCA, particularly in situations where unintended changes in foods might raise questions of harm to health or misbranding. Because most modifications would not raise such questions, the FDA did not propose to require pre-market approval for all GM foods or to require special labels.[83] The proposed rule has not been promulgated. At the same time, the FDA published a draft guidance for industry for voluntary labelling,[84] which remains in draft form.

5.2 Guidance Documents

5.2.1 Biotechnology consultation

The 1992 Policy Statement indicated that it is 'prudent practice' for developers of foods using new technologies, including new plant varieties, to consult with the FDA before commercial distribution of food or feed from new plant varieties.[85] In 1997, therefore, the FDA published a guidance document for industry, listing information (including safety and nutritional assessments) that should be submitted to the FDA and describing the procedure for consultation.[86] The document recommended initial consultations on new plant varieties

---

[81] Council for Biotechnology Information (n 79 above) 1. One scholar asserted that the FDA's use of substantial equivalence in the 1992 Policy Statement 'shifted the burden of proof back to the government for the vast majority of GM foods': TO McGarity, 'Seeds of distrust: Federal regulation of genetically modified foods' (2002) 35 University of Michigan Journal of Law Reform 403, 484.   [82] See Miller (n 78 above) 1042–3.
[83] FDA, Premarket Notice Concerning Bioengineered Foods, 66 Fed Reg 4706, 4711 (18 January 2001). Because consultation is successful and protects public health, the FDA believes that the rule is not needed: GAO (n 12 above) 44.
[84] Center for Food Safety and Applied Nutrition (CFSAN), FDA, *Guidance for Industry: Voluntary Labeling Indicating Whether Foods Have or Have Not Been Developed Using Bioengineering* (January 2001).   [85] FDA, 1992 Policy Statement (n 44 above) 22,991.
[86] CFSAN, FDA, *Guidance on Consultation Procedures—Food Derived from New Plant Varieties* (October 1997). Guidance documents do not bind the FDA or the public, nor do they establish legally enforceable rules. See E Seiguer and JJ Smith, 'Perception and process at the Food and Drug Administration: obligations and trade-offs in rules and guidances' (2005) 60 Food and Drug Law Journal 17.

prior to commercialization, other consultations as necessary, and a final consultation once the firm has documentation—including detailed safety and nutritional assessments—to show that its new product is safe. FDA scientists review the data to identify unresolved scientific and regulatory issues.[87] Consultation is voluntary, but food companies do follow the practice,[88] perhaps in part to avoid the extensive liability triggered by unsafe foods.

5.2.2 Early food safety evaluation

In June 2006, the FDA issued its *Recommendations for the Early Food Safety Evaluation of New Non-Pesticidal Proteins Produced by New Plant Varieties Intended for Food Use*.[89] Recognizing the likelihood of cross-pollination from field tests and commingling of seeds, the FDA issued the guidance to address the possibility of 'the inadvertent, intermittent, low-level presence in the food supply of proteins that have not been evaluated through FDA's voluntary consultation process' (the Biotechnology Consultation).[90] The guidance document encourages developers to submit food safety data about new proteins early (perhaps at the time of field tests), before those new proteins might 'inadvertently' enter the food supply via pollen flow or commingling.[91] Questions about the safety of a new protein should be resolved before the developer carries out any activity that could result in the protein entering the food supply. Early food-safety evaluation precedes the Biotechnology Consultation, which occurs when developers plan to commercialize new plant varieties. Developers can use data from the food safety evaluation in the later consultation.[92]

5.3 FDA Regulation

The FDA uses its authority to regulate food additives to govern GM foods. Under the FDCA, foods are adulterated if they bear or contain an added 'poisonous or deleterious substance which may render it injurious to health', an

---

[87] FDA (n 86 above) II.

[88] See L Thompson, 'Are bioengineered foods safe?' (2000) 34(1) FDA Consumer Magazine (January–February) (available at <http:\\www.fda.gov/fdac/features/2000/100_bio.html>, last accessed 18 May 2009).

[89] CFSAN, FDA, *Guidance for Industry: Recommendations for the Early Food Safety Evaluation of New Non-Pesticidal Proteins Produced by New Plant Varieties Intended for Food Use* (June 2006). The GAO recommended that the EPA post results of these evaluations on its website and that, at an early stage, the FDA and the USDA share information about GM crops that might raise health concerns or cause financial losses: GAO (n 12 above) 46.

[90] CFSAN, FDA (n 89 above) 3. By 2006, the FDA had reviewed over 60 bioengineered food products, which were deemed as safe as conventional counterparts: RE Brackett, 'Testimony on the Regulation of Dietary Supplements: A Review of Consumer Safeguards', Committee on Government Reform, US House of Representatives, 9 March 2006.

[91] CFSAN, FDA (n 89 above) 4 and 5.

[92] Ibid 6 (III.C.3). For application, see CFSAN, Office of Food Additive Safety, *US Food and Drug Administration's Statement on LLRICE 600 Series* (March 2007).

unsafe pesticide residue, or an unsafe additive.[93] A food additive is considered unsafe unless it has been granted pre-market approval or is exempt from approval.[94] To gain approval of a new food additive, the manufacturer submits a food additive petition, accompanied by studies to prove the safety of the additive. Approval requires that the FDA be convinced that the proposed use of the additive is safe.[95] Safety means 'a reasonable certainty ... that the substance is not harmful under the intended conditions of use'.[96] If the FDA finds the additive safe, it will issue a regulation to that effect; otherwise additives are considered unsafe and therefore adulterated.[97]

5.3.1 Generally recognized as safe (GRAS)

Most non-pesticidal GM foods escape this rigorous pre-market review as food additives under the so-called GRAS concept.[98] The FDCA definition of food additive excludes substances 'generally recognized as safe' (GRAS).[99] That is, a food additive is defined as a substance that is 'not generally recognized, among experts qualified by scientific training and experience to evaluate its safety, as having been adequately shown through scientific procedures ... to be safe under the conditions of its intended use'.[100] Substances that are GRAS—many ingredients from natural sources and some chemical additives—are not food additives and are therefore exempt from regulation as food additives.[101] Under FDA regulations, however, the determination that a substance is GRAS requires 'the same quantity and quality of scientific evidence as is required to obtain approval of a food additive regulation for the ingredient'.[102]

In 1992, the FDA had indicated that most GM foods will be considered GRAS because most new plant foods had been accepted widely as safe.[103] The use

---

[93] 21 USC s 342(a) (defining adulterated food). Any substance that is not 'an inherent constituent of food or whose level in food has been increased by human intervention' is 'added': FDA, 1992 Policy Statement (n 44 above) 22,989. These added substances fall under the more stringent 'may render [the food] injurious' safety standard, and the food is adulterated if there is a reasonable possibility that it will harm health. See *US v Anderson Seafoods Inc* 622 F2d 157 (5th Cir 1980); *US v Lexington Mill & Elevator Co* 232 US 399 (1914), both cited by the FDA.

[94] 21 USC s 348(a). The Food Additives Amendment, enacted in 1958, did not address foods from new plant varieties, which had been regulated under 21 USC s 342(a)(1).

[95] 21 USC s 348(b) and (c). See s 321(u): '[t]he term "safe" ... has reference to the health of man or animal'. [96] 21 CFR 170.3(i).

[97] Unapproved additives are considered adulterated under 21 USC s 342(a)(2)(C).

[98] On GRAS, see FH Degnan, 'The GRAS Concept: Ensuring Food Safety and Fostering Innovation' in FH Degnan, *FDA's Creative Application of the Law: Not Merely a Collection of Words* (Washington, DC: 2nd edn, Food and Drug Law Institute, 2006) 15. Degnan referred to Congress' enactment of GRAS as 'a display of practical judgment': ibid.

[99] For the regulatory definition of GRAS, see 21 CFR 170.30. [100] 21 USC s 321(s).

[101] FDA, 1992 Policy Statement (n 44 above) 22,989. GRAS substances are listed in 21 CFR parts 182, 184, and 186. See PM Gaynor et al, *FDA's Approach to the GRAS Provision: A History of Process* (2006).

[102] 21 CFR 170.30(b). See Degnan (n 98 above) 26–8 (reviewing litigation on the standard for a GRAS determination). [103] FDA, 1992 Policy Statement (n 44 above) 22,990.

of the GRAS concept for foods produced from biotechnology lets the agency 'take advantage of the dual purposes of the [Food Additives Amendment]—ensuring safety and fostering innovation by relying on the use of sound scientific judgment based on good science'.[104]

In a lawsuit that challenged the FDA's 1992 policy, the court held that the FDA's presumption of GRAS status for GM foods was not arbitrary and capricious.[105] The GRAS determination, the court noted, must be based on technical evidence of safety, which is generally known and accepted in the scientific community. The court stated that '[i]n an area characterized by scientific and technological uncertainty ... this court must proceed with particular caution, avoiding all temptation to direct the agency in a choice between rational alternatives'.[106]

One of the objections to the FDA's use of GRAS for GM foods is that the FDCA does not require the agency to be informed about the additions of substances that manufacturers consider to be GRAS.[107] A 1972 regulation allowed manufacturers to petition the FDA for affirmation of the GRAS status of new substances.[108] Although this regulation has not been repealed, the FDA now uses a notification procedure outlined in a 1997 regulatory proposal, which has not yet been promulgated.[109] Any person can notify the FDA of a claim that a particular use of a substance is exempt from pre-market approval as GRAS. The FDA responds to GRAS notices, but the FDA response is not approval.[110] This means that companies themselves must decide whether a new food substance is GRAS or a food additive that requires FDA approval.[111]

5.3.2 Labelling

In its 1992 Policy Statement, the FDA decided not to require labels for most GM foods.[112] The FDCA defines a food as 'misbranded' if 'its labeling is false or misleading in any particular'.[113] A factor in determining whether a label is misleading is 'the extent to which the labeling ... fails to reveal [material]

---

[104] Degnan (n 98 above) 30.
[105] *Alliance for Bio-Integrity v Shalala* 116 F Supp 2d 166, 177 (D DC 2000), citing 21 CFR 170.30(a) and (b) and deferring to the FDA's evaluation of scientific data within its area of expertise.
[106] *Alliance for Bio-Integrity v Shalala* 116 F Supp 2d 166, 177, quoting *Environmental Defense Fund Inc v Costle* 578 F2d 337, 339 (DC Cir 1978).
[107] See 21 USC s 348 and McGarity (n 81 above) 455.   [108] 21 CFR 170.35.
[109] FDA, Substances Generally Recognized as Safe, 62 Fed Reg 18,938 (17 April 1997). See FDA, *Guidance for Industry: Frequently Asked Questions about GRAS* (2004) ('*GRAS Questions*').
[110] FDA, *GRAS Questions* (n 109 above) Q 15. Between 1997 and February 2006, 193 GRAS notices had been filed: Gaynor et al (n 101 above).
[111] See VR Walker, 'Some dangers of taking precautions without adopting the precautionary principle: a critique of food safety regulation in the United States' (2001) 31 Environmental Law Reporter 10,040, 10,043.   [112] FDA, 1992 Policy Statement (n 44 above) 22,991.
[113] 21 USC s 343(a). The law requires that food labels bear the common or usual name of the food or of each food ingredient (s 343(i)), and some foods—eg a 'major food allergen', defined in s 321(qq)—require special label information (s 343(w)).

facts'.[114] The FDA has interpreted the term 'material' to refer to attributes of the food product and has required special labelling only when the absence of information could pose health or environmental risks, mislead the consumer, or allow the consumer to believe, wrongly, that one food has nutritional, functional, or other characteristics similar to another food.[115]

Prior to the development of GM foods, the FDA did not consider the methods used to develop new plant varieties to be 'material information' under the FDCA, and the development of GM foods did not change the agency's opinion.[116] Moreover, the FDA was

not aware of any information showing that foods derived by these new methods differ from other foods in any meaningful or material way, or that, as a class, foods developed by the new techniques present any different or greater safety concern than foods developed by traditional plant breeding.[117]

Therefore, the FDA does not require labelling of GM products unless, of course, the food itself has significant differences from similar foods (for example, nutritional content or an allergen).[118]

In a lawsuit mentioned above, the plaintiffs challenged the FDA's interpretation of 'material', alleging that the FDA's failure to require labelling of GM foods ignored the 'widespread consumer interest in having genetically engineered foods labeled, as well as the special concerns of religious groups and persons with allergies in having these foods labeled'.[119] The FDA did not believe the FDCA allowed it to require labels for GM foods only to respond to consumer demand. Moreover, the court doubted that the FDCA gave the agency authority to require labels solely because of consumer demand and therefore found the FDA's interpretation reasonable. Only after a material difference between products has been established can the FDA consider consumer opinion in its labelling decision. Because the FDA had determined that genetic modification does not alter foods materially—a determination to which the court granted deference—it lacked a legal basis for requiring labels, even in the face of consumer demand.[120]

---

[114] 21 USC Ibid s 321(n).

[115] Brackett (n 90 above) 5. So, eg if a tomato had an inserted peanut protein that might cause an allergic reaction, the presence of that protein would be a 'material fact,' and its omission would make the label misleading. See 21 USC s 343(a) and (w); and FDA, 1992 Policy Statement (n 44 above) 22,991.

[116] FDA, 1992 Policy Statement (n 44 above) 22,991. The FDA believes that it has 'neither a scientific nor a legal basis to require such labeling': Brackett (n 90 above) 5.

[117] FDA, 1992 Policy Statement (n 44 above) 22,991.

[118] See Brackett (n 90 above) 5. For example, the FDA advised that oil from a GM soybean with altered levels of oleic acid be given a new name.

[119] *Alliance for Bio-Integrity v Shalala* 116 F Supp 2d 166, 177 (D DC 2000) 178.

[120] Ibid 179. On labelling of biotech foods, see FH Degnan, 'Biotechnology and the Food Label' in P Weirich (ed), *Labeling Genetically Modified Food: The Philosophical and Legal Debate* (New York: Oxford University Press, 2007) 17.

Although the FDA has not required labels on GM food products, in 2001 it published a draft guidance for voluntary labelling.[121] Because consumers may be interested in whether food has been genetically modified, the agency developed guidance to help industry ensure that voluntary labelling is truthful and does not mislead consumers. The document gives examples of statements that could be used on food labels without causing the label to be misleading and the food therefore misbranded.[122]

6. State Regulation of GMOs

The Coordinated Framework 'does not contemplate the involvement of state government agencies in the regulatory process'.[123] States receive notice of regulatory activity (for example, when GM plants are moved into specific states) but play no statutory role in federal oversight of GM crops.[124] Federal regulation focuses on plant health, environmental protection, and food and feed safety; states share these concerns, but most agree that the federal government should bear primary responsibility in these areas.[125] Some state agencies would like to play a more active role in the regulatory process, perhaps in co-operation with federal agencies[126] or even by accompanying USDA officials on inspections of field-test sites. Representatives of the EPA, USDA, and FDA plan to co-operate with representatives of state governments to discuss possibilities for more state involvement.[127]

Despite the lack of a state regulatory role in the process of federal regulation of GMOs, state laws often have provisions that are relevant to GMOs. These laws help to promote specific state interests that include 'health and environmental concerns, capturing the economic benefits of biotechnology, preserving market access, and responding to citizens and stakeholders'.[128] Some general

---

[121] CFSAN, FDA (n 84 above), which reaffirmed the FDA's decision not to require labels for all GM food. [122] Ibid 3 and 4–6.

[123] NASDA and Pew Initiative on Food and Biotechnology, *Opportunities and Challenges: States & the Federal Coordinated Framework Governing Agricultural Biotechnology* (Washington, DC: Pew Initiative on Food and Biotechnology, 2006) 6.

[124] For example, the USDA provides information to states before field tests are undertaken. APHIS will work to address state concerns, requiring additional conditions for the tests, if necessary: 'U.S. Inspection System for Biotech Effective, Official Says' (testimony of David Hegwood, 17 June 2003).

[125] MR Taylor, JS Tick, and DM Sherman, *Tending the Fields: State & Federal Roles in the Oversight of Genetically Modified Crops* (Washington, DC: Pew Initiative on Food and Biotechnology, 2004) 21–2. See also D Farquhar and L Meyer, 'State authority to regulate biotechnology under the Federal Coordinated Framework' (2007) 12 Drake Journal of Agricultural Law 439.

[126] See, eg an Illinois law, 430 ILCS 95/0.01-11, which requires state agencies to co-ordinate with federal agencies for release of GMOs.

[127] NASDA and Pew Initiative on Food and Biotechnology (n 123 above) 43 and 65.

[128] Taylor, Tick, and Sherman (n 125 above) 22.

state laws apply to biotechnology. For example, many states have laws, similar to the FDCA, that prohibit the adulteration of food.[129] State laws govern the use of pesticides and protect the environment. Although these laws often predate GMOs, their provisions could authorize state regulation. In addition, a number of states have enacted measures that apply directly to GM crops.

6.1 State Authority

State police power, the traditional source of authority for state (and, by delegation, local) law and regulation to protect public health and welfare, would allow regulation of GM crops and food. The freedom of states and localities to regulate or ban GM crops and their products, however, may be limited by constitutional doctrines that govern the relationship of federal and state regulatory measures. Most important, insofar as GMOs are concerned, are congressional powers to regulate interstate commerce under the Commerce Clause[130] (and restrictions on discriminatory state regulation under the related dormant commerce clause) and to pre-empt state law under the Supremacy Clause.[131]

Federal regulation of GMOs is authorized under the Commerce Clause. Federal authority under the Commerce Clause is not always exclusive, and state laws that do not interfere with interstate commerce can coexist with federal laws. Even when Congress has not regulated, however, state laws that discriminate against interstate commerce or burden commerce unreasonably are prohibited under the dormant commerce clause.

Under the Supremacy Clause, properly enacted federal laws may pre-empt conflicting state and local measures when Congress intends to pre-empt those measures. Congress pre-empts expressly by an explicit statement of its intention to pre-empt (for example, in FIFRA)[132] and may also pre-empt by occupying an entire field of regulation. On the other hand, courts must often decide whether Congress intended its law to pre-empt (for example, if state law conflicts with federal law). The doctrine of pre-emption also allows states to pre-empt local regulation in areas where states have authority to regulate.

Although no challenge to state or local limitations on GMOs has been decided under the Commerce Clause or pre-emption doctrine, such a challenge is possible, particularly if state or local laws prevent the use of GM crops that have been authorized under federal law.[133] In fact, the PPA, under which the

---

[129] State food labelling laws that conflict with the FDCA have long been pre-empted: *McDermott v Wisconsin* 228 US 115 (1913).

[130] US Constitution, art I, s 8, cl 3. The Tenth Amendment provides that powers not delegated to the United States are reserved to the states: US Constitution, amend X.

[131] US Constitution, art VI, cl 2.

[132] A 'State shall not impose or continue in effect any requirements for labeling or packaging in addition to or different from those required' by FIFRA and its regulations: 7 USC s 136v(b).

[133] See JS Harbison, *The War on GMOs: A Report from the Front* (Fayetteville, Arkansas: National Agricultural Law Center, 2004) 20.

State Regulation of GMOs 319

USDA governs GM crops, expressly provides that 'no State or political subdivision of a State may regulate the movement in interstate commerce of any article' if the USDA has issued a regulation or order, unless state restrictions 'are consistent with and do not exceed' the federal provisions.[134] Some uncertainty exists about how this pre-emption provision, which applies to regulation connected with control of plant pests and noxious weeds, would apply to state restrictions on a GM crop for which the USDA has issued a permit.[135]

6.2 State GMO Measures

In recent years, state legislatures considered numerous measures to regulate GMOs within their territories. Although the majority did not pass, states have enacted a number of laws that govern GMOs.[136] States that have adopted these laws often desire independent authority to influence the application of biotechnology in their territories, and they may also wish to collaborate with the federal government to ensure the safety of crops and food. Moreover, a state permit requirement can provide access to detailed information on biotech crops before field tests are conducted.[137]

State laws that govern biotechnology focus on a number of issues, and most state laws support, rather than discourage, GM technology. For example, a majority of states have programmes to provide special funding to support or encourage biotechnology research or facilities.[138] Some state laws protect GM developers and producers by creating criminal or civil penalties for those who wilfully and knowingly destroy a test or research crop[139] or a field crop.[140] Others regulate GM crops in some way (for example by requiring a permit for release[141] or transport into the state,[142] or by requiring applicants for federal permits to submit a copy to the state).[143] A few states require GMs, especially seeds, to be certified or labelled.[144]

---

[134] 7 USC s 7756(b). APHIS regulations may also expressly pre-empt inconsistent state and local measures: see, eg 62 Fed Reg 19,903, 19,915 (24 April 1997).

[135] For further discussion, see Taylor, Tick, and Sherman (n 125 above) 39–40, 112–14, and 120–3.

[136] For recent data, see National Conference of State Legislatures, *Biotechnology Statutes Chart* (July 2007) (available at <http://www.ncsl.org/programs/agri/biotchlg.htm>, last accessed 25 March 2009); and Environmental Commons, *2008 Food Democracy Legislation Tracker* (July 2008) (available at <http://www.environmentalcommons.org/tracker2008.html>, last accessed 25 March 2009). [137] Taylor, Tick, and Sherman (n 125 above) 112.

[138] Eg Iowa Code s 15E.209 (special financing); and Mich Comp Laws s 211.7ii (authorizes property tax exemption for biotechnology research). [139] Eg Cal Food and Agr Code s 52100.

[140] Eg Fla Stat s 604.60.

[141] Eg Fla Stat s 581.083; and Minn Stat ss 18B.01, 18B.285 (GE pesticides), and ss 18F.01–18F.13 (GE agricultural organisms). [142] Eg Idaho Code s 22-2016.

[143] Eg Hawaii Rev Stat s 321-11.6. See also 430 ILCS 95/1-11 (Illinois Release of Genetically Engineered Organisms Act).

[144] Eg Vt Stat Ann Tit 6 ss 611, 644; and Va Code s 3.2-4008.

Some state laws protect innocent farmers when GM products are found on their land. In April 2008, Maine enacted a law that applies to farmers when GM material is found in their possession or on their property and when the presence of the GM material is *de minimis* or unintentional. The law protects farmers from liability for breach of a seed contract and damages claimed by the seed manufacturer. In addition, the Maine law directs the Department of Agriculture to enact best management practices to ensure coexistence of GM and other crops.[145] A California law enacted in September 2008 establishes a protocol for obtaining and testing crop samples to determine whether the farmer has breached a seed contract or infringed a patent. The law protects innocent farmers:

[a] farmer shall not be liable based on the presence or possession of a patented genetically engineered plant on real property owned or occupied by the farmer when the farmer did not knowingly buy or otherwise knowingly acquire the genetically engineered plant, the farmer acted in good faith and without knowledge of the genetically engineered nature of the plant, and when the genetically engineered plant is detected at a de minimis level.[146]

Many states have considered laws that impose pre-emption, in response to local prohibitions of GM crops. A few California counties, starting with Mendocino County in 2004, have banned GM plants,[147] and in March 2008, a small town in Maine enacted an ordinance prohibiting GM crops for 10 years.[148] In response, since late 2004, approximately 15 states have enacted pre-emption statutes. These prohibit local regulation of seeds[149] or prohibit local governments from regulating the development or the 'use of materials and organisms created through the application of biotechnology'.[150] Similar provisions were introduced, but not passed, in other states. One state allows producers who grow crops for food, feed, industrial, and pharmaceutical uses to create voluntary grower districts,[151] which decide what type of crops can be grown in the district.

Some states have adopted policies in response to specific issues connected with GM varieties. For example, in response to the contamination of rice by an unapproved GM variety in 2006, the California Rice Commission supported a moratorium on field testing of GM rice varieties for 2007 and future years and requested that rice varieties be tested for GM traits.[152] The Arkansas State Plant

---

[145] Maine Rev Stat Tit 7 ss 1053–4. See also s 1052 (requiring manufacturers or seed dealers to provide written instructions to growers and keep sales records).
[146] Cal Food and Agr Code ss 52300–6 (quotation, s 52305).
[147] See Harbison (n 133 above) 3. Other California counties rejected GM bans.
[148] 'Rule protects farmers from GE suits', Bangor Daily News, 10 April 2008. Early GM releases led to zoning rules, although local zoning of agricultural activities faces legal obstacles in some states: see MR Grossman, 'Biotechnology, property rights and the environment' (2000) 50 American Journal of Comparative Law (Supplement) 215, 244–6.
[149] Eg Iowa Code s 199.13A; and Kan Stat Ann s 2–1450. See also the Federal Seed Act, 7 USC ss 1551–611.   [150] Eg NJ Stat s 40:8C-2.
[151] Mo Rev Stat ss 261.256, 261.259. Districts could permit or prohibit GM and other types of crops.
[152] See California's Rice Certification Act, Cal Food and Agr Code ss 55000–108. The Rice Commission is authorized by ss 71000–138.

Board banned several rice varieties for 2007 and 2008 and required testing of seed.[153]

States legislatures continue to consider measures to govern biotechnology within their borders. Although states do not want to be 'co-equal' partners with the federal government in regulating GM crops and their products, many states would like a greater voice in the regulatory process and the opportunity to address issues of particular interest to their agricultural industries and citizens.[154]

7. Liability under State Tort Law

The federal regulatory framework for GMOs does not assign liability for damages that might be caused by GMOs.[155] If GMOs cause harm to others, state tort law principles will apply.[156] Producers or purchasers of non-GM, organic, or identity-preserved varieties may suffer losses caused by cross-pollination or commingling. Gene transmission (pollen drift) or commingling may lead to somewhat contained liability, but significant harm may result if GM 'events' not yet approved in the United States, or approved in the United States but not by its trading partners, enter the chain of commerce. If an unapproved variety commingles with other crops, buyers may consider those crops contaminated, with serious economic consequences.[157] At least six reported incidents involved unauthorized releases of GM crops.[158] Damages connected with StarLink™ and ProdiGene maize are early examples; Liberty Link rice is more recent.[159]

At the outset, however, it should be noted that the seed industry has developed stewardship programmes intended to avoid commingling and resultant legal claims. For example, both the American Soybean Association (ASA) and the National Corn Growers Association (NCGA) have effective stewardship programmes. Motivated by the threat of liability to farmers it

---

[153] Arkansas State Plant Board, 'Regulations on Plant Diseases and Pests', Circular 11 (December 2007) 43–4.
[154] NASDA and Pew Initiative on Food and Biotechnology (n 123 above) 4 and 61–5.
[155] The discussion of liability is drawn from MR Grossman, 'Genetically modified crops in the United States: Federal regulation and state tort liability' (2003) 5 Environmental Law Review 86; and MR Grossman, 'Anticipatory nuisance and the prevention of environmental and economic harms from GMOs in the United States' (2008) 18 Journal of Environmental Law and Practice 107.
[156] This chapter does not address contract or intellectual property issues associated with GM crops.
[157] See L Khoury and S Smyth, 'Reasonable foreseeability and liability in relation to genetically modified organisms' (2007) 27 Bulletin of Science, Technology and Society 215, 221.
[158] GAO (n 12 above) 90–5. These include StarLink (2000), Prodigene (2002), *Bt* 10 (2004) and Event 32 (2006) maize, and Liberty Link rice 601 and 604 (2006).
[159] See USDA, *Report of LibertyLink Rice Incidents* (October 2007); and USDA (n 34 above). On the *StarLink* litigation, see text at nn 226–37 below.

represents, the ASA asked seed companies not to sell unapproved varieties commercially and established minimum requirements for preserving the identity of unapproved GM soybeans. These steps have helped to avoid commingling of approved varieties with GM varieties not acceptable to trading partners.[160] Similarly, the NCGA has urged members to minimize trade disruptions by being aware of the approval status of maize hybrids and by selling hybrids not approved in the European Union (EU) into domestic, rather than export, markets.[161]

When harm follows failure of industry stewardship or the activities of neighbouring farmers, plaintiffs damaged by GM crops may sue in nuisance, trespass, negligence, or strict liability,[162] or a combination of these causes of action. Although tort causes of action can redress damage to the environment and harm to persons and property from GM crops, the efficacy of tort remedies will remain theoretical until courts have had the opportunity to decide cases involving damage from GMOs. The burden of proving causation of cross-pollination and other types of damage, likely with the aid of experts, may make GM damage cases expensive to litigate. Moreover, the plaintiff may be at a serious financial disadvantage when the defendant is a large corporation that has developed and marketed the GM crop.

7.1 Nuisance

As scientists were developing GM crops in the early 1980s, a legal scholar recognized nuisance as a possible remedy for actual damage (or even fear of future harm) resulting from biotechnology research. He asserted that common law nuisance could be 'redefined to cope with the risks of modern science and technology, then coupled with the use of complex injunctive decrees to provide a prophylactic means of governing hazard'.[163] Indeed, two decades later, plaintiffs used nuisance and other torts to claim property and related economic damage from GM crops.[164]

---

[160] See ASA, *Policy Resolution on Biotechnology Approvals and Minimum Requirement for Attempted Identity-Preserved Production, Harvesting, and Utilization of Biotechnology-enhanced Varieties/Hybrids that are Unapproved for Export to Major Markets* (2000) (available at <http://www.soygrowers.com/publications/minrequire-IP.htm>, last accessed 25 March 2009).

[161] NCGA, *Know Before You Grow* (linked from <http://ncga.eweb3.socket.net/biotechnology-reference-guide>, last accessed 25 March 2009) (listing GM maize varieties and their approval status in the EU and Japan).

[162] For more detailed discussion of trespass, negligence, and strict liability, see Grossman, 'Genetically modified crops' (n 155 above); DL Kershen, 'Legal liability issues in agricultural biotechnology' (2002) 10 Environmental Liability 203; and PJ Heald and JC Smith, 'The problem of social cost in a genetically modified age' (2006) 58 Hastings Law Journal 87.

[163] See BR Furrow, 'Governing science: public risks and private remedies' (1983) 131 University of Pennsylvania Law Review 1403, 1466 (referring to a private right of action).

[164] See *In re StarLink Corn Products Liability Litigation* 212 F Supp 2d 828 (ND Ill 2002).

Nuisance provides redress when a defendant's activities interfere unreasonably with the plaintiff's use and enjoyment of land, injure life or health, or interfere with public rights. A nuisance is any activity or condition that is harmful, annoying, offensive, unpleasant, disagreeable, or inconvenient to another.[165] Nuisance often focuses on reasonable use of land in light of the circumstances. That is, both plaintiff and defendant have the right to reasonable use and enjoyment of their property; the defendant cannot cause unreasonable harm to the plaintiff, and the plaintiff may have to endure some inconvenience to accommodate the defendant's legitimate land uses. The existence of a nuisance is a question of fact, and courts will balance the interests of the parties, weighing numerous factors, including the locality of the defendant's activities, in that balance.[166] In part because of this fact-based balance of competing interests, the doctrine of nuisance has suffered from 'confusions, contingencies and lack of principle'.[167]

Nuisance law involves several important distinctions. Focusing on the interest invaded, private nuisance results from interference with the plaintiff's use and enjoyment of land, while public nuisance arises from activities that interfere with land use of a large number of persons or with public rights.[168] Focusing on the defendant's conduct, nuisance can be intentional or negligent. Intentional nuisance requires that defendant's use of land caused the plaintiff to suffer substantial and unreasonable interference with use of property and that the defendant have knowledge (sometimes called 'civil intent') that its activities were substantially certain to injure the plaintiffs. Negligent nuisance, in contrast, requires proof that the defendant's activities themselves (instead of the defendant's interference with the plaintiff) were unreasonable.[169] Intentional nuisance often poses fewer obstacles to recovery, because it does not require proof that the defendant's behaviour (for example, in planting GM crops) was unreasonable. Further, some courts distinguish between nuisance *per se* and nuisance *per accidens*. Nuisances *per se* or in law exist if an activity (for example, a business prohibited by law) will be a nuisance under any circumstances, regardless of its location. In contrast, nuisances *per accidens* or in fact are activities (often lawful activities) that may become a nuisance, depending on the circumstances, the location, or the way they are carried out.[170] Remedies for nuisance include damage awards and (less often) full or partial injunctions against the defendant's activities.

---

[165] Restatement (Second) of Torts s 821A cmt b (1979) ('Restatement').
[166] LA Halper, 'Untangling the nuisance knot' (1998) 26 Boston College Environmental Affairs Law Review 89, 100.   [167] Ibid 91.
[168] Restatement (n 165 above) s 822 cmts c and d.
[169] ZJB Plater et al, *Environmental Law and Policy: Nature, Law, and Society* (New York: 3rd edn, Aspen Publishers, 2004) 106–7.
[170] SM Williams, 'The anticipatory nuisance doctrine: one common law theory for use in environmental justice cases' (1995) 19 William and Mary Environmental Law and Policy Review 223, 242 and 244.

7.1.1 Public nuisance

A public nuisance is an 'unreasonable interference with a right common to the general public'.[171] An interference with public rights is unreasonable when 'the conduct involves a significant interference with the public health, the public safety, the public peace, the public comfort or the public convenience, or ... the conduct is proscribed by a statute, ordinance or administrative regulation'.[172] In deciding a public nuisance claim, the court will often balance the value of the defendant's conduct against the seriousness of harm to the public. Most public nuisance cases are pleaded in intentional, rather than negligent, nuisance.[173] A government official normally brings a public nuisance case, but a private plaintiff may sue in public nuisance if the plaintiff has suffered a so-called 'special injury', different in kind from members of the general public.[174] Because public nuisance does not normally provide monetary damages to such plaintiffs,[175] an individual plaintiff (for example, a farmer) with crop damage from GM pollen has little incentive to claim public nuisance.

Environmental and economic consequences of GMOs may result in claims of public nuisance. Liability might arise from pollen drift that damages non-GM crops or causes out-crossing to wild relatives, commingling of GMOs with other crops after harvest that damages trade,[176] or marketing practices that threaten exports. A lawsuit might therefore seek an injunction against sale of an unapproved GM variety or even against sale of a variety not approved in importing nations, if such sales would interfere seriously with public interest in grain trade or if the seed company does not warn growers about commingling and its economic risks.[177]

Thus, the defendant in a public nuisance action involving GMOs may be a seed company, rather than an individual farmer. A defendant can be liable in nuisance 'not only when he carries on the activity but also when he participates to a substantial extent in carrying it on'.[178] This participation may include, for example, activities 'that create physical conditions that are harmful to neighboring land after the activity that created them has ceased'.[179] Accordingly, the

---

[171] Restatement (n 165 above) s 821B. 'Unlike a private nuisance, a public nuisance does not necessarily involve interference with the use and enjoyment of land': ibid cmt h.
[172] Ibid s 821B.   [173] Plater et al (n 169 above) 115.
[174] Restatement (n 165 above) s 821C and cmt d.
[175] Ibid cmt b. Citing s 821C, Furrow suggests that a plaintiff with standing who seeks only an injunction may not need to prove special injury: Furrow (n 163 above) 1440.
[176] See Kershen (n 162 above) 214.
[177] TP Redick and CG Bernstein, 'Nuisance law and the prevention of "genetic pollution": declining a dinner date with Damocles' (2000) 30 Environmental Law Reporter 10,328, 10,334–5.
[178] Restatement (n 165 above) s 834. See *In re StarLink Corn Products Liability Litigation* 212 F Supp 2d 828, 845 (ND Ill 2002) (quoting Restatement s 834).
[179] Restatement (n 165 above) s 834 cmt b. The comment gives land-based activities as examples, but its application seems to be broader, as *StarLink* indicates. See also s 843 cmt e, discussing liability of those whose activities have created a harmful physical condition.

seed company's activities could, in appropriate circumstances, constitute nuisance.

7.1.2 Private nuisance

A private nuisance is 'a nontrespassory invasion of another's interest in the private use and enjoyment of land'.[180] Private nuisances can result from physical invasions of land, but also from loud noises, vibrations, noxious smoke or fumes, offensive odors, pollution, and other unreasonable interferences.[181] They often arise between adjoining landowners, when the defendant's activities interfere unreasonably with the plaintiff's use and enjoyment of land, but a private nuisance may harm more than a single individual.

To recover in private nuisance, the plaintiff must demonstrate that he or she suffered unreasonable interference and that the defendant knew of the harm (intentional nuisance) or that the defendant's actions were unreasonable (negligent nuisance) and caused significant harm to a normal person residing in the relevant community.[182] An interference is unreasonable, for example, when 'the gravity of the harm outweighs the utility of the actor's conduct'.[183] Significant harm means 'a real and appreciable interference with the plaintiff's use or enjoyment of his land'; the harm must be 'more than slight inconvenience or petty annoyance [because the] law does not concern itself with trifles'.[184]

Landowners who suffer unreasonable interferences from GMOs may claim damages in nuisance. Airborne pollen from GM crops may be analogous to interferences caused by other airborne contaminants (for example, pesticides), which may constitute a nuisance when they cause actual harm to property or impair its use. Thus, private nuisance may offer a remedy for a plaintiff whose land has been 'contaminated' by GM pollen or seed, with resulting loss of an organic crop or decreased marketability of a traditional crop.

The plaintiff who claims intentional nuisance against a neighbouring farmer must prove that the defendant knowingly planted a GM crop without taking proper precautions to prevent pollen drift to a nearby field. The plaintiff who pleads negligent nuisance faces the additional burden of proving that defendant's activities were unreasonable. This burden may be difficult to meet if the defendant followed industry standards of stewardship, but the behaviour of a defendant who planted GM crops without observing recommended separation

---

[180] Ibid s 821D. See Kershen (n 162 above) 209–10.

[181] See R Repp, Comment 'Biotech pollution: assessing liability for genetically modified crop production and genetic drift' (2000) 36 Idaho Law Review 585, 605.

[182] Restatement (n 165 above) s 821F. A court will consider '[t]he location, character and habits of the particular community ... in determining what is offensive or annoying to a normal individual living in it': ibid s 821F cmt e.

[183] Ibid s 826. See Halper (n 166 above) 127 (noting that the 'utility of the defendant's conduct includes both its utility to the community *and* its utility to defendant').

[184] Restatement (n 165 above) s 821F cmt c.

distances and other precautions may be considered unreasonable.[185] The successful plaintiff may recover damages, but injunctive relief in private nuisance is less likely, particularly if the GM crop has been approved.

A GM seed company may also be a defendant in a private nuisance suit. A number of states have held that a manufacturer may be liable for a nuisance caused by a product in the hands of a purchaser. In the GM crop situation, '[r]esidue from a product drifting across property lines presents a typical nuisance claim. All parties who substantially contribute to the nuisance are liable'.[186] A court may find that a seed company substantially contributed to a nuisance caused by pollen drift and is therefore liable in nuisance.[187]

Moreover, a seed company's own behavior may constitute a nuisance. Although the sale of approved GM crops is not likely to be a nuisance *per se*,[188] a company that does not disclose known risks of commingling of unapproved GM seeds to producers who plan to export their crops may have engaged in an unreasonable marketing practice. Failure to prescribe proper practices, including separation distances and reserves, may also trigger liability. Under these circumstances, the farmer whose crop is contaminated by GM seed may have the right to recover in nuisance against the seed company.[189]

7.1.3 Anticipatory nuisance?

The doctrine of anticipatory (or prospective) nuisance allows a plaintiff to seek an injunction to prevent future harm from a defendant's proposed activity.[190] Both federal and state courts have held that such threatened nuisances can be enjoined.[191] Courts are reluctant to enjoin activities before harm has occurred, so the plaintiff's burden of proof is high. That is, the threatened tort must be reasonably certain to occur, and the harm involved must be significant. Some courts will enjoin only activities that would be a nuisance *per se* (and not a nuisance *per accidens*). Nonetheless, a court need not wait until an injury occurs,

---

[185] Right-to-farm laws protect some farmers against nuisance suits, particularly when a nuisance arises from changed circumstances in the surrounding area, but their applicability to growers of GM crops will depend on the specific facts and the language of the statute in the grower's state: see Heald and Smyth (n 162 above) 120–3 (concluding that right-to-farm laws are unlikely to shield GM farmers from nuisance liability).

[186] *In re StarLink Corn Products Liability Litigation* 212 F Supp 2d 828, 847 (ND Ill 2002). See Restatement (n 165 above) s 834.

[187] *In re StarLink Corn Products Liability Litigation* 212 F Supp 2d 828, 847 (ND Ill 2002). The court indicated that the facts of a nuisance suit will indicate whether the company made a substantial contribution. [188] Kershen (n 162 above) 209.

[189] Redick and Bernstein (n 177 above) 10,329 and 10,338. Seed contracts for 'identity preserved' crops often shift the risk of nuisance liability to the farmer.

[190] For an extensive analysis of anticipatory nuisance, see Grossman, 'Anticipatory nuisance' (n 155 above) and sources cited therein.

[191] For example, *Village of Wilsonville v SCA Services Inc* 426 NE2d 824 (Ill 1981), a classic anticipatory nuisance decision.

but can enjoin a 'threat of sufficient seriousness and imminence'.[192] Although either a private or a public nuisance can be enjoined, most anticipatory nuisance cases involve public nuisances.[193] Courts have enjoined anticipatory nuisances that threaten environmental harm, and the doctrine has applied to construction of large livestock feedlots.[194]

Anticipatory nuisance might offer a way to prevent threatened environmental harm and economic losses from GMOs when those harms and losses are reasonably certain to occur and will result in irreparable harm. Although scientists disagree about the possibility of environmental harm, cross-pollination of non-GM crops or wild relatives raises environmental issues.[195] Asynchronous approvals of GMOs in the United States, the EU, and other trading nations mean that pollen drift or commingling of a GM variety not approved by trading partners could threaten exports and cause widespread economic harm.[196] Incidents involving failures of containment, like the StarLink™ maize discussed below, demonstrate that economic loss can be extensive. Thus, an injunction against a threatened nuisance from a GM crop, in a situation where the threatened harm is serious and irreparable, might be an appropriate remedy. Anticipatory nuisance does not offer a remedy, however, when harm has already occurred and damages can be calculated.

7.2 Other Torts

Much discussion of potential losses from GMOs has focused on nuisance, but other tort causes of action are possible.

7.2.1 Trespass

Trespass requires an invasion of real property that interferes with the plaintiff's exclusive possession and causes damage to that property.[197] The tort of trespass occurs when the defendant, without permission, enters or causes an object to enter the land of another. The defendant's actions must be intentional, but 'it is

---

[192] Restatement (n 165 above) s 821F cmt b; and s 933 cmt b. See also WP Keeton et al, *Prosser and Keeton on Torts* (St Paul: 5th edn, West Publishing, 1984) 640–1.

[193] GP Smith II, 'Re-validating the doctrine of anticipatory nuisance' (2005) 29 Vermont Law Review 687, 721.

[194] For example, *Nickels v Burnett* 798 NE2d 817 (Ill App 2003); and *Superior Farm Management v Montgomery* 513 SE2d 215 (Ga 1999).

[195] See, eg *Geertson Seed Farms v Johanns* 2007 WL 518624 (ND Cal) (discussing the environmental and trade impacts of GM alfalfa), discussed at n 25 above.

[196] For examples of harm from recent controversies, see Grossman 'Anticipatory nuisance' (n 155 above) 147–56.

[197] See Repp (n 181 above) 600; Kershen (n 162 above) 205–7. An old requirement for trespass, that the invasion be 'direct and immediate', might not be met in the case of trespass by GM pollen, because of the passage of time (pollination occurs long after planting) and the intervening force of the wind: Heald and Smith (n 162 above) 135.

the intent to enter the land, not the intent to trespass, that is key'.[198] Intent for trespass, as for intentional nuisance, means knowledge. Of course, the plaintiff must prove that the defendant's activities (rather than other sources) caused the invasion of the plaintiff's property.

Courts have awarded damages in trespass for invasion of and damage to the plaintiff's property by airborne pollutants.[199] Even invisible particulates can cause trespass if invasion causes damage to the plaintiff.[200] Thus, GM crops that contaminate neighbouring land interfere with the neighbour's right to exclusive possession and leave the GM farmer vulnerable to a trespass claim. A farmer who knows that it is 'substantially certain' that GM seeds will escape to a neighbour's property can be liable for trespass and the resulting harm.[201] Pollen from GM crops may also constitute trespass in jurisdictions that have recognized trespass by particulates.[202] But because pollen flow is a normal part of crop production, recovery for trespass by pollen is likely to require a showing of substantial damage.[203] Damages from trespassing GMO seed or pollen may include contamination that makes the land unfit for use (for example, for organic crops) or loss of market for crops commingled with GMOs.[204]

7.2.2 Negligence

Negligence is conduct that 'falls below the standard established by law for the protection of others against unreasonable risk of harm'.[205] A claim of negligence usually requires the plaintiff to prove that the defendant had a duty to conform to a specific standard of conduct (to exercise reasonable care under the circumstances), that the defendant breached that duty, that the plaintiff suffered harm, and that the defendant's breach of duty was the factual and proximate cause of the plaintiff's injury.[206]

A plaintiff injured by GM crops, perhaps through cross-pollination of GM crops to organic crops, must prove that the defendant, rather than other producers in the vicinity, actually caused the harm. In addition, the plaintiff must prove that the defendant who planted approved GM crops failed to exercise reasonable care. It seems likely that planting directions from the seed company (for example, requirements for separation distances from other crops or a reserve area planted with conventional crops) establish a duty of care for farmers who

---

[198] JA Davies and LC Levine, 'Biotechnology's challenge to the law of torts' (2000) 32 McGeorge Law Review 221, 223.
[199] Eg *Borland v Sanders Lead Co* 369 So 2d 523 (Ala 1979).
[200] Eg *Martin v Reynolds Metals Co* 342 P2d 790 (Or 1959) (gases and particulates that settled on plaintiff's land, making land unfit for raising livestock, constituted trespass).
[201] Davies and Levine (n 198 above) 223–4.
[202] Redick and Bernstein (n 177 above) 10,336.
[203] Kershen (n 162 above) 206 (asking if foreseeable actions of insects carrying pollen from GM crops to neighboring fields make the GM farmer liable in trespass).
[204] See Repp (n 181 above) 602–5.   [205] Restatement (n 165 above) s 282.
[206] See Kershen (n 162 above) 208–10.

plant GM crops. A farmer who fails to follow appropriate practices (for example, by failing to maintain separation distances) may have acted unreasonably. In contrast, a farmer who follows the requirements imposed in connection with purchase of GM seed is less likely to have breached the duty of care.[207] Of course, the defendant's unreasonable conduct must have damaged the plaintiff's crops or real property.

Seed companies might also be defendants in negligence, but negligence does not always provide an adequate remedy. The economic loss doctrine, discussed below, precludes claims based on pure economic loss. Moreover, FIFRA may pre-empt some negligence claims based on failure to warn or another labelling claim, when a GM crop with pesticidal properties has been registered under FIFRA. That law pre-empts state 'requirements for labeling or packaging in addition to or different from' those required for pesticides by federal law.[208] State tort claims characterized as additional state law requirements, based on claims of inadequate labelling, have been pre-empted.[209] FIFRA pre-emption has also been applied to GM crops with pesticidal properties; claims based on failure to warn were pre-empted, but other tort law claims (for example, negligence *per se*) were not.[210]

7.2.3 Strict liability

Strict liability applies in situations where the defendant, even a defendant who acted with reasonable care, causes injury in the course of an activity characterized as abnormally dangerous or ultra-hazardous. To identify an abnormally dangerous activity, courts often consider factors that include the degree of risk, likelihood of serious harm, inability to eliminate risk, commonness of usage, appropriateness of activity to the area, and its value to the community.[211]

Strict liability will apply to GMOs only if planting, producing, or selling GM crops is considered abnormally dangerous. Cross-pollination and other contamination from GM crops may be analogous to pesticide drift, and most US jurisdictions have refused to impose strict liability for pesticide drift from aerial pesticide application.[212] Similarly, courts may be reluctant to impose strict liability for harm from GMOs. Planting GM crops, approved in a federal regulatory process designed to evaluate risks, is unlikely to be considered an ultra-hazardous activity. Instead, the presence of approved GM crops on millions of hectares of US farmland makes their use common and appropriate to the rural area. In contrast, however, the use of unapproved GMOs or practices

---

[207] Repp (n 181 above) 615–16.   [208] 7 USC s 136v(b).
[209] See *Bates v Dow Agrosciences LLC* 544 US 431 (2005); and *Riegel v Medtronic Inc* 552 US 312 (2008).
[210] See *In re StarLink Corn Products Liability Litigation* 212 F Supp 2d 828, 835–8, 852 (ND Ill 2002). See also *Mortellite v Novartis Crop Protection Inc* 460 F3d 483 (3d Cir 2006).
[211] Restatement (n 165 above) s 520 and cmt f.
[212] For example, *Bennett v Larsen Co* 348 NW2d 540, 553 (Wis 1984); but see *Langan v Valicopters Inc* 567 P2d 218 (Wash 1977).

that do not comply with regulations and industry guidelines could pose significant, foreseeable risk and thus be subject to strict liability, as well as negligence.

Two special situations may require different analysis. First, pollen drift or commingling of GM crops intended for pharmaceutical or industrial (rather than commodity) use may pose greater risks, especially to health. To determine whether production of pharmaceutical or industrial crops is abnormally dangerous, the court must decide whether the risk to the food supply outweighs the benefit from the GM products.[213] Secondly, in areas where organic crops predominate, cross-pollination from GM crops may be particularly feared. Although it might be argued that GMOs should be considered alien substances under the rule of *Rylands v Fletcher*,[214] a contrary argument is that the organic producer's activities should be considered 'abnormally sensitive'. In that case, strict liability may be precluded if harm 'would not have resulted but for the abnormally sensitive character of the plaintiff's activity'.[215]

A GM seed company might also be sued in strict products liability, which requires the plaintiff to establish that a defect, which made the product unreasonably dangerous, existed when the product was released into the stream of commerce. Defects may involve design, manufacturing, or inadequate instructions or warnings.[216] For example, a GM crop with a new allergen could trigger a claim in strict products liability, but proof that the GM product is defective in design or manufacture may pose a difficult burden for the plaintiff.

7.3 Limitations on Tort Liability

Some tort claims involving GMOs may be precluded by the economic loss doctrine. Courts in the United States, as in other common law countries, restrict compensation for 'pure economic loss' (that is, loss not accompanied by damage to the plaintiff's property) for 'policy reasons centered on the fear of extensive liability'.[217] Under the economic loss doctrine, '[p]hysical injuries to persons or property, and economic losses connected with those injuries, are compensable; solely economic injuries are not'.[218] The doctrine helps to maintain the distinction between tort law and contract law, protect parties' right to allocate risks by contract, and encourage assumption or allocation of risk.[219]

---

[213] Kershen (n 162 above) 208 (noting that the existence of a tolerance under the FDCA may affect the decision).
[214] (1868) LR 3 HL 330. See M Cardwell, 'The release of genetically modified organisms into the environment: public concerns and regulatory responses' (2002) 4 Environmental Law Review 156, 162.   [215] Restatement (n 165 above) s 524A; see Kershen (n 162 above) 207–8.
[216] Restatement (Third) of Torts: Products Liability, ss 1–2 (1998). On defects in food, see ibid s 7.
[217] Khoury and Smyth (n 157 above) 229. See, generally, D Gruning, 'Pure economic loss in American tort law: an unstable consensus' (2006) 54 American Journal of Comparative Law (Supplement) 188.
[218] *In re StarLink Corn Products Liability Litigation* 212 F Supp 2d 828, 838, 840 (ND Ill 2002).
[219] Ibid 839.

Applying the doctrine in a case involving alleged harm from GM maize, the July 2002 *StarLink* opinion, discussed below, indicated that the unknowing purchase of seed with the Cry9C protein from a contaminated inventory would constitute economic loss; that loss would involve the very property that was purchased, and the parties could have bargained for contractual protection. But if the plaintiffs' crops were contaminated from pollen drift from another farm or if their harvest was commingled with StarLink maize, the economic loss doctrine would not preclude recovery, because the plaintiffs' crops themselves (other property) sustained damage.[220] This reasoning is consistent with the principle stated in the Restatement (Third) of Torts: Products Liability, which indicates that 'harm to persons or property includes economic loss if caused by harm to ... the plaintiff's property other than the defective product itself'.[221]

Other decisions illustrate application of the doctrine. In a case involving damages under the *StarLink* settlement, the economic loss doctrine prevented recovery by a co-operative that marketed identity-preserved grain for growers. The co-operative never took title to the maize, which was transferred directly from the growers to the purchaser and became commingled with StarLink maize in the stream of commerce. In clear language, the settlement protected growers and grain elevators with direct physical injury (contaminated crops); therefore, the co-operative, which was neither a grower nor an elevator and never owned the crops, was not included in the settlement.[222] Commentators noted the implication of the ruling: 'that the pure economic loss doctrine bars recovery for claims related to loss of access to international markets'.[223]

Sample v Monsanto[224] involved claims in public nuisance and negligence for harms caused by Monsanto's introduction of GM maize and soybeans into the market. The plaintiffs had claimed that non-GM farmers lost revenue because of the European boycott of US maize. The court granted summary judgment to Monsanto under the economic loss doctrine. Both Sample and another farmer grew non-GM crops; some crops were tested, and others not. But neither farmer could show physical 'injury to his person or his property relating to the presence of GM corn or soybeans on the market'.[225]

7.4 An example: *StarLink*

No US court decision, to this author's knowledge, has found liability in tort for losses related to GM crops. One seminal decision, however, has suggested that

---

[220] Ibid 843.   [221] Restatement (Third) of Torts: Products Liability s 21 and cmt c.
[222] *Agra Marke Inc v Aventis CropScience USA LP* 2005 WL 327020 (ND Ill) (not reported in F Supp 2d) *3–4. Agra Marke's claim, for nearly $1 million, was for loss of the per-bushel premium it would have received for non-GM maize.
[223] SJ Smyth and DL Kershen, 'Agricultural biotechnology: legal liability regimes from comparative and international perspectives' (2006) 6(2) Global Jurist Advances 14.
[224] 283 F Supp 2d 1088 (ED Mo 2003).   [225] Ibid 1091–2.

tort causes of action may offer remedies for damages. *In re StarLink Corn Products Liability Litigation*[226] was a class action that consolidated 15 lawsuits brought by maize farmers against Aventis CropScience, the creator and manufacturer of the GM maize, StarLink, and a licensee-distributor. StarLink maize, which contains the protein Cry9C (with attributes like some human allergens), was approved for feed and industrial uses, but not for human consumption. Because of StarLink's limited registration, the EPA required special measures to prevent cross-pollination and commingling.[227] Although StarLink was planted on relatively few acres,[228] food products for human consumption were found to contain Cry9C, and numerous maize products were recalled. Other economic damage resulted; some US food producers used imported maize, and most trading partners stopped importing US maize. As a result, Aventis applied to cancel the limited registration of StarLink maize.

The plaintiffs alleged that the defendants' failure to comply with EPA requirements led to contamination of the US maize supply, increasing farmer costs and depressing maize prices. Among the 57 counts were common law tort claims in negligence, public and private nuisance, and strict liability. The plaintiffs alleged that StarLink's 'contamination of the general food corn supply' constituted a public nuisance.[229] Their private nuisance claim alleged that the defendants distributed their maize seeds 'knowing that they would cross-pollinate with neighboring corn crops'.[230] The plaintiffs' negligence claim referred to Aventis' duty to ensure that the maize did not reach human food and its breach of that duty by failing to monitor and enforce compliance with EPA requirements.[231]

On the defendants' motion, the court dismissed some of the plaintiffs' claims, but held that the case could go to trial on several tort claims, including negligence *per se* (for example, for violation of regulatory duties) and nuisance.[232] On the claim of public nuisance, the judge found that '[c]ontamination of the food supply implicates [public] health, safety, comfort and convenience', and concluded that contamination was 'an unreasonable interference with a right common to the general public'.[233] Moreover, because the plaintiffs, commercial

---

[226] 212 F Supp 2d 828 (ND Ill 2002). Cases brought by consumers who ate contaminated maize-based food were settled for $9 million. See A Harris, 'Danger uncertain, but suits multiply' *National Law Journal* (New York: 9 September 2002). The description of *StarLink* is drawn from Grossman, 'Anticipatory nuisance' (n 155 above).

[227] 212 F Supp 2d 828, 834. The plaintiffs alleged that Aventis did not follow EPA requirements, perhaps because Aventis expected StarLink to be approved for human use. The EPA no longer approves a GM crop with an incorporated pesticidal agent for feed, but not for food.

[228] The largest crop, in 2000, was 350,000 acres, less than 0.5% of US maize acreage: EPA, *Starlink TM Corn Regulatory Information* (October 2007).   [229] 212 F Supp 2d 828, 848.

[230] Ibid 844.   [231] Ibid 843.

[232] Ibid 852. FIFRA pre-empted negligence and strict liability claims based on failure to warn, however, and the court indicated that the economic loss doctrine would preclude some types of claims. See also *Sample v Monsanto* 283 F Supp 2d 1088 (ED Mo 2003).

[233] 212 F Supp 2d 828, 848.

maize farmers, relied on the integrity of the maize supply for their livelihood, they had suffered a special injury and could bring the public nuisance claim.[234] On the claim of private nuisance, the court noted that 'drifting pollen can constitute an invasion, and that contaminating neighbors' crops interferes with their enjoyment of the land. The issue is whether defendants are responsible for contamination caused by their product beyond the point of sale'.[235] Aventis, with obligations imposed by limited registration of StarLink maize, was in a position to control the nuisance; therefore, the court held that the plaintiffs had stated a valid claim for private nuisance. The court did not decide the merits of the claims. At trial, the plaintiffs would be required to meet the burden of proof for their tort and other claims. The tort claims in *StarLink* were never adjudicated, however, because the parties agreed to settle many of the issues raised in the lawsuit.

After the EPA accepted Adventis' voluntary cancellation of the registration of StarLink Aventis formed StarLink Logistics Inc to oversee efforts to monitor maize so that Cry9C could be contained and removed from the food supply. StarLink Logistics has a database with results of more than four million tests on over four billion bushels of maize in maize-handling facilities.[236] Recent studies indicate that efforts to contain StarLink have succeeded, and little Cry9C remains in the food supply. Therefore, in March 2008, the EPA recommended that the testing programme be ended.[237]

8. Conclusion

Since 1986, when the Coordinated Framework was published, numerous GM crops and their products have been commercialized. Since 1996, when farmers in six countries began planting GM crops, steady increases in GM production have occurred. In 2008, about 13.3 million farmers (most in developing countries) in 25 nations planted 125 million hectares of GM crops. The 2008 global market value (that is, sale of seed plus technology fees) of GM crops is estimated at 7.5 billion dollars and projected to be 8.3 billion dollars in 2009. Between 2009 and 2015, further increases in GM production are projected—with additional biotech-crop countries, number of hectares, types of crops and traits,

---

[234] Ibid. On a similar case in Canada, see H McLeod-Kilmurray, '*Hoffman v Monsanto*: courts, class actions, and perceptions of the problem of GM drift' (2007) 27 Bulletin of Science, Technology and Society 188. [235] 212 F Supp 2d 828, 845.
[236] EPA and FDA, EPA Draft White Paper Regarding StarLink® Corn Dietary Exposure and Risk; Availability for Comment, 72 Fed Reg 58,978, 58,979 (17 October 2007).
[237] Office of Pesticide Programs, US EPA, White Paper Concerning Dietary Exposure To Cry9c Protein Produced by Starlink® Corn and the Potential Risks Associated with Such Exposure (28 March 2008).

and a significant increase in the number of farmers planting biotech crops, especially in developing countries.[238]

Many of these crops were developed in the United States under the federal regulatory system described in this chapter. Although the basic structure of US regulation has remained constant over the past decades, new federal laws have been enacted, and the USDA, EPA, and FDA have articulated new policies and enacted new regulations. The agencies will continue to review their policies and to amend their regulations in response to the increasing number and variety of GM crops and products. These new policies and regulatory measures, along with continued co-operation among the USDA, EPA, and FDA, will help to ensure that GM crops and their food products continue to be safe to grow, safe for the environment, and safe to eat.

[238] James (n 1 above).

PART IV
INTERNATIONAL IMPLICATIONS

13
The *EC—Biotech* Decision: Another Missed Opportunity?

Joseph McMahon

1. Introduction

The Preamble to the WTO Agreement on the Application of Sanitary and Phytosanitary Measures ('SPS Agreement') notes the desire of the members to establish 'a multilateral framework of rules and disciplines to guide the development, adoption and enforcement of sanitary and phytosanitary measures in order to minimise their negative effects on trade'. It does this by listing the factors that must be taken into account in the decision by a member to adopt such measures: they must be based on the analysis and assessment of objective and accurate scientific data. This marks a significant shift from the discrimination-based approach that prevails in the General Agreement on Tariffs and Trade (GATT), especially Article XX.[1] Scott notes that this turn to science is enough for some to condemn the Agreement; but for others it is the manner in which science is used that is problematic. She continues:

With the turn to science, the WTO opened itself to charges of epistemological imperialism, and positivistic simple-mindedness. It is said to look for science-based truth where there is only disagreement, uncertainty and ignorance about potentially catastrophic risks. And it is said to do so in an institutional framework which lacks epistemic and moral authority. In doing so it has been charged with ignoring the cultural dimension of risk, and the democratic underpinnings of regulation.[2]

---

[1] According to the Preamble, the SPS Agreement is an elaboration of Art XX:b GATT. This would suggest that a breach of Art XX(b) must be established before the SPS Agreement becomes relevant, but this is not the case: see the Panel decision in *EC—Measures Concerning Meat and Meat Products (Hormones)* WT/DS26/R and WT/DS48/R, 18 August 1997, paras 8.41–2; and *Australia—Salmon* WT/DS18/R, 12 June 1998, paras 8.38–9. The text of the SPS Agreement is available at <http://www.wto.org/english/tratop_e/sps_e/spsagr_e.htm>, last accessed 9 November 2009.

[2] *The WTO Agreement on Sanitary and Phytosanitary Measures: a Commentary* (Oxford: Oxford University Press, 2007) 77.

Nowhere are these criticisms more apposite than in the emerging controversy over biotechnology. This chapter seeks to examine the interpretation and application of the SPS Agreement by the Panel in *EC—Approval and Marketing of Biotech Products* ('*EC—Biotech*') in the three areas of the Agreement that lie at the heart of the Panel's decision: the concept of a measure; the role of risk assessment; and the provisions on transparency.[3] By placing the Panel's decision in each of these areas within the evolving jurisprudence on the SPS Agreement, it will be possible to address the concerns identified by Scott.

According to the complainants in the *EC—Biotech* dispute (the United States, Canada, and Argentina), since October 1998 the European Community (EC) applied a de facto moratorium on the approval of products of agricultural biotechnology, namely food or food ingredients that contain or consist of, or are produced from, genetically modified organisms (GMOs), and GMOs intended for release into the environment. In particular, it was noted that since 1998 the EC had suspended consideration of application for, or granting of, approval for these products. Twenty-seven applications were delayed at various stages of the approval process under either Directive (EC) 2001/18 of the European Parliament and of the Council on the deliberate release into the environment of genetically modified organisms ('Deliberate Release Directive') (and, prior to its entry into force, Council Directive (EEC) 90/220)[4] or Regulation (EC) 258/97 concerning novel foods and novel food ingredients.[5] In addition, complaints were also made against various Member States who maintained marketing and import bans on products of agricultural biotechnology even though the EC had already approved such products before October 1998. In the first round of consultations, the EC stressed that it had a legitimate right to ensure that the regulatory regime allowed products to be placed on the market only after a careful risk assessment.[6] At the August 2003 meeting of the Dispute Settlement Body, the complainants requested the establishment of a Panel, a request that was blocked by the EC.[7] The Panel report was published on 26 September 2006 after many delays, and neither the EC nor the complaining parties decided to appeal against its findings.[8]

---

[3] WT/DS291/R, WT/DS292/R, and WT/DS293/R, 29 September 2006. For comments on the dispute, see, eg I Cheyne, 'Life after the Biotech Products Dispute' (2008) 10 Environmental Law Review 52; C Henckels, 'GMOs in the WTO: a critique of the Panel's legal reasoning in *EC—Biotech*' (2006) 12 Melbourne Journal of International Law 278; and O Perez, 'Anomalies in the precautionary kingdom: reflections on the GMO Panel's decision' (2007) 6 World Trade Review 265.   [4] [2001] OJ L106/1. For Council Directive 90/220, see [1990] OJ L117/15.
[5] [1997] OJ L43/1.
[6] European Commission, IP/03/859, *WTO Consultation over GMOs: EU Reiterates its Authorisation System is in Accordance with WTO Rules*, Brussels, 20 June 2003.
[7] See <http://www.wto.org/english/news_e/news03_e/dsb_18aug03_e.htm>, last accessed 8 September 2008.
[8] See <http://www.wto.org/english/news_e/news03_e/dsb_29aug03_e.htm>, last accessed 8 September 2008.

2. Scope of the SPS Agreement

By virtue of Article 1.1, the SPS Agreement applies to all SPS measures that may, directly or indirectly, affect international trade. This gives rise to three questions; What is encompassed within the expression 'SPS'? What is a measure? And how is 'affect international trade' to be interpreted?

2.1 Defining SPS

Annex A(1) defines the harm to which an SPS measure is to be addressed. It provides that an SPS measure is any measure applied:

(a) to protect animal or plant life or health within the territory of the Member from risks arising from the entry, establishment or spread of pests, diseases, disease-carrying organisms or disease-causing organisms;

(b) to protect human or animal life or health within the territory of the Member from risks arising from additives, contaminants, toxins or disease-causing organisms in foods, beverages or feedstuffs;

(c) to protect human life or health within the territory of the Member from risks arising from diseases carried by animals, plants or products thereof, or from the entry, establishment or spread of pests; or

(d) to prevent or limit other damage within the territory of the Member from the entry, establishment or spread of pests.

SPS measures are therefore directed at a particular harm; in other words, harm to human life or health, animal life or health, and plant life or health that arise from specific risks.[9] Previous Panels had concentrated on determining which paragraph of the Annex a particular measure fell under, as this would impact on the nature of the risk assessment upon which the measure must be based.[10]

In contrast, the Panel in *EC—Biotech* expended considerable energy on the interpretation of the various words used in Annex A(1). For example, 'arising from' was considered to be broad and unqualified, encompassing indirect and delayed effects, and those effects that might arise but may not do so.[11] As a

---

[9] The one exception appears to be para (d), which the Panel in *EC—Biotech* recognized to be a potentially very broad, residual category (paras 7.370–2). The Panel considered that it could extend to property damage, economic damage, and environmental damage not covered by other parts of the Annex. The latter would not cover loss of biodiversity since these are covered in other parts of the Annex.

[10] See, eg *Australia—Salmon* WT/DS18/R, 12 June 1998, para 8.31ff. See also J Pauwelyn, 'The WTO Agreement on Sanitary and Phytosanitary (SPS) Measures as applied in the first three SPS disputes—*EC—Hormones*, *Australia—Salmon* and *Japan—Varietals*' (1999) 2 Journal of International Economic Law 641.   [11] *EC—Biotech*, paras 7.225–6.

further example, a 'pest' included plants and animals that were annoying, destructive, and injurious.[12] Throughout this discussion, the Panel sought to establish a rational relationship between the risks identified in the Annex and the stated purpose of the SPS measures and, as a result, it expanded the scope of the SPS Agreement considerably. Perhaps the approach of the Panel in this case was dictated by the fact that some of the legislation at issue did not appear to be a measure applied to one of the identified risks in Annex A(1). This raises the question of what happens if a measure is adopted to protect one of the interests noted in Annex A(1), but also protects interests subject to another WTO agreement, for example the Agreement on Technical Barriers to Trade ('TBT Agreement').[13]

Article 1.3 of the TBT Agreement provides: '[a]ll products, including industrial and agricultural products, shall be subject to the provisions of this Agreement' and Article 1.5 goes on to state: '[t]he provisions of this Agreement do not apply to sanitary and phytosanitary measures as defined in Annex A to the Agreement on the Application of Sanitary and Phytosanitary Measures'. Article 1.4 of the SPS Agreement provides that '[n]othing in this Agreement shall affect the rights of Members under the Agreement on Technical Barriers to Trade with respect to measures not within the scope of this Agreement'. It is difficult, but not impossible, to envisage situations in which both the SPS Agreement and the TBT Agreement will apply to the same measure. For example, the complainants in *EC—Hormones* alleged that the measures taken by the EC were, *inter alia*, inconsistent with Articles 2, 3, and 5 of the SPS Agreement and Article 2 of the TBT Agreement. In response, the Panel found that, as the measures fell within the SPS Agreement, the TBT Agreement was not applicable to the dispute.[14]

In *EC—Biotech*, the EC argued that the approval procedures fell both within and outside the scope of the SPS Agreement. On this issue, the Panel concluded:

[W]e consider that to the extent the requirement in the consolidated law is applied for one of the purposes enumerated in Annex A(1), it may be properly viewed as a measure which falls to be assessed under the *SPS Agreement*; to the extent it is applied for a purpose which is not covered by Annex A(1), it may be viewed as a separate measure which falls to be assessed under a WTO agreement other than the *SPS Agreement*.[15]

---

[12] Ibid para 7.351.
[13] The text of the TBT Agreement is available at <http://www.wto.org/english/docs_e/legal_e/17-tbt.pdf>, last accessed 9 November 2009.
[14] *EC—Hormones* WT/DS26/R, 18 August 1997, para 8.29.
[15] *EC—Biotech*, para 7.165. The Panel noted: 'we are unaware of a directive in the *WTO Agreement* which says that a requirement can never be deemed to embody two or more distinct measures which fall to be assessed under different WTO Agreements': para 7.170. The Panel exercised judicial economy with respect to Canada's claims that various product-specific measures were inconsistent with Art III:4 GATT: para 7.2505; and concluded that Argentina had not established its claim under Art III:4: para 7.2516. The Panel also concluded that it was not necessary to address the complainant's claims under the TBT Agreement: paras 7.2528 and 7.3414.

After a detailed and extensive examination of the terms used in the four paragraphs of Annex A(1) and the types of risks covered by the EC regime, the Panel concluded that Directive 90/220 and the Deliberate Release Directive fell within the scope of the SPS Agreement.[16] To the extent that Regulation 258/97 was directed to preventing the consumers from being misled or nutritionally disadvantaged, it was held not to meet the requirements for the definition of an SPS measure; however, the other objective of the Regulation—that novel foods do not present a danger to the consumer—fell within the definition of an SPS measure.[17] It is interesting to note that there now exists the possibility that a measure pursuing multiple objectives, some falling under the SPS Agreement and others under the TBT Agreement, may be found to be contrary to the SPS Agreement yet consistent with the TBT Agreement.[18]

2.2 Defining 'measure'

Annex A(1) goes on to define 'measure' as including:

all relevant laws, decrees, regulations, requirements and procedures including, *inter alia*, end product criteria; processes and production methods; testing, inspection, certification and approval procedures; quarantine treatments including relevant requirements associated with the transport of animals or plants, or with the materials necessary for their survival during transport; provisions on relevant statistical methods, sampling procedures and methods of risk assessment; and packaging and labelling requirements directly related to food safety.

Annex B also uses the term 'regulation', and fn 5 in the Annex defines these as 'sanitary and phytosanitary measures such as laws, decrees or ordinances that are applicable generally'. One of the issues identified in the second review of the SPS Agreement has been the use of the expressions 'measures' and 'regulations', and it has been suggested that the Agreement should adopt a more consistent approach to the use of language.[19]

Using the SPS Agreement and previous GATT and WTO jurisprudence, the Panel in *Japan—Agricultural Products II* pointed out:

---

[16] Ibid paras 7.212–393. In para 7.1337, the Panel observed that the general definition in Annex A(1) is not to be applied in mechanistic fashion. It continued: 'we note that the mere fact that a measure within the meaning of the DSU meets the definition of an "SPS measure" set out in Annex A(1) does not mean that it is, *ipso facto*, subject to every provision of the *SPS Agreement* which applies to "SPS measures"'. [17] Ibid paras 7.415 and 7.416.

[18] It is also worth noting that the non-SPS objectives must be justifiable under another WTO agreement, otherwise the measure would fall foul of Art 5.6 of the SPS Agreement, which requires members to select measures that are least trade restrictive. On this point, see G Marceau and JP Trachtman, 'The Technical Barriers to Trade Agreement, the Sanitary and Phytosanitary Measures Agreement and the General Agreement on Tariffs and Trade: a map of the World Trade Organization law of domestic regulation of goods' (2002) 36 Journal of World Trade 811.

[19] See, eg G/SPS/W/158 (Canada) and G/SPS/W/168 (New Zealand).

[t]his context indicates that a non-mandatory government measure is also subject to WTO provisions in the event compliance with this measure is necessary to obtain an advantage from the government or, in other words, if sufficient incentives or disincentives exist for that measure to be abided by.[20]

The Panel in *EC—Biotech* adopted an approach similar to that it adopted in relation to paras (a)–(d) by discussing the two lists found in the final para, in other words the form of a measure (law, decree, or regulation) as opposed to the nature of a measure (requirement or procedure as illustrated by the list). A plain reading of this paragraph would suggest that this dual categorization is unduly complicated, as the Panel itself recognized that the list was non-exhaustive, so that SPS measures 'may in principle take many different legal forms'.[21] One interesting question that arose here is whether the de facto moratorium constituted a measure.

On this issue the Panel concluded that a de facto moratorium existed between October 1998 and August 2003.[22] It reached this conclusion through an examination of a considerable number of documents, placing much emphasis on the statement in June 1999 by Denmark, France, Greece, Italy, and Luxembourg that they would prevent the approval of biotech applications pending the adoption of new EC rules on the labelling and traceability of biotech products.[23] Despite the fact that the moratorium had ceased to exist after the establishment of the Panel, this did not deprive it of competence to rule on the WTO consistency of the moratorium.[24] In response to the claim by the three complainants in *EC—Biotech* that, by applying a general de facto moratorium on the approval of new biotech products, the EC had acted inconsistently with Article 5.1 of the SPS Agreement, the Panel concluded that, as the moratorium was not a measure applied to achieve the EC's appropriate level of protection, it could not be considered an SPS measure within the meaning of Article 5.1.[25] Article 5.1 refers to 'requirements and procedures', but the complaint was against the application of those requirements and procedures and the Panel refused to characterize the moratorium as an effective marketing ban on all biotech products subject to approval, a new procedure, or an amendment of the existing approval procedure. A similar conclusion was reached with respect to Article 5.6 and Article 5.5, leading the Panel to reject complaints made about the consistency of the de facto moratorium with Article 2.2 and 2.3 of the SPS Agreement.[26]

---

[20] WT/DS76/R, 27 October 1998, para 8.111.   [21] *EC—Biotech*, para 7.422.
[22] Ibid paras 7.1271–85.
[23] Ibid para 7.474. See also para 7.484, reproducing a declaration by seven Member States outlining their intention to 'take a thoroughly precautionary approach in dealing with applications and authorizations for the placing on the market of GMOs'.
[24] Ibid paras 7.1311 and 7.1312.   [25] Ibid para 7.1393.
[26] Ibid paras 7.1405 (Art 5.6); 7.1419 (Art 5.5); 7.1434 and 7.1440 (Art 2.2); and 7.1446 (Art 2.3).

As for the product-specific measures on the approval of biotech products, the United States ultimately made claims in respect of 25 applications, whilst Argentina complained about 11 applications and Canada four. Annexes A and C of the SPS Agreement were considered by the Panel to be the most relevant provisions, with the Panel noting:

> the term 'approval procedures' can be understood as encompassing procedures applied to check and ensure the fulfilment of one or more substantive SPS requirements the satisfaction of which is a prerequisite for the approval to place a product on the market.[27]

The approval procedures were thus SPS measures within the meaning of Annex A(1).[28] However, the application or operation of these procedures did not constitute an SPS measure within the meaning of that provision, and so the Panel dismissed the complainants' claims under Articles 5.1, 5.6, 5.5, 2.2, 2.3, and 7, and Annex B, using the same logic it had used when dismissing similar claims based on these articles against the general moratorium.[29]

Claims were also made against nine measures adopted by six Member States under Article 16 of Directive 90/220 (Article 23 of the Deliberate Release Directive), and Article 12 of Regulation 258/97. The Panel began its consideration of these claims by examining whether the SPS Agreement was applicable to these measures The Panel was required to determine whether the measure was applied for one of the purposes enumerated in Annex A(1) and, although Article 16 of Directive 90/220 and Article 12 of Regulation 258/97 allowed safeguard measures to be taken in case of a risk, or danger, to human health or the environment, the Panel considered it necessary to examine the nine safeguard measures at issue individually. A thorough examination of each measure, including the concerns it was intended to address, led the Panel to conclude that each qualified as an SPS measure within the meaning of Annex A(1) as far as their purpose was concerned and, as this affected international trade within the meaning of Article 1.1, they were subject to the provisions of the SPS Agreement.[30]

2.3 Defining 'Directly or Indirectly Affecting International Trade'

Article 1.1 of the SPS Agreement (rather than Annex A(1)) establishes the requirement that SPS measures will only be subject to the SPS Agreement if they affect trade, either directly or indirectly, and on this issue it has been confirmed that it is to be interpreted in accordance with the jurisprudence on Article III GATT, which contains a similar phrase. The Panel in *EC—Biotech* indicated that there was nothing to suggest a *de minimis* exception or another quantitative element.[31] Some greater guidance from the Panel would have been appreciated

---

[27] Ibid para 7.425. [28] Ibid para 7.432.
[29] Ibid Section E.3–E.8 (paras 7.1680–7.1778). [30] Ibid paras 7.2561–922.
[31] Ibid para 7.2609.

3. Risk Assessment

3.1 General

Although the SPS Agreement vests certain authority with the standards developed by international organizations, under Article 3.3 a member may decide to establish protection levels that exceed international standards if there is a scientific justification or if it determines that the standard does not meet its acceptable level of protection. The right of a member to establish its own level of sanitary protection was seen by the Appellate Body as 'an autonomous right and *not* an "exception" from a "general obligation" under Article 3.1'.[33] On the other hand, when a member decides not to use an international standard, its measure must be based on a proper risk assessment and is subject to a range of other conditions set out in Article 5. According to the Panel in *EC—Hormones*, risk assessment involves a scientific examination of data and factual studies as opposed to a 'policy exercise involving social value judgments made by political bodies', which it characterized as risk management.[34] However, the Appellate Body rejected this distinction between risk assessment and risk management by pointing out that Article 5 of and Annex A to the SPS Agreement refer to risk assessment and that the term 'risk management' is not to be found in the SPS Agreement.[35] This rejection by the Appellate Body of a role for risk management does not imply that members of the SPS Agreement are not free to adopt measures that they consider necessary to establish an appropriate level of protection. Article 2.1 guarantees the right of a member to adopt the necessary measures, provided that such measures are consistent with the SPS Agreement; in other words, the measures require a risk assessment conducted in conformity with Article 5. As the Appellate Body noted in *EC—Hormones*:

The requirements of a risk assessment under Article 5.1, as well as of 'sufficient scientific evidence' under Article 2.2, are essential for the maintenance of the delicate and carefully negotiated balance in the SPS Agreement between the shared, but sometimes competing, interests of promoting international trade and of promoting the life and health of human beings.[36]

---

[32] See, eg the failure of the Appellate Body and the Panel in *US—Upland Cotton* to define 'no, or at most minimal, trade-distorting effects or effects in production' in Annex 2 to the Agreement on Agriculture: WT/DS265/R, 8 September 2004, and WT/DS265/AB/R, 3 March 2005.
[33] *EC—Hormones* WT/DS26/AB/R and WT/DS48/AB/R, 16 January 1998, para 172.
[34] *EC—Hormones* WT/DS26/R, 18 August 1997, para 8.91ff.
[35] *EC—Hormones* WT/DS26/AB/R and WT/DS48/AB/R, 16 January 1998, para 181.
[36] Ibid para 177.

Risk assessment is defined in para 4 of Annex A to the SPS Agreement as:

The evaluation of the likelihood of entry, establishment or spread of a pest or disease within the territory of an importing Member according to the sanitary or phytosanitary measures which might be applied, and of the associated potential biological and economic consequences; or the evaluation of the potential for adverse effects on human or animal health arising from the presence of additives, contaminants, toxins or disease-causing organisms in food, beverages or feedstuffs.

As the Panel noted in *Australia—Salmon*, this provision offers two definitions of risk assessment: the risk of the entry, establishment, or spread of a disease; and the risk of associated potential biological and economic consequences.[37] Using para 4, the Appellate Body in the same dispute noted that a risk assessment under Article 5.1 must:

1. *Identify* the diseases whose entry, establishment or spread a Member wants to prevent within its territory, as well as the potential biological and economic consequences associated with the entry, establishment or spread of the diseases;
2. *Evaluate the likelihood* of entry, establishment or spread of these diseases as well as the associated potential biological and economic consequences; and
3. Evaluate the likelihood of entry, establishment or spread of these diseases, *according to the SPS measures that might be applied*.[38]

As a result of this, the risk assessment must be specific to the case at hand; a general discussion of the disease sought to be avoided by the SPS measures will not satisfy the definition of a risk assessment. According to the Appellate Body, the term 'likelihood' was synonymous with the term 'probability'. Therefore, a proper risk assessment would evaluate the probability of the entry, establishment, or spread of diseases, as well as the associated potential biological and economic consequences and the probability of entry, establishment, or spread of these diseases, according to the SPS measures that might be applied.[39]

Although Article 5.1 instructs members to take into account 'risk assessment techniques developed by the relevant international organizations', the SPS Agreement does not directly address the methodology of risk assessment. A certain degree of flexibility is therefore accorded to members in meeting the requirements of Article 5.1. As the Appellate Body noted in the *Japan—Apples* dispute: 'Members are free to consider in their risk analysis multiple agents in relation to one disease, provided that the risk assessment attributes a likelihood of entry, establishment or spread of the disease to each agent specifically'.[40] Given that Article 5.1 requires the SPS measure to be 'based on' a risk

---

[37] *Australia—Salmon* WT/DS18/R, 12 June 1998, paras 8.72 and 8.116.
[38] *Australia—Salmon* WT/DS18/AB/R, 20 October 1998, para 121 (emphasis in original).
[39] Ibid paras 123 and 124.
[40] *Japan—Apples* WT/DS245/AB/R, 26 November 2003, para 204.

assessment, the specific risk assessment logically precedes the selection of measures to address the risk that has been identified.[41] This will ensure that the measures adopted are no more trade restrictive than necessary to achieve the appropriate level of protection, as required by Article 5.6, rather than a retrospective justification of measures already adopted.

According to Article 5.2, risk assessment will take into account available scientific evidence. The importance of such evidence is reinforced by Article 2.2, which requires SPS measures to be 'based on scientific principles' and not be 'maintained without sufficient scientific evidence'. It is up to the complaining party to adduce prima facie evidence that the measure is inconsistent with Article 2.2, at which point the burden of proof shifts to allow the defending party to refute the claim. Article 2.2, according to the Appellate Body *in Japan—Agricultural Products II*, requires 'a rational and objective relationship between the SPS measure and the scientific evidence'.[42] In reviewing this relationship, which must be done on a case-by-case basis, the dispute-settlement bodies will not conduct their own risk assessment, but will instead rely on expert evidence. However, it must be noted that Article 5.2 allows the risk assessment to extend beyond available scientific evidence to a range of factors that can assess 'the actual potential for adverse effects on human health in the real world where people live and work and die'.[43]

In 2002 the Codex Alimentarius Committee on General Principles forwarded the Proposed Draft Working Principles for Risk Analysis in the Framework of the Codex Alimentarius to the Executive Committee for adoption as Draft Principles. These principles, which were adopted in 2003, apply within the framework of the Codex and not to governments; that is, they would be included in the Procedural Manual as general guidance to the Codex Alimentarius Commission and its subsidiary bodies. The Commission is currently developing Working Principles for Risk Analysis for Food Safety for Application by Governments, which will provide a framework to guide governments in the conduct of risk analysis applied to food safety issues.[44]

These principles recognize that the risk analysis process should not only be applied consistently but must be open, transparent, and documented. The structured process would incorporate the three elements of risk analysis—risk assessment, risk management, and risk communication—which would form part of an overall strategic approach to food health. It is recognized that risk assessment policy should be included as a specific component of risk management and that those responsible for risk management should establish a risk assessment policy that is systematic, complete, unbiased, and transparent. The

---

[41] *EC—Hormones* WT/DS26/R, 18 August 1997, paras 7.68–70.
[42] *Japan—Agricultural Products II* WT/DS76/AB/R, 22 February 1999, para 84.
[43] *EC—Hormones* WT/DS26/AB/R and WT/DS48/AB/R, 16 January 1998, para 187.
[44] See Codex Document CX 4/10 (CL 2004/34–GP) (available at <http://www.codexalimentarius.net/web/index_en.jsp>, last accessed 16 October 2009).

emphasis on the importance of science in risk assessment is underlined by the recommendation of a four-step process: hazard identification; hazard characterization; exposure assessment; and risk characterization. Ecological and environmental conditions, production, transport, and storage and handling practices are also to be taken into account, and any assessment should be based on realistic exposure scenarios.

Risk management is a similarly structured four-step process; preliminary risk management (in other words, such matters as identification of the food safety problems and establishment of a risk profile); evaluation of options; implementation of decisions; and monitoring and review of decisions. Although risk management should be based primarily on human health considerations, other legitimate factors relevant to risk management should be taken into account. Once again, in achieving these ecological and environmental conditions, production, transport, and storage and handling practices should be considered. It is interesting to note that the Draft Working Principles specifically recognize that the option of taking no action should also be considered. Once a decision has been taken, it should be monitored to ensure its effectiveness and there is the possibility of taking interim measures. Such measures would be based on the precautionary principle. The principles that would guide the final stage of the process of risk analysis—risk communication—suggest that the process should be consistent and transparent. It should promote awareness and understanding of the issues considered in the risk analysis process, so as to improve the overall effectiveness and efficiency of that process. Taken as a whole, the process of risk analysis should enhance the trust and confidence in food safety of all interested parties.

It must be noted that problems have arisen with the Draft Working Principles as a result of the decision to reopen them, taken at the meeting in April 2005 of the Codex Alimentarius Committee on General Principles.[45] Some members, who consider that existing documents already provide sufficient guidance on risk analysis, have constantly questioned the need for two sets of principles.[46] Nevertheless, the proposed creation of two sets of principles did help to resolve the conflict about the proper role of precaution in this context. The Draft Working Principles rely heavily on those agreed in 2003, which view precaution as an inherent element of risk management. Several members were unhappy about the use of the word 'precaution', whilst other wanted the document to specify how the principle should be used. Given this lack of consensus on the role of precaution in the Draft Working Principles (especially in

---

[45] Available at <http://www.codexalimentarius.net/web/reports.jsp?lang=en>, see ALINORM 05/28/33A, paras 31–54, last accessed 8 September 2008.

[46] Reference was made here to documents such as the Principles and Guidelines for the Conduct of Microbiological Risk Assessment (CAC/GL 30–1999) and Principles for the Risk Analysis of Foods derived from Modern Biotechnology (CAC/GL 44–2003).

3.2 Precautionary Measures

Under the SPS Agreement, there is a qualified exception to the obligation in Article 2.2 not to maintain SPS measures without sufficient scientific evidence. Article 5.7 states:

[i]n cases where relevant scientific evidence is insufficient, a Member may provisionally adopt sanitary or phytosanitary measures on the basis of available pertinent information, including that from the relevant international organizations as well as from sanitary or phytosanitary measures applied by other Members. In such circumstances, Members shall seek to obtain the additional information necessary for a more objective assessment of risk and review the sanitary or phytosanitary measure accordingly within a reasonable period of time.

In *Japan—Agricultural Products II*, the Appellate Body identified four requirements, two in each sentence, which are cumulative, for a Member adopting provisional measures using this provision.[47] The Appellate Body stated that the additional information to be sought must be 'germane' to conducting a more objective risk assessment, and 'reasonable period of time' had to be established on a case-by-case basis.[48]

It is the insufficiency of scientific evidence that triggers Article 5.7 and not, as the Appellate Body noted in *Japan—Apples*, the existence of scientific uncertainty.[49] The EC has argued for a right to take measures when the scientific information is inconclusive or uncertain using the precautionary principle.[50] In *EC—Hormones* it argued that the precautionary principle is, or has become, 'a general customary rule of international law' or at least 'a general principle of law'. In effect, the EC was arguing that the precautionary principle should be used in the interpretation of the SPS Agreement and could be used to override Articles 5.1 and 5.2. In contrast, the United States suggested that it represented more of an 'approach' than a 'principle', and Canada viewed it as 'an emerging principle of law'. On the specific question of its relationship with the SPS Agreement, the Appellate Body noted that the principle is reflected in the sixth paragraph of the Preamble and in Articles 3.3 and 5.7. However, as it had not been written into the SPS Agreement, it could not be used as a ground for

---

[47] *Japan—Agricultural Products II* WT/DS76/AB/R, 22 February 1999, para 89.
[48] Ibid paras 92 and 93.
[49] *Japan—Apples* WT/DS245/AB/R, 26 November 2003, para 184.
[50] See European Commission, *Communication on the Precautionary Principle* COM(2001)1. See also N Salmon, 'A European perspective on the precautionary principle, food safety and the free trade implications of the WTO' (2002) 27 Environmental Law Review 138; and V Hayvaert, 'Facing the consequences of the precautionary principle in European Community Law' (2006) 31 Environmental Law Review 185.

justifying SPS measures that are otherwise inconsistent with the SPS Agreement.[51] As for the status of the precautionary principle in international law, the Appellate Body stated:

> [t]he precautionary principle is regarded by some as having crystallised into a general principle of customary international *environmental* law. Whether it has been widely accepted by Members as a principle of *general* or *customary international law* appears less than clear.[52]

It concluded that it would be 'unnecessary, and probably imprudent' for it to take a position on this question. The Panel in *EC—Biotech* recalled these statements and considered that the debate on the status of the precautionary principle was still ongoing and, like the Appellate Body, it concluded it prudent not to resolve the issue of whether it is a recognized principle of general or customary international law.[53]

In *EC—Biotech*, the EC objected to the Panel considering the consistency of each of the national safeguard measures with Article 5.1, arguing that as provisional measures they should be assessed under Article 5.7 to the exclusion of Article 5.1. In rejecting this argument the Panel noted:

> it is clear that Article 5.7 is applicable whenever the relevant condition is met, that is to say, in every case where the relevant scientific evidence is insufficient. The provisional adoption of an SPS measure is not a condition for the applicability of Article 5.7. Rather, the provisional adoption of an SPS measure is permitted by the first sentence of Article 5.7.[54]

Having rejected the EC argument that Article 5.7 contains specific rules for the assessment of provisional measures, the Panel moved on to consider the second argument, that the relationship between Articles 5.1 and 5.7 is one of exclusion rather than exception. In considering this argument, the Panel examined the relationship between Article 2.2 and Article 5.7 as part of the process of responding to the EC argument that these provisions should be read in the same way as the relationship between Article 3.1 and Article 3.3. It suggested that 'Article 5.7 should be characterized as a right and not an exception from a general obligation under Article 2.2'.[55] Jurisprudence confirms that the right is neither absolute nor unqualified, and in this dispute the Panel, after noting that Article 5.7 does not provide for an exception from Article 5.1, but establishes a qualified right, concluded:

> [O]n this view of the relationship between Article 5.1 and Article 5.7, if an SPS measure challenged under Article 5.1 was adopted and is maintained consistently with the cumulative requirements of Article 5.7, the obligations in Article 5.1 to base SPS measures on a risk assessment is not applicable to the challenged measure. In such a case,

---

[51] *EC—Hormones*, WT/DS26/AB/R and WT/DS48/AB/R, 16 January 1998, para 124.
[52] Ibid para 123.   [53] *EC—Biotech*, para 7.89.   [54] Ibid para 7.2939.
[55] Ibid para 7.2969.

the complaining party would not have invoked the 'wrong' provision, in the sense that it should have brought a challenge under Article 5.7 instead of Article 5.1. The complaining party would have invoked the 'correct' provision, but the complaining party's claim under Article 5.1 could not succeed as long as the responding party complies with the requirements of Article 5.7.[56]

Therefore, the consistency of the safeguard measures with Article 5.7 would be examined within the context of the examination of the consistency of the measures with Article 5.1, especially the requirement to base SPS measures on a risk assessment. The Panel continued to conduct an initial assessment of the consistency of the national safeguard measures with Article 5.1, and then examine the consistency of the measure under Article 5.7, before reaching a final conclusion on the consistency of the measure with Article 5.1. The decision in *EC—Biotech* is interesting in that the interpretation of Article 5.7 is made with reference to Article 5.1, so 'insufficient' scientific evidence in the former is to be interpreted as not allowing the risk assessment provided for in the latter and Annex A(4).

4. Transparency

The transparency provision of the SPS Agreement is Article 7, which states: 'Members shall notify changes in their sanitary or phytosanitary measures and shall provide information on their sanitary or phytosanitary measures in accordance with the provisions of Annex B'. Annex B, entitled 'Transparency of Sanitary and Phytosanitary Regulations', lists a number of detailed requirements to promote transparency, including the publication of regulations, the existence of an Enquiry Point, and notification procedures. In this context, it is important to understand the meaning of transparency, especially as it is not defined in the SPS Agreement (or in other WTO Agreements) and yet is one of the fundamental principles of the WTO. The objective of transparency is to achieve a greater degree of clarity, predictability, and information about trade policies, rules, and regulations of members. In the context of the SPS Agreement, Annex B provides the elements of transparency, and these have been further developed in the decision of the SPS Committee on recommended procedures for implementing the transparency obligations of the SPS Agreement (under Article 7).[57]

The fundamental component of transparency under the SPS Agreement is publication of regulations. This is a general obligation on WTO members, and under paras 1 and 2 of Annex B, members are obliged to:

- ensure that all SPS measures, which have been adopted, are published promptly in such a manner as to enable interested countries to become acquainted with them; and

---

[56] Ibid para 7.3004.   [57] G/SPS/7/Rev.2.

- except in urgent circumstances, allow a reasonable interval between the publication of an SPS measure and its entry into force in order to allow time for producers in exporting countries, and particularly in developing countries, to adapt their products and methods of production to the requirements of the importing country.

A reasonable interval, according to a decision at the Doha Ministerial, is six months; however, SPS measures that liberalize trade are not to be unnecessarily delayed.[58]

According to para 3 of Annex B, the SPS Enquiry Point has a legal obligation 'to answer all reasonable questions' from interested members and provide relevant documents regarding, for example, 'any SPS measures adopted or proposed within the country' and 'risk assessment procedures, factors taken into consideration, as well as the determination of the appropriate level of SPS protection'. Members are also required to designate a single central government authority—the SPS Notification Authority—as responsible for implementing, on a national level, the notification requirements of the SPS Agreement. The SPS Notification Authority is responsible for: ensuring that proposed regulations are published early, to allow for comments; notifying other countries through the WTO Secretariat of SPS regulations; providing copies on request of proposed regulations; and ensuring that comments are correctly handled. Although members do not have to notify measures that conform to international standards, it is recognized that such notifications improve the objectives of transparency.

One aspect of the complaints in *EC—Biotech* was that the EC had failed to publish the existence of the de facto moratorium on the approval of biotech products and had thus acted inconsistently with Article 7 and Annex B of the SPS Agreement. The Panel began its consideration of this complaint by asking whether the moratorium on approvals was a generally applicable 'SPS measure' that had been adopted by the EC. The Panel had previously noted that, as the moratorium related to the application, or operation, of the EC's approval procedures, it did not constitute an SPS measure within the meaning of Annex A(1). Examining Annex B(1), the Panel noted that:

[N]either Annex B(1) nor its accompanying footnote suggests that a generally applicable measure concerning the administration, or operation, of an SPS measure—such as a measure providing for a particular operation of an SPS approval procedure—is, also, to be published.[59]

Article 7 supported this conclusion, as did the findings of the Appellate Body in *Japan—Agricultural Products II*, in which the Appellate Body considered that 'the object and purpose of paragraph 1 of Annex B is "to enable interested Members to become acquainted with" the sanitary and phytosanitary regulation

[58] WT/MIN(01)/17, 20 November 2001, para 3.2.   [59] *EC—Biotech*, para 7.1458.

adopted or maintained by other Members and thus to enhance transparency regarding those measures'.[60] Although publishing measures on the administration or operation of SPS measures would enhance transparency, the Panel in *EC—Biotech* considered that to decide so would go beyond its powers and, as a result, the EC had not acted inconsistently with Article 7 and Annex B(1) in respect to the moratorium.[61]

The Panel pointed out that Annex C also contains additional transparency requirements. In the dispute, the United States and Canada claimed that the EC moratorium on the approval of biotech products led to a failure of the EC to comply with Article 8 and Annex C(1)(a), whereby members are to ensure that procedures to check and ensure fulfilment of SPS measures are undertaken and completed without undue delay. The Panel first tackled the interpretation of 'undertake and complete', and concluded that it 'covers all stages of approval procedures and should be taken as meaning that, once an application has been received, approval procedures must be started and then carried out from beginning to end'.[62] As for the interpretation of 'undue delay', the Panel considered that this must be on a case-by-case basis, taking account of the relevant facts and circumstances. It then continued:

[W]e view Annex C(1)(a), first clause, essentially as a good faith obligation requiring members to proceed with their approval procedures as promptly as possible, taking account of the need to check and ensure the fulfilment of their relevant SPS requirements. Consequently, delays which are justified in their entirety by the need to check and ensure the fulfilment of a member's WTO–consistent SPS requirements should not, in our view, be considered 'undue'.[63]

The Panel completed its interpretation of this provision by noting that 'undue delay' applies to both the undertaking of approval procedures and the completion of those procedures. As for the application of these interpretations to the moratorium, the Panel, having examined the reasons for the moratorium as a justification for delay, concluded that they did not provide such justification; as such, there was undue delay in the sense of Annex C(1)(a).[64] In its analysis, the Panel also dismissed as general reasons for the delay the perceived inadequacy of the approvals legislation or evolving science and the application of a prudent and precautionary approach. With reference to the approval procedure for one product, MS8/RF3 oilseed rape, it dismissed as specific reasons for the delay the relationship of the approval procedure conducted under Directive 90/220 and the adoption of the Deliberate Release Directive, and the adoption of the Deliberate Release Directive.

The Panel went on to find a similar inconsistency with Article 8 and Annex C(1)(a) with respect to 24 out of 27 of the complainants' claims against various

[60] Ibid para 7.1459, citing WT/DS76/AB/R, 22 February 1999, para 106.
[61] Ibid paras 7.1461 and 7.1464.   [62] Ibid para 7.1494.   [63] Ibid para 7.1498.
[64] Ibid paras 7.1567 and 7.1568.

product-specific measures.[65] A complaint was also made by Argentina under the second clause of Annex C(1)(a) (namely that procedures are undertaken and completed in a manner no less favourable for imported products than for like domestic products). In assessing this complaint, the Panel referred to the jurisprudence on Articles III:1 and III:4 GATT for interpretative guidance, and concluded that Argentina had not proven its case on this issue.[66] The Panel also dismissed the US complaint that the moratorium was inconsistent with Annex C(1)(b) and a similar complaint against the product-specific measures brought by the United States and Argentina.[67] Argentina further failed to establish its product-specific claims under Annex C(1)(c) and (e).[68]

5. Conclusion

The findings of the Panel on transparency are to be welcomed as they may be viewed as confirming the overall purpose of the SPS Agreement: that is, to prevent the emergence of agricultural protectionism in the wake of the liberalization undertaken as a result of the implementation of the Agreement on Agriculture.[69] Likewise, the interpretation of Article 1 and Annex A may also be viewed in this positive light. The criticisms of the Panel decision have again revolved around the role of science in the SPS Agreement.

An *amicus curiae* brief submitted by five individuals challenged the Panel in *EC—Biotech* to update the existing jurisprudence on risk assessment to reflect current scholarship in both the sciences and social sciences on risk and regulation.[70] The authors suggest that it is no longer possible to separate risk assessment from risk management, this being especially so since the new Codex principles for risk analysis envisage interaction between risk assessors and risk managers, thus anticipating a role for value judgments in risk assessment. Given that such value judgments are dependent on political and cultural contexts, for example, in relation to the role of the consumer/public, it is possible for different WTO members to come to different conclusions on risk assessment. What is now being suggested is that there be a transparent risk assessment that affords opportunities for all interested parties to participate in an 'interactive' and 'responsive consultative process' where their views are 'sought' by the regulators. The conclusion offered by this *amicus curiae* brief is that '[a]n overly rigid conception of proper risk assessment and regulation in this area

---

[65] Ibid paras 7.1779–2391.   [66] Ibid para 7.2418.
[67] Ibid paras 7.1571–604 and 7.2422–74.   [68] Ibid paras 7.2487 and 7.2498.
[69] See D Roberts, 'Preliminary assessment of the effects of the WTO Agreement on Sanitary and Phytosanitary Regulation' (1998) 1 Journal of International Economic Law 377.
[70] *Amicus curiae* brief submitted by Busch, White, Jasanoff, Winickoff, and Wynne (available at <http://www.trade-environment.org/page/theme/tewto/biotechcase.htm>, last accessed 1 May 2005).

could ... undermine the legitimacy of the SPS Agreement and the WTO more generally'.[71] A further *amicus curiae* brief concluded:

[U]ncertainty is a given in any scientific inquiry, which can only establish the boundaries of existing knowledge, and this will not always trigger the need for precautionary action. However, when the available information cannot appropriately describe the risks to human, animal, or plant life or health because of the lack of understanding of events and processes, policy-makers cannot ignore the lack of quality of the scientific evidence.[72]

Whilst it may be argued that the SPS Agreement represents a suitable response to attempts by WTO members to restrict imports for traditional products, it is arguable that for innovative products, such as the products of biotechnology, where the challenges are more complex, the SPS Agreement may be inadequate. National regulations in these areas respond quite correctly to the prevailing socio-economic considerations and to concerns about sustainable development.[73] When these regulatory regimes impact on international trade, there are bound to be conflicts and, as the science is recognized as being inherently uncertain, such conflicts may lead to continuing conflict. The saga of the import of hormone-treated beef into the EC is one example of such continuing conflict.[74] What is missing from international trade conflict resolution is a mechanism to allow these differing national regulatory regimes to be reconciled. To this extent, the Panel decision in *EC—Biotech* represents another missed opportunity for the WTO.

The opportunity presented by the *EC—Biotech* dispute was to begin a process of reconciling the demands of science with the other factors that must be taken into account in any decision on risk. The Panel's approach of relying exclusively on science is, as Scott asserts, 'simple-minded'. It suggests that a simple and sufficient answer can emerge from science and that this answer should silence all public debate on the safety of biotech products. Such an approach, by ignoring 'the democratic underpinnings of regulation', cannot add to the moral authority of the WTO. The challenge for the SPS Agreement (and the WTO) is to move beyond the strict wording of the Agreement to reflect how it can accommodate the complexities arising from divergent national regulation.

---

[71] Ibid 38.
[72] *Amicus curiae* brief submitted by Center for International Environmental Law, Friends of the Earth (US), Defenders of Wildlife, Institute for Agriculture and Trade Policy, and Organic Consumers Association (US) (available at <http://www.trade-environment.org/page/theme/tewto/biotechcase.htm>, para 37, last accessed 1 May 2005). See also L Kogan, 'The precautionary principle and WTO law: divergent views towards the role of science in assessing and managing risks' (2004) 5 Seton Hall Journal of Diplomacy and International Relations 75.
[73] See, eg TD Epps, 'Reconciling public opinion and WTO rules under the SPS Agreement' (2008) 7 World Trade Review 359.
[74] See R Walker, 'Keeping the WTO from becoming the World Trans-Science Organization: scientific uncertainty, science policy and factfinding in the growth hormones dispute' (1998) 31(2) Cornell International Law Journal 280.

14
The Regulation of Genetically Modified Organisms and International Law: A Call for Generality

Duncan French

1. Introduction

Although this chapter concerns the international legal implications of the regulation of genetically modified organisms (GMOs), it does not seek to cover what has, by now, become well-trodden ground. In particular, it will provide neither an exposition of the 2000 Cartagena Protocol on Biosafety[1] nor consider in any detail the relationship between that international agreement and others,[2] specifically the World Trade Organization (WTO) Agreement on the Application of Sanitary and Phytosanitary Measures ('SPS Agreement').[3] Rather, the aim is to consider a number of discrete international legal issues arising from the regulation of GMOs, and especially how general rules of international law, together with more general trends in related legal and policy fields, are likely to prove increasingly apposite in the ongoing debate over how to manage and govern GMO activity. In particular, the chapter will consider two specific issues: first, the role of treaty interpretation in promoting synergies and reconciling apparently conflicting primary rules; and, secondly, the utility of the concept of sustainable development in determining a balanced framework for the inclusion of socio-economic considerations within GMO decision-making processes, as permitted under the Cartagena Protocol.

---

[1] Available at <http://www.cbd.int/biosafety/protocol/shtml>. See A Cosbey and S Burgiel, *The Cartagena Protocol on Biosafety: an Analysis of Results* (Winnipeg: International Institute for Sustainable Development (IISD), 2000).

[2] See A Qureshi, 'The Cartagena Protocol on Biosafety and the WTO—co-existence or incoherence?' (2000) 49 International and Comparative Law Quarterly 835.

[3] Available at <http://www.wto.org/english/tratop_e/sps_e/spsagr_e.htm>, last accessed 9 November 2009. For a comprehensive account of the SPS Agreement, see J Scott, *The WTO Agreement on Sanitary and Phytosanitary Measures: a Commentary* (Oxford: Oxford University Press, 2007).

What is interesting, in taking this broader—slightly more systemic—approach, is to realize that in many ways GMO regulation, though the subject of this chapter, is also very much a case study, in the sense that what is being considered here is the response of the international community, through law and policy, to a new challenge and the complexities arising therefrom. Thus, although context-specific to the GMO debate, the issues discussed, namely reconciling divergent norms and the incorporation of socio-economic values into legal decision-making, would arguably be equally pertinent if one were analysing any new scientific or technological endeavour. The recognition of this fact is significant for at least two reasons. The first is to highlight that, while the regulation of GMOs has shown itself to be novel in many respects, it is not unique. The second, and related, reason is to reaffirm the general capacity of the international legal system to be able to cope with and regulate even the most contentious of scientific advances; one might refer to this as the 'absorptive' capacity of the international law-making process to tackle the challenges posed by an increasingly complex world order.[4] Thus, before considering the two issues identified above, the chapter will also try to situate the GMO debate within a broader normative global context.

This chapter is purposely expansive in scope; it seeks to recognize the reality that, in regulating an issue such as GMOs, one needs to be aware not just of those rules that have been specifically adopted in response, but also the wider legal framework in which such bespoke legal regimes inevitably form only a small part. In conclusion, it will therefore argue that a legal understanding of GMOs which concentrates solely on the regulatory detail and ignores the broader context not only provides less than the whole story, but has the potential to both mislead and misunderstand the true nature of how GMOs are governed at the international level.

2. The Normative Context of GMO Regulation

It is a relatively common error to move from considering the Cartagena Protocol as the *principal* international response to the management of risks relating to GMOs to considering it as the *exclusive* international response. Certainly, the Protocol may be the most obvious manifestation of regulation at the

---

[4] Moreover, this is not just a technical question about the operational capability of the international legal system, but also a normative issue as to the significance of ensuring the necessary political will: see D French, 'International Rhetoric and the Real Global Agenda: Exploring the Tension between Interdependence and Globalization' in R Brownsword (ed), *Global Governance and the Quest for Justice* (Oxford: Hart Publishing, 2004) 136: '[t]o continue to be relevant ... the international community must endeavour to broaden the nature and scope of international law to reflect the changing nature of international society. In particular, it must be prepared to adopt structural and normative change, so as to maintain the legitimacy and role of the global order'.

The Normative Context of GMO Regulation

international level, but it is clearly not alone in the field.[5] There is an ever-growing number of legal and policy initiatives on the management of GMOs, as well as other aspects of biotechnological innovation.[6] But even with the growing range of national, regional, and global rules and other softer norms[7] in this area, there is still the reality that such instruments—however increasingly voluminous—are only a small part of a wider, more general, legal and juridical framework. This section will therefore explore two of the wider questions relating to the situation of GMO regulation within a more global approach: first, the connexions between national, regional, and global law-making; and, secondly, the fragmented nature of the international legal order, both of which have a significant, if largely unseen, impact upon the regulation of GMOs at the international level.

2.1 National, Regional, and International Law: Interface and Inter-linkages

The first issue is to consider the interaction between national, regional, and global regulation of an issue such as GMOs. While it is certainly trite to say that on a matter such as this, regulation at national and regional level is often mirrored by the development of an international normative framework in the same area, in fact this tells us very little about the nuanced connexions between the levels of governance that coexist. As was submitted by the EU[8] in *EC—Approval and Marketing of Biotech Products ('EC—Biotech')*, '[t]he move towards a strong regulatory process has not been limited to the national dimension. The international community has been working through the last two decades in order to develop a proper framework to address the specificities of GMOs'.[9] This apparent need for—and existence of—a supplementary international regulatory context to support national endeavours, though often essential, should not, however, be taken for granted.

First, the issue of cause and effect between national and international rules is often less than fully transparent, and says a lot about the nature of power in international relations. International rules can be both a response to, as well as an impetus for, the development of national regulation. Moreover, whereas states that have acted promptly will often seek to ensure that their approach provides

---

[5] See L Glowka, *Law and Modern Biotechnology: Selected Issues of Relevance to Food and Agriculture: FAO Legislative Study 78* (Rome: Food and Agriculture Organization (FAO), 2003).
[6] See, eg the 1997 Oviedo Convention on Human Rights and Biomedicine (Council of Europe, CETS No 164).
[7] In relation to food safety, the work of the Codex Alimentarius Commission in developing and co-ordinating food standards is singularly worthy of note (the Commission being jointly established by the FAO and the World Health Organization (WHO)).
[8] Although 'EU' as shorthand for 'European Union' is not technically accurate in relation to the WTO (where the membership is in fact the 'European Communities'), common sense and shifting patterns of governance suggest that this has increasingly become something of a verbal technicality.
[9] WT/DS291/R, WT/DS292/R, and WT/DS293/R, 29 September 2006, para 4.509.

the conventional model at the global level, later states—particularly developing countries—will be influenced by what already exists at the international level with respect to their own jurisdictions. The history of the negotiation and subsequent implementation of many international agreements, including arguably the Cartagena Protocol, would certainly support this supposition.[10] Thus, those states that have acted at an early stage in the process often have the advantage in subsequent regime development and the evolution of guiding norms, and this can be a significant factor in the governance of the issue as a whole.

Moreover, it must be recognized that international rules are often utilized not only to provide an overriding global framework for national regulation on an issue; they will also frequently create a prima facie template for national legislation on the same topic, thus having a substantial bearing on municipal implementation. Many developing states, for instance, which are currently finalizing their own biosafety laws, are not only using the Cartagena Protocol as a *basis* for their legislation, but are, in fact, *replicating* its provisions.[11] What might, at first glance, appear to be—and be considered to be—a positive attempt towards something approximating global harmonization, nevertheless has the real potential to prove problematic, especially in the longer term. Differences in domestic circumstances —political, institutional, socio-economic, and environmental, amongst many other factors—should not be marginalized at the expense of seeking to achieve a unified global approach. Although the opportunity costs would seem to support a 'one-size fits all' approach whenever and wherever possible,[12] domestic legislators should be cautious in importing wholesale a piece of international law into their national regulatory structures. This is particularly important where a piece of international law is in potential divergence with other extant rules of international law; a state needs to be aware of its other international commitments, before incorporating its obligations too slavishly under one treaty alone.

In addition, in considering the impact of international law on national regulation (or vice versa), it is not simply a question of identifying the appropriate legal antecedent: that is, which law preceded which? Or, more specifically, which law affected which? As already noted, it is also an issue of political influence and within that, North–South dominance. Bearing this in mind, it is worth remembering that negotiations at the international level are often not *de novo*—in the sense that this is the first attempt at regulation—between equal partners, but

---

[10] For a discussion of the history of another environmental treaty regime, see R Benedick, *Ozone Diplomacy: New Directions in Safeguarding the Planet* (Cambridge, MA: Harvard University Press, 1998, enlarged edition).

[11] See, in the case of Africa, the examples provided in Chapter 9.

[12] Such a regulatory approach would ordinarily also be encouraged by the major commercial players in the field, as this would likely accelerate the technology dissemination process. Moreover, although the perceived deficiencies in the Cartagena Protocol itself may militate against wholehearted commercial support, commercial certainty would still usually prefer a common approach.

rather that they form part of a much more complex state of political and legal interaction between states with often very discrete and distinct economic and political agendas, and which are in very different political and economic situations. We thus must constantly recognize that by acknowledging the *negotiated* nature of such global rules, we are concurrently admitting the underlying politics and politicking inherent in the negotiation process.[13] It is in this light that the negotiations that led to the Cartagena Protocol and the ongoing discussions on GMOs should be considered.

2.2 Fragmentation within the International Legal Order?

The second aspect of this broader normative context is the recognition that international rules on the regulation, management, and use of GMOs are developed alongside, and in light of, other rules of international law. However, as already noted, this neither implies nor ensures either convergence or the absence of conflict. In relation to such competing international norms, it is often the case that other international rules have the potential to affect, if not significantly impact upon, the application and/or interpretation of the more specific legal rules under discussion. As regards GMOs, for instance, the consistent shadow of various disciplines under the authority of the WTO, especially the SPS Agreement, has been a constant theme of legal debate since the idea of a biosafety treaty was first mooted. Critics point to the potential 'chilling effect' which such trade rules in particular have on the development of rules in other areas—the idea that their very existence inhibits the negotiation of sufficiently robust international treaties,[14] especially in areas of social welfare and ecological protection.[15] On the other hand, it might be suggested that such trade rules are needed to ensure that 'welfare' norms are not created which, either in their constitution or their implementation, provide unfair advantage and/or unjustified protectionism.[16] The stand-off between these two approaches is

---

[13] See, generally, S Barrett, *Environment and Statecraft: the Strategy of Environmental Treaty-Making* (Oxford: Oxford University Press, 2003).
[14] See, eg R Eckersley, 'The Big Chill: the WTO and multilateral environmental agreements' (2004) 4 Global Environmental Politics 24.
[15] For a more balanced interpretation, see A Bianchi, 'The Impact of International Trade Law on Environmental Law and Process' in F Francioni (ed), *Environment, Human Rights and International Trade* (Oxford: Hart Publishing, 2001) 105.
[16] See, eg 1991 Beijing Declaration of Developing Countries (INC Doc GE 91-704433, 24 June 1991), para 6: 'environmental considerations should not be used as an excuse for interference in the internal affairs of the developing countries, nor should these be used to introduce any forms of conditionality in aid or development financing, or to impose trade barriers affecting the export and development efforts of the developing countries'. More generally, such tension also reveals potential conflicts within and between countries (especially for developing countries) where on some matters the protection of national welfare will be considered the priority, but on other matters global harmonization will be preferred. Clearly, the issue will usually turn upon whether the matter is of export interest to the state.

perfectly captured by the Cartagena Protocol's apparently contradictory preambular references to:

<u>Emphasizing</u> that this Protocol shall not be interpreted as implying a change in the rights and obligations of a Party under any existing international agreements,
<u>Understanding</u> that the above recital is not intended to subordinate this Protocol to other international agreements.

Possible incompatibility between rules is, of course, nothing new; what perhaps is significant is the emerging realization within the international community of the importance of the issue[17] and, just as importantly, the difficulties in finding solutions, whether these be ad hoc or systemic. As will be noted below, one of the principal means of reconciliation is through treaty interpretation, although this remains a difficult and complex challenge. In addition, however, it should be appreciated that such conflicts, although invariably concerning disputes between *legal* norms, often hide less-subtle *political* motivations. States use legal claims, not in an abstract fashion, but as part of seeking to achieve their broader foreign policy objectives. In *EC—Biotech*, for instance, the dispute involved not just *legal* differences between the parties over specific aspects of the EU's GM decision-making process, but also clear differences in political, economic, and scientific opinion about the inherent benefits and risks of GM technology.[18] In this disagreement over these issues, the respective state parties utilized the available law in a way which best supported their cause and opinions. For the EU, it was the much more risk-averse Cartagena Protocol; for the United States and the other complainants, it was the SPS Agreement, predicated as it is on the need for 'sound science' and the promotion of scientific assessment over other means of assessment, political, cultural, religious, popular support, etc. Law thus becomes instrumental; its ability to be moulded to suit individual pleadings being a fundamental feature of its political utility.

Accordingly, it is into this international context characterized, *inter alia*, by multi-level governance and increasing regulatory fragmentation,[19] that the specific rules on GMOs (as already noted, largely but not exclusively to be found in the Cartagena Protocol) should, if possible, be *pulled together*[20] with other

---

[17] Report of the Study Group of the International Law Commission, *Fragmentation of International Law: Difficulties Arising from the Diversification and Expansion of International Law*,(A/CN.4/L.682, (13 April 2006) para 8: 'specialized law-making and institution-building tends to take place with relative ignorance of legislative and institutional activities in the adjoining fields and of the general principles and practices of international law. The result is conflicts between rules or rule-systems, deviating institutional practices and, possibly, the loss of an overall perspective on the law'.

[18] See, eg M Echols, 'Food safety regulation in the European Union and the United States: different cultures, different laws' (1998) 4 Columbia Journal of European Law 525.

[19] On these issues generally, see R Revesz, P Sands, and R Stewart (eds), *Environmental Law, the Economy, and Sustainable Development (The United States, the European Union and the International Community)* (Cambridge: Cambridge University Press, 2000).

[20] The colloquial nature of this phrase was purposely chosen as it was felt more appropriate than terms such as 'reconciled', since the latter do not reflect the pragmatic, disparate, disjointed, and often contentious manner in which such rules are brought together.

aspects of the international legal system. Of course, as has been forcefully pointed out, international law does not need convergence to operate reasonably successfully. As Lowe remarks, 'it is quite possible as a matter of fact for international law to stumble along without any consistent principle for resolving ... conflicts'.[21] That said, if only for the purposes of highlighting the universal nature of international law on topics such as the regulation and management of GMOs, a broader understanding is arguably extremely valuable, if not entirely essential.

3. Treaty Interpretation—the WTO's Missed Opportunity?

One of the most discussed aspects of the interplay between GMO regulation and international law is the role of treaty interpretation in providing a 'bridge' between potentially competing norms. Its relevance was highlighted very clearly in the WTO Panel's discussion of the matter in *EC—Biotech*. Of course, treaty interpretation is not the only—nor necessarily always the most effective—means of securing a reconciliation of potentially competing norms. Other methods include the establishment of rules relating to the *application* of law; for example, those dealing with successive treaties 'relating to the same subject-matter',[22] the development of judicial concepts such as *lex specialis*, and, significantly, the emergence of a hierarchy of norms, as best reflected by the elaboration of the notion of *jus cogens*, or peremptory norms of general international law. Nevertheless, as McLachlan notes, '[i]nterpretation, on the other hand, precedes all of these techniques, since it is only by means of a process of interpretation that it is possible to determine whether there is in fact a true conflict of norms at all'.[23]

One of the principal mechanisms by which treaty interpretation can assist in providing an overarching link between norms is through the full use of, and meaningful reliance on, Article 31(3)(c) of the Vienna Convention on the Law of Treaties 1969, which provides that '[t]here shall be taken into account, together with the context: ... (c) any relevant rules of international law applicable in the relations between the parties'.[24] As an aspect of interpretation, it

---

[21] V Lowe, 'Sustainable Development and Unsustainable Arguments' in A Boyle and D Freestone (eds), *International Law and Sustainable Development: Past Achievements and Future Challenges* (Oxford: Oxford University Press, 1999) 22.

[22] See, eg Art 30 of the Vienna Convention on the Law of Treaties 1969 (application of successive treaties relating to the same subject-matter).

[23] C McLachlan, 'The principle of systemic integration and Article 31(3)(c) of the Vienna Convention' (2005) 54 International and Comparative Law Quarterly 279, 286.

[24] On this issue generally, and for an alternative perspective to McLachlan, see D French, 'Treaty interpretation and the incorporation of extraneous legal rules' (2006) 55 International and Comparative Law Quarterly 281.

must be considered alongside, and integral to, Article 31(1) (the 'basic rule'[25]), Article 31(2) (the context), Article 31(3)(a) (subsequent agreement) and (b) (subsequent practice), Article 31(4) (special meaning), and Article 32 (supplementary means of interpretation). Nonetheless, as a feature of treaty interpretation, it has long since been marginalized and ignored. As Sands commented—admittedly prior to some recent judicial activity on the topic: 'what it actually means in practice is difficult to know since it appears to have been expressly relied upon only very occasionally in judicial practice. It also seems to have attracted little academic comment. There appears to be a general reluctance to refer to Article 31(3)(c)'.[26] Yet, as was noted in the 2004 Report of the International Law Commission Study Group on the matter, 'the fact that article 31(3)(c) was rarely expressly cited should not obscure its importance as a rule of treaty interpretation'.[27] And, as McLachlan continues, 'it is submitted that the principle is not to be dismissed as a mere truism. Rather, it has the status of a constitutional norm within the international legal system'.[28]

However, the question still remains: what does Article 31(3)(c) actually mean and, more importantly, what criteria must be met before it becomes operational? As the EU sought to rely extensively upon the Cartagena Protocol, this was an issue that confronted the panel in *EC—Biotech*.[29] Unsurprisingly, utilizing the rhetorical imperative of the importance of a seamless international system, the EU argued that Article 31(3)(c), as an element of the customary rules on interpretation,[30] should be considered an invaluable tool in this regard. The Panel, however, was less convinced as to its applicability.

On the other hand, before considering why the Panel ultimately rejected reliance on this provision, it is useful to look at those comments it did make which arguably contribute a positive understanding of the provision. First, the Panel noted that the reference to 'rules of international law' should be understood expansively, not only to include treaty (conventional) law and rules of

---

[25] Under Art 31(1), '[a] treaty shall be interpreted in good faith in accordance with the ordinary meaning to be given to the terms of the treaty in their context and in the light of its object and purpose'.

[26] P Sands, 'Sustainable Development: Treaty, Custom, and the Cross-fertilization of International Law' in Boyle and Freestone (n 21 above) 49.

[27] International Law Commission Study Group on Fragmentation of International Law, *Difficulties Arising from the Diversification and Expansion of International Law*', Report on Fifty-Sixth Session (2004) Supplement No 10 (A/59/10) 301.   [28] McLachlan (n 23 above) 280.

[29] See, generally, M Young, 'The WTO's use of relevant rules of international law: an analysis of the Biotech Case' (2007) 56 International and Comparative Law Quarterly 907. See, further, Chapter 13.

[30] On the incorporation of general rules of interpretation into the operation of the WTO, see Art 3(2) of the Understanding on Rules and Procedures Governing the Settlement of Disputes ('Dispute Settlement Understanding') ('in accordance with customary rules of interpretation of public international law'), as subsequently reinforced by the Appellate Body in such early cases as *US—Gasoline: Standards for Reformulated and Conventional Gasoline* in the penultimate line WT/DS2/AB/R, 29 April 1996, and *Japan—Alcohol Beverages* II WT/DS8,10–11/AB/R, 4 October 1996.

customary international law, but also general principles of law as included in Article 38(1)(c) of the Statute of the International Court of Justice.[31]

Secondly, the Panel also discussed what effect the inclusion of other 'relevant' rules would have on the interpretation of the primary rule, in this case the WTO covered agreements; or, in other words, what was meant by the phrase 'shall be taken into account'. Again, in a rather positive and expansive interpretation of the provision, the Panel was prepared to accept a relatively broad approach to this issue:

> It is important to note that Article 31(3)(c) mandates a treaty interpreter to take into account other rules of international law ('there shall be taken into account'); it does not merely give a treaty interpreter the option of doing so. It is true that obligation is to 'take account' of such rules, and thus no particular outcome is prescribed. However, Article 31(1) makes clear that a treaty is to be interpreted 'in good faith'. Thus, where consideration of all other interpretative elements set out in Article 31 results in more than one permissible interpretation, a treaty interpreter following the instructions of Article 31(3)(c) in good faith would in our view need to settle for that interpretation which is more in accord with other applicable rules of international law.[32]

This is an important statement of principle, and although stated by a WTO Panel, and thus not binding in any form on any other judicial or arbitral panel, it arguably reflects a sound approach to the issue. The argument that where, all things being equal, a treaty interpreter should 'settle for that interpretation which is more in accord with other applicable rules of international law' would seem to recognize the importance of treating disparate rules of international law holistically, whenever possible. However, as noted below, what would seem to be a very worthy objective was significantly diluted in this case by other reasoning in the Panel's decision, but nevertheless, in abstract and isolation, it represents a useful and positive development.

The Panel's view here, on the other hand, must be read in the light of *dicta* in other cases that would seem to take a more restrictive approach to the matter. An example is *Dispute Concerning Article 9 of the OSPAR Convention*,[33] which

---

[31] *EC—Biotech*, para 7.67: '[r]egarding the recognized principles of law which are applicable in international law, it may not appear self-evident that they can be considered as "rules of international law" within the meaning of Article 31(3)(c). However, the Appellate Body in *US—Shrimp* made it clear that pursuant [to that article] general principles of international law are to be taken into account in the interpretation of WTO provisions ... the European Communities considers that the principle of precaution is a "general principle of international law". Based on the Appellate Body report on *US—Shrimp*, we would agree that if the precautionary principle is a general principle of international law, it could be considered a "rule of international law" within the meaning of Articles 31(3)(c)'. Ultimately, the Panel did not decide upon the issue of the status of the precautionary principle, noting that '[s]ince the legal status of the precautionary principle remains unsettled, like the Appellate Body before us, we consider that prudence suggests that we not attempt to resolve this complex issue, particularly if it is not necessary to do so. Our analysis below makes clear that for the purposes of disposing of the legal claims before us, we need not take a position on whether or not the precautionary principle is a recognized principle of general or customary international law' (para 7.89).
[32] Ibid para 7.69.
[33] (2003) 42 ILM 1118, 2 July 2003.

related to a dispute between the Republic of Ireland and the United Kingdom on the accessibility to environmental information. The majority accepted that 'current international law and practice'[34] is admissible through Article 31(3)(c) and that, consequently, '[l]est it produce anachronistic results that are inconsistent with current international law, a tribunal must engage in *actualisation* or contemporization when construing an international instrument that was concluded in an earlier time'.[35] However, the majority was also very clear that interpretation must ultimately respect the rights of states as sovereign parties to treaties, firmly stating that '[t]he issue here is one of interpretation in good faith ... A treaty is a solemn undertaking and States Parties are entitled to have applied to them and their peoples that to which they have agreed and not things to which they have not agreed.[36]

Certainly, in *EC—Biotech*, the United States—as a non-party to either the Convention on Biological Diversity or the Cartagena Protocol—strongly resisted any interpretation that might suggest that obligations they had *not* entered into must affect the substantive interpretation of those agreements they had entered into. And, as noted below, the Panel equally shared this concern.

Beyond the specific issue of treaty interpretation, many have also raised the matter of the impact of the incorporation of extraneous legal material upon the jurisdictional scope of the dispute. As Judge Buergenthal noted in his separate opinion in *Case Concerning Oil Platforms (Islamic Republic of Iran v United States of America)*, a case before the International Court, whilst Article 31(3)(c) of the Vienna Convention 'is sound and undisputed in principle as far as treaty interpretation is concerned',[37] this neither gives it free reign to assist in the determination of inter-state disputes nor allows judicial/arbitral tribunals to stray across jurisdictional limits. Buergenthal's principal concern with introducing other rules of international law via Article 31(3)(c) was thus not the normative uncertainty that this might create per se (although clearly this mattered), but rather that it evades important questions of jurisdiction. Similarly, Judge Higgins noted that '[t]he [International] Court has, however, not interpreted [the primary treaty] by reference to the rules on treaty interpretation. It has rather invoked the concept of treaty interpretation to displace the applicable law'.[38]

Perhaps, it was part of the concern of the Panel in *EC—Biotech* that the EU, by seeking to rely on the Cartagena Protocol, was not actually wishing to use it to assist in the interpretation of the SPS Agreement, but rather in Judge Higgins' words, 'to displace' it. Ultimately, the Panel refused to rely upon the Cartagena Protocol and not accept the application of Article 31(3)(c) in this case. Its premise for so doing centred on precisely what that provision meant when

---

[34] Ibid para 101 (but only insofar as such law and practice are relevant).
[35] Ibid para 103. The tribunal, however, raised the question whether actualization was necessary 'of a treaty made scarcely ten years earlier': ibid.   [36] Ibid para 102.
[37] Judgment of 6 November 2003, Separate Opinion of Judge Buergenthal, para 22.
[38] Ibid Separate Opinion of Judge Higgins, para 49.

referring to 'applicable in the relations between the parties'. The Panel adopted a most stringent requirement in this regard. It reasoned that 'parties' in this context must invariably mean *all* parties and that, consequently, 'the rules of international law to be taken into account in interpreting the WTO agreements at issue in this dispute are those which are applicable in the relations between the WTO Members'.[39] In other words, *all* WTO Members must be bound by the other rule, if Article 31(3)(c) is to apply.

This, of course, need not necessarily be a problem when it comes to customary international law or general principles of law, but as regards conventional law, such as the Cartagena Protocol, it sets a huge obstacle in the way—all 153 current members of the WTO must also be party to the Protocol, an almost impossible hurdle in relation to multilateral treaties. The Panel left open the issue of whether Article 31(3)(c) should apply if all parties to the *dispute*, though not the WTO Agreement, were also party to the 'other' treaty. However, as this was not the case (since the United States was a non-party), the important secondary question was not decided (although on the basis of its primary reasoning that all WTO members must also be parties to the second treaty, this would also seem logically to stymie even this slightly broader understanding).

On first view, the reasoning of the Panel would appear to be relatively convincing:

[r]equiring that a treaty be interpreted in the light of other rules of international law which bind the States parties to the treaty ensures or enhances the consistency of the rules of international law applicable to these States and thus contributes to avoiding conflicts between the relevant rules.[40]

There would seem to be much merit in limiting applicability in some way such as this as it provides for greater legal certainty than might otherwise be the case and gives less room for overt judicial discretion. Moreover, as McLachlan notes, treaty interpretation is not just about dispute-settlement, since 'Article 31 is concerned with the promulgation of a general rule, which would apply to the interpretation of a treaty irrespective of whether any particular parties to it may happen to be in dispute'.[41] It is important for the parties themselves to know what their obligations and rights are, 'irrespective' of the existence of a dispute.

Nevertheless, both pragmatically and theoretically, I am not convinced that this understanding is altogether correct. On a pragmatic level, judicial practice seems to be against such an approach: even the WTO Appellate Body has seemingly taken a more selective (in other words, less-rigid) approach than the Panel requires. The Panel deals with such apparent generosity of interpretation by suggesting that the Appellate Body—most evidently in *US—Shrimp*[42]—was not actually involved in a process of treaty interpretation under Article 31(3)(c)

---

[39] *EC—Biotech*, para 7.68.   [40] Ibid para 7.70.   [41] McLachlan (n 23 above) 315.
[42] WT/DS58/AB/R, 12 October 1998.

at all, but rather was undertaking the more acceptable process of determining the 'ordinary meaning' of a treaty provision under Article 31(1). As it notes:

> other relevant rules of international law may in some cases aid a treaty interpreter in establishing, or confirming, the ordinary meaning of treaty terms ... Such rules would not be considered because they are legal rules, but rather because they may provide evidence of the ordinary meaning ... They would be considered for their informative character.[43]

According to the Panel, this is what the Appellate Body did in *US—Shrimp*: 'as we understand it, the Appellate Body drew on other rules of international law because it considered that they were informative and aided it in establishing the meaning and scope of the term "exhaustible natural resources"'.[44]

Whether this is actually the case is debatable. In that dispute, what the Appellate Body actually said was that '[t]he words of Article XX(g) ... were actually crafted more than 50 years ago. They must be read by a treaty interpreter in the light of contemporary concerns of the community of nations about the protection and conservation of the environment'.[45] On the face of this, it would certainly seem to suggest that the Appellate Body was relying upon other legal rules for more than just 'informative' evidence. In fact, what was actually happening in this case was the Appellate Body using an amalgam of a variety of very different approaches, without any real attempt to isolate individual theoretical positions.[46] This can be seen by the way in which the Appellate Body draws together conclusions on the point:

> Given the recent acknowledgement by the international community of the importance of concerted bilateral or multilateral action to protect living natural resources, and recalling the explicit recognition by WTO Members of the objective of sustainable development in the preamble of the *WTO Agreement*, we believe it is too late in the day to suppose that Article XX(g) ... may be read as referring only to the conservation of exhaustible mineral or other non-living natural resources. Moreover, two adopted GATT 1947 panel reports previously found fish to be an 'exhaustible natural resource' within the meaning of Article XX(g). We hold that, in line with the principle of effectiveness in treaty interpretation, measures to conserve exhaustible natural resources, whether *living* or *non-living*, may fall within Article XX(g).[47]

However acceptable or valid 'we believe it is too late in the day' might be as justification for treaty interpretation, it would nonetheless seem too simplistic to argue, as the Panel did in *EC—Biotech*, that the Appellate Body was merely

---

[43] *EC—Biotech*, para 7.92. Cf Young (n 29 above) 929: 'quite apart from the questionable doctrinal foundations of this reading of [Art 31(1) of the Vienna Convention] its application can lead to highly abstract notions that are de-contextualised from disputes'.
[44] *EC—Biotech*, para 7.94.
[45] *US—Shrimp* WT/DS58/AB/R, 12 October 1998, para 129.
[46] French (n 24 above) 298.
[47] *US—Shrimp* WT/DS58/AB/R, 12 October 1998, para 131.

seeking to interpret the 'ordinary meaning' of a term. Whichever position is ultimately correct in law, in fact both the Appellate Body and other WTO dispute panels[48] will continue to rely upon extraneous treaty law—with less concern for such formalities—when it assists in both their argumentation and final decision.

In a more recent dispute, for instance, the Panel in *Brazil—Measures Affecting Imports of Retreaded Tyres*[49] extensively relied upon the 1999 Basel Convention Technical Guidelines on the Identification and Management of Used Tyres (adopted by the Conference of the Parties by virtue of Decision V/26), notwithstanding the fact that not all Members of the WTO are party to the Basel Convention on the Control of Transboundary Movements of Hazardous Wastes and Their Disposal.[50] Significantly, although what were under discussion were non-binding technical guidelines, there was simply no discussion in the Panel's reasoning of the advisability or legality of relying upon this. It is strongly arguable that the issue is not whether or not the guidelines themselves are legally binding; it is whether the treaty from which they are derived is binding as per the reasoning in *EC—Biotech*. It may well be, of course, that the Panel in *Brazil—Retreaded Tyres* relied upon the Technical Guidelines just to determine the 'meaning and scope' of a provision of the WTO (as the panel in *EC—Biotech* thought permissible to determine a provision's 'ordinary meaning'). Yet, it might be wondered whether the imperative in Article 31(3)(c) of the Vienna Convention to 'take account of' would have actually resulted in anything more consequential.

To conclude, there is clearly much judicial discretion in this area of dispute-settlement, even within a relatively 'closed' sphere such as the WTO dispute-settlement mechanism. One can either ignore this discretion—or marginalize it as merely 'judicial politics'—or one can accept that such discretion does exist and, equally, must inevitably form part of the interpretative process. Of course, there are limits in what a tribunal may do; the Panel in *EC—Biotech* clearly could not apply the provisions of the Cartagena Protocol in precedence to those in the SPS Agreement. However, it seems eminently sensible to accept that, as most judicial decisions are rather imprecise affairs when it comes to how interpretation is reached,[51] it is ultimately overly formalistic to rule out—as completely as the Panel did—that such a potentially apposite treaty cannot provide any assistance. Moreover, the decision *not* to consider the Cartagena Protocol is

---

[48] As is clear from the next paragraph. [49] WT/DS332/R, 12 June 2007.
[50] Moreover, in this case, although Brazil and the EU are members of both the Basel Convention and the WTO, the United States—as third party—is not a party to the former, thus not meeting the Panel's interpretation of symmetry in treaty membership in *EC—Biotech*.
[51] G Schwarzenberger, *A Manual of International Law, Volume I* (4th edn, London: Stevens & Sons, 1960) 153: '[i]nternational courts and tribunals fight shy of laying bare the equitable and common-sense reasons on which, in fact, their interpretative work is based'.

itself an example of judicial discretion. Of course, the Panel suggested that such consideration would be appropriate if it was merely for linguistic clarification to determine the 'ordinary meaning', but, as suggested above, there is no clear red line between that and the consequence of applying Article 31(3)(c), certainly not to the extent of justifying the endorsement of the former and the condemnation of the latter.

4. Socio-economic Considerations and Sustainable Development

The second issue in this chapter concerns the appropriateness of the concept of sustainable development in determining a balanced framework for the inclusion of socio-economic considerations within GMO decision-making processes, as permitted under the Cartagena Protocol. While it might be thought that the principle of sustainable development should—and would—play a pivotal role in reconciling the Cartagena Protocol with the various disciplines under the WTO, as both refer to the principle in their respective Preambles, there is little consensus on what contribution sustainable development can actually make. Moreover, although the concept of 'mutual supportiveness' between the goals of the global trade regime, on the one hand, and multilateral environmental agreements, on the other, is increasingly accepted[52] as an inevitable—and invaluable—implication of the sustainable development debate,[53] what this supportiveness actually means remains both unclear and extremely contested.

One particular question in this broader debate—and one where sustainable development arguably has something meaningful to say—is to what extent should socio-economic considerations affect and influence GMO decision-making processes? The Cartagena Protocol controversially includes a specific (if not altogether clear) mandate to this effect. Article 26(1) notes that:

[t]he Parties, in reaching a decision on import under this Protocol or under its domestic measures implementing the Protocol, may take into account, consistent with their international obligations, socio-economic considerations arising from the impact of living modified organisms on the conservation and sustainable use of biological diversity, especially with regard to the value of biological diversity to indigenous and local communities.

This provision is supplemented by Article 26(2), which encourages the parties to 'cooperate on research and information exchange on any socio-economic impacts of living modified organisms, especially on indigenous and local

---

[52] See, eg Doha Ministerial Declaration, WT/MIN(01)/DEC/1 (20 November 2001) para 31: '[w]ith a view to enhancing the mutual supportiveness of trade and environment, we agree to negotiations, without prejudice their outcome, on ... '.
[53] See, generally, C Redgwell, 'Biotechnology, Biodiversity and Sustainable Development: Conflict or Congruence' in F Francioni and T Scovazzi (eds), *Biotechnology and International Law* (Oxford: Hart Publishing, 2006) 61.

communities'. Significantly, as noted above, Article 26 ranges beyond the international regime per se to incorporate other levels of governance, specifically 'domestic measures [taken to] implement ... the Protocol'. But the wider issue remains: to what extent should socio-economic considerations form part of what has traditionally been characterized as a primarily scientific decision-making process?[54]

Some (in particular, developed) states feared, and continue to be concerned by, the potential for protectionism and arbitrary discrimination if socio-economic considerations were permissible factors to be considered when administrations implement their GMO processes. In an area where certainty is an important—though not paramount—commercial and legal value, introducing socio-economic circumstances was felt likely merely to exacerbate the uncertainties already inherent in GMO approval mechanisms. On the other hand, many noted that the risks of GMO introduction were not simply ecological or human health-related, but also social and developmental, and could not be ignored.[55] As a 2005 report by the Executive Secretary of the Convention on Biological Diversity states:[56]

[m]any potential importers ... were concerned that these organisms could undermine the livelihoods of their farmers, and indigenous and local communities through the possible displacement of local varieties, loss of markets and employment, and posing threat to their cultural and ethical values. They argued that the introduction of living modified organisms could erode their diverse agricultural systems and undermine the value of biological diversity, its conservation and sustainable use.[57]

In between these two extremes, the wording of Article 26 emerged, reflecting the acute concern, particularly of GMO exporters, that any consideration of socio-economic factors should occur in a way that is 'consistent with ... international obligations', implicitly the provisions of the SPS Agreement and the Agreement on Technical Barriers to Trade.[58] However, the fact remains that if Article 26 is not to be regarded as a peripheral or a 'dead-letter' provision, and notwithstanding

---

[54] See also Art 31(7)(d) of Directive (EC) 2001/18 of the European Parliament and of the Council on the deliberate release into the environment of genetically modified organisms ('Deliberate Release Directive') [2001] OJ L106/1, which sets out the need to consider the 'socioeconomic implications of deliberate releases and placing on the market of GMOs'.

[55] Moreover, there are those who argue that the Cartagena Protocol does not go far enough: see International Centre for Trade and Sustainable Development (ICTSD), *Biotechnology: Addressing Key Trade and Sustainability Issues* (Geneva: ICTSD, 2006) 55: '[t]he Biosafety Protocol has been criticised, however, for not taking other important developing country concerns on board. For example ... Gopo and Kameri-Mbote (2005) argue that the focus of the Protocol is not on biosafety but rather on "bio-trade"—and thus primarily benefits biotechnology exporters without giving adequate protection to people in the developing world'.

[56] Executive Secretary of the Convention on Biological Diversity, *Socio-Economic Considerations: Cooperation on the Research and Information Exchange* (2005).

[57] UNEP/CBD/BS/COP-MOP/2/12 (24 March 2005) para 4.

[58] Available at <http://www.wto.org/english/docs_e/legal_e/17-tbt.pdf>, last accessed 9 November 2009.

the limited scope for consideration of socio-economic factors within WTO disciplines (especially within the SPS Agreement), there must be some middle ground to permit states to utilize this provision in the fullest manner possible. Many would suggest that this is simply impossible, particularly in the context of the SPS Agreement, where broadly interpreted socio-economic considerations are simply outside its scope of permissible risks.[59] Nevertheless, unless one is to give *de jure* primacy to the SPS Agreement over the Cartagena Protocol, Article 26 remains a valid—and lawful—provision of an international agreement.

These issues, though less discussed than others in *EC—Biotech*, did receive at least some attention in the submissions made by the EU, where it was observed, in particular, that:

> In their submissions, the complaining parties seek to evade or ignore the whole socio-political, legal, factual and scientific complexity of the case. The complaining parties wilfully ignore the social controversies that led to the revision of the European Communities' regulatory framework in the period 1998–2001 (a framework that is not challenged). They also ignore the scientific and regulatory debates at the international level that have taken place over the past years, including the process that led to the conclusion of the Cartagena Protocol on Biosafety. The Protocol is based on the understanding that the inherent characteristics of GMOs require them to be subject to rigorous scrutiny so as to ensure that they do not cause harm to the environment or human health, or cause socio-economic disruptions.[60]

The EU went on to note that:

> [i]n light of these risks, governments around the world, since the first commercialisation of GMOs in the early nineties, have started to address the question of how to regulate GMOs ... Often such systems are based on a precautionary approach, and decisions are sometimes made dependent on considerations other than scientific factors, such as, for instance, socio-economic considerations.[61]

These issues went unconsidered by the Panel.

Thus, within this apparent stalemate, where might one turn? Arguably, one of the most useful ways forward is to seek conceptual support for Article 26 from the principle of sustainable development which, as noted above, is expressly referred to in the Preamble of both the Cartagena Protocol and the WTO Agreement, and which provides a potentially compelling and constructive link between them. One need not look far for support for this view; the Appellate

---

[59] But see Annex A, (1)(d): 'any measure applied ... to prevent or limit other damage within the territory of the Member from the entry, establishment or spread of pests', which the Panel in *EC—Biotech* read as follows: '[t]he residual category of "other damage" is potentially very broad. In our view, "other damage" could include damage to property, including infrastructure (such as water intake systems, electrical power lines, etc.). In addition, we think "other damage" could include economic damage (such as damage in terms of sales lost by farmers). The dictionary defines the term "damage" as "physical harm impairing the value, usefulness, or normal function of something" and "unwelcome and detrimental effects", or "a loss or harm resulting from injury to person, property, or reputation"': para 7.370 (references removed).

[60] *EC—Biotech*, para 4.332.   [61] Ibid para 4.339.

Body itself has already relied upon the concept within the context of the General Agreement on Tariffs and Trade (GATT), and there is *legally* no distinction between the GATT and the SPS Agreement as annexed agreements to the WTO Agreement to justify any divergence in approach.[62] Specifically, one might note the following three excerpts from *US—Shrimp*:

> The preamble of the WTO Agreement—which informs not only the GATT 1994, but also the other covered agreements—explicitly acknowledges 'the objective of sustainable development'.
>
> This concept has been generally accepted as integrating economic and social development and environmental protection.
>
> We note once more that this language demonstrates a recognition by WTO negotiators that optimal use of the world's resources should be made in accordance with the objective of sustainable development.[63]

Of course, this tells us very little about the methodology to determine how the principle of sustainable development might be used in evaluating the 'WTO-compatibility' of Article 26, although equally such quotations also say very little about why it should permit the Appellate Body to reinterpret 'exhaustible natural resources' in Article XX(g) GATT to include living species, even though the Appellate Body said that it did. Instead, what these quotations tell us is that sustainable development has already been used at least once by the Appellate Body to provide a broader interpretation than it might otherwise have done; to say that it cannot do so in any other situation—or in a case involving the SPS Agreement—would be both legally incorrect and as much an example of judicial discretion. Importantly, one might point again to the Appellate Body's definition of sustainable development: 'integrating economic and social development and environmental protection'.[64] This seems to be a particularly clear imperative —if not exactly treaty language—to support a more holistic approach to the inclusion (note, not the pre-eminence) of socio-economic considerations.

Of course, the text of the SPS Agreement remains paramount; but, surprisingly, despite much that is said about the SPS Agreement, the language already contains some references to (admittedly a restricted list of) economic considerations.[65] Nevertheless, unlike the perceived open nature of some GATT

---

[62] *US—Shrimp* WT/DS58/AB/R, 12 October 1998, para 153: '[a]s this preambular language reflects the intentions of negotiators of the WTO Agreement, we believe it must add colour, texture and shading to our interpretation of the *agreements* annexed to the WTO Agreement' (emphasis added). [63] Ibid para 129, fn 107, and para 153.
[64] Ibid para 129 fn 107.
[65] See Art 5(3) of the SPS Agreement: '[i]n assessing the risk to animal or plant life or health and determining the measure to be applied for achieving the appropriate level of sanitary or phytosanitary protection from such risk, Members shall take into account as relevant economic factors: the potential damage in terms of loss of production or sales in the event of the entry, establishment or spread of a pest or disease; the costs of control or eradication in the territory of the importing Member; and the relative cost-effectiveness of alternative approaches to limiting risks'.

terminology, as evidenced by the Appellate Body's interpretation in *US—Shrimp*, the language of the SPS Agreement is arguably, and conspicuously, that much tighter, particularly after some of the earliest rulings on aspects of the agreement.[66] Therefore, could a party seek to rely upon, or would a Panel or the Appellate Body utilize, the Preamble to the WTO Agreement to ensure the appropriate incorporation of socio-economic considerations within an interpretation of the SPS Agreement? Current jurisprudence would seem to suggest that soft law general principles, even one contained in the Preamble such as sustainable development, would find their limit when confronted with a clearly worded, opposing, binding rule.[67] That said, conflict is rarely an inevitability; convergence—at some level—is often attainable to a greater or lesser extent. As the Appellate Body has clearly noted, 'it must add colour, texture and shading to our interpretation'.[68] Thus, the answer remains deeply uncertain, although, as observed above, the decision not to integrate is as much a policy choice as the decision to do so.

To conclude this section, it is also necessary to discuss briefly two further issues. First, the matter of the desired level of protection. Unlike the question of the legality of an SPS measure itself, the Appellate Body has asserted that determination of the appropriate *level* of protection which a Member State wishes to achieve is at the sovereign discretion of each member. As the Appellate Body stated in *EC—Measures Affecting Asbestos and Asbestos-containing Products*, in relation to Article XX(b) GATT (on which, of course, the SPS Agreement amplifies), 'it is undisputed that WTO Members have the right to determine the level of protection of health that they consider appropriate in a given situation'.[69] Logically, this would seem to suggest that not only is determining the level of protection outside the remit of the SPS Agreement, but consequently it is one that may also include socio-economic factors, where they are considered relevant.

Unfortunately, while it seems reasonably certain that members have a high degree of autonomy on this issue, there is no *carte blanche* for them to base their position principally or exclusively upon their own socio-economic situation.[70]

---

[66] See, generally, J Pauwelyn, 'The WTO Agreement on Sanitary and Phytosanitary (SPS) Measures as applied in the first three SPS disputes—*EC—Hormones, Australia—Salmon* and *Japan—Varietals*' (1999) 2 Journal of International Economic Law 641.
[67] See *EC—Measures Concerning Meat and Meat Products (Hormones)* WT/DS26/AB/R and WT/DS48/AB/R, 16 January 1998, with reference to the precautionary principle: '[it] does not, by itself, and without a clear textual directive ... [override] ... the normal (ie. customary international law) principles of treaty interpretation': para 124.
[68] *US—Shrimp* WT/DS58/AB/R, 12 October 1998, para 153.
[69] WT/DS135/AB/R, 12 March 2001, para 168.
[70] Art 5(4) of the SPS Agreement states: 'Members should, when determining the appropriate level of sanitary or phytosanitary protection, take into account the objective of minimizing negative trade effects'. *Cf* the Panel report in *EC—Measures Concerning Meat and Meat Products (Hormones) (Canada)* WT/DS48/R/CAN, 18 August 1997, para 8.169: '[g]uided by the wording of Article 5.4, in particular the words "should" (not "shall") and "objective", we consider that this

Moreover, to the extent that the objective to be attained and the means of achieving that objective are inextricably linked, it is impossible completely to divorce a member's autonomy in determining its desired level of protection from the requirements of the SPS Agreement.[71] As the Appellate Body noted (again in relation to the GATT) in *EC—Asbestos*:

[i]n addition, we observed, in [*Korea—Beef*], that '[t]he more vital or important [the] common interests or values' pursued, the easier it would be to accept as 'necessary' measures designed to achieve those ends. In this case, the objective pursued by the measure is the preservation of human life and health through the elimination, or reduction, of the well-known, and life-threatening, health risks posed by asbestos fibres. The value pursued is both vital and important in the highest degree.[72]

Although the Appellate Body is here specifically talking about the necessity of the means chosen to meet the objective pursued,[73] there is also an implicit judgment about the merits of the objective itself ('[t]he value pursued is both vital and important'). Arguably, where that value is less clear, or where it has been influenced by socio-economic, rather than life and health, considerations, it would seem likely that not only would justifying the necessity of the measure prove significantly more difficult, but a Panel or the Appellate Body might also feel obliged in such circumstances to question the objective itself.

The second issue is whether, as a second-best option, the SPS Agreement provides any exemptions, which might permit the inclusion of such socio-economic considerations, notwithstanding its substantive provisions. Article 10(3), for instance, states that:

[w]ith a view to ensuring that developing country Members are able to comply with the provisions of this Agreement, the Committee [on Sanitary and Phytosanitary Measures] is enabled to grant to such countries, upon request, specified, time-limited exceptions in whole or in part from obligations under this Agreement, taking into account their financial, trade and development needs.

Such special and differential treatment is not, of course, novel within the context of WTO law, although one should note its limitations, which are, in this case, *temporal* ('time-limited exceptions'), *personae* ('developing country Members')

provision of the SPS Agreement does not impose an obligation. However, this objective of minimizing negative trade effects has nonetheless to be taken into account in the interpretation of other provisions of the SPS Agreement.'

[71] See *Canada—Continued Suspension of Obligations in the EC—Hormones Dispute* WT/DS321/AB/R, 16 October 2008, para 685: '[a] WTO Member that adopts an SPS measure resulting in a higher level of protection than would be achieved by measures based on international standards must nevertheless ensure that its SPS measures comply with the other requirements of the *SPS Agreement*, in particular Article 5. This includes the requirement to perform a risk assessment' (references removed). [72] *EC—Asbestos* WT/DS135/AB/R, 12 March 2001, para 172.
[73] Some have referred to this as the adoption of a proportionality test within WTO jurisprudence: with reference to Art XX(b) GATT, see M Matsushita, TJ Schoenbaum, and PC Mavroidis, *WTO: Law, Practice, and Policy* (2nd edn, Oxford: Oxford University Press, 2007) 800.

and *materiae* ('financial, trade and development needs'). Such restricted exemptions therefore do little to address the EU's broader concern as to the 'socio-political, legal, factual and scientific complexity' surrounding the management of scientific risk. In pursuing perhaps a broader social agenda than the United States, the EU has found itself subject to the strictures of an Agreement which the EU itself played a key role in devising, with very limited scope for subsequent political and legal manoeuvre.

5. Conclusion

The purpose of this chapter was not to consider the relationship between the Cartagena Protocol and the SPS Agreement *in minutiae*; that has been done elsewhere. Rather, the aim was to locate the GMO issue within a broader (international) legal framework; and to consider the importance of the connexions between the bespoke regulation of GMOs with more generic matters of international law. In particular, in addressing issues of treaty interpretation and sustainable development, the chapter has highlighted that not only is it extremely *useful* to consider GMO regulation alongside general international law principles; it is, in fact, *essential*. To that extent, when framing the parameters of the GMO regime, one must always look beyond the immediate regulatory environment to the systemic overarching structure. Or, in other words, one can only truly understand the 'specific' by reference to, and within the context of, the 'general'. Thus, neither analysing the Cartagena Protocol on its own terms nor placing it alongside (usually in contradiction to) the SPS Agreement will ever give as full an answer as properly locating it within the *corpus* of international law as a whole. It is only by considering this broader canvas that one can identify what are the long-term problems—as compared to the 'huff-and-puff' of short-term politics—and begin to try to negotiate comprehensive and sustainable solutions. As an aside, this also means that there can be no such thing as 'GMO law' (or, for that matter, a 'GMO lawyer'). Specialization is fine; over-specialization in ignorance of the wider framework must be totally avoided.

Conclusion

Luc Bodiguel

A conclusion is not a summary. On the contrary, it takes the reader in new directions, along new paths which have not yet been investigated; it opens a final door before the last page is reached and leaves the door open; it surprises. With such in mind, this Conclusion will not be a summary of the previous chapters devoted to regulatory regimes for genetically modified organisms (GMOs). The idea is rather to expand the debate—to leave behind legal analysis of regulations, and instead question the role and concept of the law in the arena of GMOs.

For this purpose, we shall take two approaches. The first, open and general (maybe even subversive), and without a safety net, is intended to generate ideas for general discussion in relation to the law and GMOs. The second concerns civil disobedience in the context of GMOs, so exploring legal concepts which have been developed during the GM debate.

1. The Role of Law: Some (Subversive) Issues

The subject of GMOs often leads to binary and even Manichaean discussion, which can be summarized as follows.

A transgenic maize seed is produced and is either Angel or Demon. It is an Angel if it grows well in arid, sterile, or poor land, generating a level of agricultural production that ensures the self-sufficiency of each country. It is also an Angel if it avoids the intensive use of chemicals,[1] the main source of pollution in soil, water, and food, and a source of public health problems. It is yet again an Angel if it allows everyone access to agricultural production without imposing extra costs[2] and without compromising economic and social considerations. But it is a Demon if it is a wayward failure, developed at tremendous cost, serving only to reinforce the hegemony and profit of multinationals. It is also a Demon if 'mere puff', reducing science and expertise to a series of commercial arguments. And it is a Demon yet again if the technology is divorced from any human, social, and economic considerations, with scientific innovation as its sole objective.

These diametrically opposite approaches give rise to two questions for discussion on and about the law.

---

[1] Although there will be adverse impact upon employees in petro-chemical industries, which manufacture these products.
[2] Such as the cost of evaluations, monitoring, control, public participation, and insurance.

Question 1: is it Possible and Desirable to Move Away from This Dual Approach?

When implementing such measures as those governing risk analysis or coexistence, the legislation analysed in this work often offers less-radical solutions than a straight 'yes' or 'no'; but, for all that, it does not generally venture beyond the limits of binary discussion. In Europe, Canada, the United States, Argentina, Uruguay, or South Africa, 'yes' takes precedence. It is only in a few African countries or certain regions of Europe (which do not enjoy legal competence in GM matters) that 'no to GMOs' is sometimes the official line. On reflection, just as with the regulations themselves, it may be a very difficult exercise to avoid such dualism; and escape routes adopted have been to put off decisions, simply give opinions, and limit the level of commitment.[3]

In this light, the objective of this book has been to accommodate all viewpoints, not seeking to reassure or provide a single, easy answer to such a complex problem. On the contrary, the different chapters explore different approaches to GMOs; and, within an essentially legal framework, each contributor analyses the societal choices which have been translated into law—from bold liberalism to absolute prohibition, together with more reserved or prudent positions in between. In addition, the contributors have examined the extent to which emphasis has been placed on 'sound science', and also recurring problems of implementation. The result is necessarily a varied collection, with it often being difficult to strip out horizontal themes (such as participation, risk, governance, coexistence, and liability) and to compare regulation at different levels (whether international, regional, or national). However, four common trends can be distinguished running throughout the book: first, the importance of GMOs in public debate has generated the need for regulation, even in countries with a *laissez-faire* approach; secondly, the rule of law is systematically used as a tool for mediation between the various interests involved, the search for a common language being particularly sensitive at regional or international level; thirdly, the limits of regulation are broadly accepted (not least, through recognition that they may not be practicable or effective in light of financial cost, local opposition, and the limits of scientific knowledge); and, fourthly and most importantly, in most cases there is found to be a predetermined and dominant political choice in favour of free circulation.

Question 2: the GM Debate—a Consequence of Politicization and Appeals to a Morality not Based on Science?

This may be partly true. However, an explanation can be sought in the ethics, morality, and societal aspects of modern biotechnology, a technological process

---

[3] On the role of researchers in this context, see C Vélot, *OGM. Tout S'explique* (Athée: Editions Goutte de Sable, 2009) 21–2.

which can affect both humans and the natural environment. Indeed, Roy Lewis in *Evolution Man: Or, How I Ate My Father* has already explained the whys and wherefores of this debate.[4] Between 'the father' and 'uncle Vanya', discussion is down to earth. The former strongly prefers experimentation and continuous discovery in the name of 'evolution' of the species. He maintains this position even when it leads to the destruction of all or part of the forest and its wildlife following a conflagration caused by improper use of fire. The latter considers that man must stay in his place in the natural cycle, that he is not a distinct species, and that it would be better to continue to sleep in the trees rather than attempt to achieve emancipation through the use of hazardous processes: 'you cannot control this infernal thing you call progress'.[5] The sons provide their own input. They adopt a careful approach towards 'technological innovation' after the fire, but this is accompanied by the intention to make fire a competitive tool, an instrument for individual or collective power: '[h]ave I understood you properly father? Do you really intend to give your fire-lighting formula to any Tom, Dick or Harry in Africa?'[6] And this leads them to get rid of the father...

Progress in the face of nature, risk, the management of inventions, patents: all of this is involved; and it can soon be appreciated why the GM debate becomes primarily a question of moral or political ethics. In this context, science proves another factor in the decision-making process. As so often, however, it falls to the law to perform its essential function of integrating these ethical, moral, and political choices into each legal system. In the case of GMOs, the form and content of the law will therefore depend mainly on national or regional concepts of scientific progress and its assumed or demonstrated risks.[7] Yet, at the same time, these choices must conform to different legal systems, whose overall structure and detailed provisions extend far beyond the thorny question of GMOs (including, for example, general principles and rules of procedure). Otherwise, the regulatory framework for GMOs might prove unconstitutional or unlawful. Thus, the law does not simply translate, but also has a broader regulatory function, which civil disobedience campaigners are not slow to note.

2. GMOs and Civil Disobedience

Some people may think that there is no point in analysing the link between the destruction of GMOs and civil disobedience, simply because civil disobedience is a violation of the law and must be condemned. However, the

---

[4] R Lewis, *Evolution Man: Or, How I Ate My Father* (Paris: Actes Sud, Pocket, 2004).
[5] Ibid 61.   [6] Ibid 156.
[7] This dependence on industrial or agri-industrial choices and national cultures is particularly well-demonstrated by E Orsena, *Voyage aux Pays du Coton* (Paris: Fayard, 2006).

debate on civil disobedience[8] is still current[9] and of major importance in understanding the law and its sources. This is why great jurisprudence scholars, such as Rawls, Habermas, and Dworkin,[10] did not hesitate to address these issues and debate with their peers. In their footsteps,[11] a few reflections will be proffered on the subject, echoing the belief of Hiez that those who disobey deserve to be taken seriously, because they can claim to have thought about the law.[12]

The story might begin as follows: 'it is better to die than to disobey', in the words of Socrates to Crito.[13] Socrates based society on obedience and loyalty to law. From this point of view, obedience underpins the 'city' and disobedience can only break the links which obedience had built, and unravel or break the social contract.[14] However, other schools of thought cast civil disobedience in less-revolutionary, less-radical form; they envisage disobedience as a marginal, but effective, source of law based on general principles or values which transcend positivism. The destruction of GM crops provides an excellent forum to understand the theoretical and practical obstacles which must be conquered in order to accept such ideas. Certainly, 'transcending the law' is not something that is acceptable to the courts, which cling to a restrictive approach;[15] but there is already theoretical support for going further, which at least introduces the protestors' arguments into the legal debate (and, moreover, without compromising democracy).

2.1 A Further Thought on 'What Makes the Law'

Using a classic, positivist approach, the law is seen through a governmental lens: it is said to be outside any moral system and can only be questioned by those

---

[8] On the history of civil disobedience, see, eg C Mellon, 'Emergence de la Question de Désobéissance Civile' in D Hiez et B Villalba (eds), *La Désobéissance Civile. Approches Politique et Juridique* (Lille: Septentrion, 2008) 37.

[9] It may also be noted that civil disobedience is fast becoming an issue in relation to the construction of new coal-fired power plants: see, eg 'Gore Urges Civil Disobedience to Stop Coal Plants', 24 September 2008 (available at <http://www.reuters.com/article/newsOne/idUSTRE48N78A20080924>, last accessed 6 August 2009).

[10] J Habermas, *Droit et Démocratie* (Paris: Gallimard, 1997) 411; R Dworkin, *Taking Rights Seriously* (London: Duckworth, 1977) 206–22; and J Rawls, *A Theory of Justice* (Oxford: revised edition, Oxford University Press, 1999) 319–43.

[11] This Conclusion will seek only to trace the main themes of the issue and is in no sense exhaustive.

[12] D Hiez, 'Les Conceptions du Droit et de la Loi dans la Pensée Désobéissante' in Hiez and Villalba (eds) (n 8 above) 67.

[13] Plato, *Crito* 50a-52d; and see F Ost, 'La Désobéissance Civile: Jalons pour un Débat' in PA Perrouty (ed), *Obéir et Désobéir: le Citoyen Face à la Loi* (Brussels: Editions de l'Université de Bruxelles, 2000) 15.

[14] See F Vallançon, 'De la Désobéissance Civile' in Hiez and Villalba (eds) (n 8 above) 21.

[15] See Chapter 1 and, in particular, the judgment of the Court of Appeal in Orléans: CA Orléans, 26 February 2008, CT0028, No 07:00472.

who made it. In this context, the binding force of legal rules does not depend on its social acceptability,[16] but on its proper enactment and compliance with procedural and institutional rules. In other words, the law imposes external conduct, no matter what you feel inside, in your conscience.[17] According to this theoretical perspective, civil disobedience cannot therefore be analysed as part of the legal process or as a contribution to the law. It is a violation of rules and a refusal to follow legal and procedural channels.

However, this perception is widely questioned by jurisprudence scholars who claim that, although people must take account of legislative decisions and, more generally, the law, they can also contest the law in certain circumstances. Thus, according to Dworkin, 'if the issue is one touching fundamental personal or political rights ... a man is within his social rights in refusing to accept [the Supreme Court's] decision as conclusive'.[18] Furthermore, not only can the 'law' be contested; it is also fallible. For example, in the view of Perrouty, although procedural rationality might appear today to be the most serious candidate for providing a basis for the legitimacy of the law, it does not guarantee that the standards applied will be right. He emphasizes that what is novel (and for him this point is essential) is that it does not claim to be right. Once there is acceptance that the law is fallible, it then becomes possible to think of disobedience as a legitimating factor and give it a status which differs from that of resistance to oppression, so creating autonomous space for citizens confronting the normative claims of the law. His conclusion is that disobeying the law then no longer means calling into question the legitimacy of the law as a whole, but rather drawing on the consequences of a fundamental fallibility in order to improve the standard of collective life.[19]

As soon as these ideas are accepted, the law no longer appears as some intangible colossus, conferred with a legitimacy to the exclusion of all else. Rather, resulting from successive and evolutionary power relationships, it can amount simply to expression of the dominant 'voice'. In this sense, it enjoys only

---

[16] See, on the basis of Kelsen and Chevalier's ideas, S Chassagnard Pinet, 'La Désobéissance Face à la Normativité du Droit' in Hiez and Villalba (eds) (n 8 above) 51: '[l]la force obligatoire [des règles de droit] n'est pas conditionée par son acceptabilité sociale'.

[17] See Vallançon (n 14 above) 28–9: 'la loi impose une conduite extérieure, quoiqu'en pense en son for intérieur, la conscience'. [18] Dworkin (n 10 above) 214–15.

[19] PA Perrouty, 'Légitimité du Droit et Désobéissance' in Perrouty (ed) (n 13 above) 59: '[si] la rationalité procédurale [peut apparaître] aujourd'hui comme le candidat le plus sérieux pour fonder la légitimité du droit, ... elle ne garantit pourtant nullement aux citoyens que les normes adoptées seront justes. Mais la nouveauté, et ce point est essentiel, c'est qu'elle n'y prétend pas. Une fois posé ce constat de faillibilité, il devient possible de penser la désobéissance dans le cadre de la légitimité et de lui conférer un statut distinct de celui de la résistance à l'oppression, dégageant ainsi un espace de liberté à l'autonomie des citoyens confrontés aux prétentions normatives du droit. Désobéir à la loi ne revient désormais plus à remettre en cause la légitimité du droit dans son ensemble mais à tirer les conséquences d'une faillibilité principielle pour améliorer les normes de la vie en commun.'

procedural legitimacy; and, if other legitimating factors are entertained, it is subject to competition.[20] These other sources of law, other legal 'creative forces' ('forces créatrices de droit'),[21] can be added to the list of norms, and among them even civil disobedience. It is this concept which is adopted by anti-GM protestors.

2.2 '*Fauchage*': a Legitimate Source of Law?

Should societal action—in this case the destruction of fields of GM crops—thus become a legitimate source of law? This question raises issues not only as to 'what makes the law' but more broadly as to 'what makes democracy';[22] and it requires consideration of whether it is necessary to legitimize acts of civil disobedience and to agree to allow them, in principle, to go unpunished.

Civil disobedience was defined by Rawls as a 'public, nonviolent, conscientious yet political act contrary to law usually done with the aim of bringing about a change in the law or policies of the government'.[23] Under no circumstances is it a revolutionary act intending to use violence to challenge democracy.[24] In this sense, and this sense only, certain acts which challenge the existing law can be legitimized. The destruction of GM crops would seem to fall within this category. Although its non-violent nature may be questioned, because there may be material damage,[25] the action is normally circumscribed, deliberate, collective, public, and only targets changes to the law of GMOs. Besides, there is clear evidence that the activities of anti-GM protestors are more symbolic than determinedly destructive, a

---

[20] See Hiez (n 12 above) 81–2: '[l]e droit ne serait donc pas seulement une règle fondée en raison et de façon transcendantale ... il constitue l'affirmation d'un pouvoir. Dans ces conditions, la désobéissance n'a rien d'illégitime, au même titre que le droit n'a rien de légitime.'

[21] See Chassagnard Pinet (n 16 above); and see also R Encinas de Munagorri, 'La désobéissance civile: une source de droit' [2007] *Revue Trimestrielle de Droit Civil* (January–March) 73 (according to whom civil disobedience can be a material source of law, in the sense of Ripert's theory of 'creative forces of law').

[22] Such a question can only be asked, however, in a democratic context: see Chassagnard Pinet (n 16 above).

[23] Rawls (n 10 above) 320. This definition has been contested: see, eg J Raz, *The Authority of Law: Essays on Law and Morality* (Oxford: Clarendon Press, 1979) 262–75; but there is general agreement on Rawls' definition: see, eg Mellon (n 8 above) 43–6; and Ost (n 13 above) 6–17. On civil disobedience generally, see also, eg HA Bedau, 'On Civil Disobedience' (1961) 58 Journal of Philosophy 653; J Rawls, 'The Justification of Civil Disobedience' in HA Bedau (ed), *Civil Disobedience: Theory and Practice* (New York: Pegasus, 1969) 240; and B Smart, 'Defining Civil Disobedience' in HA Bedau (ed), *Civil Disobedience in Focus* (London: Routledge, 1991) 189.

[24] See Encinas de Munagorri (n 21 above).

[25] The requirement that the act should be non-violent may in itself be subject to considerable jurisprudential dispute (see, eg Smart (n 23 above) 202–6). It may also be noted that in *Monsanto plc v Tilly* the Court of Appeal referred to the 'somewhat euphemistically called "non-violent" action of pulling up the GM crops': The Times, 30 November 1999.

characteristic which may help distinguish them from the 'militant action and obstruction' which Rawls could not accept as civil disobedience.[26]

These issues may be illustrated by the activities of members of GenetiX Snowball, a British anti-GM organization. In *Monsanto plc v Tilly* the Court of Appeal noted that, following a membership guide on how to conduct public actions, the organization required each person not to uproot more than 100 plants.[27] Lang has advanced the view that, where protestors operate a self-imposed limitation on the means of violence employed, this allows their behaviour to remain within the definition of civil disobedience, since '[t]hat limitation makes quite clear both that violence itself is not the end of the actions of which it is part and that the agents responsible do not mean to employ it as a means of overthrowing the enforcement agencies as such'.[28] Further, and finally, it may be reiterated that in *Director of Public Prosecutions v Tilly* an offence of aggravated trespass was not made out where there was simply crop destruction.[29]

Similarly, the Charter of the *Faucheurs* is clear on the point.[30] While conscious of the risk of sanctions, the protestors are to undertake their activities in public and broad daylight ('une action publique concertée à visage découvert'); and, if they do so, it is for the general interest, because the regulatory system gives priority to private interests ('les intérêts privés'). In their view, this is not the expression of a just, democratic society, but reveals a state of 'non-law' ('non-droit'), which does not allow the possibility of any legal action, and characterizes the destruction of GM crops as a criminal act. Accordingly, in the current state of necessity, there is nothing else at their disposal to ensure the survival of democracy.[31] In other words, the law—or its 'inverted application' ('usage inversé') according to the Charter—is contested in the name of the law, which must be restored as the regulatory reference point between people and property. The actions of civil disobedience committed by the *Faucheurs* may, on their analysis, be justified on the basis of setting 'good law' against 'non-law' or 'perverted law'; and this explains why persons committing acts of disobedience reject the argument that the legitimacy of civil disobedience depends formally on the prior and ineffective application of all available democratic means of action.[32]

---

[26] Rawls (n 10 above) 322; and see also, eg Encinas de Munagorri (n 21 above). The same characteristic may preclude categorization as 'eco-terrorism': see, generally, eg S Vanderheiden, 'Ecoterrorism or justified resistance? Radical environmentalism and the "War on Terror"' (2005) 33 Politics and Society 425. Scientists, however, may take a different view: see, eg A Sample, 'Food: Scientists Want Top Security for GM Crop Tests', The Guardian, 29 July 2008.
[27] The Times, 30 November 1999.
[28] B Lang, 'Civil disobedience and non violence: a distinction with a difference' (1970) 80 Ethics 156, 158.
[29] [2001] EWHC 821 (Admin), The Times, 27 November 2001; and [2002] Criminal Law Review 128. Rather, it was necessary for there to be persons present engaging or about to engage in lawful activity, the offence also contemplating that they be intimidated or unable to proceed with what they intended.
[30] Available at <http://www.monde-solidaire.org/spip/IMG/pdf/Charte_faucheurs.pdf>, last accessed 25 June 2009.
[31] Ibid: '[d]ans l'état de nécessité actuelle où nous nous trouvons, nous n'avons plus rien à notre disposition pour que la démocratie reste une réalité'.   [32] See Ost (n 13 above) 37.

Accordingly, the *Faucheurs* meet one of the criteria of Rawls' definition very well: the act must be 'a political act not only in the sense that it is addressed to the majority that holds political power, but also because it is an act guided and justified by political principles, that is, by the principles of justice which regulate the constitution and social institutions generally'.[33]

The nature and content of this 'good law' is less obvious. Although the *Faucheurs* appeal to general legal principles, they legitimize their action by reference to socio-political concepts, such as the general interest, the duties and responsibilities of citizens, and the abuse of power by the authorities. They also invoke a baseline which is above the law,[34] together with a set of social values which stand outside the legal system. In other words, in their view, laws, principles, social values, and political claims form an indissoluble whole.[35]

Anti-GM protestors therefore rely on a sociological concept of the law. While they address internal legal conflict by opposing more or less technical legislation (for example, that governing GMOs) against general principles, overall they base their argument on two main ideas, broad in their interpretation. On the one hand, according to Bové and Luneau, the law is not limited to a succession of texts, but also takes root where people live and in the forms which they give to their life.[36] On the other hand, there is no political question which can be solved by a legal question and there is no genuine legal question which does not have moral implications.[37] Using this open approach, the *Faucheurs* reveal their vision of the law, characterized by its 'politicization' and its 'power-relationship dimension'.[38] They suggest that the judge and legislator should take the same open approach into account.

It is difficult to assess the legitimacy of these arguments. Everything depends upon where you stand. For example, in the view of the French court, it is not possible for a few people to use their personal convictions to pass judgment on the value of the interests to be defended.[39] However, others regard civil disobedience as something which has the capacity to generate law.[40] In this latter sense, it would be the power of the street ('street law') which would legitimize the actions of the

---

[33] Rawls (n 10 above) 321.
[34] Chassagnard Pinet (n 16 above) 55–7: 'un referential placé au dessus de la loi'.
[35] For those undertaking acts of civil disobedience, there is widely understood to be a potential contradiction between 'legislation' and 'the law': see Hiez (n 12 above) 83–4. Hiez adds that these are 'si différentes qu'il faut s'interroger sur l'unité qu'elles recouvrent: y-a-t-il vraiment des éléments communs aux diverses désobéissances?'
[36] J Bové and G Luneau, *Pour une Désobéissance Civique* (Paris: La Découverte, 2004) (although it may be noted that this is entitled *Pour une Désobéissance Civique*, as opposed to *Pour une Désobéissance Civile*): the law 's'enracine là où les gens vivent et dans les formes qu'ils donnent à leur vie': ibid 179–80 (adopting the thinking of D Rousseau).
[37] S Turenne, 'Le Discours Judiciaire face à la Désobéissance Civile' in Hiez and Villalba (eds) (n 8 above) 51. See also S Turenne, *Le Juge face à la Désobéissance Civile en Droit Américain et Français Comparés* (Paris II: Thèse, 2005).
[38] Hiez (n 12 above) 70 and 76: 'politisation' and 'dimension du rapport de force'.
[39] CA Orléans, 26 February 2008, CT0028, No 07:00472; and see Chapter 1.
[40] See, eg Hiez (n 12 above) 81.

civil disobedience movement. 'Social fact' ('fait social') would thus be capable of changing the law. Anti-GM organizations have understood this very well; they desire publicity for their actions even more than destruction.[41]

Theoretically legitimate, this broad approach does not necessarily act contrary to the effective functioning of the democratic system.

2.3 Civil Disobedience: Useful to Democracy?

For those who take a legalistic standpoint, civil disobedience is a violation of the law because the democratic system is organized to ensure that persons who suffer prejudice can take action in court, assuming that recourse to the courts is made available by the law. For example, in France, the appeal judges of the Court of Appeal in Orléans in 2008 adopted this stance when they held that it could not be tolerated for groups to take the law unto themselves so as to impose their views by force, when there were existing legal and democratic processes to address their claims.[42] In particular, they noted that there was a summary procedure which would have enabled the defendants to request suspension of proceedings, following which there might be withdrawal of the administrative authorization for release of the GMOs into the environment. Further, commentators have mentioned the possibility of making civil liability claims against the state,[43] or, more generally, all legal avenues for judicial review.[44]

It must be asked whether it is sufficient that these actions be available for the democratic system to retain its integrity. Moreover, there is no certainty that civil disobedience should be a last resort.[45] Indeed, it is important to recognize the democratic opportunity—apparently paradoxical—which is offered by civil disobedience, allied to public-participation mechanisms in decision-making. Thus, representative democracy must be made permeable to participative democracy, and then technical legal concepts can be opened to civil society.[46]

Many authors seem to accept such compatibility of civil disobedience with the proper functioning of democracy.[47] As stated by Ost, between somewhat

---

[41] As has been seen, in the UK case of *Monsanto plc v Tilly* The Times, 30 November 1999, the Court of Appeal saw courting publicity, and even martyrdom, as the motivating factor behind the activities of the anti-GM protestors. See also G Hayes, 'Collective action and civil disobedience: the anti-GMO campaign of the *Faucheurs Volontaires*' (2007) 5 French Politics 293; and Smart (n 23 above) 206 (who believes that advance publicity cannot be a requirement of all civil disobedience, since any requirement of fair notice might well frustrate its performance); but *cf* HA Bedau, 'On Civil Disobedience' (1961) 58 Journal of Philosophy 653, 656 (where the approach taken is that 'civil disobedience is necessarily *public*').

[42] CA Orléans, 26 February 2008, CT0028, No 07:00472.

[43] A Gossement, 'Le fauchage des OGM est-il nécessaire?' [2006] *Revue Environnement* (January) 9.

[44] P Billet, 'Fauchage d'OGM: une relaxe sans nécessité' (2006) 339 *Revue de Droit Rural* 60, 63.

[45] Rawls (n 10 above) 327–8: '[n]ote that it has not been said, however, that legal means have been exhausted. At any rate, further normal appeals can be repeated; free speech is always possible'.

[46] See, eg Mellon (n 8 above); and W Smith, 'Democracy, deliberation and disobedience' (2004) 10 Res Publica 353.

[47] See Mellon (n 8 above) 47–8.

rhetorical affirmation of a right to insurrection and a very limited recognition of the right to oppose an unlawful act of public authorities, there is the grey area of civil disobedience, which is subject to dispute; and, although, of course, civil disobedience in principle remains unlawful, this principle should not exclude a degree of administrative tolerance or judicial clemency.[48] In this sense, civil disobedience imposes a form of limitation, the ultimate expression of the extreme fragility of democracy.[49] This idea is all the more acceptable where those conducting acts of civil disobedience contest the law, since they are not outside the legal system, but at its core.[50] On the other hand, subject to there being moral authority (according to some commentators),[51] civil disobedience is variously described as providing a degree of stability for society,[52] as being capable of correcting democratic deficits,[53] as a rampart against excessive use of power (even if democratic power),[54] a factor to stimulate democratic debate,[55] a positive contribution to the evolution of debate in situations of controversy,[56] and, more widely, a form of democratic participation in the development of legal rules.[57]

---

[48] Ost (n 13 above) 20–1: '[e]ntre l'affirmation quelque peu rhétorique d'un droit à l'insurrection et la reconnaissance très limitée d'un droit à s'opposer à un agissement illégal de l'autorité publique, se développe la zone grise, ouverte à la discussion, de la désobéissance civile. Si le principe reste, bien entendu, l'illégalité de celle-ci, ce principe n'empêche cependant pas certaines manifestations de tolérances administrative ou de clémence judiciaire'.

[49] Ibid: 'la désobéissance civile prend place comme une modalité- limite, l'expression ultime de la fragilité forte qui fait la démocratie'.

[50] D Hiez and B Villalba, 'Réinterroger la Désobéissance Civile' in Hiez and Villalba (eds) (n 8 above) 11.

[51] For many jurists an appeal to morality is central to such action: see, eg Dworkin (n 10 above); and Rawls (n 10 above) 337 (where civil disobedience is seen as 'an appeal to the moral basis of civic life'). Yet it has been seen that for many protestors their primary motivation is to prevent perceived risk (notwithstanding that there may also exist strong moral undertones): see P Jones, 'Introduction: law and disobedience' (2004) 10 Res Publica 319, 322; and see also J Welchman, 'Is ecosabotage civil disobedience?' (2001) 4 Philosophy and Geography 97.

[52] Rawls (n 10 above) 336: '[a] general disposition to engage in justified civil disobedience introduces stability into a well-ordered society, or one that is nearly just'.

[53] Markovits has preferred the expression 'political disobedience' to 'civil disobedience', laying emphasis on its ability to correct democratic deficits in law and policy; and, instead of a theory of democracy-limiting disobedience, he advances 'a theory of *democracy-enhancing* disobedience or, more simply, *democratic* disobedience': D Markovits, 'Democratic disobedience' (2005) 114 Yale Law Journal 1897, 1898 and 1902.

[54] Perrouty (n 19 above) 8: 'un rampart contre les excès de pouvoirs'.

[55] Ibid: 'un facteur de "stimulation du débat démocratique"'.

[56] B Villalba, 'Contribution de la Désobéissance Civique à l'Établissement d'une Démocratie Technique. Le Cas des OGM' in Hiez and Villalba (eds) (n 8 above) 129: 'un apport positif à l'évolution des pratiques délibératives dans une situation de controverse'.

[57] Hiez and Villalba (n 50 above) 16: 'une forme de participation démocratique à la production de la norme'.

2.4 Conclusion

On these bases, should civil disobedience continue to be juxtaposed to democracy? To do so would be to forget that, although legal theory may be regarded as the product of a hermetically sealed literature, it can be brought back into the limelight through such actions. This is what we seem to learn from destruction of crops by the anti-GM movement and, more generally, every time people question the regulations which govern the use of GMOs. It is even more the case today, in that, although the law is not yet changing, the latest positions taken by the European Council, together with statements by the President of the European Commission and Commissioner Fischer Boel, seem to take into account the extent of the anti-GM movement. So is civil disobedience a source of law? Perhaps not always, but it can certainly be a paralyzing force on certain effects of the law.

Index

Aarhus Convention
  access to information 15–16, 21–2
  access to justice 15
  African Model Law 22, 236
  Almaty Amendment 19
  confidentiality 21–2
  costs of challenging non-compliance with environmental measures 21
  deliberate release 15–16, 84, 86
  Food Law Regulation 99–100
  non-governmental organizations (NGOs) 16
  nuclear power stations, consultation on 20
  protests in France 32, 34
  public authorities, confidentiality of proceedings of 21–2
  public participation 15–16, 21–2, 35, 60
  transparency 15
  United Kingdom 21
access to information
  Aarhus Convention 15–16, 21–2
  confidentiality 21–2
  European Food Safety Authority (EFSA) 90–1
  European Union 71, 80
  Food and Feed Regulation 90–1
  labelling 71
  public authorities, confidentiality of proceedings of 21–2
  public participation 15–16, 21–2
access to justice 15, 16, 21
additives 309–10, 313–15
administrative liability 216–21
adventitious presence, coexistence and
  compensation 160
  consistency 160
  economic effects 128, 132
  European Union 123–4, 126–33, 137, 139–40, 147, 159–61
  France 167, 173
  liability 153–4, 158
  organic production 147
  proportionality 160
  seeds 141
  Spain 181
Africa 227–53 *see also* **African Model Law on Safety in Biotechnology; particular countries**
  African Centre for Biosafety 230
  biotechnological capacity of countries 239–40
  Biowatch South Africa 229–30, 241–2
  Cartagena Protocol 238, 239, 251
  caution, countries which promote GMOs with 245–8, 251
  classification of countries by approach 239–50, 251
  COMESA (Common Market for Eastern and Southern Africa) 238
  commercialization 251–3
  Common Position on Biosafety (African Union) 250
  economic development 228–9
  economic integration agreements 238
  ECOWAS (Economic Community of Western African States) 238
  export markets, loss of 230–1, 252–3
  field trials 233
  food aid, declining 232, 250
  food production statistics 227–8
  food security 227–30, 233, 252–3
  intellectual property 252
  Millennium Development Goals (MDGs) 228
  multi-level approaches 233–7
  national approaches 239–50
  non-governmental organizations (NGOs) 229–30
  opposition to GM crops 229–31
  potential role of GM crops 229–32
  poverty 227–8, 231
  promote GMOs, countries which 240–8, 251
  protests 229–30
  research and development 233, 251, 253
  restrictive regulatory approach, countries with 249–50, 251–2
  risk assessment 251–2
  SADC (Southern African Development Community) 238
  seeds 229–30, 251, 252
  self-sufficiency in food 227–8
  sub-regional approaches 238–9, 251
African Model Law on Safety in Biotechnology (African Union)
  access to information 22, 236

African Model Law on Safety in
    Biotechnology (African Union) (*cont.*)
  background 233–4
  Cameroon 248
  Cartagena Protocol 234–5
  confidentiality 22, 236
  consultation 16–17, 236
  draft report 250, 253
  Egypt 244
  Ethiopia 250
  harmonization 16, 19
  information requirements 235
  insurance 237
  Kenya 247
  labelling 236
  liability 236–7
  Malawi 19, 245
  notice-and-comment procedure 16–17, 236
  precautionary principle 2, 235
  public participation 16, 19, 22, 236, 251
  remedies 236–7
  risk assessment 235–6
  significant risk 235–6
  socio-economic impacts 235–6
  South Africa 19, 22, 251
  standing 237
  strict liability 237
  sub-regional approaches 238
  sustainable development 235–6
  Tanzania 249–50
  traceability 236
agricultural tenancies 183
approvals *see* authorization and approvals
Argentina
  authorization 279, 282
  Biodiversity, Convention on 295
  CONABIA 281, 282–3
  confidentiality 283
  damages 284
  general framework 281–2
  harmonization 292
  institutions 282–5
  MERCOSUR 274–5, 277–8, 279, 281–4, 292–3, 297–8
  Miami Group 292
  precautionary principle 283
  registration 282
  research and development 283
  risk assessment 282, 284
  seeds, definition of 282
  specific regime 282–4
  toxicity tests 283
  trading partners 293
  transparency 283

Arkansas, United States 320–1
Austria 87, 116–7, 149–50
authorization
  Argentina 279, 282
  Brazil 280, 284–5
  Canada, coexistence in 263, 267
  coexistence
    Canada 263, 267
    European Union 132, 134, 164–5, 196
    France 167–9, 171–2, 174
  Commission Recommendation on coexistence 2003 51–2, 164–5
  cost-benefit analysis (CBA) 43–6, 47–53
  delay 352–3
  Deliberate Release Directive 2, 84, 85, 87–8, 90, 134, 200
  Environmental Liability Directive 211, 214, 220–1
  European Union 102–3, 111–22, 132, 134, 164–5, 196
  exclusion of liability 220–1
  externalities 47–50, 53
  food and feed, EU regulation and 87–9, 91
  France 167–9, 171–2, 174
  innovation, social value of 47
  labelling 46, 51–3
  liability 199–202
  MERCOSUR 278–81, 296
  negligence 220–1
  optimal level of pollution 46
  permits 220–1, 240–4, 302–3, 306
  post-authorization 43–6, 104, 111–21
  pre-market approvals 314–15
  private consequences 47–9, 53
  risk assessment 47–8
  Sanitary and Phytosanitary Agreement (SPS Agreement) 343, 351–3
  seeds 141
  segregation 51–2
  single authorization procedure 89–91, 96–7
  social value 47–8
  supply chain 52
  tolerance threshold 52
  traceability 46, 51–3
  United States 314
  Uruguay 278, 290–2
  willingness to pay (WTP) 49–53

bans
  Austria 87, 116–7, 149–50
  Brazil 285
  Charter of Florence 151, 174–5
  coexistence 132, 136, 138, 148–51, 161–71, 189, 191–3

Commission Recommendation on coexistence 2003 148–9
Declaration of Rennes 151
Deliberate Release Directive 148–9
European Food Safety Authority (EFSA) 149–50
European Union 132, 136, 138, 148–51, 161–2, 164
Food and Feed Regulation 148–9
France 166–7, 168, 169–71
GMO-Free Network 151
harmonization measures 150
Hungary 87, 149–50
imports 338
local authorities 169–71
police powers 170
regions 138, 161–2
safeguard clauses 148–9
Sanitary and Phytosanitary Agreement (SPS Agreement) 338, 342–3, 351–3
segregation 150
transparency 150
United Kingdom 189, 193–4
US states, regulation in 320
BAT (best available techniques) 61
BATNEEC (best available technique not entailing excessive cost) 61
biodiversity *see also* **Cartagena Protocol to Convention on Biodiversity**
Argentina 295
Convention on Biodiversity 14, 19, 35, 292, 295–7
cost-benefit analysis (CBA) 38
definition 211
environmental impact assessments 14
environmental liability 200–2, 206–7, 210–18, 222, 224
fault 214–16
MERCOSUR 292, 295–7
Paraguay 295
public participation 14, 19, 35
Uruguay 295
birds 210–11, 223
Bolivia 274, 279
Brazil
authorization 280, 284–5
bans 285
Biosafety Law 285–7
Cartagena Protocol 287
constitutional law 287, 298
CTN Bio 285–7
environmental impact assessments 280, 284–5

federal competence 284
Ideas Afines Group 292
institutions 286
labelling 285, 287
liability 284
MERCOSUR 274–5, 278–80, 284–7, 292–3, 298
moratoriums 285
National Biosafety Council (CNBS) 286
non-governmental organizations (NGOs) 286–7
opposition 280
public participation 286–7
research and development 287
risk assessment 286–7
state competence 284
state regulation 284–5
trading partners 293
traditional knowledge 285
BSE (bovine spongiform encephalopathy) 12, 43, 268–71, 272
buffer strips 157, 181–3, 195

California, United States 320
Cameroon
African Model Law 248
Cartagena Protocol 248
caution, regulations promoting GMOs with 247–8
damages 248
deliberate release 247–8
information, provision of 248
labelling 248
precautionary principle 247–8
risk assessment 247–8
Canada
approvals 263
authorization 267
BSE 268–71, 272
Canadian Food Inspection Agency (CFIA) 266–7
Canadian Wheat Board (CWB) 254–5
canola 254–5, 257–62, 273
causes of action 261–3, 269, 272
class actions 259–63, 268–72
coexistence 21, 254–73
commercialization 254–6
company responsibilities, denial of 259–62
company rights, recognition of 257–9
compensation schemes 273
Crown immunity 263, 265–6, 270–2
duty of care 261–7, 270, 272
economic loss 256, 259, 267–8, 270–3
foreseeability 263–4

Canada (*cont.*)
    government liability 263, 264–71
    Hoffman 21, 257, 259–62, 267, 271–2
    insurance 273
    liability 259–71
    Monsanto 254–62
    negligence 256, 261–72
    novel traits, plants with 266
    operational/policy distinction 265–7, 270–1
    patents 257–9, 262–3
    policy 255–7, 262, 265–8, 270–2
    proximity 261–4, 267, 270, 272
    public authorities, negligence actions against 264, 267–8
    risk creation measures 264, 272
    risk reduction measures 264, 272
    Sauer 268–72
    Schmeiser 257–9, 271–2
Cartagena Protocol to the Convention on Biodiversity
    Africa 234–5, 238, 239–41, 251
    African Model Law 234–5
    Brazil 287
    Cameroon 248
    consultation 14, 19
    damages, assessment of 212–13
    EC–Biotech 362, 364–5, 367–8
    Environmental Liability Directive 212–13
    France 32, 34
    international law 356–60, 374
    interpretation 362, 364–5, 367–8
    MERCOSUR 292, 295–6
    Miami Group 292
    national provisions 357–8
    Paraguay 289
    protests in France 32, 34
    public participation 14, 15–16, 19, 35
    socio-economic considerations 368–71
    South Africa 240–1
    sustainable development 368–71
    Transboundary Movements Regulation 81
causation
    coexistence 153
    Environmental Liability Directive 208–9, 213–16, 218
    liability 153, 188, 208–9, 213–16, 218
    limitation periods 218
    Lugano Convention on Civil Liability 208–9
    negligence 67–9
    risk assessment 67–9
    US states, regulation in 322, 328–9
causes of action
    Canada, coexistence in 261–3, 269, 272
    class actions 259–63, 269, 272
CBA *see* **cost-benefit analysis (CBA)**
Charter of Florence 151, 174–5
Chile 274
civil disobedience 377–85
    definition 380–2
    democracy 383–5
    European Union 385
    fallibility of law 379
    'fauchage' as legitimate source of law 380–3
    GenetiX Snowball 381
    'good law' 381–2
    legitimacy of law 379–83
    procedural legitimacy 379–80
    social contract 378
    sociological concept of law 382
    source of law, as 378, 380–3
    'street law' 382–3
    symbolism 380–1
civil liability
    coexistence 152–3, 158
    environmental liability 200–1, 206–7, 215–16, 223
    Kenya 247
    Lugano Convention on Civil Liability 206–9, 218
class actions
    ascertainability of members of class 260–1, 262, 268
    BSE 268–71, 272
    Canada, coexistence in 259–63, 268–72
    causes of action 259–63, 269, 272
    certification 259–61, 269, 271–2
Coasean analysis 202–4
codes of good practice 173, 190–1
coexistence *see also* **adventitious presence, coexistence and; coexistence in European Union**
    accreditation bodies for organic farming 192
    agricultural tenancies 183
    Austria 149–50
    authorization 141, 167–9, 171–2, 174, 263, 267
    bans 132, 136, 138, 148–51, 161–71, 174, 189, 191–3
    BSE 268–71, 272
    buffer strips 157, 181–3, 195
    Canada 21, 254–73
    causation 153
    causes of action 261–3, 269, 272
    Charter of Florence 151
    class actions 259–63, 268–72

code of practice 190–1
commercialization 254–6
company responsibilities, denial of 259–62
company rights, recognition of 257–9
compensation funds 153, 158, 273
COPA-COGECA 143, 153–4
Crown immunity 263, 265–6, 270–2
damages 151–4, 197
Declaration of Rennes 151
Department for Environment, Food and Rural Affairs (DEFRA) 189–90, 193
duty of care 261–7, 270, 272
economic effects 180–1, 188
economic loss 188, 256, 259, 267–8, 270–2
emergencies 168–9
European Food Safety Authority (EFSA) 149–50
European Seed Association (ESA) 141, 143, 145–6
field trials 187
foreseeability 263–4
France 166–74, 195–6
freedom of cropping 166–8, 183
GM-free zones 169–70, 187–8, 192–5
GMO-Free Network 151
government liability 263, 264–71
harmonization measures 150
Hungary 87, 149–50
information, provision of 167, 171–3, 182–5
insurance 153, 158, 273
Italy 174–9, 195, 196
labelling 130–40, 143–8, 181, 186,
liability 152–3, 158, 167, 185, 188, 259–71
local authorities 169–71
management measures 181–2
Monsanto 254–62
negligence 152, 256, 261–72
neighbours, notification of 182
Northern Ireland 189
nuisance 152
operational/policy distinction 265–7, 270–1
organic farming 146–8, 168–9, 192–4
patents 257–9, 262–3
police powers 170
policy 188–91, 255–7, 262, 265–8, 270–2
polluter pays 150
precautionary principle 167–8
professional codes of good practice 173
proportionality 169–70
protocols 191
proximity 261–4, 267, 270, 272

public authorities 182, 264, 267–8
public participation 167
purity 141–2, 182, 186
regions 138, 161–2
risk assessment 167
risk creation measures 264, 272
risk reduction measures 264, 272
safeguard clauses 148–9
Scientific Committee on Plants (SCP) 142–3
Scotland 189, 193–4, 196
seeds 141–6, 154, 190
segregation 142–3, 150, 181–4, 187, 191
Spain 130, 146, 179–88, 195–6
strict liability 153, 158
technical rules 171, 173–4
thresholds 141–8, 169, 186, 192–3
traceability 183
transparency 150, 172
United Kingdom 128, 188–96
United States 146, 147
Wales 189–90, 192–3, 196
zero tolerance 143
coexistence in European Union 118–19, 123–62, 191–4 *see also* **Commission Recommendation on coexistence 2003**
adventitious presence 123–33, 137–40, 147, 159–61
Austria 149–50, 196
authorization 132, 134, 164–5, 196
bans 132, 136, 138, 148–51, 161–2, 164
best practice, guidelines on 134–6
buffer strips 157, 181–3, 195
Charter of Florence 151, 174–5
competence of farmers, proof of 156
competition 136
confidentiality 172
consistency 160–1
Contained Use Directive 179
context 160–1
COPA-COGECA 133
cost-benefit analysis (CBA) 46, 51–2
damages 197
Declaration of Rennes 151
definition of coexistence 124
Deliberate Release Directive
  authorization 134
  confidentiality 172
  GM-free zones 187
  implementation 154, 165, 195
  labelling 129, 138, 192
  safeguard clause 131, 148–9
  Spain 179
  traceability 129

coexistence in European Union (*cont.*)
  economic impact 132, 137, 139–40, 154–6, 165, 197
  economic loss 132, 137, 165, 206
  ECOSOC 137
  Environmental Liability Directive 152
  European Commission 124, 131–45, 154–60, 166, 188–94
  European Food Safety Authority (EFSA) 149–50
  European Union 46, 51–3
  Fischler Communication 132–4, 136
  Food and Feed Regulation 138, 148–9
  France 166–74, 195–6
  Friends of the Earth Europe (FOEE) 125, 137–8, 140
  gene flow, limiting 135, 139
  Germany 128
  GM-free zones 136, 187, 196–7
  GMO-free Europe 125
  GMO-Free Network 151
  Greenpeace 138
  guidelines for Member State strategies 134–6, 164, 195
  harmonization 150
  herbicides 123
  horizontal elements 127–8
  Hungary 87, 149–50
  implementation of Member State measures 154–8, 163–97
  imports 125
  indicative catalogue of measures 136
  insurance 134, 158
  Italy 174–9, 196
  labelling 124, 129–40, 147–8, 156, 192
  legislation of Member States 154–5, 157, 161
  liability 135, 137, 141, 151–4, 158–9, 197
  management measures 134–5, 139, 173–4, 180–4, 186–7
  market access problems 128
  multi-level governance 118–21
  national measures 134–6, 154–9, 163–97
  national registers of GM crops 156
  neighbours, notifying 135, 136
  New Case Studies. Joint Research Centre of European Commission 138–40
  non-governmental organizations (NGOs) 123, 137–8
  on-farm measures 156–7
  organic production 146–8
  pesticides 123
  policy 130–40
  polluter pays 150
  practicality 160
  precaution and prevention 130, 196–7
  precautionary principle 130
  principles of coexistence 160–2
  Products Liability Directive 152
  proportionality 156, 160–1
  purity standards 134
  regional bans 138, 161–2
  registers of GM crops 156
  reports 138–9, 154–8, 166
  research 138–40, 161
  risk assessment 132, 165
  risk management 133, 137
  safeguard clauses 131, 148–9
  seeds 136–7, 139–46, 157
  segregation 134–5, 136, 150, 157, 159, 195
  sources of admixture 135, 139
  Spain 130, 146, 179–88, 195–6
  stakeholder involvement 134
  StarLink maize 128
  strict liability 137
  subsidiarity 133, 165
  thresholds 129–31, 133–5, 137–40, 147–8, 160, 164
  traceability 129–30, 133–4
  Traceability and Labelling Regulation 129
  training of farmers, proof of 156
  transparency 134, 150
  United Kingdom 128, 188–96
  United States 127–8, 146, 147
  vertical elements 127–8
  voluntary agreements 135, 136
  zero tolerance 127, 160
COMESA (Common Market for Eastern and Southern Africa) 238
comitology procedure 85–6, 107, 109–14, 122
commercialization
  Africa 240, 251–3
  Canada, coexistence in 254–8
  Egypt 243
  South Africa 240
  United States 299
Commission *see* **Commission Recommendation on coexistence 2003; European Commission**
Commission Recommendation on coexistence 2003
  authorization 51–2, 164–5
  bans 148–9
  best practices 124
  France 195
  implementation measures 154–9
  Issue by European Commission 133–8, 164

Index

Italy 178
Spain 180–4, 187, 195
United Kingdom 191–5
compensation
  adventitious presence 160
  Canada, coexistence in 273
  coexistence 153, 158
  compensation schemes 273
  liability 153, 158
competence of farmers, proof of 156
competent authorities
  consent 82, 85–6
  Contained Use Directive 82
  Deliberate Release Directive 85–6
  European Union, multi-level governance in 106–11
  Food and Feed Regulation 107
  national bodies 107–11
  notification 85, 106
  reports 85
  risk assessment 108
  role 82
  Uruguay 290–1
competition 294
confidentiality
  Aarhus Convention 21–2
  access to information 21–2
  African Model Law 22, 236
  Argentina 283
  deliberate release 22, 172
  public authorities 21–2
  South Africa 241
  Tanzania 249
conservation sites
  Environmental Liability Directive 210–12
  property rights 205–6
  sites of special scientific interest (SSSIs) 204–6, 211–12, 215, 223–4
constitutional law 175, 177–8, 287, 298, 318
consultation 18–21
  African Model Law 16–17, 236
  Cartagena Protocol 14, 19
  Deliberate Release Directive 17
  Environmental Impact Assessment Directive 20
  Food and Feed Regulation 91
  nuclear power stations, consultation on 20
  risk assessment 59, 70
  United Kingdom 18, 19–20
  United States 312–13
consumer preferences 22–5
consumer surplus 39–40
contained use
  annual reports 83

  classification 82–3
  competent authorities, consent of 82
  Contained Use Directive 80, 81–3, 179, 199
  definition of contained use 82
  Environmental Liability Directive 214
  liability 199
  premises, notification of 82–3
  risk assessment 82–3
  South Africa 240–1
  Spain 179
COPA-COGECA 133, 143, 153–4
cost-benefit analysis (CBA) 37–53
  aim and scope 38–46
  altruistic value 42
  application of CBA 47–53
  authorization measures 43–6, 47–53
  bequest value 42
  biodiversity 38
  coexistence 46, 51–3
  consumer surplus 39–40
  economic analysis 45, 53
  existence value 42
  externalities 38–40, 47–50, 53
  innovation 37–8, 47
  intrinsic value 43
  labelling 46, 51–3
  marginal external cost (MEC) 40, 44
  novel issues 43–6
  optimal level of pollution 44, 46
  option value and 'quasi-option value' 42
  post-authorization measures 43–6, 51–3
  precautionary principle 45
  presentation 39–43
  private consequences of authorization 47–9, 53
  'revealed preferences' approach 40–1
  risk assessment 37, 39, 44–5, 47–8, 53
  segregation costs 51–2
  social value 44–5, 47–8
  'stated preferences' approach 41
  supply chain 52
  tolerance threshold 52
  traceability 46, 51–3
  United Kingdom 18
  use value and non-use value 41–2
  willingness to pay (WTP) 39–43, 49–53
  World Trade Organization (WTO) 38, 44, 53
cost-internalization 201–2
costs of challenging non-compliance with environmental measures 20–1
Council of EU 109–10, 112–13
criminal offences
  Environmental Liability Directive 215

criminal offences (*cont.*)
    France 29–35
    protests 26–35
    sites of special scientific interest
        (SSSIs) 215
    United Kingdom 26–9, 36
Crown immunity 263, 265–6, 270–2
culture 295
customary international law 362–3, 365

damages
    Argentina 284
    assessment 212–13, 217
    Cameroon 248
    coexistence 151–4, 197
    Environmental Liability Directive
        212–13, 217
    European Union 197
    nuisance 323, 326
    remediation 217
    South Africa 241
    StarLink litigation 321, 332
    trespass to land 328
    United States 320–3, 326, 328, 332
Declaration of Rennes 151
deliberate release *see also* **Deliberate Release**
        Directive
    Aarhus Convention 15–16
    Cameroon 247–8
    confidentiality 22
    Environmental Liability Directive 2, 214
    European Union 22
Deliberate Release Directive 80, 83–7
    access to information 84, 86
    authorization 2, 84, 85, 87–8, 90,
        134, 200
    bans 116–7, 149–50
    coexistence
        authorization 134
        confidentiality 172
        GM-free zones 187
        implementation 154, 165, 195
        labelling 129, 138, 192
        safeguard clause 131, 148–9
        Spain 179
        traceability 129
    comitology procedure 85–6
    Community procedure 109
    competent authority, consent of 85–6
    confidentiality 172
    consultation with public 17
    Contained Use Directive 84
    environmental impact assessment 84
    Environmental Liability Directive 214, 220

European Food Safety Authority (EFSA)
        85–6, 87
    field trials 83–4
    Food and Feed Regulation 85, 90–1, 105–6
    France 31–2, 33–4
    GM-free zones 187
    implementation 154, 165, 195
    Italy 176, 178
    labelling 86, 129, 138, 192
    liability 199–200
    multi-level governance 103–8, 111, 116–17
    new information 86
    non-governmental organizations
        (NGOs) 17
    notifications 84, 85
    placings on the market 84–7
    precautionary principle 65, 66, 84, 86, 98
    protests in France 31–2, 33–4
    public participation 17, 66
    reports 85
    risk assessment 56, 65, 66, 90–1, 104–8,
        114–15, 200
    safeguard clause 86–7, 114, 131, 148–9
    Sanitary and Phytosanitary Agreement
        (SPS Agreement) 338, 341, 343
    single authorization procedure 90
    Spain 179, 187
    traceability 129
    transparency 66, 99
    United Kingdom 192
    World Trade Organization (WTO) 86–7
democracy 383–5
demonstrations *see* **civil disobedience; protests**
deterrence 201–2
developing countries *see* **Africa; MERCOSUR;**
        particular countries (eg Brazil)
    food security 13
    Sanitary and Phytosanitary Agreement
        (SPS Agreement) 373–4
    socio-economic considerations 373–4
    sustainable development 373–4
discounts 25
discrimination 369
duty of care
    Canada, coexistence in 261–7, 270, 272
    US states, regulation in 328–9

EC law *see* **European Union**
EC-Biotech
    Cartagena Protocol 362, 364–5, 367–8
    international law 357, 360
    Sanitary and Phytosanitary Agreement
        (SPS Agreement) 338, 339–45,
        349–54

socio-economic considerations 370
sustainable development 370
treaty interpretation 362, 364–8
economic development 80, 228–9
economic impact 132, 137, 139–40, 154–6, 165, 197
economic integration agreements 238
economic loss
  Canada, coexistence in 256, 259, 267–8, 270–3
  coexistence
    Canada 256, 259, 267–8, 270–3
    European Union 132, 137, 165, 206
    Spain 188
  European Union 132, 137, 165, 206
  nuisance 327
  policy 267–8
  Spain 188
  United States 327, 329, 330–1
ECOSOC (Economic and Social Committee) 137
ECOWAS (Economic Community of Western African States) 238
EFSA *see* **European Food Safety Authority (EFSA)**
Egypt
  African Model Law 244
  commercialization 243
  field trials 243
  import permits 243–4
  labelling 243
  promoting GMOs, regulations 243–4
  research and development 233, 243
  seeds 243
emergencies 115, 168–9
environmental impact assessments (EIAs)
  Biodiversity, Convention on 14
  Brazil 280, 284–5
  Deliberate Release Directive 84
  Environmental Impact Assessment Directive 20
  United Kingdom 19–20
  United States 303–6
environmental liability *see also* **Environmental Liability Directive**
  authorization 199–202
  biodiversity, damage to 200–2
  civil liability 200–1, 206
  Coasean analysis 202–4
  conceptual framework 199–206
  Contained Use Directive 199
  cost internalization 201–2
  Deliberate Release Directive 199–200
  deterrence 201–2
  field trials 199–200
  foreseeability 200, 202
  labelling 199
  negligence 200–1
  polluter pays 201, 207, 212–14, 219–21
  post-release measures (*ex post*) 199, 201
  pre-release measures (*ex ante*) 199
  property rights 202–6
  public liability regimes 200–1, 204
  remediation 201–2
  remedies 200–1, 204–5
  resource allocation 201–4
  risk assessment 199–200, 202
  sites of special scientific interest (SSSIs) 204–6
  strict liability 200–1, 207, 209, 222
  utilitarianism 204–5
Environmental Liability Directive
  administrative liability 216–21
  application 209–13
  authorization 211, 214, 220–1
  background 206–7, 213
  basis for liability 213–16
  biodiversity, damage to 206–7, 210, 211–18, 222, 224
  Cartagena Protocol to Convention on Biodiversity 212–13
  causation 208–9, 213–16, 218
  civil liability 206–7, 215–16, 223
  coexistence 152
  compensatory remediation 217
  conservation sites 210–12
  contained use 214
  coverage 207, 221
  criminal offences 215
  'damage', definition of 210–13
  damages, assessment of 212–13, 217
  definition of environmental damage 210–13
  Deliberate Release Directive 214, 220
  deliberate releases 2, 214
  'environmental', definition of 210
  European Commission 214
  exclusion of liability 219–21
  fault 214–16, 220–1
  Food and Feed Regulation 221
  foreseeability 216
  habitats, damage to 17, 209–12, 215–17, 223
  Habitats Directive 210–11, 215, 223–4
  hazardous activities 206, 213–16
  implementation 207, 221
  individual claims 219

Environmental Liability Directive (cont.)
  insurance 222
  internalization costs 207, 213
  land damage 210
  liability 152
  limitation of actions 217–18
  Lugano Convention on Civil Liability 199
  Natural England, consent of 211–12
  negligence 215–16, 220–1
  non-governmental organizations (NGOs) 17, 219
  nuisance 216
  operational activities, fault-based liability for 215
  organic farming, nuisance and 216
  permit defence 220–1
  polluter pays 201, 207, 212–14, 219–21
  precautionary principle 220
  pre-release measures 201
  primary remediation 217
  property rights 223
  public liability 207
  remediation 207, 212–14, 216–23
  risk allocation 214, 222
  risk assessment 199, 219, 220
  sites of special scientific interest (SSSIs) 211–12, 215, 223–4
  species, damage to 17, 210
  standing 219
  state of the art defence 219–21
  strict liability 201, 207, 209, 213–16, 222
  territorial extent 224
  unknown risk 198–9
  water damage 210
  Wild Birds Directive 210–11, 223
Ethiopia 250
European Commission *see also* Commission Recommendation on coexistence 2003
  coexistence 124, 131–45, 154–60, 166, 188–94
  comitology 107, 109–12
  Consumerchoice report 24–5
  Environmental Liability Directive 214
  implementation of member state measures 154–8, 163–97
  indicative catalogue of measures 136
  Life Sciences and Biotechnology Communication 79
  multi-level governance 104–22
  New Case Studies. Joint Research Centre of European Commission 138–40
  reports 138–9, 154–8, 166
  seeds 143–5
  supermarkets and retailers 24–5
European Convention on Human Rights 32, 34
European Food Safety Authority (EFSA)
  access to information 90–1
  bans 149–50
  coexistence 149–50
  Deliberate Release Directive 85–6, 87
  European Union 101, 107–9, 111, 113–15, 117
  Food and Feed Regulation 90–1
  Food Law Regulation 81
  multi-level governance 101, 107–9, 111, 113–15, 117
  networking 111
  political decision-making 109
  risk assessment 107
  safeguard clause 149
  scientific opinions 108–9
European Seed Association (ESA) 141, 143, 145–6
European Union 79–100 *see also* Deliberate Release Directive; Environmental Liability Directive; European Commission; Food and Feed Regulation
  access to information 71, 80
  adventitious presence 123–4, 126–33, 137, 139–40, 159–61
  authorization 87–9, 102–3, 111–22
  bans 87, 116–7, 132, 136, 138, 148–51, 161–2, 164
  BSE 88
  centralization 107–8, 111, 121–2
  civil disobedience 385
  coexistence 118–21, 132, 134, 164–5, 196
  comitology 107, 109–14, 122
  competent authorities 106–8
  confidentiality 22
  Contained Use Directive 80, 81–4, 179, 199
  Council 109–10, 112–13
  de-centralization 118–19
  decision-making process 105–11
  dispersal of authority 105–11
  divergent opinions, transparency of 108–9, 111
  ECOSOC (Economic and Social Committee) 137
  economic development 80
  economic impact 132, 137, 139–40, 154–6, 165, 197

Index

Environmental Impact Assessment Directive 20
European Food Safety Authority (EFSA) 101, 107–9, 111, 113–15, 117
  experts, networking of 118–19
  Food Law Regulation 94–100
  free movement of goods 120, 121–2
  Habitats Directive 119–20
  hierarchy 101, 111–13, 121–2
  insurance 2
  internal market 118–22
  labelling 71, 80–1, 88–9, 91–4, 129
  MERCOSUR 292–5, 296
  minimizing admixture, guidelines on 2
  multi-level governance 101–22
  national competent bodies 107–11
  networking 111, 118–19
  new governance 110–11
  notification requirements 88, 106
  Novel Food and Food Ingredients Regulation 88, 338
  novelty of scientific information 116–17
  organic products 4, 93–4, 146–8, 192
  policy on GMOs 79–100
  precautionary principle 65, 72–5
  Products Liability Directive 152
  public participation 17, 23–5, 35, 36, 104, 111
  regulatory committee, opinions of 109–10
  regulatory context 102–5, 113
  risk 56, 65–6, 71–5, 79–80, 88, 104–8, 114, 117
  safeguard clauses 71–3, 114–16, 118
  scientific authority, centralization of 107–8
  scientific information 116–18
  simplified procedure 88
  socio-economic considerations 71, 370, 373–4
  sustainable development 370, 373–4
  technical standards and regulations 120
  threshold for cross-contamination 4, 93–4, 129–30, 142–6, 147–8, 192
  Traceability and Labelling Regulation 80–1, 89, 91–4, 129
  transparency 108–9, 111
exclusion of liability 219–21
experts, networking of 118–19
export markets to, loss of 230–1, 252–3

Faucheurs Volontaires 29–34, 36, 380–3
fault 214–16, 220–1

field trials
  Africa 233
  Deliberate Release Directive 83–4
  Egypt 243
  environmental liability 199–200
  notification 302
  permits 302–3
  South Africa 240
  Spain 187
  standards 302
  United States 300–3, 313, 320–1
Fischler Communication 132–4, 136
food additives 309–10, 313–15
food aid 2–3, 232, 246, 250
Food and Feed Regulation 80, 89–91
  access to information 90–1
  authorization 85, 91
  coexistence 138
  competent authorities 107
  consultation 91
  Deliberate Release Directive 85, 90–1, 105–6
  emergency measures 115
  Environmental Liability Directive 221
  European Food Safety Authority (EFSA) 90–1
  labelling 89, 91
  monitoring reports 91
  multi-level governance 105–9, 115
  notification requirements 89
  precautionary principle 91
  processing aids 96–7
  scientific opinions, reasons for 108
  single authorization procedure 89–91
Food Law Regulation 94–100
  access to information 99–100
  civil society 99–100
  convergence of principles 97–100
  European Food Safety Authority (EFSA) 81
  information for consumers 99–100
  integrated approach 94–7
  labelling 95–6, 99
  policies, integrated approach towards 94–6
  precautionary principle 97–9
  products, integrated approach towards 96–7
  risk assessment 98–9
  risk management 98
  single authorization procedure 96–7
  social value 95
  sustainable consumerism 95
  traceability 96–7, 98, 99
food security 1–2, 227–30, 233, 252–3

398   Index

foreseeability 200, 202, 216, 263–4
France
  Aarhus Convention, protests and 32, 34
  adventitious presence 167, 173
  authorization 167–9, 171–2, 174
  bans 166–7, 168, 169–71
  Cartagena Protocol, protests and 32, 34
  coexistence 166–74, 195–6
  Commission Recommendation on coexistence 2003 195
  criminal offences, protests and 29–35
  Deliberate Release Directive 31–2, 33–4, 172
  emergencies 168–9
  European Convention on Human Rights 32, 34
  European Union 166–74, 195–6
  Faucheurs Volontaires 29–34, 36, 380–3
  freedom of cropping 166–8
  French Charter for the Environment 32, 34
  GM-free zones 169–70
  information, obligations to provide 167, 171–3
  labelling threshold 167
  liability 167
  local authorities 169–71
  media 12, 29
  necessity, defence of 30–5
  non–governmental organizations (NGOs) 16
  organic producers 168–9
  police powers 170
  precautionary principle 32, 34, 167–8
  prevention 167–72
  professional codes of good practice 173
  proportionality 169–70
  protests 12, 29–35
  public participation 167
  publicity 12, 29
  risk assessment 167
  Spain 195–6
  technical rules 171, 173–4
  threshold 169
  transparency 172
free movement of goods 120, 121–2
freedom of cropping 166–8, 183
Friends of the Earth 25–6, 125, 137–8, 140

GATT (General Agreement on Tariff and Trades) 343–4, 353, 371–3
gene flow, limiting 135, 139

general principles of environmental law 63, 97, 362–3
GenetiX Snowball 381
Germany 87
GM-free zones
  coexistence 136, 169–70, 176, 178, 187–8, 192–7
  Deliberate Release Directive 187
  European Union 136, 196–7
  France 169–70
  Italy 176, 178
  Scotland 192–3
  Spain 187–8
  United Kingdom 192–5
  voluntary zones 194
  Wales 192–3
GM Nation 11–13, 18, 23, 25, 59
GMO-free Europe 125
GMO-Free Network 151
'good law' 381–2
government liability 263, 264–71
Greece 87
Greenpeace 25–9, 138

habitats
  Environmental Liability Directive 17, 209–12, 215–17, 223–4
  Habitats Directive 119–20, 210–11, 215, 223–4
  Natura 2000 211, 224
  remediation 223
HACCPs (hazard analysis and critical control points) methods 93
harmonization 16, 19, 150, 292–8
hazardous activities 206, 208, 213–16
health and safety 69
herbicides 123
Hungary 87, 149–50

Ideas Afines Group 292
imports
  bans 339
  Egypt 243–4
  European Union 125
  permits 243–4
  Sanitary and Phytosanitary Measures Agreement (SPS Agreement) 338
indigenous and traditional knowledge 285, 368–9
information *see also* access to information; confidentiality
  African Model Law 235
  Cameroon 248

Index

Deliberate Release Directive 86
France, coexistence in 167, 171–3
Spain, coexistence in 182–5
injunctions 323, 324, 326–7
insurance
  African Model Law 237
  Canada, coexistence in 273
  coexistence 134, 153, 158, 237
  Environmental Liability Directive 222
  European Union 2
  liability 153, 158, 222
  necessity 32–3
intellectual property 252, 257–9, 262–3
internal market 118–22
internalization costs 207, 213
international law 355–74
  Cartagena Protocol 356–60, 374
  EC-Biotech 357, 360
  fragmentation within international legal order 359–61
  national regulation 357–9
  normative context 356–61
  political motivations 360
  public participation 14–17
  regional regulation 357–9
  Sanitary and Phytosanitary Agreement (SPS Agreement) 360, 374
  socio-economic considerations 368–74
  sustainable development 368–74
  treaty interpretation 360, 361–8, 374
interpretation *see* **treaty interpretation**
isolation *see* **segregation**
Italy, coexistence and
  central government 174–9, 195, 196
  Charter of Florence 174–5
  Commission Recommendation on coexistence 2003 178
  constitutional law 175, 177–8
  Deliberate Release Directive 176, 178
  European Union 174–9, 196
  GM-free zones 176, 178
  management rules 178–9
  national legislation 176–9, 195, 196
  regions 174–9, 195, 196

joint and several liability 209
judicial review, costs of 21
jury trials 27–9
justice *see* **access to justice**

Kenya
  African Model Law 247
  caution, regulations promoting GMOs with 245–7
  civil liability 247
  food aid from United States 246
  labelling 247
  National Biosafety Authority (NBA) 246–7
  research and development 233
  risk assessment 247
  socio-economic development 245–7

labelling
  access to information 71
  adventitious presence 167
  African Model Law 236
  authorization 46, 51–3
  Brazil 285, 287
  Cameroon 248
  coexistence 143–8
    adventitious presence 167
    Deliberate Release Directive 129, 138, 192
    European Union 124, 129–31, 133–5, 137–8, 140, 156, 192
    policy 130–40
    Spain 181, 186
    totally GM-free 181
  cost-benefit analysis (CBA) 46, 51–3
  de minimis 86
  Deliberate Release Directive 86, 129, 138, 192
  Egypt 243
  European Union
    coexistence 124, 129–31, 133–5, 137–8, 140, 156
    thresholds 124, 129–31, 134–5, 138
  food and feed, EU regulation and 88, 89, 91
  Food Law Regulation 95–6, 99
  France, coexistence in 167
  Kenya 247
  liability 158, 199
  Malawi 245
  MERCOSUR 296
  optimal level of pollution 46
  organic production 147–8
  pesticides 308
  policy 130–40
  post-authorization 46, 51–3
  precautionary principle 99
  public participation 12
  risk assessment 71, 75
  seeds 143–5
  South Africa 242
  Spain 181, 186

labelling (*cont.*)
  Tanzania 249–50
  thresholds 124, 129–31, 134–5, 138, 167, 192
  totally GM-free 181
  Traceability and Labelling Regulation 80–1, 89, 91–4, 129
  United Kingdom 12, 190, 192
  United States 12, 308, 312, 315–17, 319
  Uruguay 292
  voluntary labelling 317
law, role of 375–7
  morality, ethics and politics 376–7
  politics 376–7
  rule of law 376
legitimacy of law 379–83
liability *see also* environmental liability; Environmental Liability Directive; strict liability
  adventitious presence 153–4, 158
  African Model Law 236–7
  Brazil 284
  Canada 259–71
  causation 153, 188
  civil liability
    coexistence 152–3, 158
    environmental liability 200–1, 206–7, 215–16, 223
    Kenya 247
    Lugano Convention on Civil Liability 206–9, 218
  coexistence 152–3, 158
  compensation funds 153, 158
  COPA–COGECA 153–4
  damages 151–4, 197
  economic loss 256, 259, 267–8
  Environmental Liability Directive 152
  European Union 135, 137, 141, 151–4, 158–60, 197
  France 167
  government 263, 264–71
  insurance 153, 158
  joint and several liability 209
  Kenya 247
  Lugano Convention on Civil Liability 206–9, 218
  negligence 152
  nuisance 152
  principles of liability 263–8
  Products Liability Directive 152
  seeds 154
  South Africa 241
  Spain 185, 188
  US state law 299, 320, 321–33

LibertyLink rice 321–2
limitation of actions 217–18
low-level gene mixing 305, 313
Lugano Convention on Civil Liability 206–9, 218
  causation 208–9
  continuous or series of occurrences, time limits and 218
  Environmental Liability Directive 199
  hazardous activities 206, 208
  joint and several liability 209
  limitation periods 218
  ratification 207–8
  remediation 208, 209
  tolerable levels of damage 208
Luxembourg 87

Maine, United States 320
Malawi
  African Model Law 19, 245
  labelling 245
  liability 245
  National Biotechnology Policy 244
  promoting GMOs, regulations 244–5
  public participation 245
  research and development 244
  risk assessment 244–5
management *see also* risk management
  coexistence 134–5, 139, 178–9, 181–2
  European Union 134–5, 139
  Italy 178–9
  MERCOSUR 297
  risk assessment 55–6, 63–4
  Spain 181–2
marginal external cost (MEC) 40, 44
marginal willingness to pay (MWTP) 39–43
media 11–12, 27–8, 29
MERCOSUR 274–98
  arbitration 277
  Argentina 274–5, 277–84, 292–8
  authorization 278–82, 296
  biodiversity 292, 295–7
  Biotechnology Group 296
  Bolivia 274, 279
  Brazil 274–5, 278–80, 284–7, 292–3, 298
  Cartagena Protocol 292, 295–6
  Chile 274
  Commercial Commission 276
  common frontier tariff 274
  Common Market Group 276
  competition 294
  Council of the Common Market 276

Index

customs union 274
Declaration on Biodiversity Strategy 297
differences in production 278–80
economies, size of 275
European Union 292–5, 296
external market 293–4
Framework Agreement on the Environment 294–5
frontiers 275
harmonization of regulation 292–8
Ideas Afines Group 292
increase in use 278
institutions 276–7
integration process 274–5, 276, 294–5
intellectual property 279
interests, difficulties in reconciling 292–3
internal market 293–4
labelling 296
legal order 276–7, 297–8
management tools for biosafety 297
members 274–5
Miami Group 292
multidisciplinary bodies 281
national implementation 277
normative position 294–5
Paraguay 274–5, 278, 279, 287–9, 293, 295, 298
partial-scope agreements 276–7
precautionary principle 283, 295
recent developments 296–7
regulatory framework 279, 280–97
risk assessments 282, 284, 295–6
risk management 297
sanctions 277
single market 293–6
statistics 278
transcription into national law, provisions not requiring 276–7
Uruguay 274–5, 277–8, 279, 290–3, 295, 298
Miami Group 292
Millennium Development Goals (MDGs) 228
Monsanto 254–62
morality, ethics and politics 376–7

national measures *see* particular countries (eg France)
Natura 2000 211, 224
Natural England 211–12
necessity, defence of 27–35
negligence *see also* causation; economic loss
authorization 220–1

Canada, coexistence in 256, 261–72
coexistence 152
duty of care 261–7, 270, 272, 328–9
environmental liability 200–1, 215–16, 220–1
foreseeability 200, 202, 216, 263–4
nuisance 323
proximity 261–4, 267, 270, 272
public authorities 264, 267–8
risk assessment 67–9
United Kingdom 67–9
United States 323, 328–9, 330, 332
neighbours, notifying 135, 138, 182
networking 108, 111, 118–19
***New Case Studies*. Joint Research Centre of European Commission** 138–40
New Zealand, public participation in 18–19
non-governmental organizations (NGOs)
Aarhus Convention 16
access to justice 16
Africa 229–30
Brazil 286–7
coexistence 123, 137–8
Deliberate Release Directive 17
developing countries 13
Environmental Liability Directive 17, 219
European Union 66, 123, 137–8
France 16
Friends of the Earth 25–6, 125, 137–8, 140
Greenpeace 25–9, 138
protests 25–7
public participation 24
rights of NGOs 16
risk assessment 59, 71
Northern Ireland 189
notice-and-comment procedure 16–17, 236, 242
notification
competent authorities 85, 106
Deliberate Release Directive 84, 85
field trials 302
food and feed, EU regulation and 88, 89
multi-level governance in EU 106
neighbours, notifying 135, 138, 182
premises for contained use 82–3
Sanitary and Phytosanitary Agreement (SPS Agreement) 350–3
nuclear power stations, consultation on 20
nuisance
anticipatory nuisance 326–7
coexistence 152
competing interests 323, 324

nuisance (*cont.*)
  damages 323, 326
  definition of nuisance 323, 324, 325
  economic loss 327
  Environmental Liability Directive 216
  foreseeability 216
  injunctions 323, 324, 326–7
  intention 323
  negligence 323
  precautions 325–6
  private nuisance 323, 325–6, 332–3
  public nuisance 323, 324–5, 327, 332–3
  reasonableness 323, 324, 325–6
  remedies 323, 324, 326–7
  seeds 324–6
  StarLink litigation 327, 332–3
  US states 332–7, 331–3
  use and enjoyment of land, interference with 323, 324, 325

operational/policy distinction 265–7, 270–1
organic production
  accreditation bodies 192
  adventitious presence 147
  certification 147
  coexistence 168–9, 192
  emergencies 168–9
  Environmental Liability Directive 216
  European Union 4, 146–8, 216
  France 168–9
  freedom of cropping 168
  growth in production 146
  International Federation of Organic Agricultural Movements (IFOAM) 146
  labelling 93–4, 147–8
  marketing 93–4, 147–8, 168
  nuisance 216
  Spain 146
  thresholds 93–4, 147–8, 169, 192–3
  United Kingdom 192–4
  United States 146–8

Paraguay
  Biodiversity Convention 295
  Biosafety Commission 287–8
  Cartagena Protocol 289
  COMBIO 287–9
  Ideas Afines Group 292
  intellectual property 279
  legislation 298
  MERCOSUR 274–5, 278, 279, 287–9, 293, 295, 298
  public participation 287–8

  risk assessment 288–9
  SENAVE 288
  trading partners 293
participation *see* consultation; public participation
patents 257–9, 262–3
permits
  Environmental Liability Directive 220–1
  field trials 302–3
  imports 243–4
  South Africa 240–1
  United States 302–3, 306
pesticides
  Advisory Committee on Pesticides (ACP) 70
  coexistence 123
  European Union 123
  labelling 308
  plant-incorporated protectants (PIPs) 307, 308–9
  registration 308–9
  residues 309–10
  tolerances 309–10
  United States 307, 308–10, 313–14, 329
phytosanitary measures *see* Sanitary and Phytosanitary Measures Agreement (SPS Agreement)
police powers 170, 318
policy
  Canada, coexistence in 255–7, 262, 265–8, 270–2
  coexistence
    Canada 255–7, 262, 265–8, 270–2
    European Union 130–40
    United Kingdom 188–91
  economic loss 267–8
  European Union 79–100, 130–40
  Food Law Regulation 94–6
  integration 94–6
  operational/policy distinction 265–7, 270–1
  United Kingdom 188–91
  United States 300–1, 305, 310–12, 318–20, 329, 334
polluter pays
  bans 150
  coexistence 150
  environmental liability 201, 207, 212–14, 219–21
  exclusion of principle 219–21
  remediation 219–21
  risk assessment 61–2
  state of the art defence 219–20
poverty 227–8, 231

Index

precautionary principle
  African Model Law 2, 235
  Argentina 283
  Cameroon 247–8
  coexistence 130, 167–8
  cost-benefit analysis (CBA) 45
  Deliberate Release Directive 65–6, 84, 86, 98
  Ethiopia 250
  European Union 65, 72–5, 130
  Food and Feed Regulation 91
  Food Law Regulation 97–9
  France 32, 34, 167–8
  general principles of environmental law 63, 97
  knowledge 64–5
  labelling 92, 99
  MERCOSUR 283, 295
  protests in France 32, 34
  response to 64
  risk assessment 61–6, 70, 72–5, 98–9
  risk management 63–4
  Sanitary and Phytosanitary Agreement (SPS Agreement) 348–9
  science, relationship with 64–5
  South Africa 241
  timing 64
  traceability 92, 98
  trigger 64
  World Trade Organization (WTO) 63, 65
premises for contained use, notification of 82–3
procedural legitimacy of law 379–80
ProdiGene maize 321
product liability 152, 330
professional codes of good practice 173
property rights 202–6, 223
proportionality 156, 160–1, 169–70
protectionism 368
protests
  Aarhus Convention 32, 34
  Africa 229–30
  Cartagena Protocol 32, 34
  civil disobedience 377–85
  criminal offences 26–35
  Deliberate Release Directive 31–2, 33–4
  democracy 383–5
  European Convention on Human Rights 32, 34
  fallibility of law 379
  Faucheurs Volontaires 29–34, 36, 380–3
  France 29–35
  French Charter for the Environment 32, 34
  Friends of the Earth 25–6
  GenetiX Snowball 381
  'good law' 381–2
  Greenpeace 25–9
  jury trials 27–9
  justification, defence of 27–9
  legitimacy of law 379–83
  necessity, defence of 27–9, 30–5
  non-governmental organizations (NGOs) 25–7
  precautionary principle 32, 34
  present and imminent danger 30–2, 33–5
  public participation 25–35, 36
  publicity 27–8, 29
  sanctions 29–30
  social contract 378
  sociological concept of law 382
  source of law, civil disobedience as a 378, 380–3
  'street law' 382–3
  supermarkets 25–6
  symbolism 380–1
  trespass to land 27–8
  United Kingdom 26–9
  United States 26
proximity 261–4, 267, 270, 272
public authorities
  Aarhus Convention 21–2
  Canada, coexistence in 264, 267–8
  confidentiality 21–2
  negligence 264, 267–8
  Spain 182
public liability regimes 200–1, 204, 207
public participation 11–36 *see also* consultation
  Aarhus Convention 15–16, 21–2, 35, 60
  access to information 15–16, 21–2
  African Model Law 16, 19, 22, 236, 251
  Biodiversity, Convention on 14, 19, 35
  Brazil 286–7
  Cartagena Protocol 14, 15–16, 19, 35
  coexistence 167
  consumer preferences 22–5
  costs of challenging non-compliance with environmental measures 20–1
  Deliberate Release Directive 17, 66
  developing countries, food security for 13
  effectiveness of participation 19–22
  European Union 17, 23–5, 35, 36, 104, 111
  Food Law Regulation 99–100
  formal public influence 14–22
  France 12, 29–35, 36, 167
  GM Nation 11–13, 18, 23, 25, 59

public participation (*cont.*)
  informal public influence 22–35
  international level 14–17
  labelling 12
  Malawi 245
  media 11–12
  national level 17–19
  New Zealand 18–19
  non-governmental organizations (NGOs) 35
  Paraguay 287–8
  preferences of consumers 22–5
  protests 25–35, 36
  risk assessment 58, 59–61, 66, 70, 75
  South Africa 242
  supermarkets and other retailers 24–5, 36
  Tanzania 249
  transparency 111
  United Kingdom 11, 12–13, 19–21, 25, 26–9, 35–6, 60, 70
  United States 12–13, 17–18, 23, 304
  Uruguay 290
publicity 11, 12, 27–8, 29

regions
  Africa 238–9, 251
  African Model Law 238
  bans 138, 161–2
  coexistence 138, 161–2, 174–9, 195, 196
  European Union 138, 161–2
  international law 357–9
  Italy 174–9, 195, 196
remediation
  damages, assessment of 217
  environmental liability 201–2, 207, 212–14, 216–23
  Habitats Directive 223
  Lugano Convention on Civil Liability 208, 209
  polluter pays 219–21
  risk allocation 222
  state of the art defence 219–21
  United States 305
remedies *see also* damages
  African Model Law 236–7
  compensation 153, 158, 160, 273
  environmental liability 200–1, 204–5
  injunctions 323, 324, 326–7
  nuisance 323, 324, 326–7
research and development
  Africa 233, 251, 253
  Argentina 283

Brazil 287
coexistence 138–40, 161
Egypt 233, 243
European Union 138–40, 161
Kenya 233
Malawi 244
South Africa 233, 240, 251
Uruguay 290
resource allocation 201–4
risk and risk assessment 54–76
  Advisory Committee on Pesticides (ACP) 70
  Africa 235–6, 240–5, 247, 249, 251–2
  African Model Law 235–6
  allocation of risk 214, 222
  analysis 55–6, 64–5
  Argentina 282, 284
  authorization 47–8
  BAT 61
  BATNEEC 61
  Brazil 286–7
  Cameroon 247–8
  Canada, coexistence in 264, 272
  causation 67–9
  coexistence 132, 165, 167
  consultation 59, 70
  Contained Use Directive 82–3
  cost-benefit analysis (CBA) 37, 39, 44–8, 53
  courts 67–76
  definition of risk assessment 345
  Deliberate Release Directive 56, 65, 66, 90–1, 104–8, 114–15, 200
  engaging with risk 55–8
  Environmental Liability Directive 199, 214, 219, 220, 222
  environmental risk assessment 56, 69–70
  Ethiopia 250
  European Food Safety Authority (EFSA) 107
  European Union 65–6, 71–5, 79–80, 98–9, 104–8, 114, 117, 132, 165
  food and feed, EU regulation and 88
  Food Law Regulation 98–9
  France 167
  GM Nation 59
  health and safety 69
  Kenya 247
  knowledge 54
  labelling 71, 75
  liability 199–200, 202
  limits of risk-based approaches 61–2
  Malawi 244–5
  management 55–6, 63–4
  MERCOSUR 282, 284, 295–6

Index

negligence 67–9
networking 108
non–governmental organizations (NGOs) 59, 71
Paraguay 288–9
political decision-taking 56
polluter pays 61–2
precautionary principle 61–6, 70, 72–5, 98–9
prevention of environmental damage 61–2
public participation 58, 59–61, 66, 70, 75
risk management 344, 346–7, 353
risk society 54–5
Royal Commission on Environmental Pollution (RCEP) 70
Sanitary and Phytosanitary Agreement (SPS Agreement) 338, 344–51, 353–4
science 56–8, 70, 346–7, 348–50
significant risk 235–6
social dimensions 58, 71
South Africa 240–2
strict liability 214
Tanzania 249
United Kingdom 67–71
United States 313
Uruguay 290
risk management
coexistence 133, 137
European Union 98, 133, 137
Food Law Regulation 98
MERCOSUR 297
precautionary principle 63–4
risk assessment 344, 346–7, 353
Sanitary and Phytosanitary Agreement (SPS Agreement) 344, 346–7, 353

SADC (Southern African Development Community) 238
safeguard provisions
bans 87, 116–7, 149–50
coexistence 131, 148–9
Deliberate Release Directive 86–7, 114, 131, 148–9
European Union 71–3
Food and Feed Regulation 148–9
multi-level governance in European Union 114–16, 118
Sanitary and Phytosanitary Agreement (SPS Agreement) 349–50
Sanitary and Phytosanitary Agreement (SPS Agreement) 337–54
appropriate level of protection 372–3
approval procedures 343, 351–3
bans 338, 342–3, 351–3
cost-benefit analysis (CBA) 53
definition of sanitary and phytosanitary measures 339–40
Deliberate Release Directive 338, 341, 343
developing countries 373–4
'directly or indirectly affecting international trade', definition of 343–4
EC–Biotech 338, 339–45, 349–54
GATT 343–4, 353, 371
import bans 338
international law 360, 374
interpretation 371–2
marketing bans 338, 342
measure, concept of a 338, 341–3
notification requirements 350–3
Novel Food and Food Ingredients Regulation 338
Preamble 337
precautionary measures 347–50
precautionary principle 348–9
publication of regulations 350–2
risk assessment 338, 344–51, 353–4
risk management 344, 346–7, 353
safeguard measures 343, 349–50
science 337, 346–7, 348–50, 353–4, 360
scope 339–44
socio-economic considerations 369, 371–4
special and differential treatment 373–4
sustainable development 369, 371–4
Technical Barriers to Trade Agreement (TBT Agreement) 340–1
transparency 338, 350–3
science
centralization of scientific authority 107–8
decision-making 70
distrust 57–8
European Food Safety Authority (EFSA) 108–9
European Union, multi-level governance in 107–8, 116–18
evidence 346–7, 348–50
law and science 70
political decision-making 109
precautionary principle 64–5
reasons for opinions 108
risk assessment 56–8, 70, 346–7, 348–50
Sanitary and Phytosanitary Agreement (SPS Agreement) 337, 346–50, 353–4, 360
United Kingdom 18
Scotland 3, 189, 193–4, 196
seeds
Africa 229–30, 251, 252

seeds (*cont.*)
  Argentina 282
  authorization 141
  coexistence 141–6, 154, 190
  COPA–COGECA 143
  definition 282
  Egypt 243
  European Commission 143–5
  European Seed Association (ESA) 141, 143, 145–6
  European Union 136–7, 139–46, 157
  labelling 143–5
  legislation, proposals for 143–4
  liability 154, 330
  nuisance 324–6
  purity 141–2
  Scientific Committee on Plants (SCP) 142–3
  segregation 142–3
  thresholds 141–6
  United Kingdom 190
  United States 321–2, 324–6, 330
  zero tolerance 143
segregation
  authorization 51–2
  coexistence 134–5, 136, 142–3, 150, 157, 159, 181–4, 187, 195
  cost-benefit analysis 51–2
  distances 157
  European Union 134–5, 136, 157, 159, 195
  post-authorization 51–2
  seeds 142–3
  Spain 181–4, 187
  United Kingdom 191
sites of special scientific interest (SSSIs)
  environmental liability 204–6, 211–12, 215, 223–4
  property rights 205–6
social contract 373
socio-economic considerations
  African Model Law 2, 235–6
  appropriate level of protection 372
  Cartagena Protocol 368–71
  developing countries 373–4
  discrimination 369
  EC–Biotech 370
  European Union 71, 370, 373–4
  GATT 371–3
  indigenous and local communities 368–9
  Kenya 245–7
  precautionary principle 2
  protectionism 368

risk assessment 58, 71
  Sanitary and Phytosanitary Agreement (SPS Agreement) 369, 371–4
  sustainable development 368–74
  Technical Barriers to Trade Agreement (TBT Agreement) 369
  World Trade Organization (WTO) 369–74
sociological concept of law 382
source of law, civil disobedience as 378, 380–3
South Africa
  African Model Law 19, 22, 251
  Biowatch South Africa 229–30, 241–2
  Cartagena Protocol 240–1
  commercialization 240
  confidentiality 241
  contained use 240–1
  damages 241
  field trials 240
  labelling 242
  legislation 240–3
  liability 241
  notice-and-comment procedure 242
  permits 240–1
  precautionary principle 241
  promoting GMOs, regulations 240–3
  public participation 242
  research and development 233, 240, 251
  risk assessments 240–2
  substantial equivalence test 241
Spain, coexistence and
  adventitious presence 181
  agricultural tenancies 183
  application of measures 181–4
  autonomous communities 179–80, 182, 184–5, 187–8, 196
  breach of rules 185
  buffer zones 181–3
  Commission Recommendation on coexistence 2003 179–83, 187–95
  competence 185
  Contained Use Directive 179
  controversial issues 186–8
  Deliberate Release Directive 179, 187
  economic effects 180–1, 188
  economic loss 188
  European Union 130, 179–88, 195–6
  farmers, obligations on 181–4
  field trials 187
  France 195–6
  freedom of cropping 183
  General State Administration 179–80
  GM-free zones 187–8

increase in production 179
information, provision of 182–5
labelling threshold 181, 186
legislation 179–85
liability 185, 188
management measures 181–2
monitoring and control 184–5
National Biosafety Commission 180
neighbours, notification of 182
organic production 146
producers or sellers 183
public authorities 182
registers 184–5
reports 184–5
segregation 181–4, 187
thresholds 186
traceability 183
SPS *see* Sanitary and Phytosanitary Agreement (SPS Agreement)
standing 219, 237
StarLink litigation
coexistence 128
damages 321, 332
escapes 2, 13
European Union 128
nuisance 327, 332–3
US states, regulation in 321, 331–3
state of the art defence 219–21
state sovereignty 2–3
'street law' 382–3
strict liability
African Model Law 237
coexistence 137, 153, 158
environmental liability 200–1, 207, 209, 222
European Union 137
hazardous activities 213–16
public liability 200–1
risk allocation 214
seeds 330
US states, regulation in 329–30
subsidiarity 133, 165
supermarkets and retailers 24–6
sustainable development
African Model Law 235–6
appropriate level of protection 372–3
Cartagena Protocol 368–71
consumerism 95
developing countries 373–4
discrimination 369
EC–Biotech 370
European Union 370, 373–4
Food Law Regulation 95
GATT 371–3

indigenous and local communities 368–9
international law 368–74
mutual supportiveness 368
protectionism 368
Sanitary and Phytosanitary Agreement (SPS Agreement) 369, 371–4
socio-economic considerations 368–74
Technical Barriers to Trade Agreement (TBT Agreement) 369
World Trade Organization (WTO) 369–74

Tanzania
African Model law 249–50
confidentiality 249
labelling 249–50
National Biosafety Framework (NBF) 249–50
public participation 249
restrictive national regulations 249–50
risk assessment 249
Technical Barriers to Trade Agreement (TBT Agreement) 340–1, 369
technical standards and regulations 120
thresholds
coexistence 129–31, 133–5, 137–48, 160, 164, 169, 186
cost-benefit analysis (CBA) 52
cross-contamination 4
European Union 4, 129–31, 133–5, 137–40, 160, 164
France 169
labelling 124, 129–31, 134–5, 138, 167, 192
organic production 93–4, 147–8, 192–3
post-authorization 52
seeds 141–6
Spain 186
traceability 129–30, 134
United Kingdom 192–3
time limits 217–18
traceability
access to information 97
African Model Law 236
authorization 46, 51–3
coexistence 129–30, 133–4, 183
cost-benefit analysis (CBA) 46, 51–3
Deliberate Release Directive 129
Food Law Regulation 96–7, 98, 99
HACCPs (hazard analysis and critical control points) methods 93
optimal level of pollution 46
post-authorization 46, 51–3
precautionary principle 92, 98
production methods 92

traceability (*cont.*)
  Spain 183
  thresholds 129–30, 134
  Traceability and Labelling Regulation 80–1, 89, 91–4, 129
  unique identifiers 92
traditional knowledge 285, 368–9
training of farmers, proof of 156
Transboundary Movements Regulation 81
transparency
  Aarhus Convention 15
  Argentina 283
  bans 150
  coexistence 134, 150, 172
  Declaration of Rennes 151
  Deliberate Release Directive 66, 99
  divergent opinions 108–9, 111
  European Union 108–9, 111, 134
  France 172
  public participation 111
  Sanitary and Phytosanitary Agreement (SPS Agreement) 338, 350–3
treaty interpretation 360, 361–8, 374
  see also **international law**
  application of laws 361
  Cartagena Protocol 362, 364–5, 367–8
  customary international law 362–3, 365
  EC–Biotech 362, 364–8
  extraneous legal material 364, 367
  general principles of law 362–3
  good faith 364
  judicial discretion 365–8
  relevant rules, other 363–4
  Vienna Convention on the Law of Treaties 361–2, 364, 367
  World Trade Organization (WTO) 361–8
trespass to land 27–8, 327–8

United Kingdom
  Aarhus Convention 21
  access to justice 21
  accreditation bodies for organic farming 192
  bans 189, 193–4
  code of practice 190–1
  coexistence 128, 188–96
  Commission Recommendation on coexistence 2003 191–5
  consultation 18, 19–20
  cost-benefit analysis (CBA) 18
  criminal offences, protests and 26–9
  Deliberate Release Directive 192
  Department for Environment, Food and Rural Affairs (DEFRA) 189–90, 193

  Environmental Impact Assessment Directive 20
  European Commission 188–9, 191, 192, 194
  European Union 128, 188–96
  evidence-based policy-making 18
  GM Dialogue 18
  GM-free zones 192–5
  GM Nation 11–13, 18, 23, 25, 59
  Greenpeace 26–9
  health and safety 69
  implementation measures, form of 190–1
  judicial review, costs of 21
  jury trials 27–9
  justification defence, protests and 27–9
  labelling 12, 190, 192
  media 11
  necessity, protests and 27–9
  negligence 67–9
  Northern Ireland 189
  nuclear power stations, consultation on 20
  organic farming 192–4
  policy 188–91
  protests 26–9, 36
  protocols 191
  public participation 11, 12–13, 19–21, 25, 26–9, 35–6, 60
  publicity, protests and 27–8
  risk assessment 67–71
  Royal Commission on Environmental Pollution (RCEP) 70
  Royal Society (UK). *Reaping the Benefits* 1–2
  scientific review 18
  Scotland 189, 193–4, 196
  seeds 190
  segregation 191
  supermarket retailers 25, 36
  thresholds 192–3
  trespass to land 27–8
  Wales 189–90, 192–3, 196
United States 299–334 *see also* **US states, regulation in**
  advance notice of proposed rule-making 309
  APHIS (Animal and Plant Health Inspection Service) 301–7
  approvals 314
  Biotechnology Regulatory Services (BRS) 301, 304
  coexistence 127–8, 146, 147
  commercialization 299
  consultation 312–13
  consumer preferences 23
  Coordinated Framework 1986 300–1, 333–4

Index

data requirements 304, 309, 313
Delaney Clause 309–10
Department of Agriculture (USDA) 300–7, 334
deregulation, petitions for 303–4
early food safety evaluation 313
environmental impact assessment 303–6
Environmental Protection Agency (EPA) 300–1, 305, 307–10, 334
European Union 127–8
federal agencies 300
Federal Insecticide, Fungicide, and Rodenticide Act (FIFRA) 307, 308–9
federal laws 299–317, 334
federal policy 300–1, 305, 310–12, 334
field trials 300–3, 313
food additives 309–10, 313–15
food aid 246
Food and Drug Administration (FDA) 300, 304, 307, 310–17, 334
Food, Drug, and Cosmetic Act (FDCA) 307, 309–17
GRAS (generally recognized as safe) 314–15
guidance documents 312–13
increases in production 1, 299, 333–4
labelling 12, 308, 312, 315–17
liability under state law 299
low-level gene mixing 305, 313
non-regulated status 17–18, 303–4, 306–7
notification 302, 312, 315
Office of Science and Technology Policy (OSTP) 300–1, 307
organic production 146, 147
permits 302–3, 306
pesticides 307, 308–10, 313–14
plant–incorporated protectants (PIPs) 307, 308–9
Plant Protection Act 17–18, 301–7
policy 300–1, 305, 310–12, 334
pre-market approvals 314–15
pre-market safety review of new foods 310–12
protests 26
public participation 12–13, 17–18, 23, 304
registration of pesticides 308–9
remediation 305
risk assessments 313
StarLink litigation 2, 13, 23
statistics 1, 299, 333–4
substantial equivalence principle 311–12
UN Food and Agriculture Organization (FAO) 311
World Health Organization (WHO) 311

Uruguay
authorization 278, 290–2
Biodiversity, Convention on 295
CERV 290–1
competent authorities 290–1
harmonization 292
labelling 292
legal regime 290–2, 298
maize 291
MERCOSUR 274–5, 277–8, 279, 290–3, 295, 298
Miami Group 292
National Biosafety Cabinet 292
public participation 290
research and development 290
risk assessment 290
trading partners 293
US states, regulation in 299, 317–33
Arkansas 320–1
authority 318–19
bans 320
California 320
causation 322, 328–9
certification 319
Commerce Clause 318–19
constitutional law 318
cooperation with federal agencies 317
Coordinated Framework 317
damages 320, 321–2, 328, 332
de minimis rule 320
duty of care 328–9
economic loss 329, 330–1
federal agencies 317
federal authority 317
federal regulation 317, 321
Federal Insecticide, Fungicide, and Rodenticide Act (FIFRA) 329
field trials 320–1
interstate commerce 318–19
labelling 319
liability 299, 320, 321–33
LibertyLink rice 321–2
limitations on tort liability 330–1
Maine 320
negligence 328–9, 330, 332
nuisance 322–7, 331, 332–3
pesticides 329
pharmaceutical or industrial use 330
police powers 318
policy 320–1, 330
pre-emption doctrine 318–20, 329
ProdiGene maize 321
product liability 330
protective measures 319

US states, regulation in (*cont.*)
  rice 320–1
  seeds 321–2, 330
  StarLink litigation 321, 331–3
  state measures 319–21
  stewardship programmes 321–2
  strict liability 329–30
  Supremacy Clause 318
  tort law, liability under state 321–33
  trespass 327–8
utilitarianism 204–5

Vienna Convention on the Law of Treaties 361–2, 364, 367
voluntary agreements 135, 136, 194

Wales 189–90, 192–3, 196
water damage 210

Wild Birds Directive 210–11, 223
willingness to pay (WTP) 39–43, 49–53
World Trade Organization (WTO) *see also* Sanitary and Phytosanitary Agreement (SPS Agreement)
  cost-benefit analysis (CBA) 38, 44, 53
  Deliberate Release Directive 86–7
  precautionary principle 63, 65
  socio-economic considerations 369–74
  sustainable development 369–74
  Technical Barriers to Trade Agreement (TBT Agreement) 340–1, 369
  treaty interpretation 361–8
  United States 311

Zambia 2–3, 232
zero tolerance 127, 143, 160